GLOBAL INTERACTIONS

YEAR 11

Grant Kleeman
David Hamper
Helen Rhodes

Pearson Australia
(a division of Pearson Australia Group Pty Ltd)
459–471 Church St, Level 1, Building B, Richmond, Victoria, 3121
www.pearson.com.au

First published 2024 by Pearson Australia

Portfolio Manager: Simon Tomlin
SME/Senior Publisher (Humanities): Niki Horin
Programme Manager: Michelle Thomas
Production Editor: Jaimi Kuster
Editorial Development Manager: Leanne Peters
Development Editor: Fiona Doig
Copyeditor: Kirsty Hine
Proofreader: Margaret Trudgeon
Indexer: Max McMaster
Rights & Permissions Editor: Jes Senbergs
Designer: Anne Donald
Typesetter: Integra Software Services
Publishing Services Analyst: Jit-Pin Chong
First Nations content reviewer: Gary Passmore
Illustrators: QBS; Bruce Rankin
Printed in Singapore by Markono/03 (2025)

A catalogue record for this book is available from the National Library of Australia

ISBN 978 0 6557 1305 0
Pearson Australia Group Pty Ltd ABN 40 004 245 943

Acknowledgement of Country
Pearson acknowledges the Traditional Custodians of the lands upon which the many schools throughout Australia are located.

We respect the living cultures of Aboriginal and Torres Strait Islander peoples and their ongoing connection to Country across lands, sky, seas, waterways and communities. We celebrate the richness of Indigenous Knowledge systems, shared with us and with schools Australia-wide.

We pay our respects to Elders, past and present.

Content warning
Some of the images used in Global Interactions Fourth Edition might have associations with deceased First Nations Australians. Please be aware these images might cause sadness or distress in Aboriginal and/or Torres Strait Islander communities.

Dedicated to:
Ollie Taylor Fuller, Gemma Louise Rose Fuller, Jaxon John Ray Fuller, Charlotte Teresa Rose Kleeman & Raphael James Noel Kleeman

Zoe Elizabeth Hamper & Phoebe Anne Hamper

Eloise Christie Brook Rhodes, Xander Louis Nathan Smith, Felicity Grace Brook Rhodes, William Woodsell Brook Rhodes & Beatrice Brooke Ruth Smith

It's not 'our' world! We are its custodians. The world belongs to our children, our grandchildren, our great-grandchildren and all future generations. Protecting this treasured place is our obligation, our legacy, our gift to the future.

—Grant Kleeman

Global Interactions

The *Global Interactions Fourth Edition* series is written for the NSW Geography 11–12 Syllabus (2022), for implementation from 2024. *Global Interactions 11* has been developed to support teaching and learning of the contents of the NSW Geography Syllabus (Year 11). It has been written by one of Australia's most experienced and accomplished team of geography educators. It provides comprehensive coverage of the syllabus's core topics and its various options. It also provides a detailed guide to the geographical investigation students are required to undertake and the tools and skills they are expected to master.

Student book

- Chapter and unit structure closely reflect the NSW Geography 11–12 Syllabus (2022)
- Full-colour text with engaging and highly visual design
- Dynamic, contemporary and relevant textual examples, images, graphs and maps
- Topic-based units written in accessible language with clear and concise explanations of key terms and concepts
- A variety of learning and comprehension activities for regular revision and consolidation
- Spotlight features and case studies that describe and encourage in-depth investigation, and address the syllabus's focus on 'place'
- Front-of-chapter glossary for easy reference, with select extended glossaries for additional support
- Written by an experienced author team:
    - **Grant Kleeman** (lead author) is one of Australia's leading geography educators. He is an experienced teacher educator, geography teacher, author, curriculum consultant and examiner. Grant has been closely involved in the development of the geography curriculum at the state and national levels for more than three decades. He is the coordinating author of numerous texts, including the Pearson Geography NSW series.
    - **David Hamper** is an experienced geography educator and is currently Deputy Principal at Hills Grammar School, Sydney. David has also contributed to several geography texts, including the Pearson Geography NSW series.
    - **Helen Rhodes** is an experienced geography educator and is currently a Master Assisting in Geography at Shore, North Sydney. Helen has also contributed to several geography texts, including the Pearson Geography NSW series.

The Publishers wish to thank Rod Lane and Barbara Rugendyke for their contribution to *Global Interactions 1 Second Edition*.

Reader+

Pearson Reader+ allows you to use *Global Interactions* online or offline, anywhere and anytime. The Reader+ edition retains the look and integrity of the printed book and includes additional resources:

- two digital-only chapters: G1 and G2
- For teachers: Teaching and Learning Program
- For students: Digital resources, downloadable worksheets and interactive activities to support understanding of key geographical tools, including graphs, maps and statistics
- Access to both student and teacher resources

Contents

How to use

Chapter glossaries

Chapter glossaries provide an easy reference for more than 350 key terms. Glossary words are marked in bold in the main text. Glossaries appear at the front of chapters, with select extended glossaries included for chapters requiring greater vocabulary support.

Chapter glossary

atmosphere the gaseous layer surrounding Earth that acts as a barrier between space and the surface

biosphere the living components of an ecosystem; the flora and fauna

fieldwork observing and investigating in the real world

geographical thinking making sense of the world by viewing it through a geographical lens; conceptual understanding developed through the study of geography

geosphere all the rocks that make up Earth's crust; also called the lithosphere

human geography the study of the dynamics of cultures, societies and economies

hydrosphere the interconnected system of water in the atmosphere and lithosphere, including the oceans, ice caps, rivers and groundwater

pedosphere the outermost layer of Earth made of soil

perspectives the viewpoints, theories or worldviews that shape people's values; those deeply held beliefs about what is important or desirable

physical geography the study of the natural features of Earth's surface

transferable skills skills that can be applied in a range of contexts

Case studies

Three levels of case studies—smaller ones within units, unit-long case studies, and full chapter case studies—allow students to dip in or deep-dive into a specific event or location. Unit-level and smaller case studies offer a good introduction to a specific event or place, and are written to extend students' knowledge and understanding. Case study chapters offer a far deeper exploration. See page 1 for more about the case study chapters.

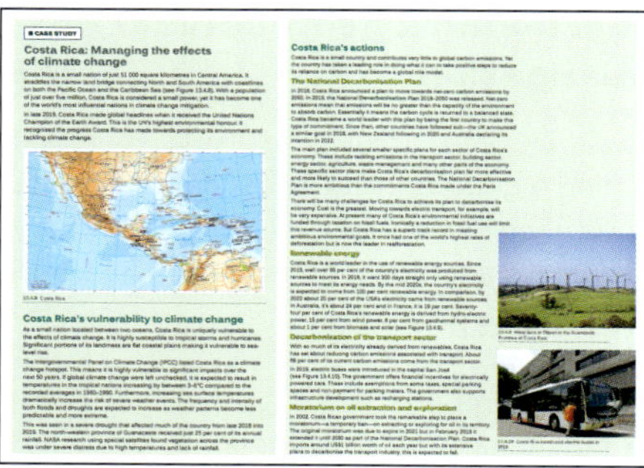

CASE STUDY

Costa Rica: Managing the effects of climate change

Costa Rica's actions

The National Decarbonisation Plan

Costa Rica's vulnerability to climate change

Skills builders

Skills builders are embedded in selected units and concentrate on key geographical skills and concepts.

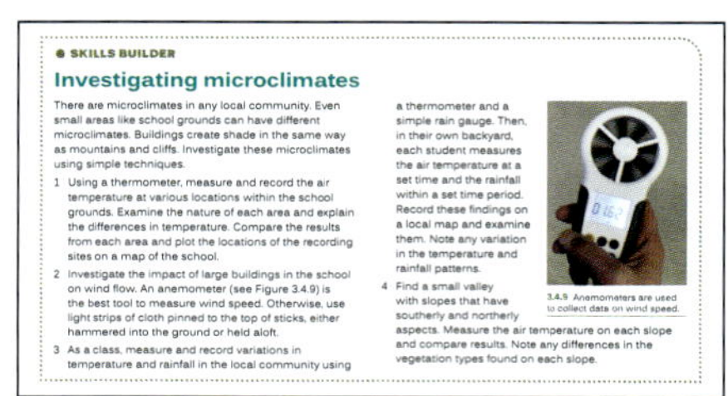

SKILLS BUILDER

Investigating microclimates

There are microclimates in any local community. Even small areas like school grounds can have different microclimates. Buildings create shade in the same way as mountains and cliffs. Investigate these microclimates using simple techniques.

1 Using a thermometer, measure and record the air temperature at various locations within the school grounds. Examine the nature of each area and explain the differences in temperature. Compare the results from each area and plot the locations of the recording sites on a map of the school.
2 Investigate the impact of large buildings in the school on wind flow. An anemometer (see Figure 3.4.9) is the best tool to measure wind speed. Otherwise, use light strips of cloth pinned to the top of sticks, either hammered into the ground or held aloft.
3 As a class, measure and record variations in temperature and rainfall in the local community using a thermometer and a simple rain gauge. Then, in their own backyard, each student measures the air temperature at a set time and the rainfall within a set time period. Record these findings on a local map and examine them. Note any variation in the temperature and rainfall patterns.
4 Find a small valley with slopes that have southerly and northerly aspects. Measure the air temperature on each slope and compare results. Note any differences in the vegetation types found on each slope.

3.4.9 Anemometers are used to collect data on wind speed.

Spotlights

Spotlights offer real-life contexts for key ideas, concepts or issues discussed in the unit. They are designed to develop students' knowledge and understanding of the concepts and processes that are central to the study of geography at this stage of learning.

SPOTLIGHT

Cruising sunk by COVID-19

In its 2019 annual report, the Cruise Line International Association (CLIA) predicted that 32 million people would take a cruise in 2020. There was good reason for this—there had been continual growth since 2009, when 17.8 million people cruised, a 44 per cent growth in a decade.

The 2020 reality was all major cruise lines posted revenue declines of at least 75 per cent. Outbreaks of COVID-19 onboard some cruises saw restrictions on cruise ship movements in many locations. CLIA estimates that the industry lost 520 000 jobs in 2020 as cruise ships were left with skeleton crews of engineers and sailors just to maintain them.

In a bid to stimulate their economies, many governments around the world announced large scale spending on infrastructure during the pandemic. This saw a significant increase in demand for steel. Many cruise ships became worth more as scrap metal. Consequently 2020 saw the scrapping of dozens of previously popular cruise ships, many long before their anticipated end of life (see Figure 15.4.4).

15.4.4 Cruise ships at a scrap yard in Turkey being broken down for scrap steel

Activities

Activities have been created to cater for the full range of student abilities. Many activities are based on the stimulus material presented and aim to facilitate the development of the skills used by geographers.

Using a ladder approach (lower to higher order), three different sets of activities:

- guide students towards an understanding of the content specified in the syllabus
- extend students beyond the text and involve them in a variety of learning experiences
- engage students in higher order tasks, applying inquiry skills and tools.

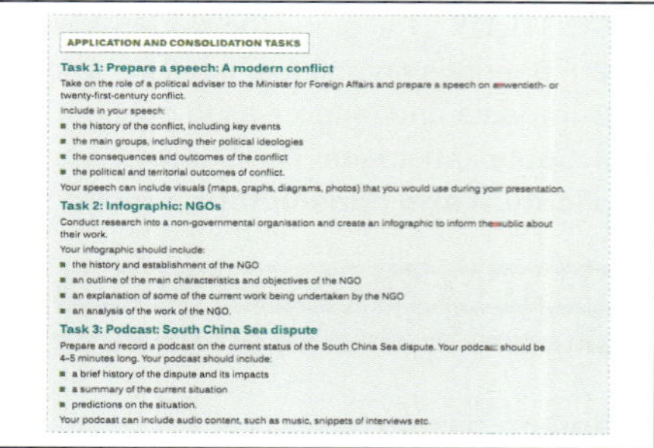

APPLICATION AND CONSOLIDATION TASKS

Task 1: Prepare a speech: A modern conflict

Take on the role of a political adviser to the Minister for Foreign Affairs and prepare a speech on a twentieth- or twenty-first-century conflict.

Include in your speech:
- the history of the conflict, including key events
- the main groups, including their political ideologies
- the consequences and outcomes of the conflict
- the political and territorial outcomes of conflict.

Your speech can include visuals (maps, graphs, diagrams, photos) that you would use during your presentation.

Task 2: Infographic: NGOs

Conduct research into a non-governmental organisation and create an infographic to inform the public about their work.

Your infographic should include:
- the history and establishment of the NGO
- an outline of the main characteristics and objectives of the NGO
- an explanation of some of the current work being undertaken by the NGO
- an analysis of the work of the NGO.

Task 3: Podcast: South China Sea dispute

Prepare and record a podcast on the current status of the South China Sea dispute. Your podcast should be 4–5 minutes long. Your podcast should include:
- a brief history of the dispute and its impacts
- a summary of the current situation
- predictions on the situation.

Your podcast can include audio content, such as music, snippets of interviews etc.

READER+ RESOURCES

Visit the digital toolkit in the Resources folder of Reader+ for student support with key geographical tools, including reading and creating graphs and maps, and understanding statistics. Also, access blank maps and graphic organisers.

To locate these resources in the eBook, go to your Reader+ edition on pearsonplaces.com.au and select VIEW MORE.

To access these resources when you are in the eBook, select the folder icon on the navigation pane.

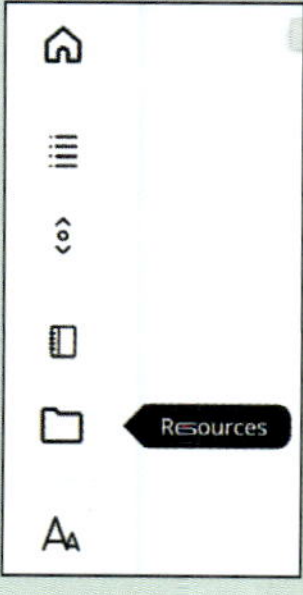

Using Global Interactions 11

Structure

The text is organised into four sections, corresponding with the four focus areas of the Year 11 Geography Syllabus.

1 Earth's natural systems

This section focuses on key processes, cycles and circulations of the atmospheric, hydrological, geomorphic and ecological systems that form Earth's integrated system. It explores Earth's uniqueness and diversity, and land cover change. Three case study chapters investigate land cover change—a topic prominent in the Year 11 syllabus. Each case study chapter investigates one type of land cover, its nature and extent, the land cover changes, the natural and human factors involved, and management practices.

Students are provided with three case study chapters:

- Deforestation
- Glaciers and ice sheets
- Desertification (digital-only chapter)

2 People, patterns and processes

This section explores the extent of human activity on Earth's surface, the processes of our increasingly integrated world, and patterns of human populations and resource consumption. Three case study chapters investigate the role of different human processes in changing environments and shaping the unique character of places, with students developing an understanding through the study of one specific human process at work.

Students are provided with three case study chapters:

- Place and cultural change
- Political power and contested spaces
- Human resilience in diverse environments (digital-only chapter)

3 Human–environment interactions

This section explores the effects of human interactions with the environment on Earth's natural systems over time, and the impacts of humans on land use and land cover. Three case study chapters investigate human–environment interactions to help develop an understanding of the natural and human components and their interactions with each other and the implications of these, with students developing understanding through the study of one specific human–environment interaction.

Students are provided with three case study chapters:

- Climate Change
- Contemporary hazard: Bushfires
- Contemporary hazard: COVID-19

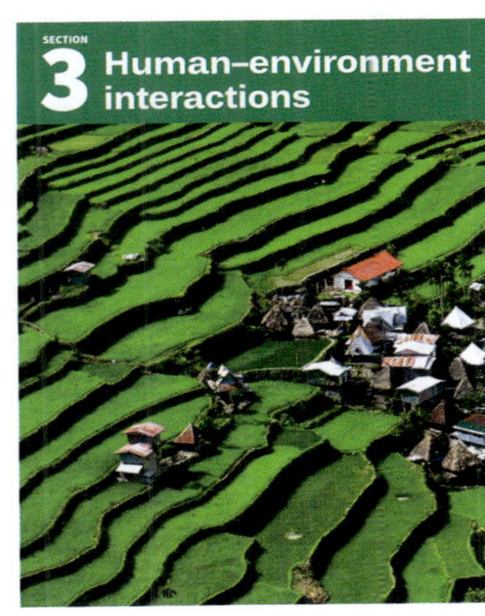

4 Geographical investigation

This section focuses on the geographical investigation students are required to undertake in Year 11.

CHAPTER

1 Geography: An introduction

The Royal Geographical Society (UK) sees geography as building a unique bridge between the social sciences and natural sciences. They say, 'Geography ... makes a vital contribution to our knowledge of the rapidly changing environmental and social challenges facing us and how we should tackle them.'

The Australian Academy of Sciences describes geography as a 'wide-ranging and dynamic discipline concerned with exploring issues affecting the wellbeing of people and places'. It states geography '... provides an understanding of the diversity of environments, places and cultures on this planet, and the inequalities within and between places'. On the utility of geography, the Academy notes '... applying geographical understandings to contemporary issues allows us to integrate knowledge about the natural world, society and the humanities through the perspective of space, place and the environment'.

This introductory chapter offers students a range of important insights into the nature of the discipline of geography—its intrinsic and utilitarian value, its key concepts, its focus on inquiry-based learning and its transferable skills, geographical perspectives and the role of fieldwork.

Our daily lives are interwoven with geography. Each of us lives in a unique place and in constant interaction with our surroundings. Geographic knowledge and skills are essential for us to understand the activities and patterns of our lives and the lives of others.

Gilbert M. Grosvenor, former president and chairman of the National Geographic Society

1.0.1 A computer-generated topographic map

Chapter glossary

atmosphere the gaseous layer surrounding Earth that acts as a barrier between space and the surface

biosphere the living components of an ecosystem; the flora and fauna

fieldwork observing and investigating in the real world

geographical thinking making sense of the world by viewing it through a geographical lens; conceptual understanding developed through the study of geography

geosphere all the rocks that make up Earth's crust; also called the lithosphere

human geography the study of the dynamics of cultures, societies and economies

hydrosphere the interconnected system of water in the atmosphere and lithosphere, including the oceans, ice caps, rivers and groundwater

pedosphere the outermost layer of Earth made of soil

perspectives the viewpoints, theories or worldviews that shape people's values; those deeply held beliefs about what is important or desirable

physical geography the study of the natural features of Earth's surface

transferable skills skills that can be applied in a range of contexts

UNIT 1.1

The nature of geography

Geography is the study of places and the relationships between people and their environments. Geographers explore both the physical properties of Earth's surface and the human societies spread across it. They also examine how human culture interacts with the natural environment and the way that locations and places can have an impact on people. Geography seeks to understand where things are found, why they are there, and how they develop and change over time.

National Geographic Society

Geography prepares young people with the knowledge, skills and understanding to make sense of their world and to face the challenges that will shape our societies and environments at the local, national and global scales.

Dr Rita Gardner CBE, former director, Royal Geographical Society

Geography explains the past, illuminates the present and prepares us for the future. What could be more important than that?

Sir Michael Palin, former president, Royal Geographical Society

Geography is, in the broadest sense, an education for life and living. Learning through geography … helps us all to be more socially and environmentally sensitive, better informed, and more responsible as citizens and employees.

Royal Geographical Society

… the fundamental question of geography is about how humans shaped Earth's surface and how we, in turn, are shaped by the ways in which we have shaped Earth's surface. So, for me, geography was just a set of tools that allowed me to ask these kinds of questions and to try to think through them.

Trevor Paglen, American artist, geographer and author

Geography is a dynamic subject. It explores the complexity of our world. It offers an understanding of the diversity of environments, places and cultures on this planet, and the inequalities within and between places.

Applying geographical understandings to contemporary issues allows us to integrate knowledge about the natural world, society and the humanities through the perspective of space, place and the environment.

In studying geography, students gain the opportunity to investigate geographical issues and phenomena at a variety of scales and contexts. What makes geography unique is the integrated study of Earth's physical and cultural geographies.

Each definition on the left emphasises the relevance of geography.

Physical and human geography

Geography is the only subject that spans both physical and social sciences. Traditionally, it is divided into two parts: **physical geography** and **human geography**, though each is rarely studied in isolation. As a physical science, geography covers the natural environment and the effects of human activities on it. As a social science, it focuses on the spatial aspects of people, communities and cultures. Environmental geographers work across this division, studying the interrelationships between human and physical geography.

● SPOTLIGHT

Physical geography

Physical geography is the study of Earth's natural surface features, to better understand the dynamics of landscapes and environments (see Figures 1.1.1 to 1.1.4). In doing so, geographers study one or more of the following: the **atmosphere** (air and gases surrounding Earth), **hydrosphere** (oceans, rivers and glaciers), **pedosphere** (soils), **geosphere** (rocks) and **biosphere** (flora and fauna). Physical geography is sometimes called earth sciences.

1.1.1, 1.1.2, 1.1.3 and 1.1.4 Physical geography studies Earth's natural surface features.

● SPOTLIGHT

Human geography

Human geography focuses on the dynamics of cultures, societies and economies, and how networks of people and cultures are globally distributed (see Figures 1.1.5 to 1.1.8). Human geographers investigate and find solutions for issues related to climate change, economic development, environmental management, population and cultural change, and urban, rural and regional planning. When studying conflicts and tensions, they identify solutions with environmental sustainability and social justice in mind. Their role in environmental management, urban planning, and processes of social and economic change is crucial.

1.1.5, 1.1.6, 1.1.7 and 1.1.8 Human geography focuses on population, culture, society and economic dynamics.

Activities

Acquiring and processing geographical information

1. Define geography.
2. Explain the difference between physical geography and human geography.

Applying and communicating geographical understanding

3. Read the quotes about geography at the beginning of this unit. As a class:
   - **a** discuss the observations made about the relevance of geography in the contemporary world.
   - **b** explain why you have elected to study geography in your senior school years.
4. To what extent do your reasons reflect the points raised by the quotes?
5. Write your own definition of geography. Share it with your classmates. What elements do the definitions have in common? As a class, formulate a collective definition.
6. Read the definitions of geography at the beginning of this unit again. Which one best reflects the definition your class developed?
7. Read the list of key features of geography in the definitions the text provides. Are there any more features that you can add?

UNIT 1.2

Becoming a geographer

Why study geography?

In an interconnected world, it is vital for students to deepen their knowledge and understanding of why the world functions the way it does. It requires appreciating the nature and extent of the connections between people, places and environments over time and space. In acquiring geographical knowledge and a unique spatial perspective, geography students can engage with some of the great challenges facing humanity.

Consider these attributes of geography.

- Orientates students to the future, empowering them with the knowledge, understanding and skills needed to shape a socially just and sustainable world.
- Inspires curiosity and wonder about the diversity of the world's places, peoples, cultures and environments.
- Provides structure for exploring, analysing and understanding the characteristics of places. It encourages questioning why the world is the way it is and reflection on personal relationships and responsibilities for it.
- Teaches students to respond to questions geographically—to plan inquiries, collect, evaluate, analyse and interpret information, and respond to what is learnt. It also develops broad skills, capabilities and dispositions helpful in life.
- Fosters an appreciation and respect for social, cultural and religious diversity and different **perspectives**, and an understanding of ethical research principles. It encourages skills in teamwork, problem-solving, critical thinking and creativity.
- Encourages regional and global citizenship, making students capable of active and ethical participation.

SPOTLIGHT

Transferable skills

Geography students develop **transferable skills** they can apply in a wide range of jobs:

- planning and conducting inquiries using ethical research principles
- collecting, evaluating, processing and analysing information (e.g. from field observation, data collection, mapping, monitoring, remote sensing, case studies and reports)
- spatial and visual representation and interpretation skills (cartographic, diagrammatic, graphical, photographic and multimodal forms)
- problem-solving, critical thinking and creativity using evidence and logic
- teamwork and time management skills
- being able to respond to and act on the findings of an inquiry or investigation
- developing the transferable skills known as the '4 Cs':
    - critical thinking: understanding problems through carefully analysing data
    - communication: using written, oral and visual forms
    - collaboration: working with experts from other fields to find solutions to problems
    - creativity: drawing on many sources for information and collating it to creatively solve issues.

Thinking geographically

Geographical thinking is when geographers make sense of the world around them by viewing it through a 'geographical lens' (see Figure 1.2.1). This lens consists of the conceptual knowledge and understanding developed by geographers when they integrate information from different sources and use geographical tools and skills to help them inquire about and interpret what they discover.

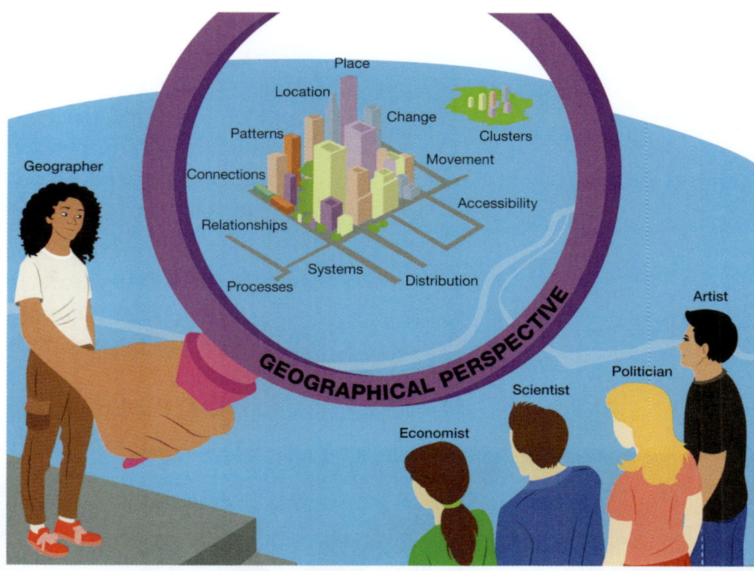

1.2.1 The geographical thinking lens

Geographical concepts

Concepts are abstract ideas that help simplify and understand a complex world. The building blocks of thoughts and beliefs, they shape how we perceive and interpret the world. They are the key to geographical literacy. Once mastered, geographical concepts will forever shape the way we see, interpret and react to those places, issues and events we encounter and engage with throughout our lives.

Conceptual understanding develops all through life, starting with early childhood experiences, like home and school. It expands into adulthood through personal experience to a global understanding of similarities and differences of places, even those not directly experienced.

Geography draws on seven key concepts: place, space, environment, interconnection, sustainability, scale and change (see Figures 1.2.2 and 1.2.3). Considered 'big ideas', these concepts can be applied across the subject to identify a question, guide an investigation, organise information, suggest an explanation or assist in decision-making. They underlie the geographical methods used to investigate and understand the world.

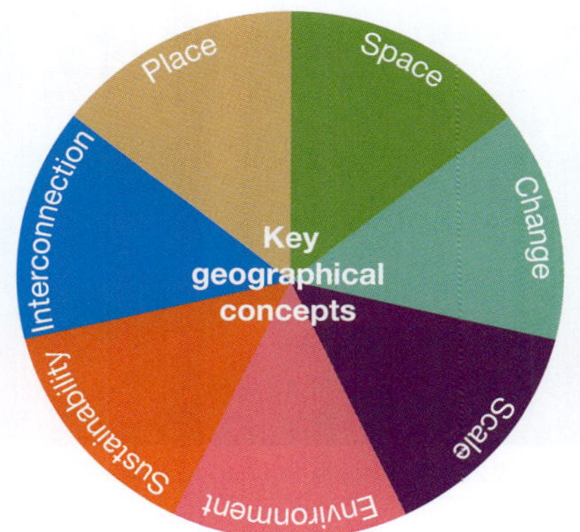

Perspectives in geography

Perspectives are the bodies of thought, theories or worldviews that shape people's values—those deeply held beliefs about what is important or desirable. They are determined by perceptions (how people view and interpret things) and viewpoints (what they think) on issues and events. Understanding different perspectives helps one person critique another's insight into the world. Building the capacity to appreciate others' perspectives takes time.

Concept	Definition
Place	Any part of Earth's surface identified and given meaning from local to global. Places may be perceived, experienced, understood and valued differently. They can be described in terms of their location, shape, boundaries, features and environmental and human characteristics. Some characteristics are tangible (e.g. landforms, people), others intangible (e.g. scenic quality and culture).
Space	The three-dimensional surface of Earth on which everything is located and across which people, goods and information move.
Environment	The totality of our living and non-living surroundings, classified as natural, managed (e.g. farmland) or constructed.
Scale	Differences in spatial level, from personal to local, national, regional and global. Used to examine geographical phenomena and problems.
Sustainability	The ongoing capacity of an environment to meet present needs without compromising future generations' ability to meet their needs.
Change	Any act or process through which something becomes different. Being aware of change over time and space is important. Change helps us understand what is happening around us, and to see the world as dynamic.
Interconnection	Geographical phenomena are not isolated, but connect through various processes, across space (e.g. environmental, people movement, trade and investment flows, goods and service purchases, cultural influences, the exchange of ideas and information, political power, and international agreements).

1.2.2 and 1.2.3 Seven key geographical concepts

Perspective is determined not just by physical position, but also by a person's mindset or worldview, which gives order and meaning to their surroundings. These viewpoints are adopted through social positioning or belonging to a particular group. Exploring different perspectives builds knowledge and enhances understanding of geographical issues and phenomena. For example, reflect on the perspective of an impoverished farmer who needs to clear forest trees for agriculture. Now consider a conservationist who is concerned with forest preservation. Valuable insights emerge.

People perceive, describe, explain and evaluate geographical issues and phenomena differently. But perspectives are not just personal. A person's worldview is influenced by their gender identity, ethnicity, cultural background, socioeconomic status, sexuality, indigenous heritage and more.

Geographical inquiry and skills

1.2.4 Fieldwork is used to gather data.

An inquiry-based approach to learning is central to studying geography. It involves acquiring, processing and communicating geographical information. Such an approach develops the skills needed to explain spatial patterns, evaluate consequences and manage complex places and environments. **Fieldwork** and tools, including mapping and spatial technologies, are fundamental (see Figure 1.2.4). Geographical inquiries help students develop and apply inquiry skills, including:

- asking distinctly geographical questions
- planning an inquiry and evaluating information sources
- collecting, processing, analysing and interpreting information
- reaching conclusions based on evidence and logical reasoning
- evaluating and communicating findings
- reflecting on the inquiry and responding, through action, to what was learnt.

The relationship between geographical inquiry and geography tools and skills

Geographical inquiries involve individual or group investigations, starting with geographical questions. A series of steps follows—collecting, evaluating, analysing and interpreting information. Conclusions are developed and actions are proposed. Geographical tools and skills are the techniques that geographers use in such investigations. These tools and skills are typically grouped under five subheadings aligned to the stages of a complete investigation:

- observing, questioning and planning
- collecting, recording, evaluating and representing
- interpreting, analysing and concluding
- communicating
- reflecting and responding.

The tools students use in Years 11 and 12 include maps, graphs and statistics, spatial technologies, visual representations and fieldwork.

● SPOTLIGHT

Fieldwork

Fieldwork enables students to apply classroom-based learning to real-world geography. It means observing geographical features and processes, spatial patterns and relationships, and geographical issues outside the classroom (see Figure 1.2.5). The key features of fieldwork involve:

- connecting learning experiences to the outside world
- developing geographical thinking skills
- undertaking tasks individually or in groups
- exploring different perspectives on many geographical issues
- gathering information and fostering understanding by observing geographical processes
- fostering active learning to sharpen perception by encouraging observation, problem identification, and questioning and testing hypotheses
- engaging in larger research projects that offer freedom to decide what to investigate and where
- employing a range of methodologies, including surveys (e.g. face-to-face interviews and web-based questionnaires), observation and experiments
- using fieldwork to collect primary data from external sources.

Students in Year 11 undertaking fieldwork should be able to:

- determine the fieldwork's inquiry questions with teacher guidance
- decide which methodologies and data sources to employ
- analyse the data collected and independently make decisions or reach conclusions
- reflect on the data-collection methods used and the data's validity.

1.2.5 Drones are used in fieldwork.

Activities

Acquiring and processing geographical information

1. Define transferable skills. Identify which can best be applied in the career you want to choose.
2. State what is meant by the terms geographical thinking and geographical lens. What does the lens consist of?
3. What does the mastery of geographical concepts enable us to do?
4. Explain what is meant by the term inquiry-based approach. What are the inquiry skills developed in geography?

Applying and communicating geographical understanding

5. Reorder the points under 'Why study geography?' to best reflect your motivation to study geography.
   - **a** Identify which points are of intrinsic nature (of value by its very nature) or are more utilitarian (or functional).
   - **b** Discuss how perspective influences your new order.

SECTION

1 Earth's natural systems

Earth is constantly evolving. Understanding its complexity means studying its component parts, their interactions, functions and anticipated changes. The interconnectedness of its natural systems is best approached through its four components—the atmosphere, hydrosphere, lithosphere and ecosphere. Each is a vast reservoir of material, with flows of matter and energy passing between them.

Geographers use the systems concept to break down complex phenomena into smaller components. This enables the planet to be viewed as a system of separate but interacting parts. The oceans, atmosphere, landmasses, lakes and rivers, soil, plants and animals can be studied independently, yet each remains interdependent. This systems-based approach focuses on the processes, circulations and cycles, within and between the parts.

Earth is a closed system because only energy—not matter—is transferred or exchanged with its surroundings. Energy from the Sun enters and leaves Earth's system as radiation. Being a closed system has two vital implications. First, the amount of matter in a closed system is fixed. Earth's natural resources are all that are available and any wastes remain within the planet's system. Second, changes made in any part of the system will eventually affect the other parts.

Content focus

In this section, students have the opportunity to investigate Earth's natural systems and develop an understanding and appreciation of the wonder, complexity and dynamics forming the integrated Earth system. In doing so, they analyse natural processes and cycles to develop an appreciation of the connections between the atmosphere, lithosphere, biosphere and hydrosphere. They investigate land cover change and have the opportunity to focus on either deforestation, melting glaciers and ice sheets or desertification.

The authors also address the role of humans as a key element of Earth's natural systems and the extent to which people destabilise natural processes and contribute to environmental change on a range of scales. In doing so, we embrace an anthropogenic perspective, one that holds that Earth's most recent geological period (the Anthropocene) is human influenced and that there is now overwhelming evidence that atmospheric, geologic, hydrologic, ecological and other Earth system processes all bear the impact of humans.

In this section

CHAPTER

2 The uniqueness and diversity of Earth

Sir David Attenborough, the famous British naturalist and broadcaster, described the natural world as 'the greatest source of excitement; the greatest source of visual beauty; the greatest source of intellectual interest. It is the greatest source of so much in life that makes life worth living'. He believes, 'It's surely our responsibility to do everything within our power to create a planet that provides a home not just for us, but for all life on Earth.'

This chapter introduces Earth as a source of wonder, spiritual connection and universal value. The focus is on inspirational landscapes, biodiversity hotspots, great wildlife migrations and the global commons. It examines alternative worldviews, Aboriginal and Torres Strait Islander peoples' connection to Country, and the overview effect. Landscapes explored are land and ocean environments, ice sheets and drainage basins.

The more clearly we can focus our attention on the wonders and realities of the universe about us, the less taste we shall have for destruction.

Rachel Carson, American marine biologist and author

2.0.1 The 12 Apostles, Port Campbell National Park, Victoria

Chapter glossary

anthropocentrism the worldview that humans are the central or most significant beings in the world

biodiversity hotspots areas that are highly endangered centres of biodiversity

bycatch the incidental capture of non-target species, including dolphins, marine turtles and seabirds

Country often used by Aboriginal and Torres Strait Islander peoples to describe a specific place to which they are connected. The concept of Country is complex and incorporates cultural, physical, linguistic and spiritual features

denudation the sum of weathering, mass-wasting and erosional processes that result in the progressive lowering of Earth's surface

drainage basin an area drained by a river and all of its tributaries

ecocentric a worldview explicitly recognising the intrinsic value of Earth's ecosystems and their biological and physical elements

ecosystem services the outputs, conditions or processes of natural systems that directly or indirectly benefit humans or enhance social welfare

endemic species restricted to a geographical area, and which do not occur naturally in any other part of the world

evaporation the process by which water changes from a liquid to a gaseous state

firn compacted snow that will eventually become glacial ice

geomorphology the scientific study of the origin and evolution of features created by physical, chemical or biological processes operating at or near Earth's surface

global commons global resources; Earth's shared natural resources, such as the oceans, the atmosphere and Antarctica.

High Seas those parts of the world's oceans beyond the exclusive economic zones of countries

intrinsic value the environmental perspective that nature has value, independent of potential human uses

land lores the complete body of knowledge, customs, rules and traditions regarding the land

littoral rainforest a forest distinguished by its proximity to the ocean (generally <2 km) with a closed canopy

tectonic relating to the structure of Earth's crust

topographical the forms and features of land surfaces

watershed the dividing ridge between drainage areas

UNIT 2.1

Nature as a source of wonder

Did you know?

Lions dominate their environment; they are the only predators on the savanna capable of killing Africa's largest herbivores, including elephants and giraffes (see Figure 2.1.1).

2.1.1 Male lions roar to show their power.

The natural world fascinates us—whether we simply glance at passing clouds or stand in awe viewing a spectacular vista. Engaging with nature provides solace, feelings of ease and being grounded. The naturalist, John Muir wrote, 'When we try to pick out anything by itself, we find it hitched to everything else in the universe.' Indigenous peoples have a spiritual and cultural attachment to the natural world.

Inspirational landscapes

Geographers often ask, what makes a landscape inspirational? Studies reveal many people prioritise naturalness, especially landscapes featuring unique landforms, vegetation and spectacular water features (see Figure 2.1.2). Inspirational landscapes provoke deep feelings of connection to a place. This preference often goes beyond an aesthetic response. Some people find greater inspiration in landscapes dominated by cultural features, often because they have spent time there and developed a strong emotional attachment to it. There are often enduring historical cultural associations for indigenous peoples.

The perception of landscapes is highly individual. Being inspired by nature can promote creativity, reflection and action to protect a place. Everyone views the world through a lens shaped by their life experiences. This is what is meant by a person's perspective or worldview.

2.1.2 Many consider the spectacular northern lights (Aurora borealis) inspirational, seen here at Kirkjufell, Iceland.

● **SPOTLIGHT**

Victoria Falls

Some landscapes are so breathtakingly beautiful that they enrich and inspire people. Lying on the Zimbabwe and Zambia border, Victoria Falls is the world's largest waterfall by water volume (see Figure 2.1.3). Known as *Mosioa-Tunya*, or 'smoke that thunders', it describes the experience encountered by visitors. As a World Heritage site, holding outstanding universal value to humanity, the falls are protected.

2.1.3 Victoria Falls on the Zambezi River

Biodiversity hotspots

Biodiversity hotspots are the places richest in biodiversity—there are over 30 recognised worldwide. At high risk for destruction, the classification means a region has lost at least 70 per cent of its original natural vegetation, usually due to human activity such as deforestation. While biodiversity hotspots make up less than 3 per cent of the land surface, they represent 44 per cent of the world's plant species and 35 per cent of land vertebrate species. Most plants in such hotspots are **endemic**—found nowhere else.

Some examples of biodiversity hotspots are:

- the Tropical Andes Biodiversity Hotspot is the world's most diverse, harbouring about one-sixth of the world's plant species.
- the islands of the New Zealand archipelago evolved in isolation, giving rise to many unique species. Over 90 per cent of its insects and 80 per cent of its plants are only found there.
- the mountains of Southwest China are home to unique temperate coniferous forests found in the river valleys of the Tibetan Plateau.
- the Philippines has more than 7000 islands, with larger islands holding more unique species than most countries. This includes 600 recorded indigenous bird species.
- the unique rainforests of Belize (see Figure 2.1.4) are home to the world's highest mammalian endemism and the second highest amphibian, bird, reptile and non-fish vertebrate endemism.

Australia's two biodiversity hotspots are southwestern Australia and the Eastern Australian temperate forests (see box, Spotlight: Eastern Australian temperate forests).

Many biologically diverse environments are being eliminated, degraded and fragmented by human activity. This includes tropical forests, wetlands and coral reefs that would otherwise see new species emerge and evolve. Also, the **ecosystem services**, on which all life depends, are being disrupted.

2.1.4 The rainforests of Belize are a biodiversity hotspot.

SPOTLIGHT

Eastern Australian temperate forests

The Eastern Australian temperate forests extend along the Great Dividing Range, from the east coast of New South Wales to the South Coast, and into the south-eastern corner of Queensland. Within the dry and wet sclerophyll eucalyptus forests (see Figure 2.1.5) lie pockets of ecologically significant subtropical and **littoral rainforests**, warm temperate rainforest, deciduous rainforest and cool temperate rainforest. Some eucalyptus forests have been cleared for development, agriculture or logging, but much is protected in national parks. The Gondwana Rainforests of Australia hotspot is within a World Heritage Area. It contains the world's most extensive areas of subtropical rainforest and Antarctic beech cool-temperature rainforest, and large areas of warm temperate rainforest. Primitive plants and animals, from which other life evolved, live here.

2.1.5 Wet sclerophyll eucalyptus forest in the Eastern Australian biodiversity hotspot

Wildlife migrations

Animal migrations can cover staggering distances and scales. The longest-distance movements of animals are usually seasonal, but can also be linked to the local availability of food or mating. Such behaviour occurs in all major animal groups, including birds, mammals, fish, reptiles, amphibians, insects and crustaceans.

Of the world's 10 000 bird species, around 1800 migrate long distances each year. Migrations are often seasonal, with feeding and breeding in high northern latitudes during summer, and migrating south for winter. The Arctic tern (a tiny 120 grams) migrates further than any animal—up to 80 000 kilometres per year from its Arctic breeding grounds to Antarctica. Migrations reveal the interconnectedness of the world's ecosystems and habitats.

● **SPOTLIGHT**

Australia's great whale migrations

Humpback whales migrate along Australia's east coast (see Figure 2.1.6) from the icy Southern Ocean, where they develop an insulating layer of fat by consuming vast amounts of food, principally krill. Lacking this fat, newborn whales need warmer water while they build up their fat stores from their mothers' rich milk. Humpbacks head north in May and June to the waters off North Queensland to breed, returning south with newborns from August to November. Once hunted, humpbacks are now protected, with their numbers rising from 5000 to 135 000. However, entanglement in fishing nets and collisions with ships remain threats. Southern right whales and sometimes blue whales, minkes and orcas (killer whales) also migrate.

2.1.6 Humpback whale breaching off the coast of NSW

Activities

Acquiring and processing geographical information

1. Briefly outline the importance of the natural world and all its wonders.
2. Explain what makes a landscape inspirational. Why is it considered important?
3. Explain what biodiversity hotspots are. Why are they considered important?

Applying and communicating geographical understanding

4. Select one of the biodiversity hotspots that is not used as an example in the text. Research why that place was designated a hotspot.
5. Study the box, Spotlight: Eastern Australian temperate forests. Why are these forests worthy of protection?
6. Study the text about wildlife migrations. Investigate one animal migration and share your findings with your class.

UNIT 2.2

People's connection to the natural world

There are many ways to view our connection with the natural world. One way is by relating to something greater—aesthetically, emotionally and spiritually. Being immersed in nature provides a range of mental and physical benefits. In an increasingly urbanised, technological world, many people feel separated from nature.

Worldviews

The ways people relate to the natural world range from those best described as **ecocentric** to those that are **anthropocentric**. These represent the two ends of a continuum of environmental worldviews.

Ecocentric worldviews hold that humanity is part of nature and that all forms of life have inherent value. Humans are custodians who preserve biodiversity and help it to function. Every organism can evolve and adapt to changing environmental conditions. Resources are limited, so economic growth is only encouraged when it is environmentally beneficial.

Anthropocentric (or human-centred) worldviews hold that elements of the natural world have value only because of their usefulness to people. That is, humans are viewed as environmental masters rather than custodians. Earth's resources are seen as being unlimited. If shortages do arise, substitutes can be found. It is argued that fossil fuels are the key to sustained economic growth and advances in human wellbeing. Those who do believe the planet is warming will argue that technological advances will fix the problem.

Aboriginal and Torres Strait Islander Peoples' connection to Country

Across the globe, many indigenous peoples developed deep spiritual links with their traditional lands over thousands of years. In Australia, Aboriginal and Torres Strait Islander peoples believe the land owns the people and every aspect of life is connected to the land. In their society, life originated in the **land lores** and they are governed by it (see Figure 2.2.1).

2.2.1 Narrawallee Inlet is part of the traditional Country of the Jerrinja People.

Aboriginal and Torres Strait Islander peoples' identity and sense of belonging is deeply connected to **Country**. All aspects of their identity—family, language, culture and spirituality—are connected to their Country. They are part of the land, for it is the land (their mother) who gave birth to them and they have a responsibility to care for and look after her until they finally return to her. Many non-Indigenous Australians may find this strong connection to Country a little difficult to comprehend.

Today, the responsibility to care for Country remains strong for many Aboriginal and Torres Strait Islander peoples. The health of the land and sea is central. Plants, animals and ecosystems are deeply valued. Using wild foods and traditional medicines helps pass on cultural knowledge. Their connection is spiritual and cultural, obliging them to caretake cultural sites, including sacred sites, archaeological sites, waterholes and burial grounds. The land and water are central themes in Aboriginal and Torres Strait Islander art, including theatre, dance, music and painting.

The 'overview effect': Seeing Earth from space

2.2.2 Earthrise viewed from the Moon during the Apollo 8 mission, 1968

Once a photograph of Earth, taken from outside is available ... a new idea as powerful as any in history will be set loose.

Fred Hoyle, English astronomer

In December 1968, the crew of Apollo 8 turned the spacecraft's camera back toward Earth (see Figure 2.2.2). The images it beamed back to mission control forever changed the way people thought about their planet. Set against the vastness of space, the blue, pearl-like sphere—now home to more than 8 billion people—looked small and very vulnerable. Astronauts seeing Earth from space talk about an overview effect—an experience that has transformed their perspective of the planet and humanity's place on it. They perceive it as our shared home, without boundaries between nations or species.

It suddenly struck me that that tiny pea, pretty and blue, was Earth. I put up my thumb and shut one eye, and my thumb blotted out the planet Earth. I didn't feel like a giant. I felt very, very small.

Neil Armstrong, American astronaut

Activities

Acquiring and processing geographical information

1. How can noticing and appreciating the natural world benefit humans?
2. Distinguish between ecocentric and anthropocentric worldviews. Outline the characteristics of each.
3. Explain why Country is important to Aboriginal and Torres Strait Islander peoples.
4. Why do Aboriginal and Torres Strait Islander peoples have such a sense of responsibility towards their Country?
5. Explain what is meant by the overview effect.

Applying and communicating geographical understanding

6. Write a paragraph describing an experience through which you gained a new appreciation and understanding of the value of the natural world.

UNIT 2.3

The universal value of Earth's environments

Intrinsic value is a concept that describes nature as having a value independent of human use, reflecting an ecocentric worldview. It gives nature, ecosystems and landscapes value, even if they do not benefit humans. Nature can have an aesthetic, emotional and/or spiritual value and, in some instances, an economic value, such as being a tourist destination.

The global commons

Many global resources are land-based and subject to the territorial claims of nations. Other resources (including the atmosphere, the open ocean, outer space and the polar lands, especially Antarctica) are part of what is called the **global commons**, a term describing resources that no individual or country owns or has legal responsibility for. They are shared resources.

It is especially challenging to effectively manage global commons because governance structures and management systems must be designed to address often conflicting public and private interests.

The ocean commons

The **High Seas** belong to everyone. Covering 60 per cent of the ocean's surface, these largely unprotected waters support an amazing array of marine creatures and are a vital source of food. Countries govern oceans within their exclusive economic zones (EEZ)—generally 370 kilometres from shore. The High Seas sit outside any authority.

The UN Convention on the Law of the Sea (UNCLOS) is the principal international treaty regarding the sea. Initiated in 1956 and legally in force since 1994, its rules govern territorial boundaries (22 kilometres from shore), resource management and the rights of states within their exclusive economic zones. The International Tribunal for the Law of the Sea can resolve disputes. Most countries, except the United States, have ratified UNCLOS.

The ocean commons are also home to major fisheries, with more than 3 million vessels catching 78.8 million tonnes of seafood in 2020. With fish populations rapidly declining, fisheries have collapsed. Many believe industrial-scale fishing is unsustainable. There are environmental impacts on non-targeted marine life, called **bycatch**. There is an urgent need for better management.

2.3.1 Fish stock depletion is only one of the issues impacting the ocean commons.

Fishing (see Figure 2.3.1) and pollution are the main causes of declining ocean health. Entanglement in abandoned plastic and nylon nets are a major threat to marine life. Widespread overfishing and microplastics are disrupting food chains, endangering ecosystems and human health.

Better management means fishing sustainably. This involves imposing fishing quotas, eliminating destructive and illegal fishing practices, establishing protected fish-breeding areas, restoring collapsed fisheries, educating stakeholders and the public, and developing sustainable fishing certification programs.

● SPOTLIGHT

The polluted atmosphere commons

The atmosphere has no territorial boundaries and cannot be exploited as a sovereign right by individual countries. The atmosphere belongs to no-one and everyone. It is a global common.

The atmosphere is a complex dynamic natural system that supports all life (see Chapter 3). Unfortunately, it receives an array of airborne pollutants that can harm life, including people and food crops (see Figure 2.3.2). They can be chemicals, solid particles, liquid droplets or gases, and be natural or human-made. Reducing pollutants is a massive global challenge because it is impossible to restrict them to the place they originate from or to contain them within country borders. The most damaging is carbon dioxide (discussed in detail in Chapter 13).

International action has had some success in mitigating the impacts of some pollutants. The Montreal Protocol on Substances that Deplete the Ozone Layer is one. Adopted in 1987, the Protocol had some success in reducing damage to the ozone layer. The ozone layer is the shield protecting Earth's surface from harmful ultraviolet (UV) radiation levels. It is the only United Nations treaty to be ratified by every country. Its landmark multilateral environmental agreement regulates the production and consumption of nearly 100 human-made chemicals that damage the ozone layer.

2.3.2 Pollutants entering the atmosphere can have a global impact.

Activities

Acquiring and processing geographical information

1. What is meant by the term intrinsic value?
2. Describe what the term global commons refers to. Why does the protection and management of the global commons prove so difficult?

Applying and communicating geographical understanding

3. Study the text on the ocean commons.
   a. Describe what constitutes the High Seas.
   b. Outline the regulatory framework and institutions that govern the ocean commons.
   c. Outline the issues impacting the wellbeing of the ocean commons. How are these issues being addressed?
4. Study the box, Spotlight: The polluted atmosphere commons.
   a. Explain why it is so difficult to manage the impacts of pollution on the atmosphere.
   b. Outline the forms atmospheric pollution can take.
   c. Describe how humans have sought to mitigate the impacts of atmospheric pollution.

UNIT 2.4

Earth's landscapes and biophysical features

Our planet yields a spectacular variety of natural landscapes and biophysical features. They include all landmasses and oceans, terrestrial and marine environments, geomorphic landscapes, permanent ice sheets and drainage basins. The interacting components of the planet's natural systems produce this diversity—the atmosphere, hydrosphere, geosphere and biosphere.

The processes creating and transforming Earth's features operate on broad time and spatial scales (see Figure 2.4.1). Some occur abruptly and result only in local changes. Examples include storms, floods, landslides, earthquakes and volcanic eruptions. Other processes are extremely slow and almost impossible to see. Examples of places associated with **tectonic** processes and the rock cycle include landmasses and oceans, terrestrial and marine environments, geomorphic landscapes, ice sheets and glaciers, and drainage basins.

2.4.1 Rivers and waterfalls in southern Iceland—a geomorphic landscape shaped by hydrological processes

Landmasses and oceans

A landmass is a large area of land where Earth's crust extends above sea level. There are seven continental landmasses. Largest to smallest, these are Asia, Africa, North America, South America, Antarctica, Europe and Australia (see Figure 2.4.2). They account for 148 647 000 square kilometres of land, or 29.1 per cent of the planet's surface (510 065 600 square kilometres). They make up most, but not all, of the land surface. The rest consists of non-continental islands.

The world's oceans make up most of the hydrosphere. They cover 361.9 million square kilometres (70.8 per cent) of Earth's surface, with a total volume of roughly 1332 million cubic kilometres (see Figure 2.4.3). The Pacific Ocean is the largest, covering more than 30 per cent of the planet. With a surface area of more than 155 million square kilometres, the Pacific basin is larger than the landmass of all the continents combined.

Oceans provide a range of environmental services that sustain life. They produce more than half the oxygen we breathe and, as a vital part of the water cycle, they provide most of the precipitation that sustains our water supply. They are also an important source of food. Every year, humans take 100 million tonnes of seafood from the oceans. Figure 2.4.4 shows the ecosystem and economic services provided by the world's oceans.

Continent	Area (km^2)	Percentage of total landmass
Asia	44 579 000	29.5
Africa	30 370 000	20.4
North America	24 709 000	16.5
South America	17 840 000	12.0
Antarctica	14 000 000	9.2
Europe	10 180 000	6.8
Australia	8 600 000	5.9

2.4.2 The relative size of continents as a percentage of terrestrial surface

Did you know?

There's no precise definition for a continent. With an area of about 7.5 million square kilometres, Australia is the smallest recognised continent. Greenland doesn't count as a continent and is considered the world's largest island. It extends over 2.1 million square kilometres and is part of North America. The cut-off for a continent lies somewhere between Australia and Greenland.

Continent	Area (km²)	Percentage of total landmass	Average depth (m)	Length of coastline (km)
Pacific Ocean	168 723 000	46.6	3970	135 663
Atlantic Ocean	85 133 000	23.5	3646	111 866
Indian Ocean	70 560 000	19.5	3741	66 526
Southern Ocean	21 960 000	6.1	3270	17 968
Arctic Ocean	15 558 000	4.3	1205	45 389

2.4.3 The relative size of oceans, average depth and coastline length

Ecosystem services

- Oxygen produced via the process of photosynthesis
- Carbon dioxide absorption
- Climate moderation
- Nutrient recycling
- Water purification
- Biodiversity and habitats
- Reduced storm damage (coastal wetlands, barrier islands and mangrove forests)

Economic services

- Food
- Energy from waves and tides
- Pharmaceuticals
- Transport routes
- Recreation and tourism
- Employment
- Minerals

2.4.4 The ecosystem and economic services provided by oceans

● **SPOTLIGHT**

The role of oceans in regulating climate

The most vital role of oceans is keeping our planet warm. They store solar radiation, distribute heat and moisture around the globe, and drive weather systems. They absorb most solar radiation, especially in tropical and equatorial waters. The oceans release this heat slowly, moderating temperatures.

Oceans help distribute heat around the globe via **evaporation**. As water molecules are heated, they break away from the water body and move into the atmosphere, increasing the air temperature and humidity. Rain and storms develop and are carried by trade winds. Almost all rain falling on land originates in the oceans. The tropics are particularly wet because the heat absorption, and therefore the rate of evaporation, is highest in these regions.

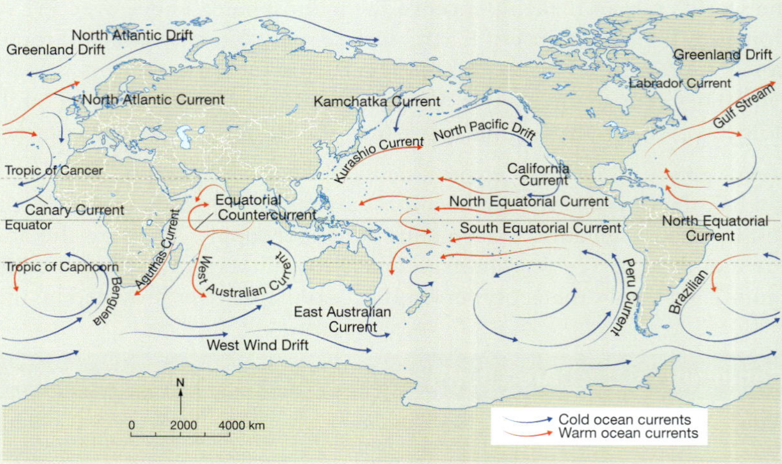

2.4.5 Circulation flows in the world's oceans

Ocean currents drive weather patterns. These are movements of water in a continuous flow created largely by prevailing surface winds, but also temperature and salinity variations, planetary rotation and tides. The currents typically flow clockwise in the Northern Hemisphere and anticlockwise in the Southern Hemisphere. Often tracing coastlines, these currents transport warm water and precipitation from the Equator toward the poles and cold water from the poles toward the Equator (see Figure 2.4.5). This regulates global climate, balancing the uneven distribution of solar radiation over the surface. Without these currents, temperatures would be far hotter at the Equator and colder at the poles, making the planet largely uninhabitable.

Oceans play a critical role in climate change. They have absorbed 93 per cent of the extra energy generated by the greenhouse effect, with warming observed at depths of 1 kilometre. This has led to changed ocean currents, increased stratification within water bodies (because of less mixing of water), oxygen depletion and changes to the distribution of marine plant and animal species. At the same time, weather patterns are changing, with extreme events increasing in frequency and severity.

Terrestrial and marine environments

Terrestrial environments are land-based ecosystems consisting of biotic (living) and abiotic (non-living) components in an area. Examples include deserts, tropical rainforests, deciduous forests, grasslands, tundra and taigas. The environment in a particular place depends on the interaction between temperature, precipitation, soil type and amount of light received.

Marine environments include the abyssal plain and the continental shelf. The abyssal plain is the deep ocean floor (3000–6000 metres below sea level) covering over half of Earth's surface. Continental shelves are found off the coast of the continents and are home to coral reefs, seagrasses and kelp forests. Mangroves, salt marshes, mudflats, and rocky and sandy shores are found where terrestrial and marine environments meet.

Marine ecosystems can be highly productive. Coral reefs, estuaries, salt marshes and mangrove forests teem with life (see Figure 2.4.6). Others, like the abyssal plain, are largely devoid of life, the few life forms that do live there are in constant darkness. Unable to photosynthesise, these organisms feed on dead organic detritus falling from above.

2.4.6 Coral reef, Raja Ampat, Indonesia. While reefs occupy less than 0.1% of the world's ocean area, they are home to at least 25% of all marine species.

Geomorphic landscapes

Geomorphic landscapes result from a distinctive set of tectonic and geological processes, and include volcanic, riverine (see Figure 2.4.7), arid and coastal landscapes. Such landscapes are created by physical, chemical or biological processes. The study of their origin and evolution is called **geomorphology**.

Large-scale **topographical** features are the product of the interaction of surface and subsurface processes. Mountain ranges are uplifted by tectonic processes and worn down and shaped by the processes of **denudation** (weathering and erosion) and the deposition of eroded materials. Small-scale features develop in response to the balance between uplift, erosion, deposition and subsidence.

2.4.7 The Grand Canyon in Arizona, USA, is a geomorphic landscape. The layers of rock reveal two billion years of geological history. The 446 km long, 29 km wide, and up to 1.8 km deep canyon was formed by the Colorado River cutting down through layer after layer of rock as the Colorado Plateau was slowly uplifted by forces deep within Earth.

Ice sheets and glaciers

Glaciers are slowly moving rivers of ice that move downslope under the influence of gravity and their own mass. They form where the snow accumulation exceeds its ablation over many years, often centuries. Glaciers can move quite fast—in some instances up to 1 kilometre a year.

An ice sheet is defined as any mass of glacial ice over 50 000 square kilometres in size. There are only two ice sheets, one in Antarctica (see Figure 2.4.8) and the other in Greenland. Combined they contain 99 per cent of the world's freshwater. An ice sheet extending beyond the coastline over the ocean is called an ice shelf. A mass of glacial ice covering an area under 50 000 square kilometres is called an ice cap, and a series of connected ice caps is an ice field.

Ice sheets form as snow accumulates year on year and is ultimately compressed to form ice or **firn**. Layers of firn eventually build up on top of each other. When the ice grows thick enough (about 50 metres), the firn fuses into a huge mass of solid ice. Ice sheets tend to be slightly dome-shaped and spread out from their centre, a process caused by the weight of the ice. Ice sheets and glaciers are studied in detail in Chapter 6.

2.4.8 The ice sheet over the Antarctic continental land mass

Did you know?

Coastlines don't indicate the actual boundaries of the continents. Continental areas are defined by their continental shelves. This is a gently sloping area extending outward from beaches far into the ocean. It is both part of the ocean and part of the continent.

● SPOTLIGHT

Largest ice sheet

At over 400 kilometres long and 2500 metres thick, the world's largest glacier, or ice stream, is the Lambert Glacier in Antarctica. It can move up to 1200 metres a year.

The Antarctic ice sheet is the largest accumulation of ice on Earth (see Figure 2.4.9). It covers more than 14 million square kilometres and averages over two kilometres thick. It contains about 30 million cubic kilometres of water, enough to raise global sea levels by 60 metres. The Greenland ice sheet is only 1.7 million square kilometres.

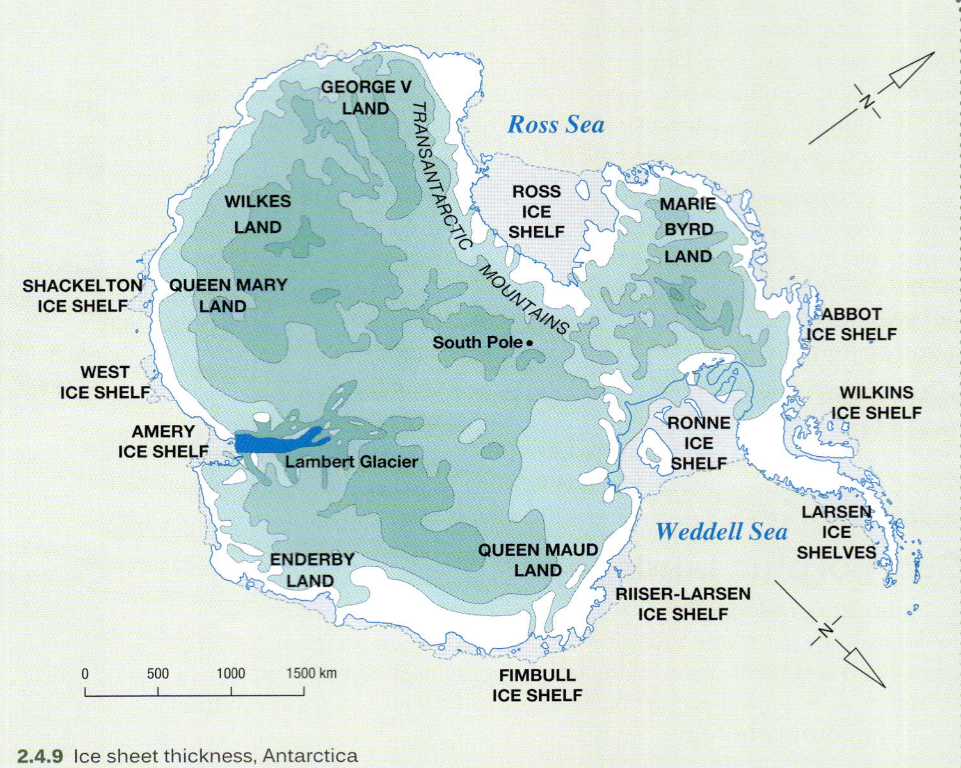

2.4.9 Ice sheet thickness, Antarctica

2.4.10 Aerial view of Meghna River and the Ganges Delta, India, the world's largest delta

Drainage basins

Running water has shaped almost every land surface, even where rainfall is low and infrequent. Landforms resulting from river and stream activity are part of complex drainage systems. These develop when precipitation collects and moves downslope to the nearest sea or ocean, or even inland depressions such as Lake Eyre in South Australia.

Drainage systems play a critical role in the hydrological cycle: water evaporates from water bodies (primarily oceans) and some of it falls as precipitation onto land. Some infiltrates the ground, while some flows across the surface. Streams—water flowing downslope in a well-defined channel—erode the land, and transport and deposit sediment (alluvium) on their way to the sea (see Figure 2.4.10).

Water moving downslope in thin sheets is called overland flow. This eventually concentrates in well-defined channels to become streamflow. Run-off is a combination of both types.

Every stream has its own **drainage basin**—the area contributing water to the stream. Australia's largest drainage basin, the Murray–Darling, covers 1 million square kilometres (14% of the landmass). **Watersheds** separate adjacent drainage basins which are often nested in a hierarchy. The Mississippi River system in the USA has 10 major tributaries, each with its own catchment (see Unit 3.8).

The world's largest drainage basins

1 The Amazon Basin, South America, covers over 7 million square kilometres; the 6575-kilometre-long Amazon and its tributaries drain into the Atlantic Ocean.

2 The Congo River Basin, west-central Africa, covers 3.4 million square kilometres and is home to vast tropical forests.

3 Nile River Basin, North Africa, covers 3.25 square kilometres. Stretching 6695 kilometres, the Nile is the world's longest river and flows into the Mediterranean Sea.

4 The Mississippi River catchment covers all or part of 32 states of the USA and two Canadian provinces, and area of over 3.2 million square kilometres (see Spotlight box in Unit 3.8).

Activities

Acquiring and processing geographical information

1 Account for the diversity of Earth's distinctive biophysical features.

2 Define terrestrial environment. What factors determine the nature of the environment found in a particular place?

3 Outline the diversity of marine environments. Which are the most productive?

4 Outline the key processes responsible for the formation of geomorphological landscapes.

5 Explain what an ice sheet is. Name Earth's two ice sheets. How much water do they store?

6 Explain what a drainage basin is. What is a watershed?

Applying and communicating geographical understanding

7 Study the photos in Unit 2.4. As a class, discuss how the photographs reflect Earth's diverse and distinctive biophysical features. What other types landscapes and features can you identify?

8 Study Figure 2.4.4. Write a report outlining the ecosystem services supplied by the world's oceans.

9 Study Figure 2.4.2. Construct a column graph illustrating the relative size of Earth's continental landmasses.

10 Study Figure 2.4.3. Construct a column graph illustrating the relative size of Earth's oceans.

11 Study Figure 2.4.1. As a class, brainstorm the geomorphic process responsible for the development of the Southern Region of Iceland.

APPLICATION AND CONSOLIDATION TASKS

Task 1: Report writing: Oceans and global warming

Write a report on oceans and global warming. It should include the following:

- The role of ocean currents in regulating global weather patterns (see Figure 2.4.5)
- The impact of climate change on the world's oceans
- The impacts that are currently being observed and the predicted impacts.

Your report can include visuals (maps, graphs, diagrams, photos).

Task 2: Annotated visual display: Drainage basins

Investigate one of the world's largest drainage basins and create an annotated visual display (AVD). Select one of the following:

- The Congo Basin
- The Amazon Basin
- The Murray–Darling Basin

Your AVD should include information that outlines:

- the hydrological features of the basin
- the types of environments it supports.

You should include visuals (maps, graphs, diagrams, photos).

CHAPTER

3 Earth's natural systems at work

Earth's natural **systems** are incredibly complex and highly integrated. To make them easier to investigate and understand, geographers divide these processes into four readily identifiable components or spheres. These are:

- the atmosphere—the combination of gases and particles surrounding the globe
- the hydrosphere—the interconnecting system of water in the atmosphere and lithosphere
- the **lithosphere** (or geosphere)—Earth's solid outer shell
- the **ecosphere** (or biosphere)—Earth's surface zone and atmosphere in which all life exists.

Each of these helps sustain the amazingly diverse life on Earth, providing all the elements necessary. Interactions between the systems that make up Earth's biological and physical environment are vital to life on the planet. Such interactions occur on local and global scales.

This chapter explores the characteristics of Earth's natural systems and the factors affecting their functioning. The focus is on key processes, cycles and circulations connecting these systems.

Those who contemplate the beauty of the Earth find reserves of strength that will endure as long as life lasts. There is something infinitely healing in the repeated refrains of nature—the assurance that dawn comes after night, and spring after winter. In nature nothing exists alone.

Rachel Carson, *Silent Spring*, author and pioneer of the environmental movement

3.0.1 Volcanic eruptions occur when molten rock reaches Earth's surface.

Chapter glossary

albedo the ability of a surface to reflect light

asthenosphere a layer of partially molten material within the upper part of the mantle, on which the tectonic plates move

atmospheric pressure systems areas of the atmosphere with relatively high or low barometric pressure, often referred to as highs and lows

barometric pressure the pressure of the atmosphere

becalmed to stop through lack of wind, particularly ships

biological productivity the amount of biomass produced in an ecosystem

biomass the combined weight of the organisms living in an area

catchment a drainage area, especially of a river

condensation the transformation of gas or vapour into a solid or liquid

conduction energy transfer from one material to another by direct contact

continental drift a former theory that all the continents were once joined together in a giant 'supercontinent', which subsequently split apart, with the continents drifting away from one another until they reached their present locations

continentality the degree to which the weather in an area is impacted by being an inland region compared to a coastal region

convection currents movements within a medium (e.g. air or water) caused by differences in temperature

Coriolis force a force that results from Earth's rotation, deflecting moving objects to the left in the Southern Hemisphere and the right in the Northern Hemisphere

depression an area of low atmospheric pressure

dew point the temperature at which a sample of air will have a relative humidity of 100 per cent

ecosphere the living part of the planet that consists of a thin layer extending from just above Earth's surface to just below it; the combined ecosystems of the planet

ecotone a transition between two plant communities

edaphic related to soil

epicentre the point from which an earthquake seems to radiate out

erosion the act of the Earth's surface being worn away, by wind, water, glaciers, waves, etc.

fossil fuels energy sources formed in past geological times from organic materials; examples include coal, petroleum and natural gas

gradational processes the processes by which Earth's surface gets flattened

greenhouse effect the increase in temperature on Earth created by the Sun's radiant heat passing through the atmosphere, in the same way the glass of a greenhouse creates a warmer temperature inside it

This chapter's glossary words list continues on page 77.

UNIT 3.1

Earth's natural systems and their functioning

Earth's natural systems are incredibly complex, but some important universal factors affect how they function. These impact how natural systems operate and interact in different places on Earth's surface. They include latitude, altitude and the roles played by continents, oceans and seasons (see Figure 3.1.1).

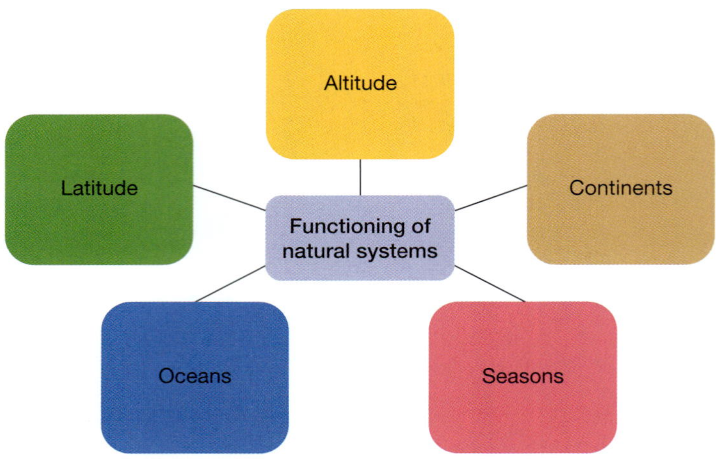

3.1.1 Factors affecting the functioning of natural systems

Latitude

There is a relationship between latitude and temperature. Temperatures are typically warmer towards the Equator and cooler towards the poles. This is the result of different latitudes receiving different amounts of solar radiation. Places at, or close to, the Equator receive more direct solar radiation than places near the poles. There is a roughly gradual temperature gradient from the low to the high latitudes. Other factors complicate this pattern, including altitude, ocean currents, ice and snow coverage, and **precipitation**. See Units 3.3 and 3.4.

Altitude

Earth and its atmosphere are warmed by incoming solar radiation, specifically the infrared component. It warms up the land and oceans, which, in turn, warm up the air in the atmosphere. The atmosphere warms from the bottom up, so the air is usually at its warmest at the surface and cools as altitude increases. Although local factors may influence the rate of cooling, temperatures typically drop at a rate of 1°C per 100 metres. More broadly put, temperatures fall 9.8°C every 1000 metres. See Units 3.3, 3.4 and 3.8.

Continents

Remoteness from the oceans and oceanic air affects the climate of any one place. How much of an impact this has is called **continentality**. Winds and air masses of moderate temperature that originate over oceans and move onshore reduce the differences in winter and summer temperatures in coastal areas of continental landmasses. Continents' interiors are too distant to experience the ocean's moderating effect. As a result, continental interiors have greater climatic seasonal differences in temperatures. There are also greater daily variations in temperature. Typically, daylight hours are warmer and the nights are cooler than coastal areas, which gain and lose heat at a slower rate than inland. Distance from the sea also impacts precipitation. Continental interior climates tend to be subhumid to arid, as oceans are primary sources of atmospheric moisture. See Unit 3.4.

Oceans

Any point on Earth's surface is subject to the sea's influence. How much oceans impact a particular place is termed **oceanity**. It is the opposite of continentality. Places close to large water bodies have smaller temperature ranges than places well inland. The temperature of the water body also influences precipitation. Warmer water bodies are associated with higher rates of evaporation. Areas adjacent to these waterbodies receive greater precipitation than places inland. Cold ocean temperatures are associated with low rates of evaporation and less precipitation. See Units 3.4 and 3.5.

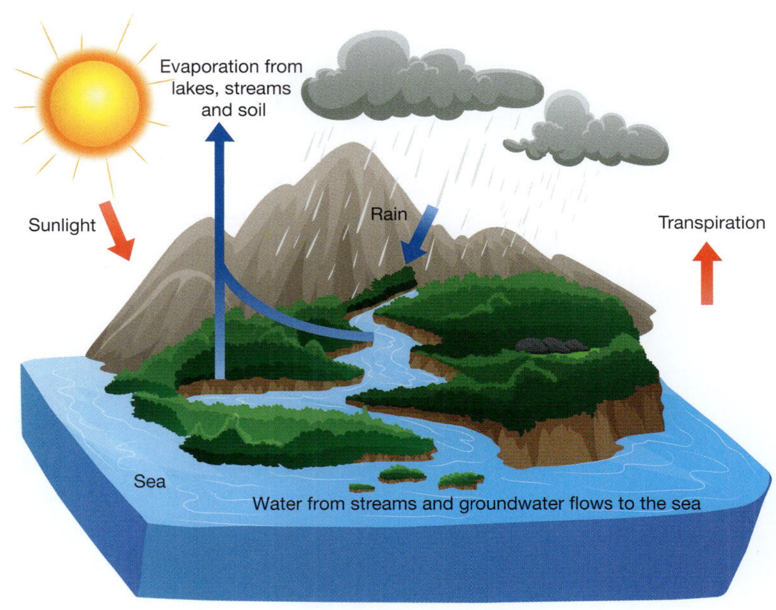

3.1.2 An island can be conceived of as an 'open system'. Energy from sunlight and rainfall reaches the island from external sources. The energy leaves the island via radiation. The water either evaporates, transpires through leaves or drains into the sea.

Seasons

Seasons occur at specific regular intervals of less than a year, typically quarterly—spring, summer, autumn and winter. **Seasonality** refers to the climatic variations caused by the tilt of Earth's axis. Throughout the year, different parts of Earth receive the Sun's most direct rays. When the South Pole tilts towards the Sun, it's summer in the Southern Hemisphere. When the North Pole tilts towards the Sun, it's winter in the Southern Hemisphere. See Unit 3.4.

The system concept

Applying the system concept involves breaking down complex problems into smaller, more easily studied, pieces. Systems vary in complexity and size, and one can be part of a larger system. A leaf is a system that is part of a larger system (a tree), in turn itself part of a larger system (a forest). To be able to study a system it must have a boundary separating it from everything else. The boundary's nature varies, leading to three basic types of systems:

- **Isolated systems** prevent the exchange of matter and energy with their surroundings (these do not exist in the natural world).
- **Closed systems** permit the exchange of energy but not matter.
- **Open systems** can exchange both energy and matter (see Figure 3.1.2).

Earth as a closed system

Earth is a closed system with energy coming in as solar radiation and leaving as infrared radiation (see Figure 3.1.3). Although tiny amounts of hydrogen do escape, the planet's mass is so vast this is inconsequential. This is important because the amount of matter in a closed system is fixed—Earth's resources, such as minerals and **fossil fuels**, are finite and wastes must remain within closed systems. Whenever changes are made in one part of a closed system, they eventually affect all other parts.

In contrast, Earth's system's smaller parts are open systems because energy and matter are exchanged (see Figure 3.1.4). These include the atmosphere, hydrosphere, geosphere and ecosphere—all are open systems. As such, they are interconnected, so understanding their interactions is a major challenge in physical geography. It can help geographers to predict likely responses to disturbance.

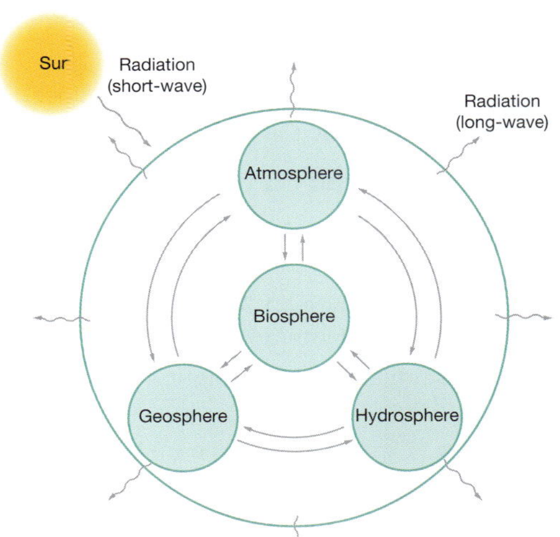

3.1.3 Earth as a closed system—the Sun's energy reaches Earth and is eventually radiated back into space; Earth's component systems are open systems.

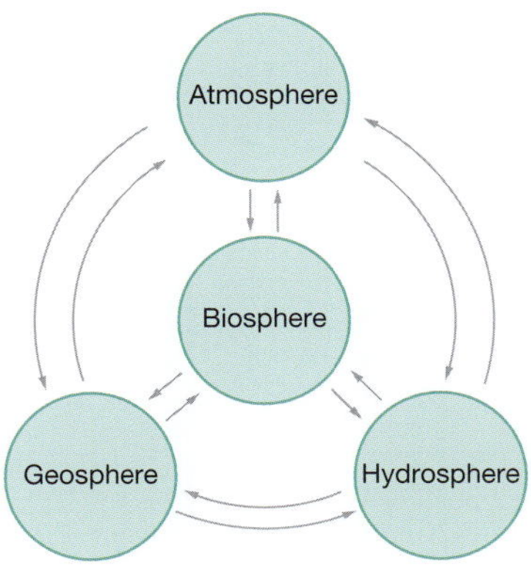

3.1.4 Earth as a system of interacting parts—each represents a reservoir of matter and each arrow represents a flow of energy or materials.

● SPOTLIGHT

The scientific method

The scientific method can be used to study and understand Earth's physical geography. Its key steps are:

- **observing** through gathering observable, measurable evidence
- **developing a hypothesis** which is an unproven explanation of why something happens
- **testing hypotheses** by comparing observations and measurements to predictions
- **creating a theory** by testing and proving hypotheses. A generalisation is made when the results match predictions
- **formulating a law** is creating a statement to explain that an aspect of the natural world consistently and predictably happens in the same way
- **re-examining** hypotheses, theories and laws must be carried out whenever new evidence is found.

Did you know?

Physical geography (sometimes called Earth sciences) assumes events in the natural world follow cause-and-effect patterns. These patterns can be understood through observations, measurements and experimentation.

Geography's inquiry-based methodology is derived from this scientific method, but has been expanded to include social, cultural, economic and political phenomena and issues. This methodology will be revisited in greater detail in Chapter 16.

The use of the scientific method by geographers demonstrates why the subject can be viewed as making a significant contribution to science, technology, engineering and mathematics (STEM).

Threats to system stability

The related issues of rapid population growth, habitat destruction (principally caused by deforestation and desertification), global warming and pollution of all types threaten to destabilise the environmental systems on which life on Earth depends.

Activities

Acquiring and processing geographical information

1 Identify and define the four spheres that interact with parts of Earth's system.

2 Explain the systems concept.

3 Distinguish between isolated, closed and open systems.

4 Explain why Earth can be defined as a closed system.

Applying and communicating geographical understanding

5 Study Figure 3.1.1. Identify the five factors affecting the functioning of natural systems. Which natural system is most affected?

6 Study Figure 3.1.3. Write one or two paragraphs outlining how Earth operates as a system of interacting parts and as a closed system.

7 Study the box, Spotlight: The scientific method and then complete the following tasks:

   a Explain what is meant by the term scientific method.

   b Identify the steps involved in the scientific method.

   c State what a scientific law is.

UNIT 3.2
Atmospheric systems

The atmosphere is the gaseous layer surrounding Earth (see Figure 3.2.1). It forms the transition between the planet's surface and the vacuum of space. Consisting of a mixture of gases: principally nitrogen, oxygen, carbon dioxide and water vapour, the atmosphere extends some 500 kilometres above the surface. The lower level is the **troposphere**. The troposphere houses the climate system that maintains life-sustaining conditions. Above it, the **stratosphere** lies 12 to 48 kilometres above Earth. The stratosphere insulates the **ozone layer**, which protects life on the planet by filtering harmful ultraviolet radiation emitted by the Sun.

3.2.1 Earth's atmosphere from space

When Earth formed around 5 billion years ago, it did not have an atmosphere. Massive volcanic eruptions released gases. As they cooled, the atmosphere developed. Their composition was probably 80 per cent water vapour, 12 per cent carbon dioxide, 7 per cent sulphur dioxide and 1 per cent nitrogen plus other trace gases. The oxygen necessary for life to flourish was missing. The incoming ultraviolet (UV) radiation was too intense for life forms to develop. These high UV levels eventually declined. This triggered chemical reactions which are believed to have kickstarted the evolution of life.

Formation of the atmosphere

When small organisms broke down carbon dioxide to use the carbon, oxygen was emitted into the atmosphere as a waste product. The increasing concentrations of oxygen absorbed greater amounts of UV radiation. So Earth's surface became warm and rich enough for life forms to flourish. Oxygen and nitrogen together now make up 90 per cent of the atmosphere's volume. The remainder comprises other gases (such as ozone and carbon dioxide) plus pollutants and particulates. See Figure 3.2.2 and Unit 3.4.

3.2.2 The composition of Earth's atmosphere

Component	Composition by volume (%)	Importance
Permanent gases: ■ **nitrogen** ■ **oxygen**	78.09 20.95	■ Nitrogen plays an important role in the growth of plants. ■ Oxygen is a product of photosynthesis, so it is reduced as the world's forests are diminished.
Variable gases: ■ **water vapour** ■ **carbon dioxide** ■ **ozone**	0.2–4.0 0.03 0.00006	■ Water vapour reflects and absorbs incoming radiation and provides the moisture for cloud formation and precipitation. ■ Carbon dioxide absorbs heat radiated from Earth's surface. ■ Ozone absorbs harmful incoming UV radiation.
Other gases: ■ **argon** ■ **helium** ■ **krypton** ■ **neon**	0.93 traces traces traces	
Particulates: ■ **dust**	traces	■ Dust absorbs and reflects incoming solar radiation and provides the condensation nuclei required to form clouds; it is added to the atmosphere by volcanic eruptions and wind erosion.
Pollutants: ■ **methane** ■ **nitrogen dioxide** ■ **sulphur dioxide**	traces traces traces	■ Methane contributes to global warming. ■ Sulphur dioxide mixes with atmospheric moisture to form acid rain; it is added to the atmosphere by industrial processes, power generation and vehicle exhaust.

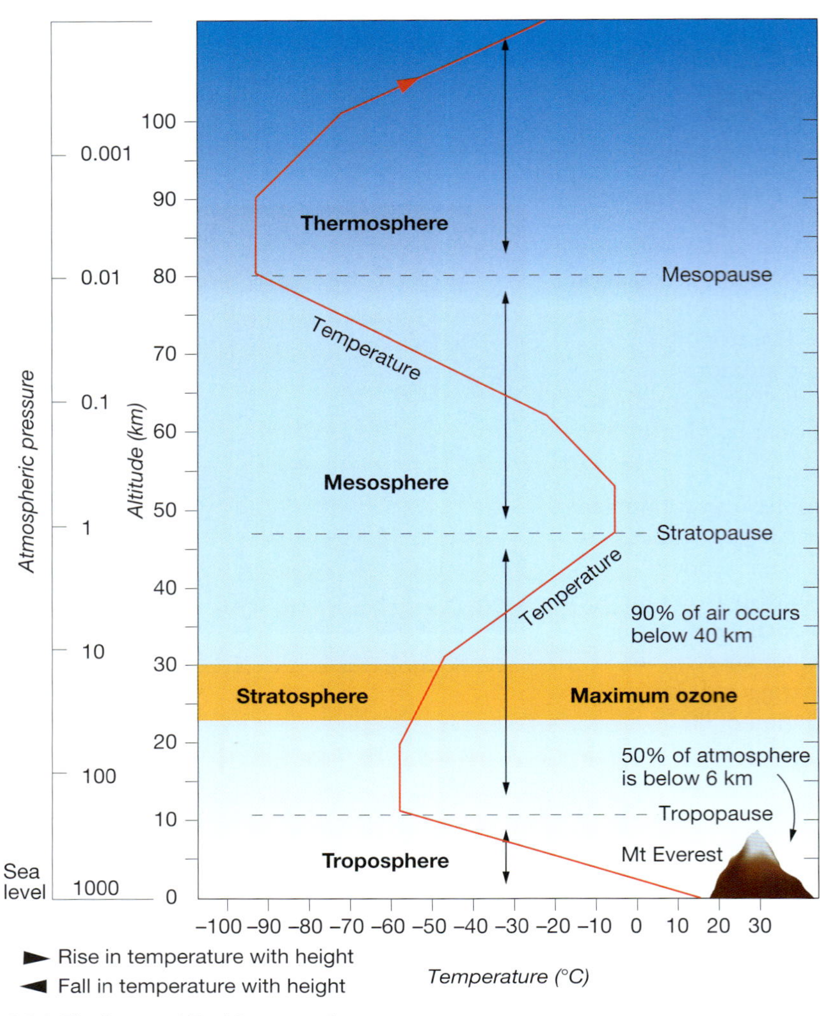

3.2.3 The layers of Earth's atmosphere

The atmosphere's composition has changed over time due to significant events during Earth's history. For example, the impact of a huge asteroid around 65 million years ago altered the atmosphere's chemical composition.

Structure of the atmosphere

Earth's atmosphere has a layered structure based primarily on temperature characteristics. See Figure 3.2.3.

The troposphere

The **troposphere** extends up from ground level to about 8 to 10 kilometres above the poles and 15 to 17 kilometres above the Equator. The top is called the tropopause. It forms the boundary between the troposphere and the stratosphere. Temperatures at the tropopause are very low: –40°C to –80°C. Heat radiating from Earth's surface warms the troposphere's air.

Almost all weather and related processes occur in the troposphere (see Figure 3.2.4). It contains all the weather systems that produce precipitation, surface winds and other climatic variables that impact living things.

The stratosphere

The stratosphere lies between 10 and 50 kilometres above Earth's surface. It is relatively calm compared to the troposphere. Lower stratosphere temperatures are fairly constant. Upper stratosphere temperatures increase with altitude, reaching about 0°C at the top. Solar radiation is absorbed, causing this warming.

3.2.4 Most weather systems are found within the troposphere.

The mesosphere

The **mesosphere** extends about 80 kilometres above the surface. Temperatures fall rapidly with elevation because there is no water vapour, cloud or dust to absorb incoming radiation. Temperatures are as low as –90°C and wind velocities (speeds) are as high as 3000 kilometres per hour, the strongest in the atmosphere.

The thermosphere

Within the **thermosphere**, the atmosphere becomes more tenuous or thinner with elevation. Temperatures rise rapidly, reaching as high as 1500°C. This is due to increasing concentrations of atomic oxygen in the atmosphere, which, like ozone, absorbs incoming UV radiation.

Activities

Acquiring and processing geographical information

1 Define the atmosphere, in your own words.

2 Explain how the atmosphere was formed.

3 Outline how oxygen began to accumulate in the atmosphere.

3 Describe the troposphere. Why is it so important to humans?

4 Outline the composition of the atmosphere.

5 Discuss the factors that determine how much incoming solar radiation occurs at particular locations on Earth's surface.

Applying and communicating geographical understanding

6 Using Figure 3.2.2, write a report on the role different gases play in the atmosphere.

7 Study Figure 3.2.3. Write a report outlining variations in temperature at each layer of the atmosphere. Explain why temperatures in the upper stratosphere and above begin to warm.

UNIT 3.3
Atmospheric processes

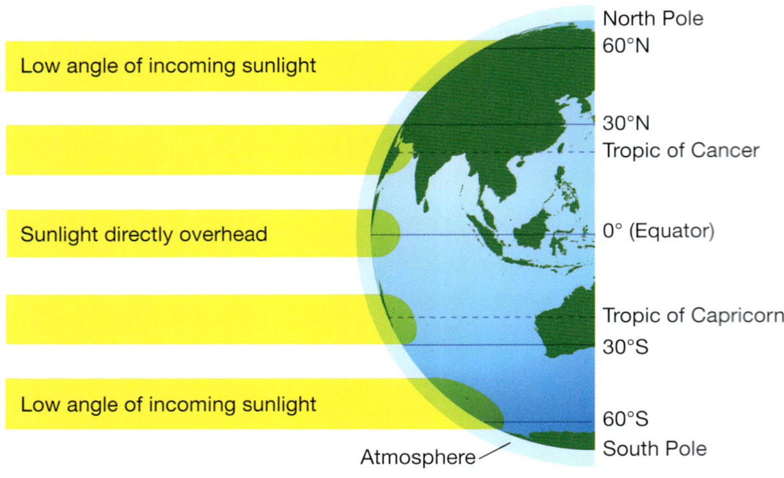

3.3.1 Latitude largely determines how much heat from the Sun reaches Earth's surface.

Energy in the atmosphere

The Sun is Earth's main source of energy. Earth receives this energy as incoming short-wave solar radiation, which controls the planet's climate, weather and water cycle. Through its conversion into chemical energy via the process of photosynthesis in green plants, it supports all life.

The amount of incoming solar radiation received (**insolation**) at a particular location on Earth's surface is determined by:

- **distance from the Sun:** Earth's elliptical orbit can result in a 6 per cent variation in the amount of insolation it receives at different places and times
- **latitude:** incoming solar radiation has twice the area to heat at 60° north and 60° south as it does at the Equator. So average temperatures are lower at higher latitudes (see Figure 3.3.1)
- **length of day and night:** Earth's axis tilts at a 23.5° angle, creating seasonal variations in the length of day and night. Shorter days mean less opportunity to absorb incoming solar radiation.

The natural greenhouse effect

Not all incoming solar radiation reaches the ground. Most is absorbed, reflected and scattered as it passes through the atmosphere. Some is absorbed by ozone, water vapour, carbon dioxide and particles of dust and ice present in the atmosphere. Some is also reflected into space from clouds and, to a lesser extent, from Earth's surface.

The ratio between incoming radiation and the amount reflected into space is called the **albedo**, expressed as a percentage. Different surfaces affect the albedo considerably (see Figures 3.3.2 and 3.3.3). Human activities also affect it. Clearing and overgrazing land often remove vegetation, which increases the albedo. This can make cloud formation and precipitation less likely, increasing the risk of desertification. Desertification is studied in more detail in Chapter G1.

Land surface	Albedo effect
Oceans	Less than 10%
Dark soils	Less than 10%
Grasslands	25%
Deserts	40%
Snow and ice	85%

3.3.2 The albedo effect varies depending on the surface.

3.3.3 Grasslands have low albedo.

Incoming solar radiation is diverted by molecules of gas in the atmosphere. This scatters the Sun's energy in all directions, with some radiation reaching Earth's surface as diffuse radiation. Only 24 per cent of this radiation reaches the surface directly. Another 21 per cent arrives as diffuse radiation. Incoming radiation is transformed into heat energy at the surface. It heats the ground, radiating long-wave or infrared energy back into the atmosphere. In the atmosphere, 94 per cent of this energy is absorbed by water vapour and carbon dioxide, creating a natural **greenhouse effect** (see Figure 3.3.4).

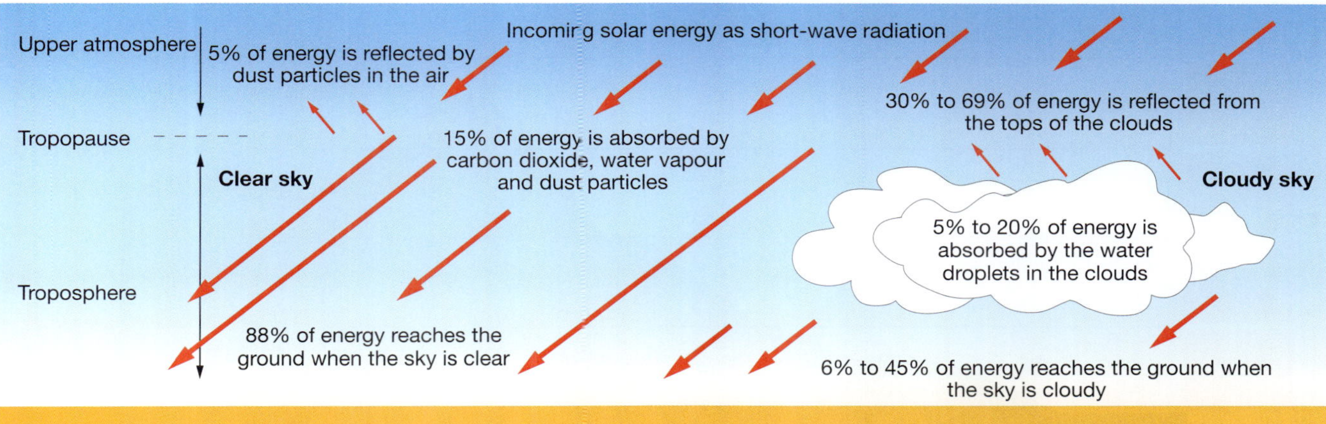

3.3.4 The absorption and scattering of insolation

The global heat budget

Earth's temperature remains relatively constant from year to year. This means there is a balance between incoming insolation and outgoing **terrestrial radiation**. This global heat budget has two key features. The first is a net gain in radiation everywhere on the surface except the poles, which have a high surface albedo. The second is a net loss in radiation throughout the atmosphere.

After accounting for incoming and outgoing radiation, a net surplus of heat remains between 35° south and 40° north of the Equator and a net deficit poleward from these latitudes. The larger land masses in the Northern Hemisphere create this latitudinal difference.

There is a positive heat balance within the tropics and a negative heat balance in the polar regions and at high altitudes. This imbalance causes heat from the tropics to transfer towards the poles and to higher altitudes. Without it, the tropics would 'overheat'. Two types of transfer take place:

- **horizontal transfers:** Wind and ocean currents transfer heat from the tropics towards the poles. Winds account for 80 per cent of the heat transfer and ocean currents 20 per cent
- **vertical transfers:** Heat is transferred to the atmosphere from the surface by terrestrial radiation, **conduction**, **convection currents** and latent heat transfers via **condensation**. Without this transfer, the surface would get hotter and the atmosphere colder.

Altitude and lapse rates

Air temperature declines with higher altitudes in the troposphere. This decline is called the **lapse rate** and it greatly influences the lithosphere's impact on atmospheric processes. Despite the Sun's extreme heat, most heat reaching the surface is from solar radiation, which gets absorbed by the land surface. This effectively heats the troposphere from the ground up. Air conducts heat poorly. So the higher the altitude in the troposphere, the colder the air becomes as it rises further from the heated surface.

On average, air temperature decreases by approximately 6°C per 1000 metres of elevation. By applying the lapse rate, if the air temperature at sea level is 30°C then at the peak of Mt Everest, which is at 8848 metres above sea level, it would be –24°C (see Figure 3.3.5). The actual lapse rate varies between places due to humidity. When applied at a given place and time, this is called the environmental lapse rate.

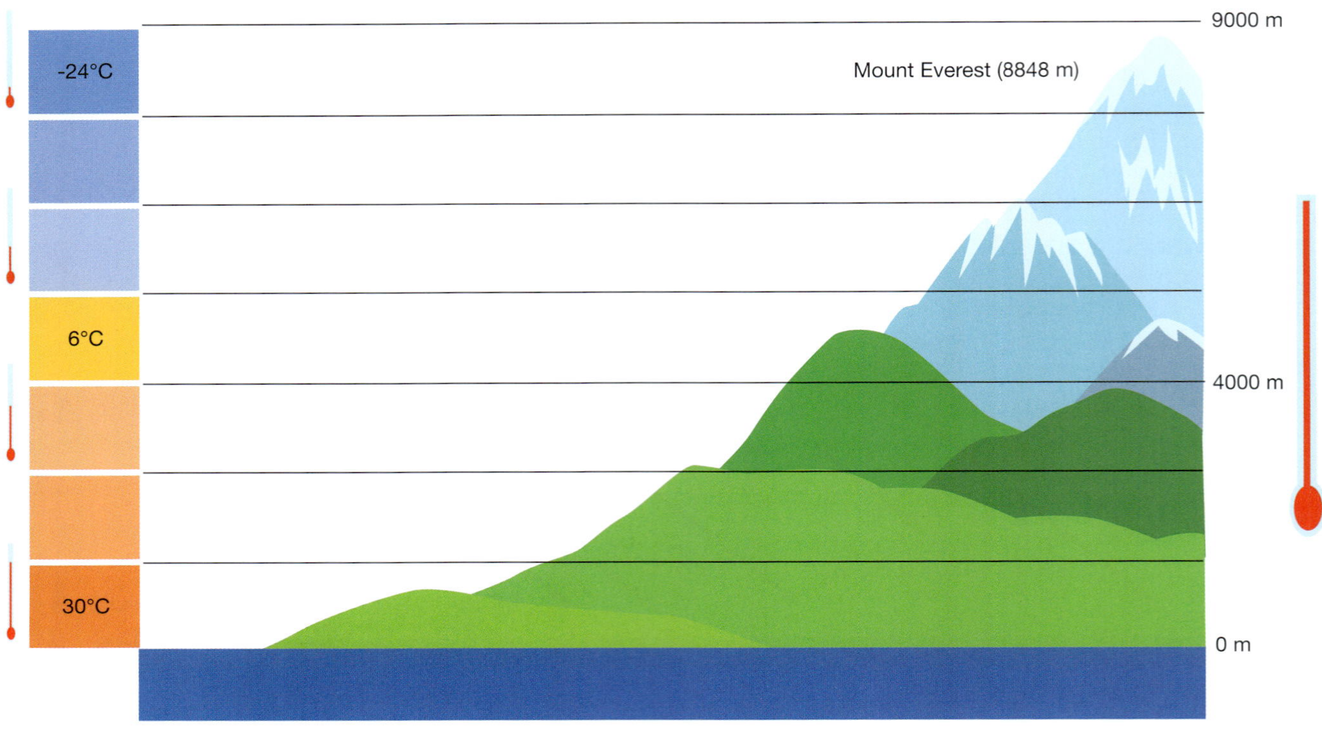

3.3.5 Lapse rates relate to the influence of altitude on temperature (applying average lapse rates).

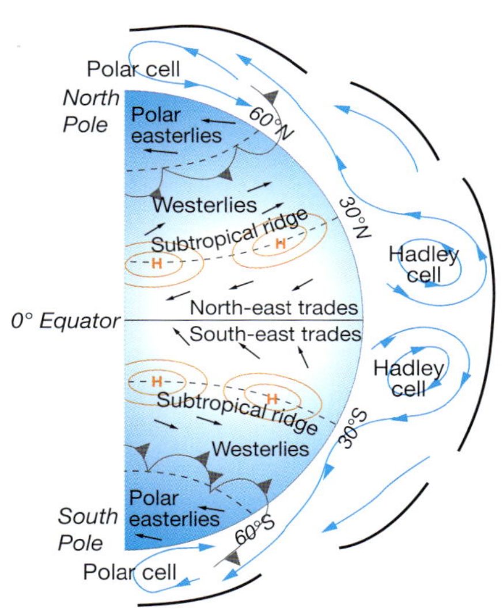

3.3.6 Movement of winds around Earth

Movement of air

The air within the troposphere never rests. The Sun heats the equatorial regions faster than the polar regions. This causes air movements within the troposphere at a global scale, which produce the distinctive climatic features experienced on the surface. Figure 3.3.6 shows the global pattern of atmospheric circulation.

At the Equator, heated air rises. When it reaches the tropopause, it moves towards the poles. At 30° north and south of the Equator, a body of descending air creates the subtropical high-pressure belt, bringing clear skies and dry, stable conditions. On the surface, some descending air moves back towards the Equator. The cells created by this circulation pattern are called Hadley cells.

On the polar side of the Hadley cells, the remaining air moves towards the poles. When this warmer air meets the cold air at the polar front, it is pushed under and uplifts the warmer air. This forms an area of low pressure called a polar cell. The resulting polar front marks the boundary between the warm tropical air masses and the cold polar air masses. The frontal zone is distinct and moves as a non-continuous band around Earth.

The upper troposphere contains narrow bands of extremely fast-moving air called jet streams. They help rapidly transfer energy and can exceed 230 kilometres per hour. Military and commercial aircraft take advantage of these winds—eastward flights are much faster than westward flights. Jet streams are found in both hemispheres, with one circling each of the two poles (the polar jet streams). A second jet stream encircles the tropics (called the subtropical jet streams), shown in Figure 3.3.6.

Surface winds

Surface winds are caused by air movements from high-pressure areas to low-pressure areas. Highly variable in speed and direction, these winds have an overall global pattern.

Hadley cells produce descending air in the mid-latitudes on either side of the Equator (see Figure 3.3.6). This air moves back towards the Equator as a surface wind. Air moving towards the Equator is deflected to the left in the Southern Hemisphere and to the right in the Northern Hemisphere by the **Coriolis force**. These deflections give the winds a direction—from south-east to north-east in the Southern Hemisphere and from north-east to south-west in the Northern Hemisphere. The surface winds are called the south-east trade winds and the north-east trade winds, respectively.

Air moving away from the Equator toward the poles is also deflected to the left in the Southern Hemisphere and to the right in the Northern Hemisphere. On the polar side of the Hadley cells, descending air moves toward the poles and is also deflected. This forms a band of strong winds in each hemisphere, dubbed the 'Roaring Forties' in the south as they persist from 40° to 49° south.

In the equatorial region, air rises and there are no strong horizontal surface winds, creating the 'equatorial doldrums'. Relatively calm weather also occurs below the descending air of the polar ends of the Hadley cells. According to legend, early mariners applied the term 'horse latitudes' to these commonly **becalmed** zones because their ships' drinking water supplies would run low, and to save water animals were thrown overboard.

Pressure systems

Atmospheric pressure systems are areas of the atmosphere with relatively high or low **barometric pressure**, often called 'highs' and 'lows'. Large areas of permanent high and low pressure exist in bands around Earth. They influence world climate patterns and ocean currents (see Figure 3.3.7).

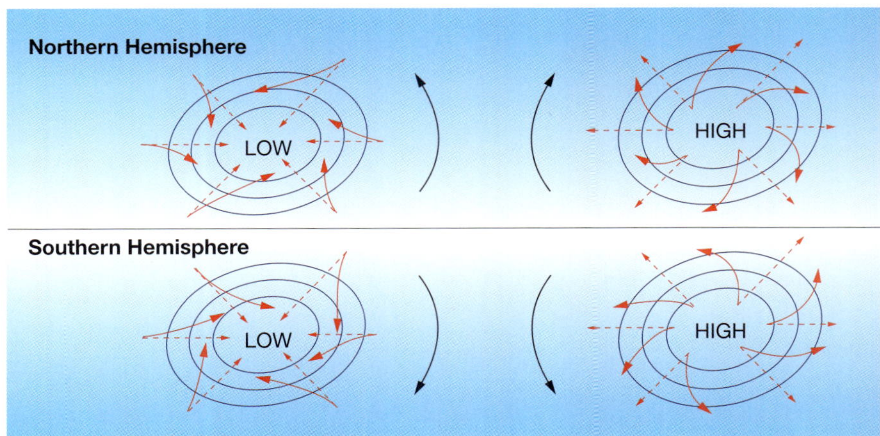

3.3.7 Patterns of air movement in pressure cells

Did you know?

The location of pressure systems largely determines wind direction. In the Southern Hemisphere, winds blow in an anticlockwise direction from areas of high pressure. Winds from low pressure systems blow clockwise towards the centre.

High pressure systems

A high pressure system, or anticyclone, is a large body of sinking air that produces an area of high pressure on Earth's surface. The air is sourced from the upper atmosphere where there is very little moisture. Because descending air warms, weather conditions remain dry. When the pressure gradients remain gentle, wind velocities stay low. (Pressure gradients are shown by the isobars on a synoptic chart.) The weather in regions dominated by high pressure systems features stable atmospheric conditions, clear skies and gentle winds.

Low pressure systems

A low pressure system, often called a cyclone, is a large body of rising air that produces an area of relatively low atmospheric pressure. As the air's source is close to the surface, there may be a lot of moisture. As moist air rises it cools rapidly and condenses, potentially increasing precipitation. Wind velocities are relatively high because the pressure gradients are relatively steep (these synoptic charts show the isobars close together.) The weather in areas dominated by low pressure systems is characterised by unstable atmospheric conditions, cloudy skies, rain and strong winds.

Tropical cyclones are intense low pressure systems that develop over warm tropical oceans. They are often erratic and unpredictable. Once a cyclone crosses a coastline it

Did you know?

The terms hurricane, typhoon and tropical cyclone are used for the same type of storm. Hurricanes are storms that form in the Gulf of Mexico, the northern Atlantic and eastern Pacific. Tropical cyclones occur in the southern Pacific and Indian oceans. Typhoons occur in the north-western Pacific Ocean.

turns toward the pole. Its intensity then reduces and it develops into a rain **depression**. To form and move, cyclones need a continuous supply of heat and moisture to maintain the necessary rising thermal currents (see Figure 3.3.8). A huge amount of moisture is required to provide the latent heat that drives the cyclone. Once released by condensation, the heavy rain associated with cyclones falls. At the cyclone's centre is the 'eye'. Stretching 30–50 kilometres wide, it is typified by subsiding air, light winds, clear skies and relatively high temperatures. The descending air around the eye warms quickly, increasing the cyclone's intensity.

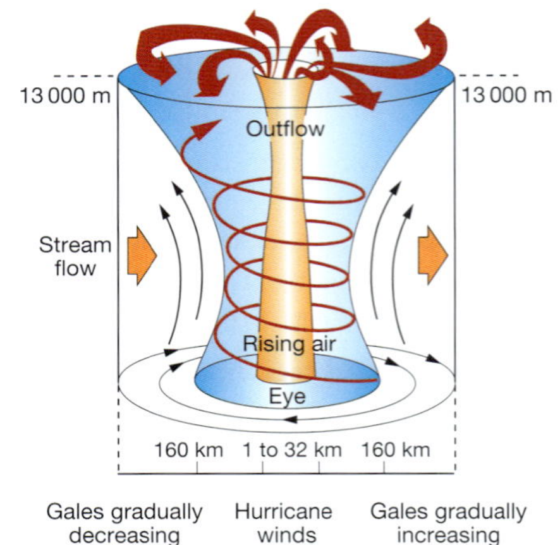

3.3.8 Formation of a cyclone

● **SKILLS BUILDER**

Interpreting synoptic charts

A synoptic chart, or weather map, is a record of the atmospheric conditions experienced at a particular time in a particular place (see Figure 3.3.9). It illustrates air pressure, the location of air masses and frontal activity, cloud cover extent, wind speed and direction, and rainfall distribution. This information can help predict temperature, humidity, ocean conditions and the likely sequence of weather in coming days.

Synoptic charts display several features.

Isobars: the lines joining places of equal barometric pressure, measured in units called **hectopascals** (hPa). Checking the value of adjacent isobars reveals the atmospheric (or barometric) pressure of a particular location. This process is similar to using contour lines to estimate height above sea level.

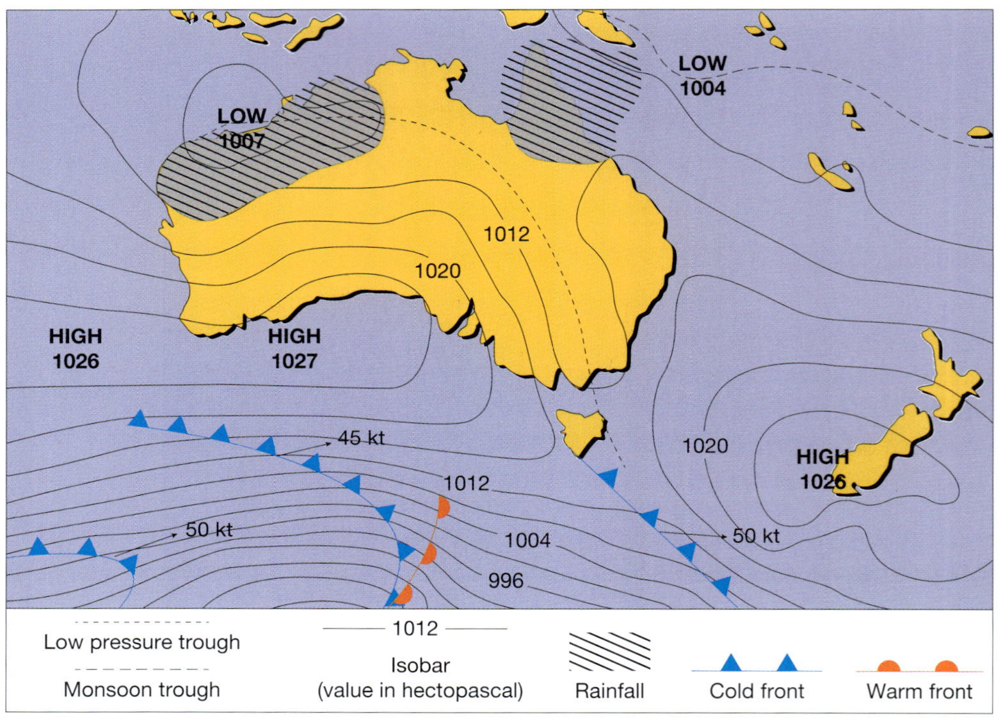

3.3.9 Synoptic chart of typical Australian summer weather patterns

Pressure systems: defined by the patterns formed by isobars. High pressure systems exhibit barometric pressure increasing towards the centre of a set of enclosed isobars. Low pressure systems show barometric pressure decreasing towards the centre.

Cold front: forms when a mass of cold air moves in and displaces warmer air. The uplift of warm, moist air increases the chance of rain.

Warm front: forms when a mass of warm air moves in and displaces cold air.

Wind speed: the closer the isobars, the greater the wind speed.

Wind direction: the direction wind comes from, so a wind coming from the south is southerly.

Rainfall: areas in which rainfall has occurred in the previous 24 hours are highlighted by shading.

Thunderstorms

Three conditions are required for thunderstorms to develop: a source of moist air, an unstable atmosphere and a mechanism to initiate their development. Moist air is important because it condenses to form cloud. As condensation takes place, heat energy is released. This makes the rising air more buoyant and promotes further cloud growth. Unstable atmospheric conditions allow developing clouds to rise to great heights. The mechanisms initiating thunderstorms are fronts, troughs and regions of low pressure (see Figure 3.3.10). Landform features, such as mountain ranges, may also promote storm development.

Storm severity depends largely on the rate at which the air rises. Severe storms occur when the atmosphere is very unstable and the upward air movement is very rapid. Sometimes several storm cells, at different stages of their life cycle, are found within the same storm system. These 'multicellular' storms produce hail, strong winds and possible flash flooding. In some cases, severe storms develop into long-lived thunderstorms called supercells. A supercell is a rare but unusually severe type of thunderstorm. Its structure, behaviour, intensity and longevity are quite different from those of ordinary thunderstorms. Supercells can yield very large hail, extraordinary wind gusts and heavy rainfall.

3.3.10 Thousands of storms occur each year in NSW. Over 100 are classified as 'severe' and can produce hailstones over 2 cm in size, wind gusts over 90 km/h and flash flooding.

Activities

Acquiring and processing geographical information

1. Describe the concept of insolation. Explain the factors determining its rate.
2. Define the term albedo.
3. Outline the operation of the global heat budget.
4. Differentiate between horizontal and vertical heat transfers.
5. Explain the processes leading to the creation of Hadley cells.
6. Why can planes flying in an easterly direction travel faster than those flying west?
7. Describe the relationship between atmospheric pressure and the movement of air across Earth's surface.
8. Outline the conditions that lead to thunderstorms.

Applying and communicating geographical understanding

9. Examine Figure 3.3.2. Briefly explain the relationship between land surface type and albedo.
10. As a class, discuss the importance of heat transfers to Earth. What would happen if these transfers ceased?
11. Using Figure 3.3.5 and the related text, describe the concept of a lapse rate.
12. Using diagrams, briefly explain the difference between high and low pressure cells.
13. Collect a series of Australian synoptic charts over five or more consecutive days. Trace the passage of high and low pressure systems across the continent. Note the associated weather conditions. Write two or three paragraphs explaining what occurred over the days.

UNIT 3.4
Global climate patterns

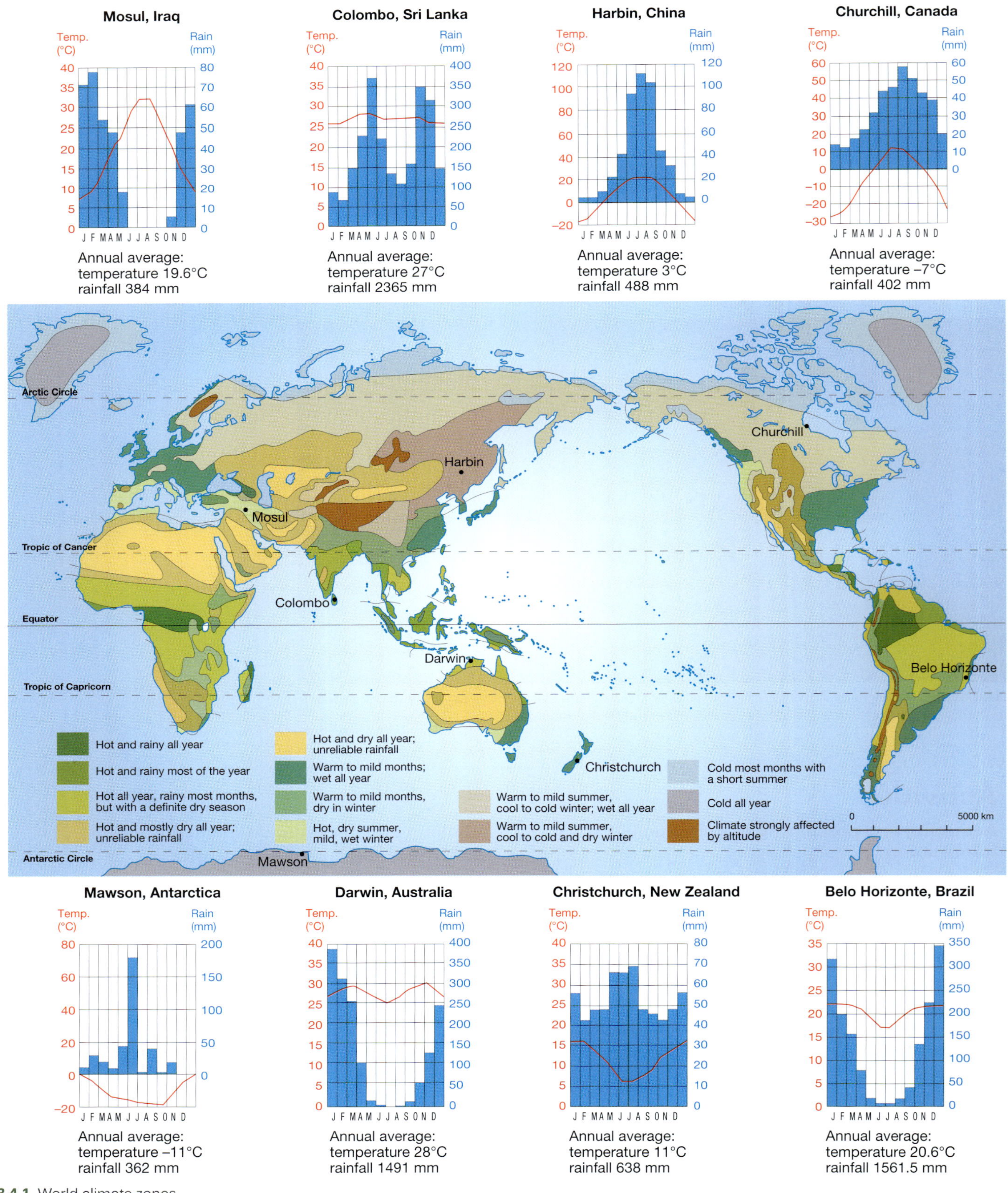

3.4.1 World climate zones

There are five main factors contributing to the patterns of climate across the globe. The first is how much insolation varies between locations as it hits the surface. The second is Earth's revolution around the Sun and its rotation on its axis. The last three relate to the parts of the atmosphere, how continents and oceans are distributed, and topography (see Figure 3.4.1).

Insolation variations

The amount of incoming solar radiation reaching Earth's surface varies with latitude (see Figure 3.4.2). Near the Equator, the Sun's rays are almost vertical, so they are concentrated on a smaller surface area. Closer to the poles, rays strike at an angle, spreading them over a larger area. This means there is less solar energy per unit area.

The relatively large input of heat near the Equator warms the air. This warmer air rises because it has a lower density than cold air. It moves northward and southward, toward the poles, carrying heat away from the Equator. At higher latitudes it cools and begins to descend. These cool air masses then flow back towards the Equator as surface winds, to fill the void left by the rising air. This circular pattern of air movement moderates global air temperatures.

Latitude	Insolation percentage
0°	100.0
10°	98.6
20°	94.5
30°	88.0
40°	79.2
50°	68.5
60°	57.0
70°	47.4
80°	43.0
90°	41.6

3.4.2 Variations in solar insolation due to latitude

Earth's revolution and rotation

At the Equator, temperatures remain fairly constant throughout the year. Elsewhere, average temperature patterns vary with the seasons. This is because Earth's axis is tilted at 23.5°. During its revolution, when the South Pole tilts toward the Sun, its rays strike the Southern Hemisphere more directly and intensely than in winter, when the South Pole is tilted away. So, the Southern Hemisphere experiences its summer while the Northern Hemisphere experiences its winter (see Figure 3.4.3)

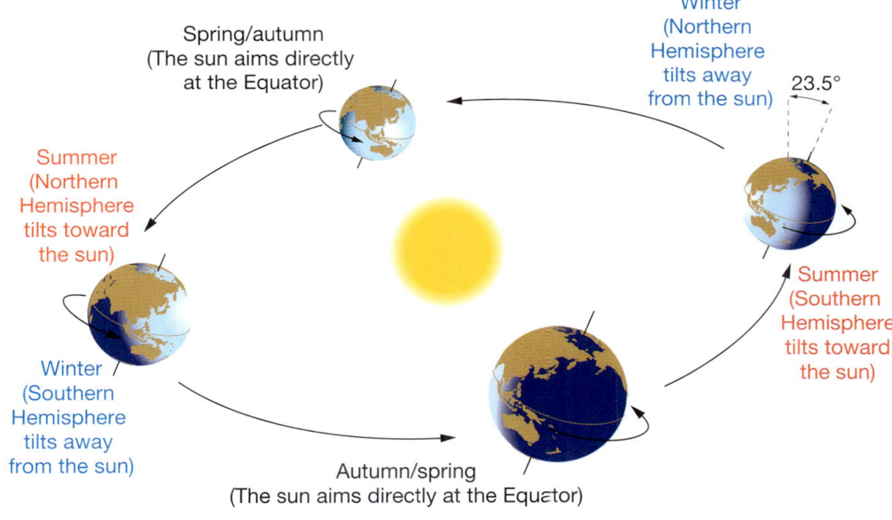

3.4.3 How Earth's revolution around the Sun influences climate

The forces generated by Earth's rotation break up the general pattern of air circulation from the Equator to the poles. This creates six separate belts of moving air: three to the north of the Equator and three to the south. Earth's revolution around the Sun and its angle also cause the length of night and day to vary. Longer days mean that more incoming solar radiation hits the surface. Therefore more solar energy is available to plants, increasing their production.

The atmosphere's components

Earth's temperature and climate are largely determined by the atmosphere's structure (detailed in Unit 3.2). In the troposphere, carbon dioxide and water vapour are important in regulating temperatures. They let in the Sun's radiant energy, but prevent some of it from escaping back into space. This creates a 'natural greenhouse effect'. Without it, Earth would be a cold and lifeless planet with an average atmospheric temperature of −18°C.

Distribution of the continents and oceans

Ocean currents also influence climate, especially in coastal areas. Warm ocean currents tend to have higher evaporation rates. Hence, coastal areas generally get more precipitation, especially where the prevailing wind blows onshore. Cold ocean currents have relatively low levels of evaporation and low moisture in the air above them. These currents tend to flow along the western coastlines. This is why more—and larger—deserts lie adjacent to the western coastlines of continents (see Figure 3.4.4).

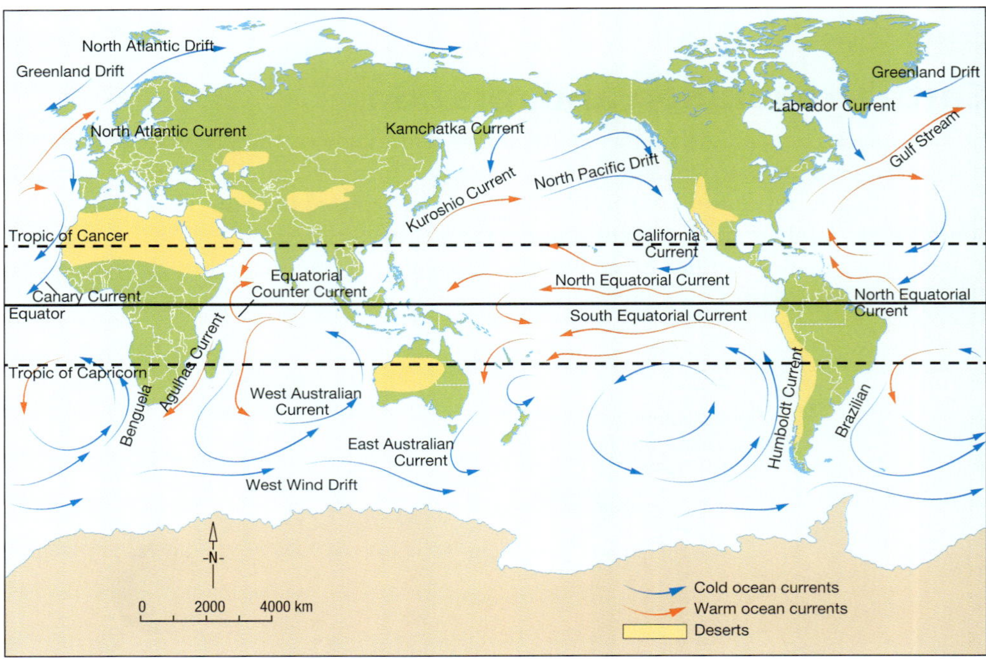

3.4.4 Main surface currents in the world's oceans and proximity of desert

● **SPOTLIGHT**

The Humboldt Current

The influence of ocean currents on climate can be seen by comparing the Humboldt, a cold ocean current, with the Gulf Stream, a warm ocean current. Also called the Peru Current, it flows northward from Antarctica into the eastern Pacific Ocean, along the western coastline of South America. The cold, slow-moving waters allow for little evaporation, creating an extremely arid landscape. Stretching 1600 kilometres along the coastline, the Atacama Desert (see Figure 3.4.5) is one of the world's driest places, which limits life in this desolate area. In contrast, the ocean here supports highly abundant marine life. This is because the current pushes nutrient-rich waters up to the surface, a process called upwelling.

3.4.5 The Atacama Desert is an extremely dry environment influenced by the cold waters of the Humboldt Current that run along its coastline.

Topography

Altitude means that mountainous areas tend to be cooler than adjacent lower altitudes. Despite being near the Equator, the summit of Mt Kilimanjaro, Africa's highest mountain, is typically covered with snow. Human-induced climate change, however, means its snow cap is less permanent now.

Mountain ranges also interrupt the movement of prevailing winds and moisture-laden air. When moist air is forced to rise, it cools and expands. This makes it lose most of its moisture as either rain or snow, on the mountain range's windward side (the side exposed to the wind). This process produces the rain-shadow effect, or orographic rainfall (see Figure 3.4.6).

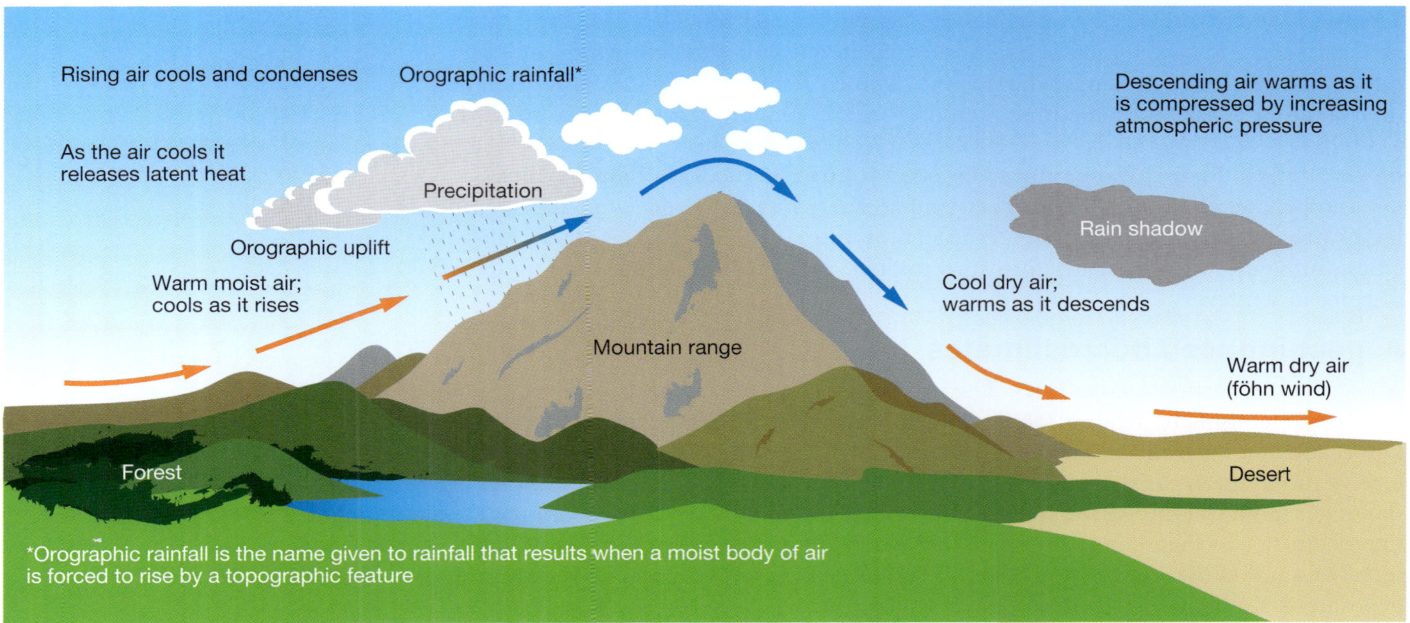

3.4.6 Orographic rainfall shows the influence of topography on climate

● **SPOTLIGHT**

The Western Ghats

On India's western edge, the Western Ghats mountain range acts as a barrier to the moisture-rich monsoons that flow across the coastline from the Indian Ocean. As the moisture-laden clouds rise over them, rains fall on the seaward western side. This creates a lush, highly productive environment (see Figure 3.4.7). The vast, dry Deccan Plateau stretches out on the leeward, or eastern, side. Here, rainfall is far lower and the biophysical environment stands in stark contrast to that on the western side (see Figure 3.4.8).

3.4.7 The Western Ghats' western side is lush with plentiful water.

3.4.8 The hot and dry Deccan Plateau is in the rain shadow on the eastern side of the Western Ghats.

As drier air flows down the leeward slope (the side sheltered from the wind) it is compressed and becomes warmer. Thus, deserts and semi-arid areas are often found on the leeward side of large mountain ranges. The rain-shadow effect impacts places like the semi-arid regions adjacent to the Himalayas in Asia and the Andes in South America.

Microclimates

Significant climatic differences often occur over relatively small areas. The study of these climatic differences is called microclimatology. Many occur naturally, resulting from differences in factors such as aspect and type of land surface.

The most widely studied microclimates are large urban centres. They feature hard and dark surfaces, and have high volumes of dust due to air pollution and wind tunnels—canyons between tall buildings where wind channels at higher velocities. Together, these alter the local, and sometimes regional, climate. Building materials and roads are less reflective. During the day these surfaces absorb and store heat, which is released slowly at night. Heat is also produced by other sources, particularly motor vehicles, but also people. Hard surfaces like roads and concrete increase rainfall run-off. With so little surface to cool the air through evaporation, this creates urban heat islands. Large cities can be up to 5°C warmer than surrounding rural areas in summer, and up to 2°C warmer in winter.

Aspect-induced microclimates

Aspect is the direction a slope faces. The northerly aspect faces the Equator in the Southern Hemisphere, while a southerly aspect faces it in the Northern Hemisphere. Direct sunlight is greater on these slopes than those facing the poles. Slopes facing east receive more sun in the mornings when temperatures are usually lower. Western-facing slopes get more sun during the hottest times. Shaded slopes have a cooler, moister microclimate (see Figure 3.4.10). Farms and settlements in mountainous regions are usually found on the sunnier side of the valley. Vegetation patterns also vary with exposure to sunlight— pockets of rainforest along Australia's eastern seaboard are often found on slopes that have a southerly or easterly aspect.

SKILLS BUILDER

Investigating microclimates

There are microclimates in any local community. Even small areas like school grounds can have different microclimates. Buildings create shade in the same way as mountains and cliffs. Investigate these microclimates using simple techniques.

1 Using a thermometer, measure and record the air temperature at various locations within the school grounds. Examine the nature of each area and explain the differences in temperature. Compare the results from each area and plot the locations of the recording sites on a map of the school.

2 Investigate the impact of large buildings in the school on wind flow. An anemometer (see Figure 3.4.9) is the best tool to measure wind speed. Otherwise, use light strips of cloth pinned to the top of sticks, either hammered into the ground or held aloft.

3 As a class, measure and record variations in temperature and rainfall in the local community using a thermometer and a simple rain gauge. Then, in their own backyard, each student measures the air temperature at a set time and the rainfall within a set time period. Record these findings on a local map and examine them. Note any variation in the temperature and rainfall patterns.

4 Find a small valley with slopes that have southerly and northerly aspects. Measure the air temperature on each slope and compare results. Note any differences in the vegetation types found on each slope.

3.4.9 Anemometers are used to collect data on wind speed.

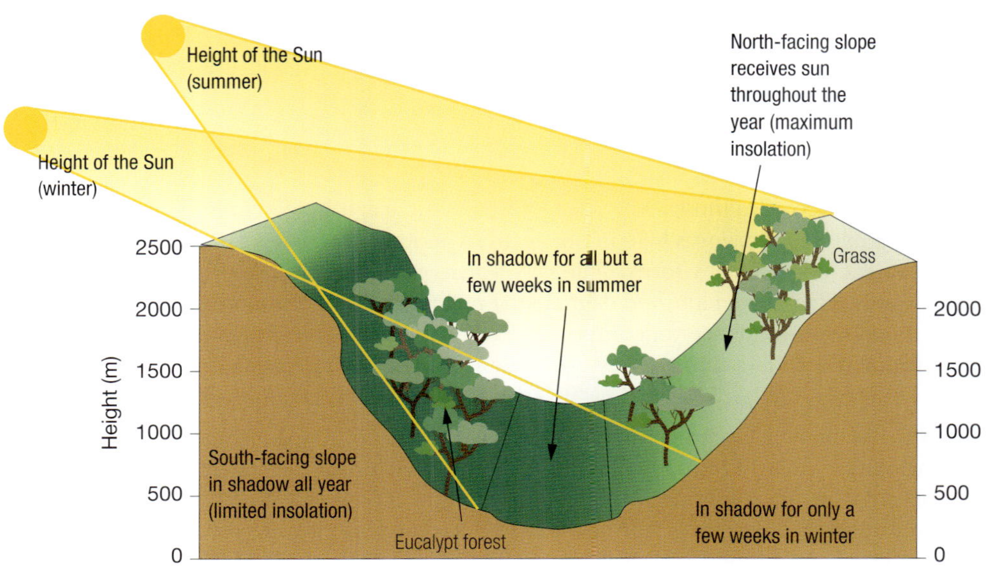

3.4.10 The impact of aspect on climate

Activities

Acquiring and processing geographical information

1. Explain how variations in insolation reaching Earth's surface affect climate.
2. Outline how Earth's revolution around the Sun and its daily rotation on its axis affect climate.
3. Describe the effect of the composition of the atmosphere on climate.
4. Describe the impact of topography on climate and rainfall.
5. Write your own definition of a microclimate.
6. Describe how large urban centres can influence climate.
7. Outline the meaning of aspect.

Applying and communicating geographical understanding

8. Study Figure 3.4.1.
   a. Comment on the longitudinal spread where climates are hot and wet for most of the year.
   b. Describe the distribution of hot and dry climates.
   c. Describe the distribution of places with hot, dry summers and mild, wet winters.
   d. Name the type of climate experienced by these geographical locations: Northern Australia, Sub-Saharan Africa, Tasmania, Alaska, Northern India, United Kingdom, Papua New Guinea and Northern Russia.
9. Examine Figure 3.4.4. Write a report describing the patterns of Earth's ocean currents and the relationship to deserts.
10. Write a paragraph outlining the impact of the Humboldt Current on the adjacent landmass.
11. Using the Western Ghats as an example, write a report on the relationship between topography, the atmosphere and the biophysical environment.
12. As a class, discuss strategies to reduce the impact of urban areas on climate. Brainstorm ideas and record these in a graphic organiser. Evaluate the ease of implementing the various strategies.
13. Examine Figure 3.4.10.
    a. Describe the relationship between altitude and shadow formation.
    b. How do you think the environment of the valley floor would differ from that at the top of the slope? Explain why.

UNIT 3.5

Hydrological systems

Although water covers almost 70 per cent of Earth's surface, it is not a limitless resource. Every drop existing today was here in one form or another on the day Earth came into being. Water comes in many forms. The salty oceans contain more than 97 per cent of all Earth's water. They help dilute and degrade the pollutants flowing into them. The next largest component of surface water is polar ice caps and ice floes (2.7 per cent). An important influence on the world's climate, they provide essential habitat to many unique life forms.

The 0.3 per cent left is the fresh water we see daily. Rivers, streams and creeks transport **run-off** that flows from rain, snow and ice melt, storms or floods. Together with **groundwater**, run-off is also a vital source of fresh water for domestic, industrial and agricultural uses. Wetlands and marshes hold unique ecological value. They cleanse our drinking water (by removing sediments and contaminants), provide long-term storage for the slow uptake of groundwater (recharge) and protect us from floods and droughts.

The water cycle

Water is always moving. Its state and geographical location are always changing. Consider Earth's 4.6-billion-year history as a 24-hour clock. At 2 am the planet was still being born. Heat from atomic decay in the pressurised core, combined with the heat generated by continual meteorite impacts, fused it into a ball of **magma**, rich in iron. The heat in the interior drove oxygen and hydrogen atoms from the materials in which they were contained, and the newly formed molecules surfaced in upwellings of magma. Steam rose from the crust as lava and emerged through surface fractures. Rising water vapour cooled and clouds soon enveloped the planet. Violent electrical storms churned through the atmosphere.

When the surface had cooled sufficiently, the rains came. The first droplets fell at about 2.30 am. The storms continued until around 3.05 am, having lasted for about 112 million years. The result was a planet mostly covered by a shallow sea, muddy with sediment washed from the land—land with no plants and no **soil**.

The same water is still circulating today, driven by the Sun's energy. At a global level, the hydrological (water) cycle is a closed system. No water is added and none is taken away. What changes is its distribution, its geographical location and sometimes its form (see Figure 3.5.1).

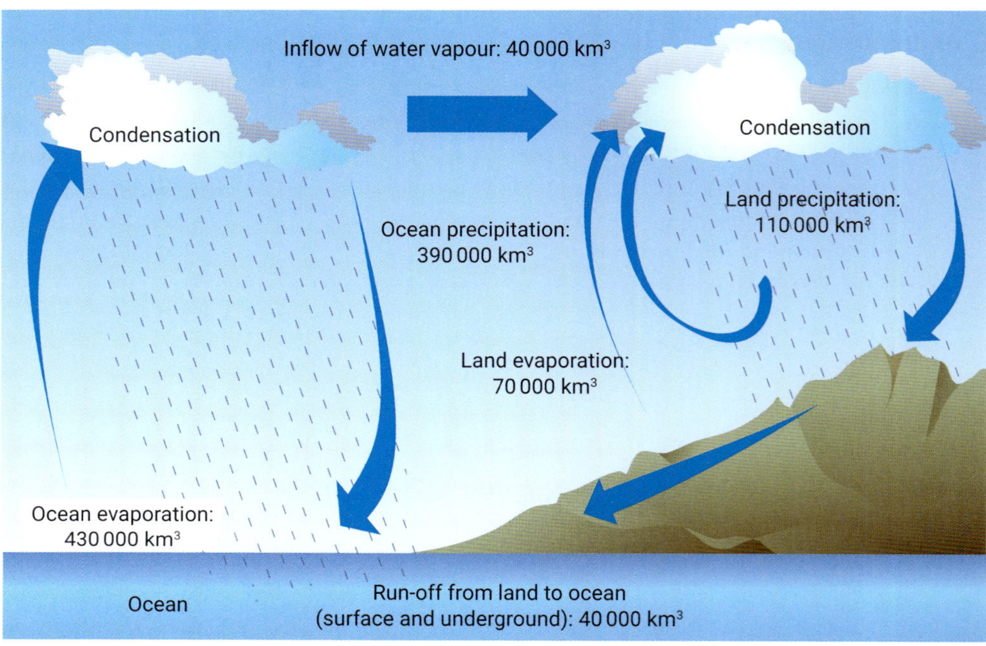

3.5.1 Vast quantities of water are transferred within the water cycle.

About 430 000 cubic kilometres (km^3) of water evaporates annually from the surface of the oceans. Another 70 000 cubic kilometres evaporates from the land. This gives a total of approximately 500 000 cubic kilometres.

Since the volume of water vapour in the atmosphere remains relatively constant, the same amount is precipitated back to the surface. However, its distribution varies. Precipitation over oceans is roughly 390 000 cubic kilometres, while precipitation over land is around 110 000 cubic kilometres. However, its combined evaporation is 70 000 cubic kilometres. This gives a net transfer from the oceans to the land of 40 000 cubic kilometres a year. This transfer sustains all the terrestrial life within the ecosphere.

However, far more water exists than the amount transferred annually (see Figure 3.5.3). Water is stored in many forms and for varying periods. Just over 2.6 per cent is fresh water—that is, free of salt (see Figure 3.5.2).

Storage	State	Volume (km^3)	Percentage of fresh water	Percentage of total water on Earth
Ice caps and glaciers	Solid	282 000 000	76.5	2.14
Groundwater				
≤ 800 m deep	Liquid	3 740 000	10.1	0.27
800–4000 m deep	Liquid	4 710 000	12.8	0.34
Lakes	Liquid	125 000	0.34	0.00091
Soil moisture	Liquid	69 000	0.19	0.005
Atmospheric vapour	Gas	13 500	0.037	0.01
Rivers and marshes	Liquid	1500	0.0004	0.0091
Total sources		36 895 000	100	2.665

3.5.3 Global distribution of fresh water in its various states

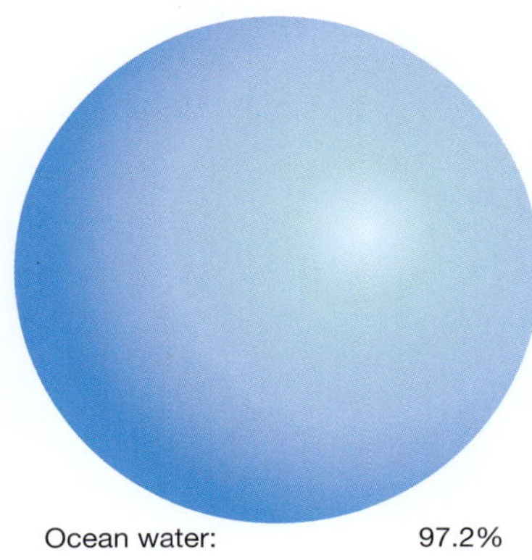

3.5.2 Where water is stored

Key processes in the water cycle

Figure 3.5.4 summarises the key processes involved in the water cycle. Each of these processes works in concert with the others in the operation of the cycle.

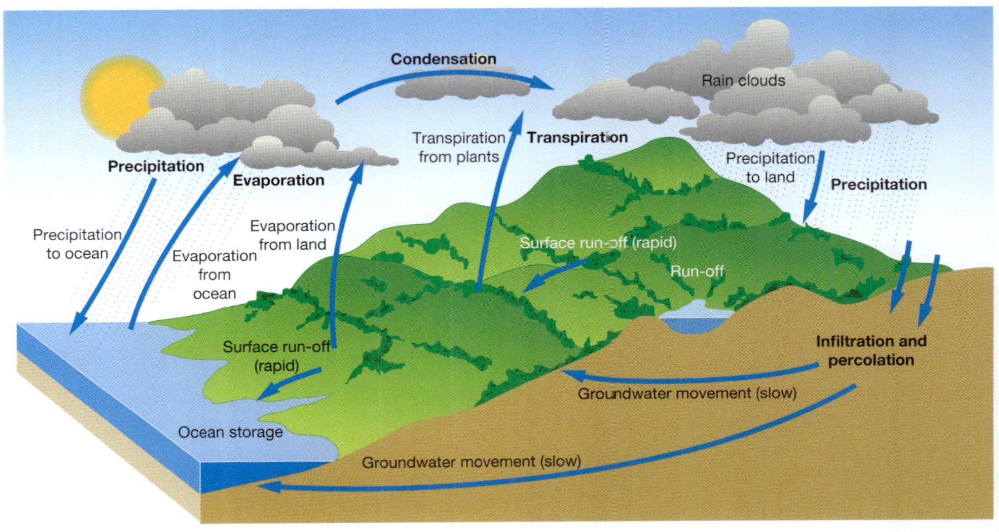

3.5.4 Key processes in the water cycle

● **SKILLS BUILDER**

Interpreting proportional circles

Proportional circles, as in Figure 3.5.2, show the relative size of the units being measured by the area of each circle. Each circle's size must be in proportion to the total value it represents. Proportional shapes are sometimes used instead, such as squares, rectangles, cubes, triangles, symbols or the shape of nation-states.

Condensation

Water changes from a gaseous state back into a liquid state in a process called condensation. It can only take place when there are solid objects on which moisture can be deposited. Dust particles in the atmosphere act as condensation nuclei—there would be no condensation without them. Condensed water may be visible as cloud, dew or mist.

The rate of condensation depends on the original air temperature (which influences the air's capacity to hold water), the amount of moisture or humidity in the air, and the rate of atmospheric cooling.

Condensation is a warming process because the latent heat absorbed by water during evaporation returns to the atmosphere. It means air can move heat as well as moisture from one place to another. The more moisture in the atmosphere, the more energy it can store and transfer as heat. The air needs to be cooled to a certain temperature before it becomes saturated (100 per cent humidity). This is called its **dew point**. When the dew point is below freezing, water vapour changes directly to ice or frost, a process called sublimation.

Evaporation

Evaporation is the physical process by which moisture is lost directly to the atmosphere. This can be due to the Sun's heat on a body of water or the effects of air movements. Its rate depends on temperature, wind velocity, the level of humidity and hours of sunshine. Evaporation is a cooling process because it involves a transfer of heat energy from the body of water to the atmosphere.

Transpiration

Transpiration is a biological process in which water is lost via the pores in leaves (see Figure 3.5.5). Its rate depends on time of year, the type and amount of vegetation and the length of the growing season.

Precipitation

Precipitation results when large masses of moist air are cooled rapidly below dew point. With continued condensation, water droplets or ice crystals become larger and heavier. Eventually, they become so heavy the atmosphere cannot support them. They fall as rain, sleet, snow and hail.

Infiltration

Following precipitation, soil becomes moist and absorbent. Water entering it passes through successive layers of the soil profile, a process known as **percolation**. How quickly this occurs depends on how much water the soil already holds. Also relevant is its porosity and structure, as well as the type and number of plants covering it. The speed at which water passes through soil is called its **infiltration** capacity. Infiltrating water eventually collects above an impermeable layer of rock or fills up all pore spaces. This forms a saturation zone—its upper level is called the **water table**. Water may gradually be transferred horizontally as groundwater flow or base flow.

3.5.5 Mists over forests are caused by transpiration.

Activities

Acquiring and processing geographical information

1 Describe the main states in which water can be found.

2 Write your own definition of the water cycle.

3 Describe the process of condensation.

4 Differentiate evaporation from transpiration.

5 Outline the forms that precipitation takes.

6 Explain the process of infiltration.

Applying and communicating geographical understanding

7 Study Figures 3.5.2 and 3.5.3.

   a List all the water storages that are part of the atmosphere, the lithosphere and the ecosphere.

   b Explain how water reaches the atmospheric store.

   c Explain why some parts of the lithosphere contain water and others do not.

   d How does water reach the ocean from the lithosphere?

8 Study Figure 3.5.3.

   a In which form is water most often found?

   b Where is fresh water most common?

   c Where is water found in a gas form?

   d Calculate the amount of water currently found in ice caps and glaciers.

9 Study Figure 3.5.4. Using facts, briefly report on the water cycle's operation.

Global water budget

Earth's **water budget** is the sum of the inputs, outputs and net changes in the hydrosphere over a period. The global water cycle contains around 1385 million cubic kilometres of water. Less than 3 per cent is fresh water. The world's population requires only a fraction of this—less than one per cent. Roughly 3000 cubic metres of renewable water supply is available per person every year, compared with the average use of 750 cubic metres.

The location, quantity and quality of water resources depend largely on the relationship between climate (precipitation and evaporation) and biophysical features, such as geology (see Figure 3.5.6). For example, South America has high rainfall in equatorial areas and high run-off, so it has large river systems and plentiful water supplies, particularly in the east where the mighty Amazon Basin is found (see Figure 3.5.7).

Continent	Area (km^2)	Average rainfall (mm)	Average run-off (mm)	Average run-off (km^3)	Percentage (run-off)
Asia	45 000 000	600	290	13 000	48
Africa	30 300 000	690	260	7900	38
North America	20 700 000	660	340	6900	52
South America	17 800 000	1630	930	16 700	57
Antarctica (and Greenland)	17 100 000	150–200*	160*	2800*	*
Europe	9 800 000	640	250	2500	39
Australia	7 700 000	465	57	440	12

* Data uncertain

3.5.6 Average annual rainfall and run-off for each continent

With mid-latitude continents, the pattern is quite different. Australia has the lowest rainfall and run-off proportional to area of all the continents. It also has the lowest percentage of run-off proportional to rainfall. Here, evaporation and transpiration are so high that the amount of rain ending up as streamflow is meagre. This leaves the continent with few major river systems. It is one of the reasons why the Central Australia is so dry. Comparing discharge rates—the amount of water a river releases at its mouth—highlights this. The discharge rate of South America's Amazon River is 180 000 cubic metres per second, while Australia's Murray River has a rate of 470 cubic metres per second.

3.5.7 High rates of rainfall around the equatorial Amazon River feed its vast quantities of water.

Variations in the water budget

The total amount of water within the water cycle is fixed, but it is unevenly distributed. The amount of rain and the seasons it occurs in varies between locations. Continental water budgets vary as a result. A continent's size and location impact on the interactions between the hydrosphere and lithosphere, affecting the general atmospheric circulation. Such interactions affect the land climate. Globally, broad patterns of vegetation result. Locally, they influence how much water a place has for plants to grow. In equatorial areas, where high rainfall dominates, plant life is more prolific.

There is a strong link between precipitation, evaporation rates and run-off (see Figure 3.5.8). It determines if rivers can form and how easily people can access water. The type of intensive agriculture that occurs in Europe is not possible in low rainfall areas like Australia. Here, grazing and less intensive farming dominate. Also, consider the size of the continent when looking at its data. While Europe has a relatively low amount of precipitation by area compared to Asia, it also has a much smaller landmass—9.9 million square kilometres compared to Asia's 44.6 million square kilometres. Evaporation rates also have an impact. They are high in Africa despite high precipitation rates, so there is far less run-off. This limits how much water enters rivers and lakes.

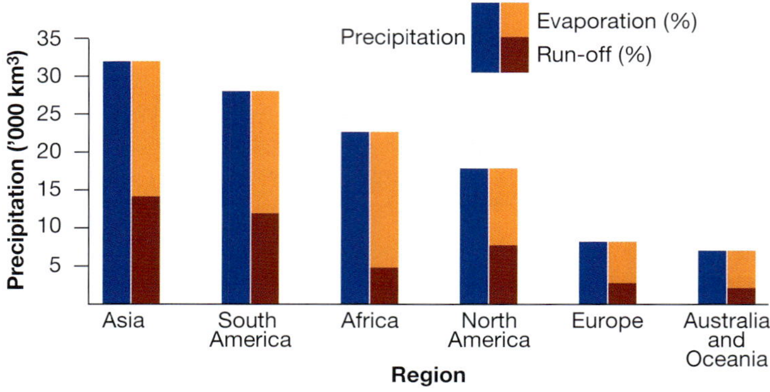

3.5.8 Annual precipitation, evaporation and run-off rates around the world. These patterns vary not only between locations but also over time.

3.5.9 The thin band of green winding through the desert demonstrates the power of water to bring plant life to the banks of Egypt's Nile River.

River catchments

Rivers play a crucial role in the water cycle, as they are one of the main ways in which water is moved across the land's surface. They transfer water from areas with a net surplus in their water budgets to areas that have less water (see Figure 3.5.9). The water flowing down a river system comes from a **catchment**. Its size and the amount of rainfall that falls in it determine the flow and discharge rate of the river. A catchment is more than its rivers; it includes lakes and other natural water storage such as wetlands.

River catchments are separated by a ridge of land called a watershed. When rain falls at higher points in the natural landscape it is divided between two river systems. The watershed marks the boundary between river catchments (see Figure 3.5.10). Rivers are generally fed by a network of smaller rivers or streams, called tributaries. One example is the Murray–Darling catchment. It covers over 1 million square kilometres, the most significant river system in Australia.

A catchment is an open system. It forms part of the water cycle. Its inputs are precipitation (mainly rain and snow) and its outputs are water lost from the system through flow to the sea, evaporation, transpiration (from plants) or human use. Within the system, some of the water is stored, either in lakes or in the soil.

The stored water otherwise passes through a series of transfers. First it infiltrates the ground, becoming groundwater. This groundwater may pass into a different catchment in a process called throughflow. Hence, the system involves far more than the flow of water. It also transports nutrients, soil and organic material. This means rivers are far more than systems of the hydrosphere—they also create crucial changes in the lithosphere and ecosphere.

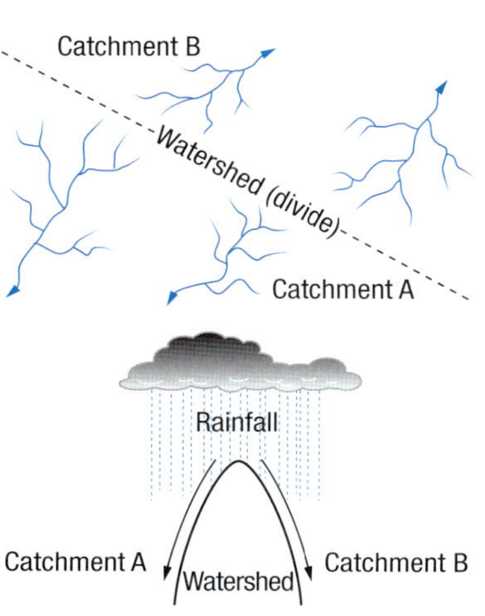

3.5.10 Two river catchments and the watershed dividing them

● **SPOTLIGHT**

Canada's watersheds

Canada is one of the wettest countries, with five main watersheds (see Figure 3.5.11). Each watershed is defined by a mountain range, with the Rocky Mountains marking the largest catchment area. On the eastern side, the rivers flow to the Arctic Ocean or Hudson Bay. The Mackenzie River—the longest in Canada at over 4200 kilometres—is the main river here. Its catchment is also the largest in Canada and the tenth largest in the world. The river arises from the mountains where the Columbia Icefield's meltwaters feed it (see Figure 3.5.12). It is a massive area of ice—around 325 square kilometres—that receives up to 7 metres of precipitation annually, nearly all of it falling as snow.

Over 30 per cent of Canada lies within the Hudson Bay Basin, a vast area of 819 000 square kilometres. Technically part of the Arctic Ocean, this basin is considered a separate watershed and drains the Nelson, Churchill and La Grand rivers. It is surrounded by land, except for a few straits and channels that flow into the Arctic Ocean, making it almost an inland sea. Hudson Bay Basin's ecosystem is complex and its waters play a key role in its functioning.

Most of the remaining watersheds in the eastern part of Canada drain into the Atlantic Ocean from catchments in a region known as the Maritimes. A small area drains southward into the Mississippi River basin, which flows through the USA before entering the Gulf of Mexico.

3.5.11 Canada's five watersheds

3.5.12 Icefield meltwaters feed Canada's Mackenzie River

Hydrological functions

Rivers are drainage systems that transport water (under the influence of gravity) from higher points to lower points of the topography. These systems are generally fed by precipitation. River systems do not always drain into oceans. Some rivers drain into lakes—such as Lake Eyre in South Australia—or inland seas, such as the Caspian Sea in Central Asia.

Geomorphic functions

River systems move and redistribute the products of **weathering**. Carrying these materials in water erodes the riverbed and banks, which can be deposited. A large river channel with an especially high flow can shift huge boulders downstream. Sand generally stays within the river channel, while finer silts and clays can reach the ocean. Sediments can also be deposited onto floodplains. They help nourish these soils. As a river moves across the flatter coastal plain it loses velocity, often branching out to form vast deltas (see Figure 3.5.13).

3.5.13 The Ganges River delta in the Bay of Bengal at the end of its 2700 km journey from the Himalayas

Catchment run-off

The catchments that collect water for rivers can be viewed as simple input–output models. The inputs are generally precipitation, but may also include water introduced from other catchments or sources.

Much of the water infiltrating the ground surface is taken up by vegetation. Other water eventually passes into the river system via groundwater flows. Water passing deeper into aquifers may be lost from the catchment. The rest passes as overland flow or run-off to the river. Water can also bring inputs—eroded material, nutrients and organic matter. Sometimes groundwater can add high levels of salts. All this depends on the nature of the rainfall and land management in the catchment. Substantially, the losses come from two sources—evaporation from land, vegetation and water surfaces, and transpiration from growing vegetation. The relationship between the inputs (precipitation) and their pathway can be expressed as:

inputs = run-off + evaporation + transpiration + groundwater + loss to other catchments + change in soil/surface storage

Alternatively, the relationship can be expressed in the following form.

run-off = inputs − evaporation − transpiration − groundwater − loss to other catchments − change in soil/surface storage

How much run-off (and the sediment and other materials carried by it) enters a river depends on the nature of the rainfall and the soils, vegetation cover, natural processes and land management. Vegetation, for example, can create ponding (or water retention) over the catchment's surface. This slows down the surface water flow, allowing more surface water to infiltrate the soil.

Both the nature of the soils and the vegetative cover of catchments will influence the volume and rate of run-off. In this way, lithospheric and biospheric conditions play key roles. If soils are sandy, much of the water will infiltrate the soils. The presence of dense ground-cover vegetation will also increase infiltration. In contrast, impermeable clay-rich soils with little vegetation cover, or soils saturated from previous rainfall, will yield the most run-off and the highest flood peaks.

The influence of vegetation

Vegetation has many roles in a catchment. It modifies local climatic conditions and intercepts rainfall. This helps to protect the soil surface from more intense precipitation. It also acts as a barrier to overland flow by:

- reducing the velocity of run-off, which in turn reduces surface **erosion**
- causing ponding, which provides more time for infiltration to occur.

An established stand of vegetation helps infiltration because a more permeable layer of organic material can develop. It also helps to provide a network of fissures, or cracks, through the soil profile. The decay of old root systems causes this.

Activities

Acquiring and processing geographical information

1 Describe the composition of the global water budget.
2 Outline the factors that determine the location, quantity and quality of water resources.
3 Explain the absence of very large river systems in Australia.
4 Explain why continental water budgets vary.
5 Define the terms catchment and watershed.
6 Outline the role of the geomorphic functions of rivers.
7 Explain how river deltas form.
8 Describe how sediment transfers by rivers benefit the ecosphere.
9 Describe the influences of run-off on rivers.
10 Explain the role of soil in determining run-off rates.
11 Explain how vegetation affects run-off.

Applying and communicating geographical understanding

12 Study Figure 3.5.6. Calculate the differences between Australia and South America in terms of average rainfall and run-off.

13 Study Figure 3.5.8.

a Which continent receives the most precipitation?

b Which of the continents featured has the least precipitation?

c Which continent has the highest rate of run-off?

d Which continent has the greatest rate of evaporation?

e What are the implications of this data for agriculture in Africa?

14 Review the box, Spotlight: Canada's watersheds. Write a short report explaining the role of the lithosphere in creating watersheds.

15 Explain the relationships between river inputs and outputs.

Floods

A river floods when it discharges more water than its channel can hold. The extra water spills over the banks and onto the floodplain. There are different types of floods.

- **Slow-onset floods:** these develop over several days and last for a week or longer. They can cause stock and crop loss, and road damage.
- **Rapid-onset floods:** these are faster and occur in the mountain headwater areas of large rivers. Often only lasting one or two days, they are often very destructive.
- **Flash floods:** these are caused by intense storms dropping large amounts of water within a short period, and there may be little warning. They can peak within minutes.

Two major factors affect the rate of river discharge and how it impacts on the scale of a flood—river catchment (covered above) and precipitation. Precipitation determines the amount of water in the system in several ways:

- **Type of precipitation:** if the precipitation falls as snow, it may be days, weeks or even months before it melts and runs off into waterways. If it melts rapidly and the soil underneath is frozen, the large amounts of run-off can cause flash flooding.
- **Intensity of rainfall:** during storms, if the rainfall is too intense the ground might not be able to absorb all the water. Excess run-off flows quickly overland and into waterways. A rapidly increasing discharge can lead to flash flooding.
- **Frequency of rainfall:** the infiltration rate is affected by how much water is already in the soil. If it is still wet from earlier rain, it will quickly become saturated. If it has been able to dry out, it will be able to absorb new rainfall. This is why floods are more likely to occur after long periods of wet weather.

The nature of the river catchment also impacts flooding. There are many ways it determines what happens to the water once it is on the ground:

- **Rock type:** permeable rocks, such as sandstone, readily absorb water and reduce the amount of run-off. Flooding is more common in catchments with impermeable rocks such as granite.
- **Soil:** sandy soils have large air spaces that water can infiltrate. Soils with a high clay content are less able to absorb water. Clay particles are very small and fit together tightly. It is difficult for water to infiltrate these tiny spaces, causing it to accumulate on the surface.
- **Slope gradient:** on gently sloping land, rainfall has time to infiltrate before it runs off. On steep slopes, rainfall runs off before it can be absorbed.

● **SPOTLIGHT**

New South Wales floods 2022

After years of severe drought, weather patterns across Australia's east shifted markedly during 2021–2022. Record-breaking rainfall brought major flood events (see Figure 3.5.14). The Wilsons River flows through Lismore, a town in northern NSW. With floods being part of the town's history, many homes were built above the record peak 1974 flood height of 12.15 metres.

3.5.14 Lismore city in flood, February 2022

In late February 2022, after months of higher-than-average rainfall, northern NSW was hit with a massive rain event. At first Wilsons River seemed to peak well below the 1974 flood levels. Within hours this was revised after a report that the flood height was above 13 metres. But it still hadn't reached its peak—a massive 14.4 metres. It inundated hundreds of homes and businesses across Lismore. Subsequent months saw many smaller floods, further impacting the town and the community's ability to recover. Many months and even years later, hundreds of residents were still living in temporary housing.

- **Vegetation:** vegetation slows the flow of water across the land surface and encourages infiltration. This reduces the risk of flooding.
- **Land use:** agricultural land uses can affect the rate of run-off. Ploughing parallel to the contours of the land (contour ploughing) can reduce run-off. Soil compacted by stock and farm machinery can do the opposite—it forms an impermeable layer, encouraging run-off. Urban buildings, paving and roads form an impermeable layer that precipitation rapidly runs off. Drainage systems guide this run-off into local waterways.

3.5.15 Extreme drought in the Entrepeñas reservoir, Spain

Drought

Droughts are prolonged periods of below-average rainfall (see Figure 3.5.15). Regions with low average rainfall are more likely to experience droughts than places with high average rainfall. Droughts inhibit the growth of crops and natural vegetation.

Severe droughts can also contribute to land degradation. Wind erosion takes place when bare, dry soil is no longer bound together by tree roots, grasses and other plants. Topsoils are the richest soils, found near the surface where **humus** accumulates. They are critical for plant growth. This layer is most susceptible to erosion from wind and water. Even drought-breaking rains can erode the soil by washing away unprotected topsoil.

The effects of drought on the people in an area vary with their socioeconomic situation. In one community a drought may simply lead to restrictions on car-washing, garden sprinklers or topping up swimming pools. In another, it can result in famine and death.

Activities

Acquiring and processing geographical information

1 Outline the different types of floods.
2 Explain how rock and soil type can impact floods.
3 Describe the role of gradient on floods.
4 Define the term drought.
5 Describe the impact of droughts.

Applying and communicating geographical understanding

6 Write a paragraph explaining how interactions within the biophysical environment affect the global water budget.
7 Using examples from the text, outline the impacts of floods on communities.

UNIT 3.6

Geomorphic systems

Soaring mountains, deep gorges and rugged coastlines tell us of the tug of opposing forces: tectonic and gradational. **Tectonic forces** are caused by Earth's internal energy and upheave the land surface, creating new landforms. Without them, Earth's surface would have been reduced to flat, featureless plains long ago by the relentless **gradational processes**. These are weathering, **mass movement** and erosion. These processes are caused by gravity and the Sun's radiant energy wearing down and smoothing the surface. The Snowy Mountains of south-eastern Australia were once 5000 metres higher than they are today. Millions of years of gradation reduced them to their current height and rounded form.

Earth's surface seems solid and unmoving. The lithosphere is the solid, outer portion of Earth's rigid upper mantle. But deep within the planet's interior are convection currents. They are so powerful and their movement so great that the lithosphere is broken up into a dozen or so vast sections. These are called **tectonic plates**. They move slowly atop the **asthenosphere**. Ocean basins and continents are located on them.

Much of Earth's geological activity takes place in the lithosphere at the plate boundaries. The plates move very slowly—about as fast as fingernails grow. Over geological time the movements become significant. Land masses split or joined together as the tectonic plates shifted around. This changed the size, shape and location of the continents. Their slow movement across the surface is called **continental drift** (see box, Spotlight: Continental drift).

● **SPOTLIGHT**

Continental drift

Published in 1915, Alfred Wegener's theory of continental drift explained the similarity of shoreline edges on either side of the Atlantic Ocean. He matched fossils and rocks from two vastly separated continents. He believed these continents shifted from a former 'supercontinent' named Pangaea. It consisted of Laurasia (Asia, Europe, Greenland and North America) and Gondwana (Australia, Antarctica, Africa, India and South America), which began separating about 200 million years ago. Each part slowly drifted to its present location. If put back together, the broken pieces fit like a vast jigsaw puzzle (see. Figure 3.6.1).

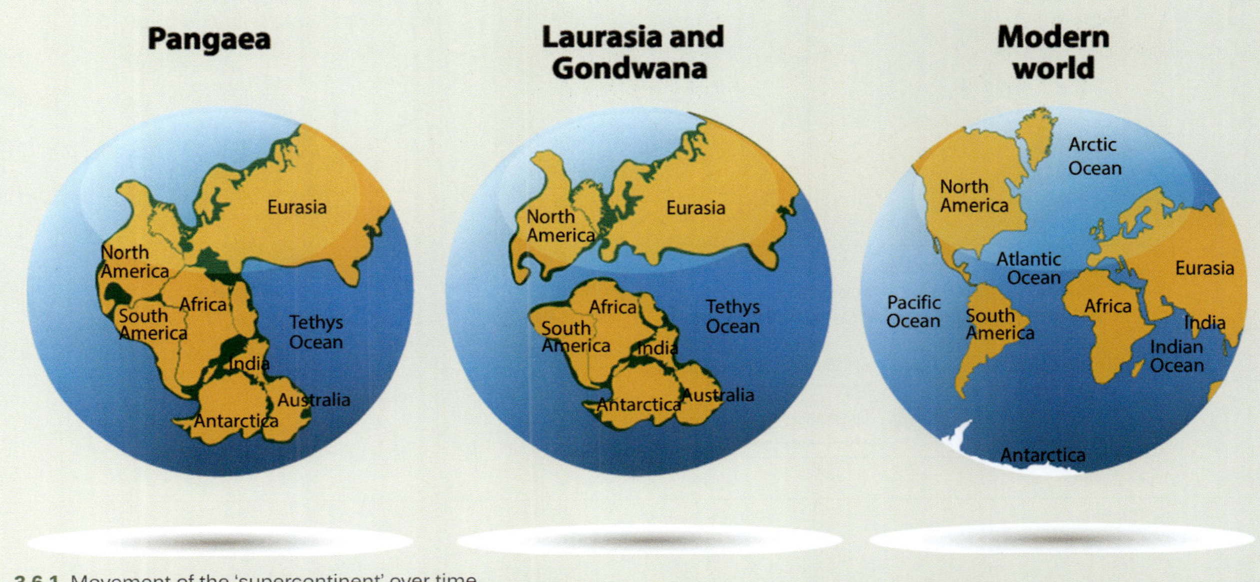

3.6.1 Movement of the 'supercontinent' over time

Plate tectonics

A plausible mechanism to explain the continents' movement was discovered relatively recently. Geological investigations of the mid-Atlantic Ocean revealed that the sediments near the mid-Atlantic ridge are thin and relatively young. Those progressively further away are broken and older. This means new land is being created under the ocean, forcing the continents further apart as it emerges.

This process is called **plate tectonics** and the theory is based on:

- identifying the asthenosphere as a layer of weak, plastic-like material that acts as a lubricating layer
- identifying the cold, rigid, lithospheric plates moving like great rafts over the fluid asthenosphere
- the idea that if new oceanic crust is being continuously created along the mid-ocean ridges, either Earth must be expanding and the oceans must be getting larger, or an equal amount of old crust must necessarily be destroyed in order to maintain a constant size.

A convection current in the asthenosphere rises, separates and forms a mid-ocean ridge—as seen in Figure 3.6.2. The convection current sinks again where the oceanic plate meets the continental plate. As it sinks, it carries the solid rocks of the crust downwards into Earth's interior. Here, they melt back into the asthenosphere. The places where this downward movement occurs are called subduction zones.

Figure 3.6.2 shows five major lithospheric plates and some minor ones. NASA scientists measured the velocities of plate movements using satellites and lasers. Their calculations confirm the plates move 1.5 to 7 centimetres per year. Australia continues to drift north by 6.7 centimetres each year.

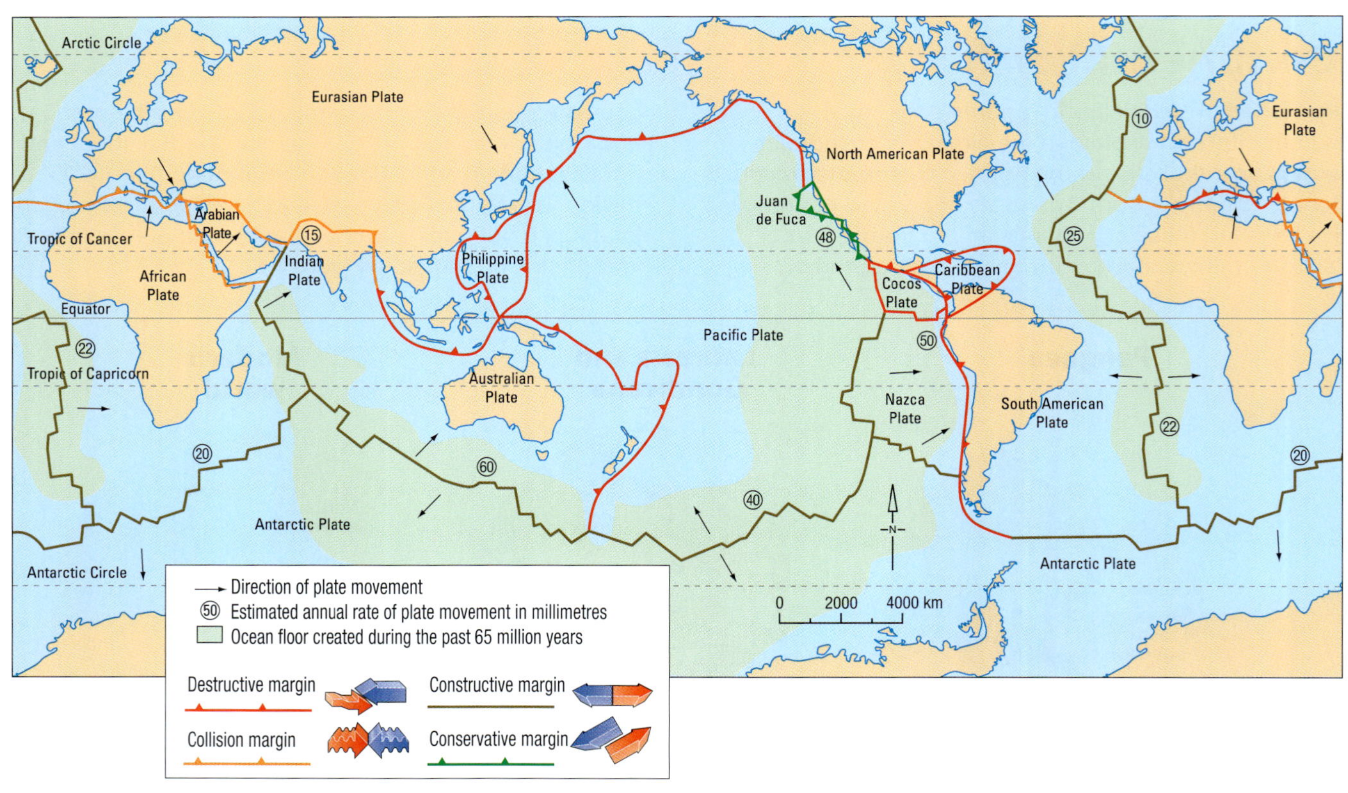

3.6.2 Earth's tectonic plates

Plate motion and boundaries

Convection currents within Earth's mantle were long seen as the cause of plate motion. Rising magma at the diverging mid-ocean ridges is believed to push the plates apart horizontally. Mantle convection depth remains unknown. Research suggests it is shallow and largely confined to the asthenosphere, rather than extending further into the mantle.

The lithospheric plates are rigid slabs of rock that move individually. As a unit, their inner parts are relatively inactive. Almost all surface tectonic and volcanic activity occurs at the plate boundaries. Here, they diverge, converge or slide past one another. Each type of boundary produces distinctive and recognisable landform features, like the mountains in Figure 3.6.4.

Diverging plate boundaries occur where the plates separate and new crust is created (see Figure 3.6.3). As the plates move apart, molten material fills the gap, forming a mid-ocean ridge. These ridges extend for some 65 000 kilometres throughout all oceans. The Hawaiian Islands in the mid-Pacific are the peaks of a mid-ocean ridge.

Converging plate boundaries are the points where plates collide. The process of sea-floor spreading increases Earth's land surface area. To balance this process, plates must collide. The old lithosphere is destroyed in the subduction zones—an area where two tectonic plates come together, one overriding the other. There are three types of collisions.

- **Colliding oceanic and continental plates** occur when the dense ocean floor is subducted (or forced down) beneath the less dense continental plate. This forms a deep trench. The descending slab of the sea floor melts and then erupts in a chain of volcanoes along the edge of the continental plate. Over hundreds of millions of years, the Andes Mountains of South America were formed in this way.
- **Colliding oceanic plates** occur when oceanic plates collide, and the plate that is older, colder and therefore denser, subducts under the other, forming a trench. This movement weakens the crust, and molten rock escapes to the surface. It forms islands of volcanic mountains. Chains of active volcanoes form on the landward side, parallel to the deep-sea trench. The Pacific basin has many such island arcs, including the Aleutian Islands, the Philippines and Japan.
- **Horizontal movements** can cause plates to move horizontally past each other along a single vertical fault line. When the built-up pressure is released this causes earthquakes. The San Andreas Fault in California is a major fault system arising from movement between the Pacific and North American plates.

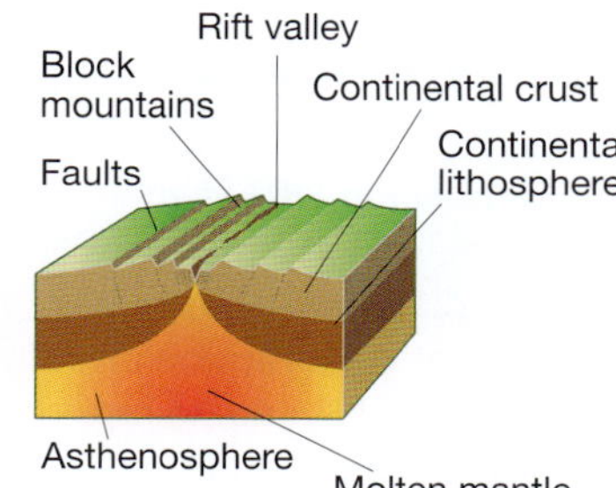

1 Plates break apart, forming rift valley, faults and block mountains (e.g. African Rift Valley)

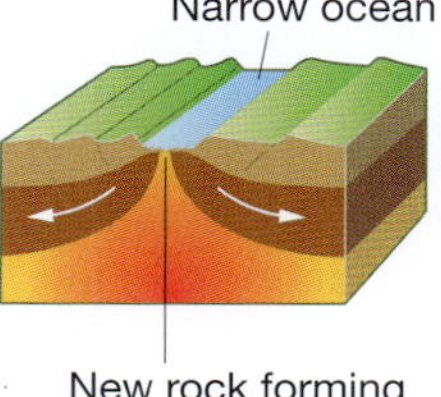

2 Plates move apart and new rocks are formed on the ocean floor (e.g. Red Sea)

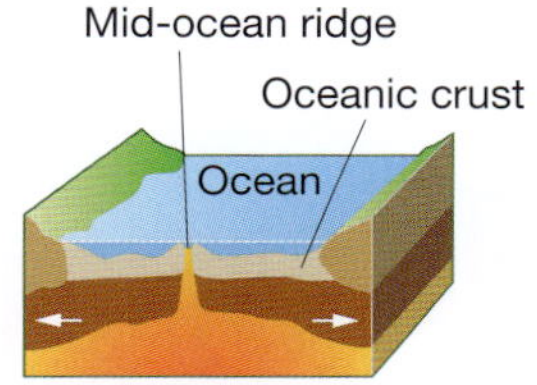

3 With further movement, the sea floor spreads to form major oceans (e.g. Atlantic Ocean)

3.6.3 Divergence and convergence of tectonic plates

3.6.4 The Himalayas are continually being pushed up by the Indian Plate ramming into the Eurasian Plate.

> **Did you know?**
>
> The Tambora eruption in Indonesia in 1815 is the largest in modern history. It killed 120 000 people. It was the only eruption in at least 1000 years to rate seven on the logarithmic Volcanic Explosivity Index. The eruption cast a pall of ash worldwide, lowering global temperatures so much (over 2.8°C) that 1816 became the 'year without a summer'. About 300 centimetres of snow fell in Canada's Quebec City in midsummer that year. The eruption is credited as the motivation for the bicycle's invention, as the cost of horse feed rose and many horses died.

Mountain building

The world's great mountain ranges are produced through plate tectonics. They form at the advancing edges of continental plates. As these plates collide, the crust thickens to absorb the impact. Pressure is exerted and layers of rock are compressed. They are forced up, folding and faulting as they go.

As continental movements tend to be very slow, the mountain chains they produce are dominated by folded rock strata (layers of rock). These are classified as fold mountains. Folds are wave-like patterns in Earth's crust. Mountains can also be created by faulting and volcanic activity. Faults are fractures in the rock structure. On a large-scale, faulting creates landforms such as rift valleys and block mountains.

Earthquakes

Earthquakes are a series of rapid vibrations caused by sudden movements in the crust. When rock strata are put under great stress by plate movements, they usually fracture or fault. The energy released by these sudden movements travels through the ground in waves. These waves spread out from the focus (the point where the earthquake began) and **epicentre** (the point on the surface directly above the focus). Earthquakes occurring in heavily populated areas can cause widespread devastation (see Figure 3.6.5). Most loss of life is caused by collapsing buildings, fires and tsunamis.

The world's most powerful earthquake had a magnitude of 9.5. It struck southern Chile in 1960, leaving 4485 people dead or injured and 2 million people homeless. The port of Puerto Saavedra was destroyed in the ensuing tsunami. It caused US$550 million damage in Chile alone. Five-metre waves hit the coasts of Japan and the Philippines, killing another 170 people. A day later, Volcán Puyehue in Chile's lake district spewed ash 6000 metres into the air. Its eruption lasted several weeks.

Volcanoes

Volcanic eruptions occur when molten rock reaches Earth's surface, through cracks or faults in the underlying rock structures. Once at the surface, the molten material (lava) cools and hardens. Over time, successive layers of lava and volcanic ash may build up a volcanic cone. A caldera, or crater, forms when a particularly violent eruption blasts away the top of an existing volcanic cone. The world's most active volcanoes lie in volcanic belts coinciding with zones where tectonic plates collide and fracture.

3.6.5 Earthquake devastation in Aleppo, Syria, in February 2023. The quake killed over 50 000 people and injured over 100 000 across Syria and Türkiye.

Volcanic eruptions can produce landforms rapidly and spectacularly. Past activity created many of the largest mountains and plentiful oceanic islands. Roughly 1500 potentially active volcanoes have at least 500 million people living nearby. Major eruptions occurred at Mount St Helens in the United States (1980), Mt Pinatubo in the Philippines (1991) and Mount Merapi in Indonesia (2010). Ash clouds can damage aeroplanes with jet engines and have frequently disrupted aviation.

The annual level of volcanic activity is surprisingly high. In 2022 alone there were 80 volcanic eruptions. Large-scale eruptions can cause short-term climate change. Millions of tonnes of volcanic ash and smoke can be released into the atmosphere, limiting the sunlight that reaches the surface.

Activities

Acquiring and processing geographical information

1. Outline Wegener's theory of continental drift. What evidence did he draw on to support this theory?
2. Identify the discoveries that helped revolutionise our understanding of movements within the lithosphere.
3. Explain why the discovery of the asthenosphere is so significant to the theory of plate tectonics.
4. State what is regarded as the cause of plate motion.
5. Describe the distinctive features produced at each type of plate boundary.
6. Describe the relationship between the location of the world's great mountain ranges and continental plates.
7. Explain the differences between mountains formed by folding and those formed by faulting.
8. Identify where most active volcanoes are found.
9. Explain how volcanic eruptions create short-term changes in climate.

Applying and communicating geographical understanding

10. Study Figures 3.6.3 and 3.6.4.
    - a Using an atlas, name the mountain ranges formed by the collision of the following crustal plates:
        - i the Indian and Eurasian plates
        - ii the South American and Nazca plates.
    - b Using an atlas, identify islands in the Atlantic Ocean that are part of the mid-ocean ridge.
    - c Identify two continents that have deep ocean trenches offshore due to subduction.
11. Study Figure 3.6.3. Write a report outlining:
    - a the process of plate divergence
    - b the process of plate convergence.

UNIT 3.7

Geomorphic processes

3.7.1 Colorado River's Horseshoe Bend in Arizona, USA, is shaped by gradational forces.

Tectonic forces constantly disturb the surface of the land. They create new landscapes that make the surface uneven. Gradational processes are the opposite; they smooth out surfaces. These geomorphic processes include weathering, erosion and deposition. When newly formed land features are exposed to the atmosphere and gravity, they swiftly come under attack by these gradational processes. Exposed rock material is fragmented by weathering. It can be detached by mass movement and erosion. Ultimately it is deposited at lower elevations as debris. Gradational processes act on the surface, wearing down high places and filling in low places (see Figure 3.7.1).

Soil formation

Soils teem with hidden life: billions of micro-organisms, bacteria and fungi. They are vital for the conversion of inorganic and organic material into the nutrients plants need. Soils develop because the lithosphere interacts with the other components of the biophysical environment. They are a complex mix of inorganic minerals (usually in the form of clay, sand and silt), air, water and organic matter (see Figure 3.7.2).

The profile of a soil

Soils are developed by physical, chemical and biological processes. These include rock weathering and vegetation decay. For soils to develop, two processes must take place. First, water moving down through cracks in the rock strata must cause physical and chemical changes in the original material. Second, living organisms in the soil must bring about further change. Where the presence of water and organisms is limited, soil development will be poor.

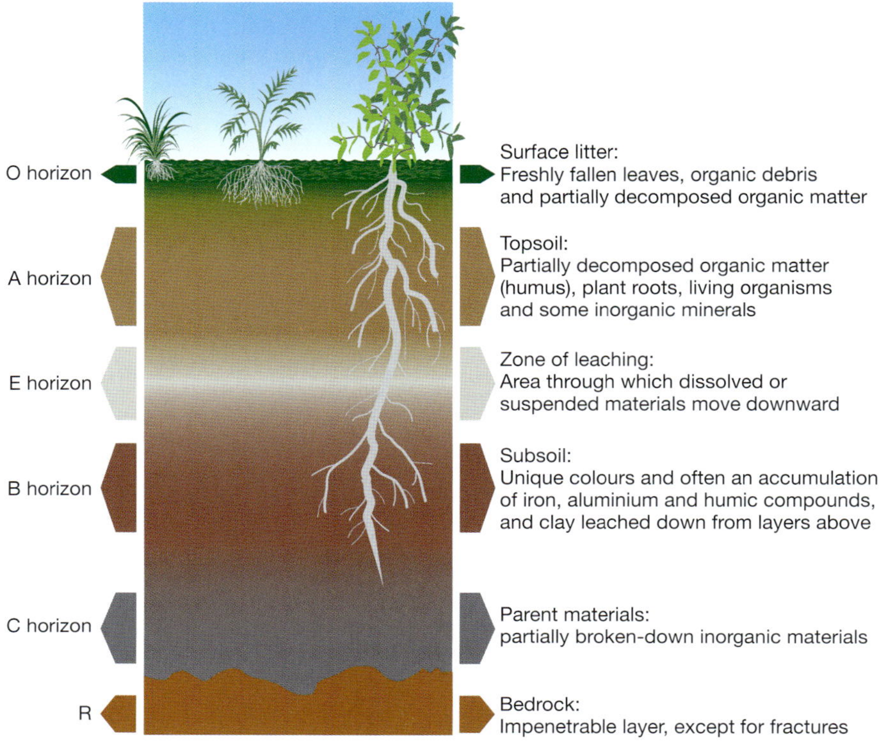

3.7.2 A generalised soil profile

Soils typically comprise several horizons roughly parallel to the surface. Together, these form the soil profile. Each horizon has a distinct colour, thickness, texture and composition. The number of horizons varies depending on the soil type. In a generalised profile, the top horizon—Horizon A—is where organic material is most prevalent (see Figure 3.7.2). This topsoil is ideal for plant growth because it has more nutrient-rich humus. It is also the type of soil most easily removed by wind and water erosion.

Weathering

Weathering is the physical disintegration and chemical decomposition of rocks and minerals by atmospheric and biological agents. It loosens surface material, making it more readily removed. It is a prerequisite for many other types of erosion. The process is often unobtrusive. Physical weathering slowly breaks up the rocks of the land surface into smaller particles, but does not change their chemical composition. Chemical weathering involves the decomposition of rocks, primarily by means of exposure to water, oxygen and carbon dioxide. Most of the original minerals formed deep in the crust are changed by chemical weathering into new compounds. These compounds are usually stable in the atmosphere and the conditions (temperature and air pressure) found at the surface. The substantial age and stability of the Australian continent produced a great thickness of rock that has been modified by the process of weathering. In some areas, the modified rock runs as deep as 100 metres. See Figure 3.7.3 for the types of weathering.

Physical weathering	
Type	**Description**
Unloading	When once deeply buried rocks have the weight and pressure of overlying material removed. The rocks expand and may crack along joints parallel to the surface and peel off in sheets.
Frost action	When water held in the tiny fissures of rocks freezes, it expands by about 10% of its volume, forcing cracks apart. Recurrent freezing and thawing, especially daily, enlarges such cracks.
Organic action	Germinating seeds and roots wedging in rock cracks exert pressure, causing the cracks to widen. Burrowing animals, such as rabbits and earthworms, allow air to penetrate further. This may indirectly accelerate weathering.
Some experts now disregard temperature change, or thermal expansion and contraction of rock, as significant weathering processes. Others argue it is vital in certain contexts—the extreme heat from bushfires can dramatically affect rock exteriors. Biological activity is now considered to be the most influential weathering process. Plant roots can cause physical and chemical weathering.	

Chemical weathering	
Oxidation	Minerals react with oxygen to form oxides and, as water is also usually involved, hydroxides. Iron is the most common element in the process. Rocks containing iron appear to be rusting as their surface decays.
Solution	The process by which weak acids dissolve minerals in rocks and distribute them in solution. Limestone is susceptible to attack from carbonic acid, formed when carbon dioxide is dissolved in rainwater. Limestone caves and karst landscapes result from this process. Karst is shaped when the soluble layers of bedrock dissolve, usually carbonate-based rocks such as limestone or dolomite.
Hydration	Rocks expand as their minerals chemically combine with water. Clay minerals have a vast capacity for such water uptake. This results in 'physio-chemical weathering'.
Organic acids	Produced when water combines with the decaying organic material in humus. The acids attack—and weather—rocks as they seep through the soil.

3.7.3 Types of weathering

Erosion and deposition

The land's surface is shaped continuously by erosion. The eroded material is transported and deposited, ultimately ending up on the lowest place—the ocean floor. Wind, water and glacial ice are the agents that carry it. Processes such as the movement of water, wind and ice shape the landscape on the surface, with fluvial process (running water) being the most important. Wave action is dominant in shaping coastlines.

Landforms shaped by water

Water is a significant agent of weathering, but its impact is greatest in erosion. The fluvial processes of streams and rivers dislodge rock particles and carry them away as sediment. Running water and surface run-off shapes into rivulets and streams that collect this loose material as its sediment load. Running water is most effective in arid and semi-arid regions (see Figure 3.7.4), or where human activities have damaged the protective vegetation that binds the soil. On these exposed landscapes any rain swiftly forms rivulets, eroding deep gullies and eventually the rocks and sediments in rivers that the water flows over. Several processes are involved:

- **Hydraulic action:** the power of flow, where surging water drags along beds and banks. It can excavate vast amounts of the loosest alluvial material (sand, silt and gravel), especially during floods.
- **Abrasion:** occurs when rock particles carried by the current strike channel walls and chip off rocks and banks.
- **Corrosion:** caused by the chemical solution of material from rocks exposed to solvents carried in the river.

Coastal landscapes shaped by waves

The ocean's waves, tides and currents relentlessly transform coastlines, making them the most varied and rapidly changing part of the lithosphere. Rivers, glaciers and even daily weather conditions also shape coastal environments. Such terrains extend as far inland as the saltwater, seaspray and windblown sands can reach. They extend into the ocean as far as waves can move the material on the seabed.

Desert landscapes shaped by water

While desert rain is rare, it can be heavy, causing flash flooding, particularly in areas with little vegetation. Surface run-off is rapid, eroding large amounts of material. It is channelled into dry riverbeds, or wadis, that cut through plateaus, forming canyons. When plateaus are eroded, this leaves isolated mesas and buttes. Water carries the eroded material via wadis onto lowlands, creating alluvial fans. These spread across the desert basin or bolson, where the wind can shape the fine particles into dunes. Playa lakes form when water flows into depressions. It eventually evaporates, forming salt or clay pans. Inselbergs are resistant rock masses that are exposed when surrounding rock erodes, such as Uluru in Central Australia.

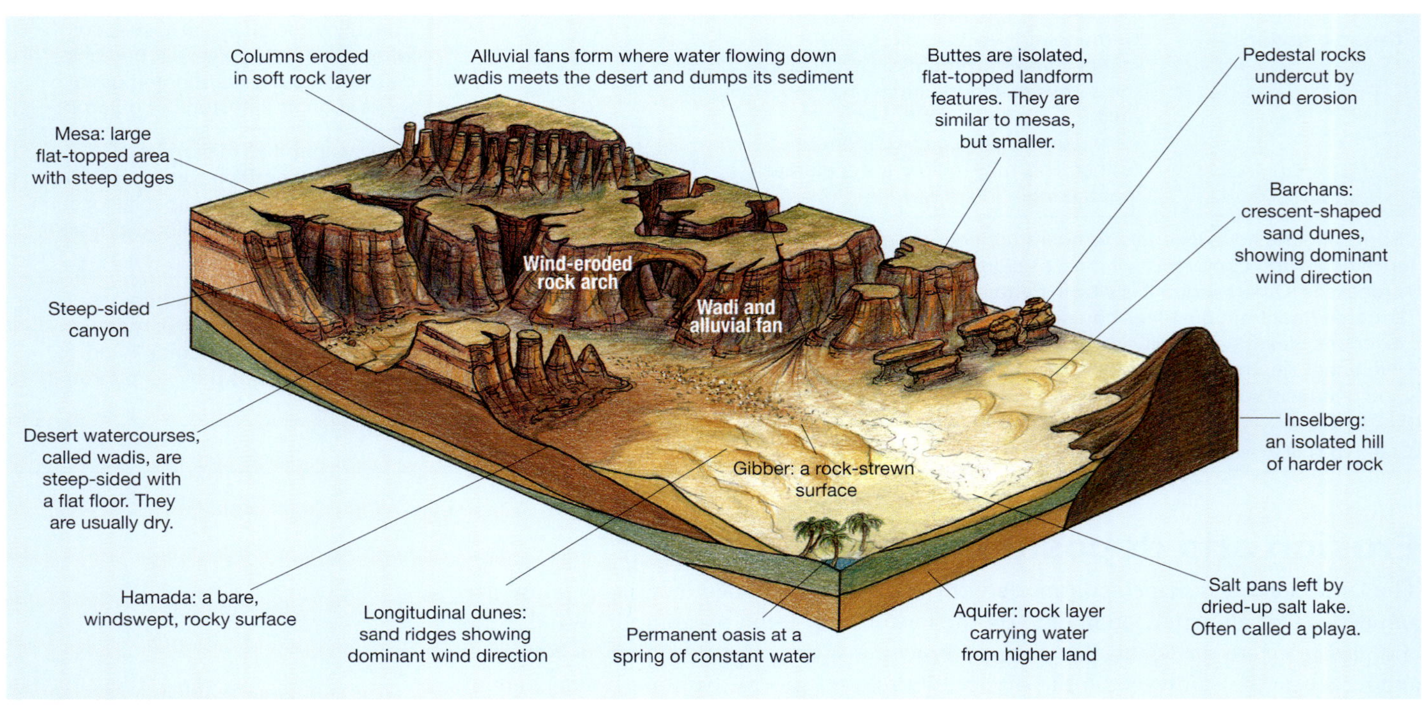

3.7.4 Landforms of the desert

3.7.5 Sand dunes of the Sahara Desert around Merzouga, Morocco

Landforms shaped by wind

Wind most effectively shapes landforms in areas with little or no vegetation, such as arid areas, or areas degraded by human activity. A process called abrasion picks up weathered materials to create distinctively sculptured, large rock formations like pedestal rocks. On a broad scale, wind erodes finer material, in a process called deflation. Deflating exposed rock creates rock-strewn reg or desert pavements. Sand dunes form when windblown material accumulates. Dune shape varies with wind strength, how much sand is available and the vegetation type and extent. Dunes are named for their shape and direction (see Figures 3.7.5 and 3.7.6).

Landforms shaped by ice

Glaciers are huge masses of ice made of long-compacted snow (introduced in Chapter 2). Due to its increasing weight, gravity forces the compacted glacial ice to move outwards and downwards (see Figures 3.7.7 and 3.7.8).

Figure 3.7.7 shows the ablation zone where Bear Glacier is losing ice. Broken pieces look like glass shards in the water. Upslope from the lake, the glacier's foot is riddled with crevasses. These are cracks in the ice caused by the glacier's movement over a rough surface. As it moves, the glacier picks up dirt and debris from the rocks it passes. When two glaciers merge, as they have in Bear Glacier, the dirt and debris they carry form parallel stripes, or medial **moraines**, on the surface of the ice.

Ice sheets are layers of ice covering relatively flat landscapes. Like glaciers, they have great erosional and depositional power. They 'pluck' rock fragments from the ground and embed them in the base of the ice, constantly grinding and scratching the land's surface as they move. This process, called abrasion, creates distinctive erosional features. Ice sheets occur in Antarctica and Greenland.

Distinctive U-shaped valleys form when active glaciers deepen and widen old river valleys. Hanging valleys mark where smaller tributary glaciers enter. Valley glaciers are found in the Himalayas, the Andes and New Zealand.

Moraines form when deposits of soil and rock embedded in glacial ice are eventually deposited. They include lateral moraines (sides of glaciers), medial moraines (from merged glaciers) and terminal moraines, the remaining rock debris deposited at the glacier's end.

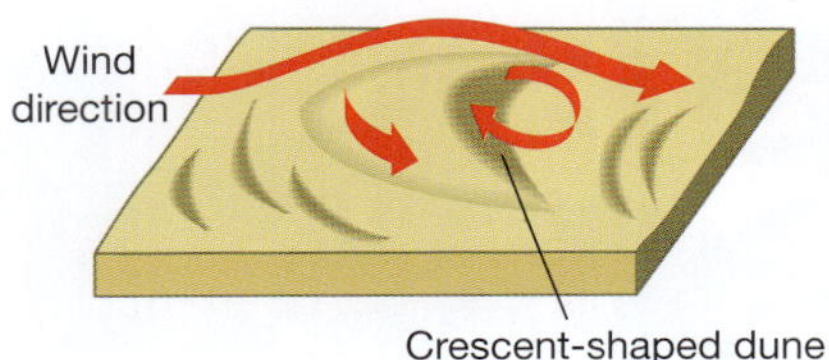

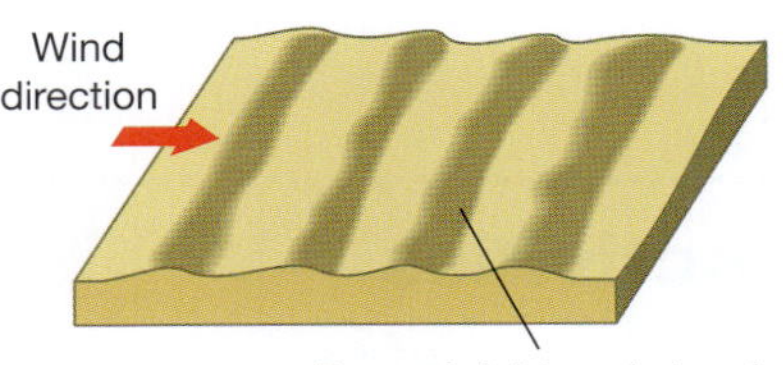

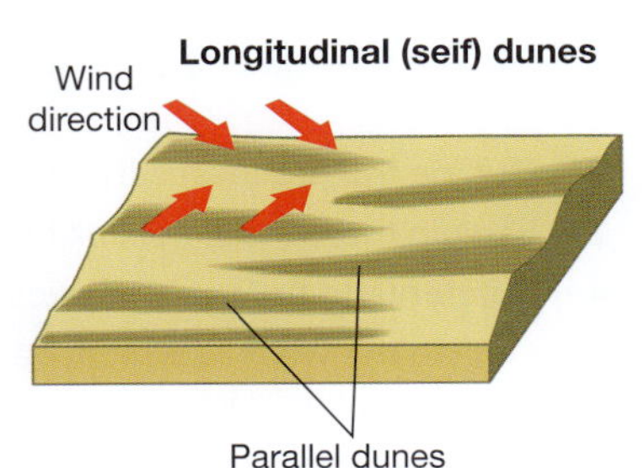

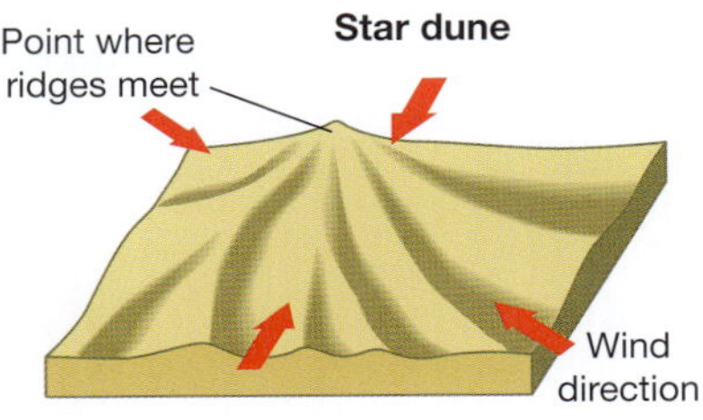

3.7.6 Different types of common sand dunes

3.7.7 Aerial view of Bear Glacier, Alaska, USA

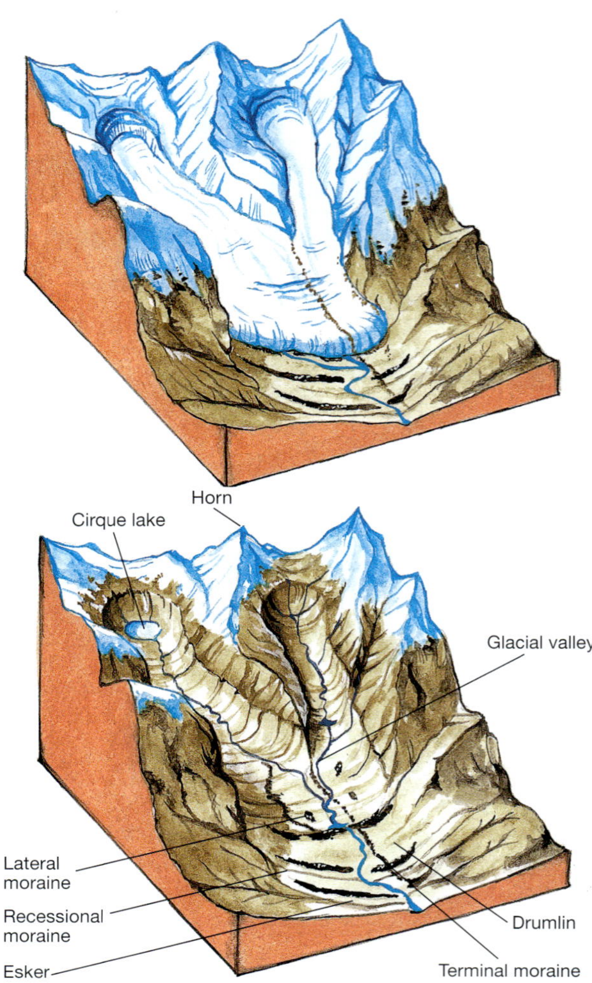

3.7.8 The development of glacial landform features

Activities

Acquiring and processing geographical information

1. Explain the role of gradational forces in shaping Earth's surface.
2. Outline the processes that lead to soil formation.
3. Explain the concept of soil horizons.
4. Differentiate between physical and chemical weathering.
5. Outline the ways in which weathered material can be moved downslope.
6. Identify the circumstances under which water erosion is most effective.
7. Explain how material in streams is transported.
9. Explain what abrasion is. How do glaciers abrade the surface of the land?

Applying and communicating geographical understanding

10. Study Figure 3.7.4 and select one landform feature on it. Investigate the processes responsible for its formation.
11. Briefly explain the processes that form the various types of sand dunes.
12. Briefly explain the formation of pedestal rocks.
13. Draw a block diagram of a glacial landscape. Label the landform features.

UNIT 3.8
Ecological systems

Ecological systems relate to all living things on Earth, both plant (flora) and animal (fauna). Collectively, Earth's living component is called the ecosphere (or biosphere) (see Figure 3.8.1). The ecosphere's functioning relies on interactions with the non-living components of the environment. For example, the soils, atmosphere and water cycle all influence, and are influenced by, the ecosphere.

3.8.1 The ecosphere of Ngorongoro Crater, Tanzania, includes flamingos, zebras and wildebeests.

Ecosystems

Ecology is a science that examines the interactions between organisms and their living (biotic) and non-living (abiotic) environment. The key is interactions. Groups of organisms interact with each other and their biophysical environment. Collectively, they form an ecological system or ecosystem. An ecosystem is an identifiable system of interdependent relationships between living organisms and their biophysical environment. Ecosystems are dynamic—constantly changing and adapting. Different types of ecosystems can be detected by noticing their usual interaction patterns.

The size of an ecosystem can vary dramatically, from a small pond to a vast area of rainforest or even an entire ocean. Large or small, ecosystems rarely have distinct boundaries. This can complicate their management when there are defined boundaries such as national parks or international borders. Individual ecosystems blend into adjacent ecosystems via a zone of transition or **ecotone** (see Figure 3.8.2). An ecotone contains organisms common to both ecosystems, but may also have organisms unique to that area. This means an ecotone often has greater biodiversity than its surrounding ecosystems.

3.8.2 This forest ecosystem gives way to an alpine ecosystem, marking an ecotone.

Classifying ecosystems

Ecosystems are usually classified by their dominant feature. They are named for their climate, physical features or vegetation. Climate is the key to the type of ecosystem that occurs in a location—temperature and precipitation are critical to plant growth. Polar regions and deserts are ecosystems named for their climatic conditions. Others are classified by their dominant physical feature, such as a mountain ecosystem. Another may be classified by its dominant vegetation, such as rainforest. Figure 3.8.3 shows how the climate conditions of temperature and precipitation in a location determine the different types of ecosystems that arise. It highlights the complex interactions taking place between the atmosphere, hydrosphere and ecosphere.

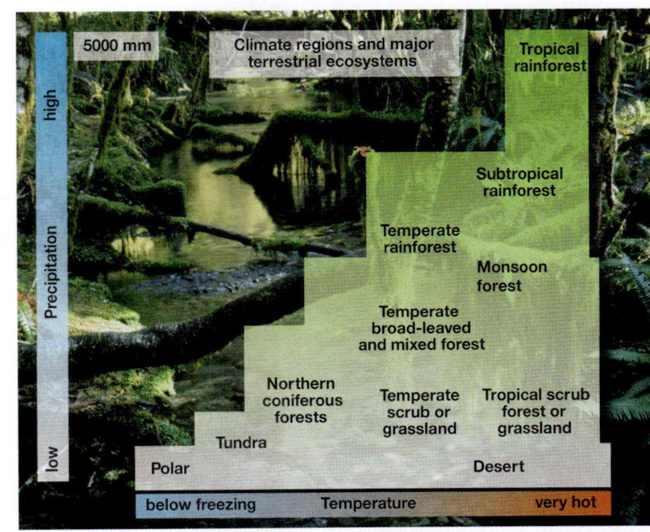

3.8.3 The relationship between climate and ecosystems

Did you know?

Mediterranean vegetation is called chaparral in North America, matorral in Chile and maquis in the Mediterranean. It is also found in southern Western Australia and South Africa.

Terrestrial ecosystems

Land-based ecosystems (e.g. forests and deserts) are called terrestrial ecosystems (see Figure 3.8.4).

The differences between terrestrial ecosystems are due to varying climatic factors. Latitude is, in turn, key in determining these climatic conditions. So, a pattern in ecosystem type and latitude emerges. This is linked to how much insolation the surface receives at the locations where latitude is the key determinant. Figures 3.8.5 and 3.8.6 demonstrate the links between climate, latitude and terrestrial ecosystem type (e.g. the world's tropical rainforests are found around the Equator). The weather is hot and rainy all year round because this latitude receives the maximum insolation. These forests are the most complex and productive of all land ecosystems.

Ecosystem	Characteristics
Polar	■ Permanent ice cap, up to 5 km deep in places ■ No plant growth, no animal life beyond the coast
Tundra	■ Usually covered with ice and snow, permanently frozen subsoil ■ 1–3 month growing season ■ Treeless, shrubby or mat-like vegetation ■ Most extensive in the Northern Hemisphere
Northern coniferous forest (taiga)	■ Long winters with deep snow, short summers with long, warm days ■ 3–4 month growing season ■ Dominated by conifer trees, thick layer of needles on the ground ■ Occurs on large continental landmasses
Temperate grassland	■ Erratic rainfall, fires occur ■ Dominated by grasses and annuals (single growing season plants) ■ Often exploited for grazing sheep and cattle
Temperate broad-leaved and mixed forest	■ Warm, mild growing season varying with latitude; moderate precipitation throughout the year; large seasonal differences and changes in day length; rich topsoil ■ Some trees evergreen, some deciduous; well-developed understorey
Mediterranean-type vegetation	■ Long, hot, dry summers, mild winters with reliable rainfall; growth often stops in summer drought ■ Open forest with stunted tree growth; woodland and shrubland; tough evergreen leaves, often spiny
Desert	■ Very little rain; true desert has less than 100 mm precipitation per year and arid areas less than 250 mm; high summer daytime temperatures (often >37°C); large temperature difference between day and night ■ Widely scattered shrubs, water-conserving plants and non-drought-adapted ephemerals (which grow and set seed quickly on rare occasions when water is available); some very dry, sandy deserts have almost no plant growth ■ Generally between 20° and 35° north and south of the Equator
Tropical grassland (savanna)	■ Low rainfall but seasonal heavy storms can occur; frequent fires; thin soil ■ Grasses with scattered clumps of trees, grading into either open plain or woodland
Tropical scrub forest	■ Rainfall not abundant, high evaporation ■ Thorny shrubs and trees ■ Grades into tropical grassland and savanna
Monsoon forest	■ In the tropics, but with distinct wet and dry seasons ■ Trees less closely spaced than in rainforest, many trees shed their leaves in the dry season
Tropical rainforest	■ Warm and humid with no true seasons (averages 25°C); frequent rain; little change in day length; growth throughout the year; infertile clay soil ■ Closed canopy with little understorey; diverse plant species competing for light; trees often large-trunked with buttress roots; many vines and epiphytes (plants that grow on other plants); little leaf litter
Mountain	■ Increasing altitude produces a decrease in temperature, similar to the effect of increasing latitude ■ Vegetation types vary with altitude, if too high, trees do not grow and vegetation resembles tundra

3.8.4 Characteristics of the major terrestrial ecosystems

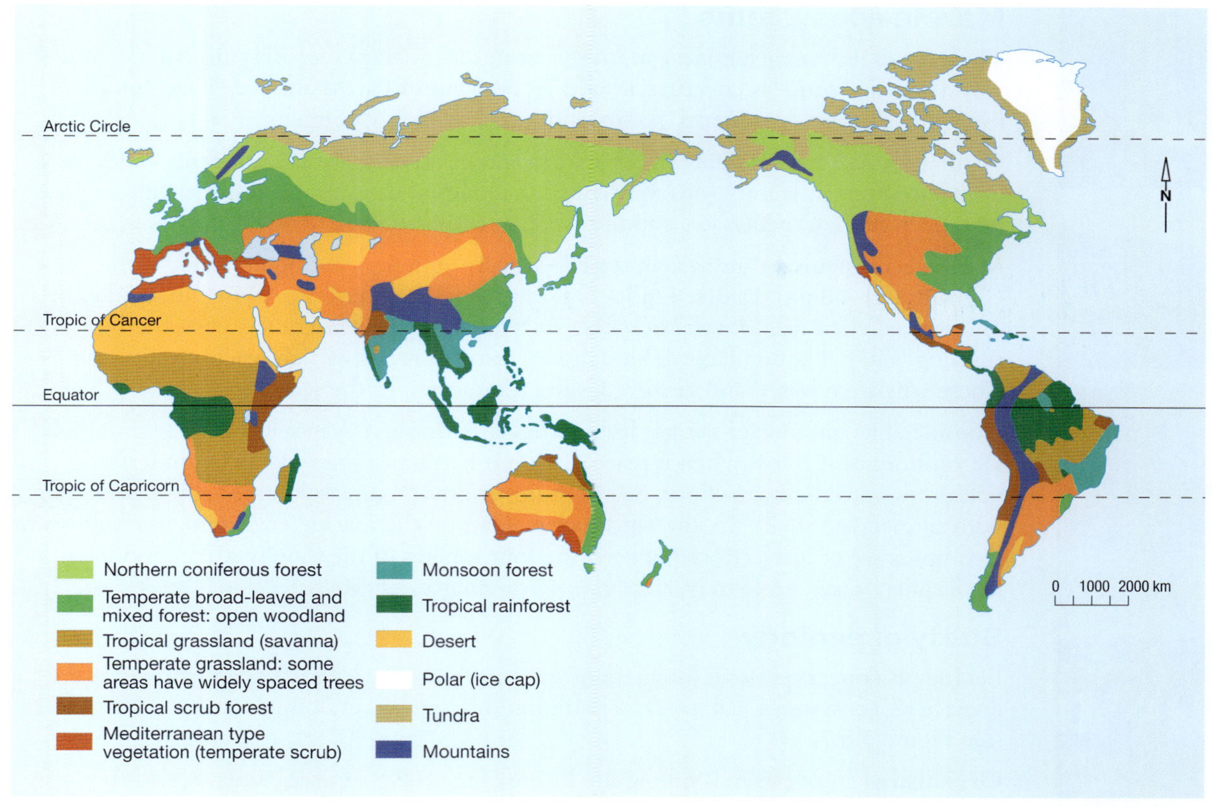

3.8.5 The world's major terrestrial ecosystems

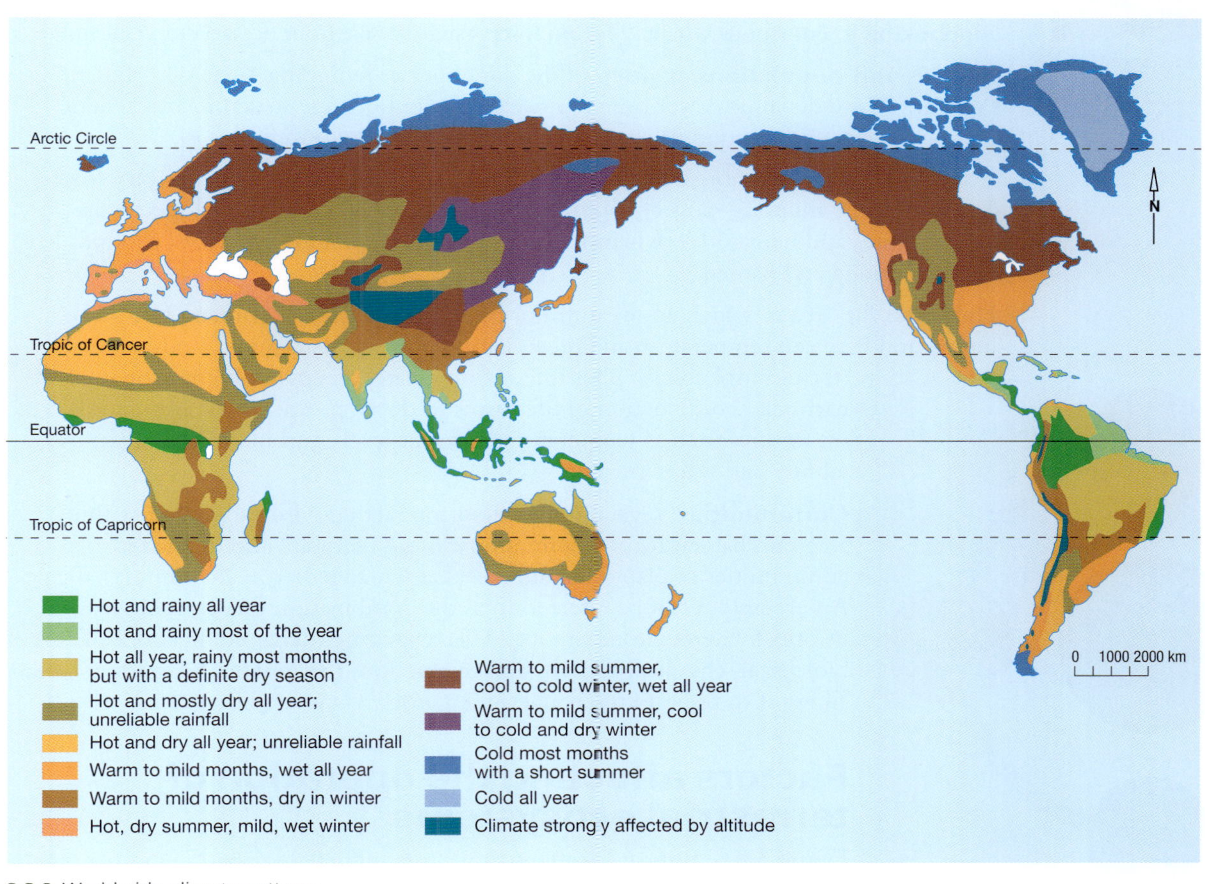

3.8.6 Worldwide climate patterns

3.8.7 The five levels of ecological organisation

Marine ecosystems

The ocean's dominance is clear in images from space. Marine ecosystems encompass the world's oceans and seas covering around 70 per cent of Earth's surface. They comprise the world's largest ecosystems. As on land, they are highly complex.

Naturally, there are differences both within and between marine ecosystems. These arise from variations in salinity (dissolved salts), the amount of nutrients dissolved in the water, how deep sunlight penetrates and the average temperature.

Marine ecosystems are always salt water—usually 96.5 per cent liquid water with chloride and sodium dissolved in it. This salt comes from rivers carrying huge quantities of dissolved chemicals. River rocks that once contained salt become weathered by running water. Their chemicals are released into the rivers, which react, creating salt. These salts are released and dissolved, then transported to the sea.

Considerable amounts of salt get lost, particularly through evaporation. The wind blows salt onshore or it is lost when it reacts with hot basalt lava spewed out by underwater volcanoes onto the ocean floor. Sea creatures consume some salt, especially tiny diatoms with silica-based shells. So, the salt water making up the marine environment results from a complex set of interactions between the four spheres of the biophysical environment. Each sphere plays a role in the marine environment's delicate balance of salt.

Study of ecology

Ecology is concerned with interactions occurring at five levels of organisation: ecosphere, ecosystems, habitats and communities, species and populations, and organisms (see Figure 3.8.7).

Organisms: an organism is simply any form of life. There are many ways to classify organisms. The simplest distinction is between producers (plants), consumers (most animals) and decomposers (such as bacteria that feed on dead animal and plant matter). Plants range from single-celled phytoplankton to huge trees. Animals range from microscopic zooplankton to enormous whales. Decomposers range in size from bacteria to giant fungi.

Species and populations: a group of the same species living together is a population. Populations are dynamic. Over time their size, distribution, age structure and genetic make-up adapt and respond to changing environmental conditions.

A species is one type of organism able to reproduce its own kind. Estimates vary from 5 million to 30 million—possibly 50 million—species exist. Most animal species are insects, mites and nematodes (worms). So far, around 1.7 million species of all kinds have been identified and named.

Habitats: the area in which an organism or population lives is its habitat. The interaction between temperature and precipitation plus soil determines land-based (terrestrial) habitats. This interaction creates an environment allowing a particular combination of life forms to develop. An aquatic habitat is based on features such as temperature, nutrient levels, turbidity (light intensity), salinity and water currents.

Communities: several populations interacting with each other within a certain habitat form a community. Ecosystems are also defined as communities of plants and animals that live together in a common habitat An ecosystem can be referred to as the combination of a community and its non-living environment an—ever-changing (dynamic) network of biological, chemical and physical interactions that sustains a community and allows it to respond to changes in environmental conditions.

Climate

Biotic factors

Factors affecting Ecosystem Operation

Topography

Soils

3.8.8 Factors affecting terrestrial ecosystem operation

Factors affecting the operation of terrestrial ecosystems

Figure 3.8.8 summarises the main factors influencing ecosystem operation. These four factors work in concert with each other.

Climatic factors

The key climatic factors affecting ecosystem operation are: precipitation, temperature, sunlight and wind. Water is essential to all living organisms. Precipitation is the most significant climatic factor. Plants require water to photosynthesise and reproduce. As plants constantly lose water via transpiration, they need to receive at least the same amount from the environment.

Precipitation has two key impacts on ecosystem functioning:

1 The availability of water. This includes the type and amount (e.g. is it steady rain or does it come in short intense bursts during storms, resulting in high run-off levels?).

2 Its timing. Does it coincide with the growing season of the vegetation, is it concentrated in a short wet season or is it spread out over the entire year?

3.8.9 The stinging tree (*Dendrocnide moroides*) has large leaves. It grows below the canopy in Queensland's tropical rainforests where light is scarce.

Temperature is another key climatic factor impacting ecosystems operation. For most plants the ideal range for photosynthesis is 10–35°C. Above 35°C will cause too much moisture loss. Below 10°C can make it difficult for plants to grow.

Adequate sunlight is vital for ecosystems to function. Photosynthesis is the single most important biological process for green plants. Leaves act like tiny solar cells collecting the Sun's energy. They combine it with carbon dioxide and water to create nutrients. It means the intensity of sunlight determines how much energy is available to plants. In dark environments, such as under a rainforest tree canopy, plants have evolved to capture all available light. Some have huge leaves (see Figure 3.8.9), others climb trees to reach the light.

Wind is the fourth main climatic factor affecting ecosystem functioning. Its impact on vegetation is usually mild, although intense storms might cause massive damage. Where winds are consistently strong, such as along coastlines and on mountain ridges, plants may hug the ground to avoid being damaged. Wind can also limit the moisture available to the ecosystem by increasing evaporation. Many plants rely on wind to disperse their seeds or pollen.

Topographic factors

Topography relates to the shape of the land and is a part of the lithosphere. Vegetation patterns are affected indirectly by the gradient, aspect and altitude of the slope, with slope the most vital element of topography. All these factors influence ecosystem development (see Figure 3.8.10). They also modify the climate and impact the soils.

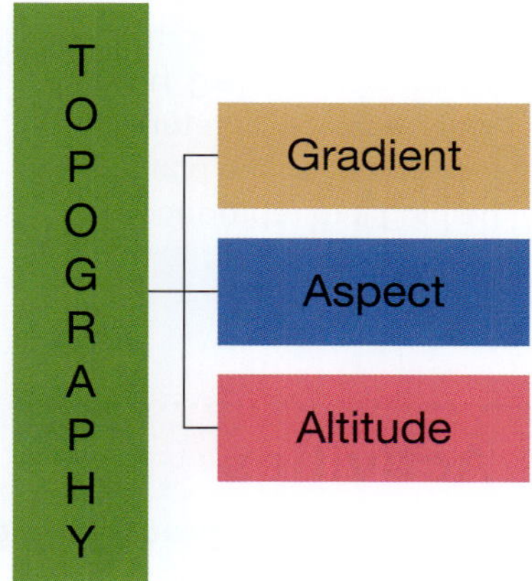

3.8.10 Topographic factors that affect ecosystems

Gradient: the steepness of the slope, which impacts surface stability. For example, on very steep slopes soil is more easily eroded by water. This makes it susceptible to mass movement such as landslides. This can strip the slopes, leaving little soil available to support life.

Aspect: the orientation, or direction, the slope is facing. It impacts sunlight and temperature conditions. North-facing slopes in the Southern Hemisphere favour plant growth over those facing south (especially in winter). These slopes receive the most direct insolation.

Altitude: as elevation increases, there are corresponding decreases. These occur in soil depth and air temperature. Altitude also increases exposure to wind. Collectively, these limit the conditions for plant growth. As altitude increases there is a noticeable decrease in the number of plant species. It also decreases plant height, density, growth rate and growing season length.

In very high mountains, vegetation shows a distinct vertical zoning. As latitude increases and elevation rises, there is a noticeable shift to cooler vegetation types, along with a decrease in community diversity (see Figure 3.8.11). The best-known zoning is the tree line—the altitude above which trees no longer grow. The tree line height is influenced by latitude. The tree line closer to the Equator has a much higher altitude than the one closer to the poles.

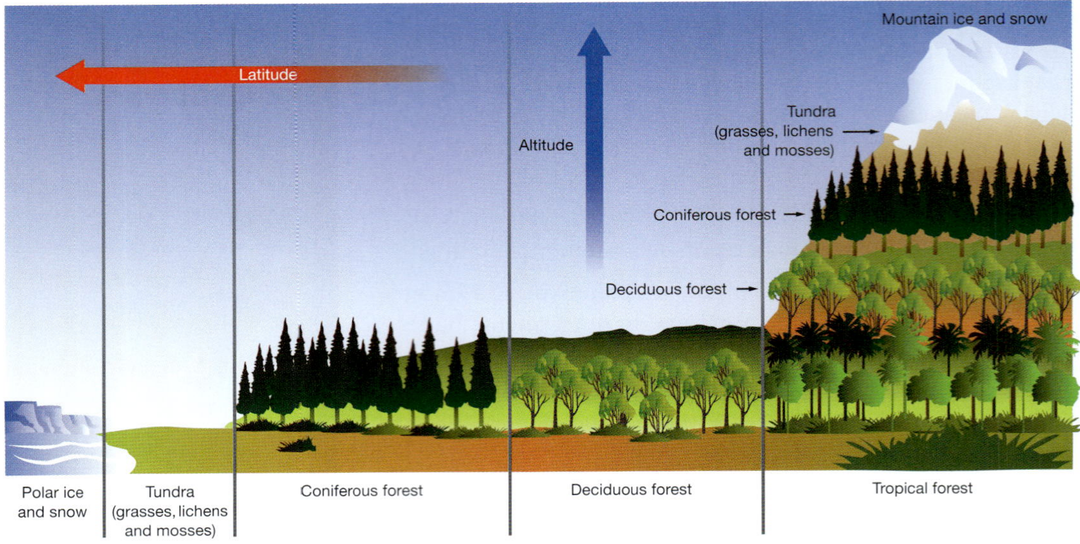

3.8.11 Relationship between latitude, altitude and the tree line

Did you know?

Highly beneficial relationships can occur between species. Many plants evolved to rely on animals to help them reproduce. Most notable is the role bees play in pollination. Such relationships are often symbiotic—both species benefit (e.g. bees receive nectar in return for pollinating the plant, helping it to reproduce).

Soil factors

Aspects associated with soil are called **edaphic** factors. Most plants rely on soil as a growth medium. Soil provides the essential moisture, nutrients and oxygen. Vegetation, in turn, impacts the soil's characteristics—decaying plants add organic material. Soil and vegetation, then, have an intimate mutual relationship in ecology. Soils with high levels of humus benefit plants by stimulating their growth. This in turn stimulates further organic material deposits into the soil. A positive feedback loop is created. It is equally true that poor soils usually limit vegetation growth. This creates limited organic deposition into the soil, producing a negative feedback loop. If the rich topsoil is lost from a site, it may be unable to produce crops or pasture. It can become prone to desertification.

Biotic factors

Biotic factors are aspects of ecosystem functioning. They connect the elements of the biosphere–the living components—including flora and fauna. A biome is a large group of plants and animals. It is not a random collection but a highly complex, interactive community. Within each ecosystem, plant and animal species form a multitude of relationships. The biotic factors are those arising from these intertwining relationships.

Activities

Acquiring and processing geographical information

1 Explain what an ecosystem is.
2 Distinguish between terrestrial and aquatic ecosystems.
3 Describe the ecological terms of community, population and organism.
4 Define the term ecotone.
5 Explain the influence of the availability and timing of precipitation on an ecosystem.
6 Describe the role of temperature on ecosystem operation.
7 How do gradient and aspect influence ecosystem operation?
8 What is the tree line?

Applying and communicating geographical understanding

9 Select two of the ecosystems listed in Figure 3.8.4. Compare their location and climatic conditions.
10 Write a paragraph explaining the role of climate in ecosystem development using Figures 3.8.5 and 3.8.6
11 Create a glossary for the terms aspect, gradient and altitude.
12 Examine Figure 3.8.11. Write a short report describing the link between latitude, altitude and the tree line.

Energy flows and nutrient cycling

The productivity of an ecosystem can be expressed in two ways:

Biomass—the mass of new living matter produced per square metre of land (or within a volume of water) per unit of time. The greater the biomass, the more productive the ecosystem is said to be.

Energy flow—the amount of energy (in kilojoules) that is 'locked into' all the organisms in an area per unit of time.

Both rates depend on the quantity of available energy and the nutrients in the environment. Add to this mix the efficiency with which energy and matter are incorporated into producers and passed up the food chain or food web.

Figure 3.8.12 compares the productivity of major ecosystems. Ecosystems with low productivity (such as deserts) require larger areas to be sustainable. Highly productive ecosystems, such as tropical estuaries (coastal rivers), require only small areas.

3.8.12 The relative productivity and size of key ecosystems. This graph shows the vast differences in how much life different ecosystems can support.

Biomass

Biomass describes the total weight or number of organisms in an area. For example, the biomass of a forest can be estimated by considering the trees and other plants, along with the animals living in it. Scientists do this by identifying a number of sample areas (such as 10-square-metre plots) from across an ecosystem. They calculate the biomass for these small plots, then use the results to draw conclusions for the whole ecosystem. The amount of biomass produced in an ecosystem is its **biological productivity**.

The way the four spheres of the biophysical environment—the atmosphere, hydrosphere, lithosphere and ecosphere—interact with each other is crucial in determining an ecosystem's biological productivity. For example, the low rates of precipitation in desert ecosystems greatly limits plant growth. This impacts soil production, further limiting plant growth. In turn, this limits the diversity, size and number of animal species. Hence, deserts overall have low biological productivity.

Conversely, highly productive ecosystems, such as tropical rainforests (see Figure 3.8.13), have plentiful precipitation and high temperatures, along with solar energy. These all produce excellent growing conditions. Due to the vast size of tropical rainforests, their highly complex **nutrient cycles** and numerous species, they yield a staggering 10 per cent of the world's biomass production.

3.8.13 The Amazon Rainforest is one of the most biologically productive ecosystems.

Energy flows

The Sun's energy powers ecosystems. This energy flows through the ecosystem as it transfers between organisms. Unlike many of Earth's processes, this energy flow is not a cycle. Instead it is transferred through a chain, from producers to consumers. Producers—plants—use solar energy to produce more energy. Consumers—animals—use the energy stored in plants, either by eating the plant or eating an animal that ate the plant. As energy flows through the food chain, at each link some energy is lost in the form of heat into the atmosphere.

Geographers categorise species into **trophic levels** (see Figure 3.8.14). Energy flows from one level to another. As energy is lost between each level, there is less energy available for each subsequent level. Hence each trophic level is smaller than the level below it.

As primary producers, plants occupy the first trophic level so they are the most numerous. They add—or produce—energy in the system but take none out as their energy source is from the Sun. They expend little energy to obtain nutrients.

Primary consumers are herbivores, animals that exclusively gain their energy from eating plants. Secondary consumers gain their energy from eating the primary consumers.

Tertiary consumers are at the top of the food chain because they are not eaten by others. Sometimes they are called apex predators due to their position on food chain. They gain their energy from eating the secondary consumers. Tertiary consumers use large amounts of energy to hunt for food. Ecosystems with low levels of energy cannot support many tertiary consumers.

Food chains also help recycle nutrients from producers and consumers back into the system for future use. Decomposers like fungi, worms, beetles and termites complete this role. They break down organic materials allowing them to recycle.

It is useful to understand the simple transfer of energy in an ecosystem as a food chain or a pyramid. Yet the relationships between producers, consumers and decomposers in any ecosystem is far more complex than in the highly simplified Figure 3.8.14. Organisms in a natural ecosystem are usually part of an intricate network of interacting food chains, called a food web. Figure 3.8.15 shows a simplified food web of Antarctica. There can be multiple levels of consumers and some species have diverse diets (such as penguins). Others (such as baleen whale species) have very limited diets. Despite being tertiary consumers they consume species towards the bottom of the food chain.

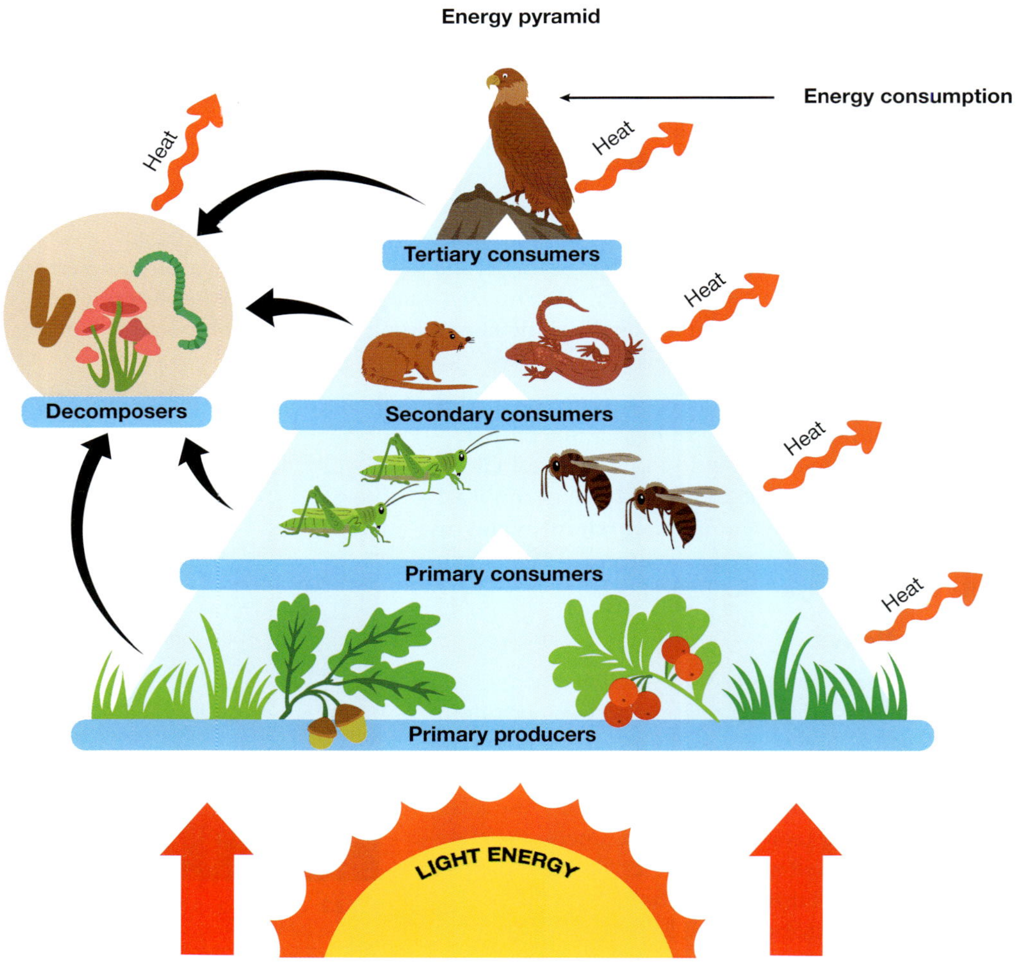

Energy pyramid

Energy consumption
Heat
Tertiary consumers
Secondary consumers
Primary consumers
Primary producers
Decomposers
LIGHT ENERGY

3.8.14 The transfer of energy within an ecosystem

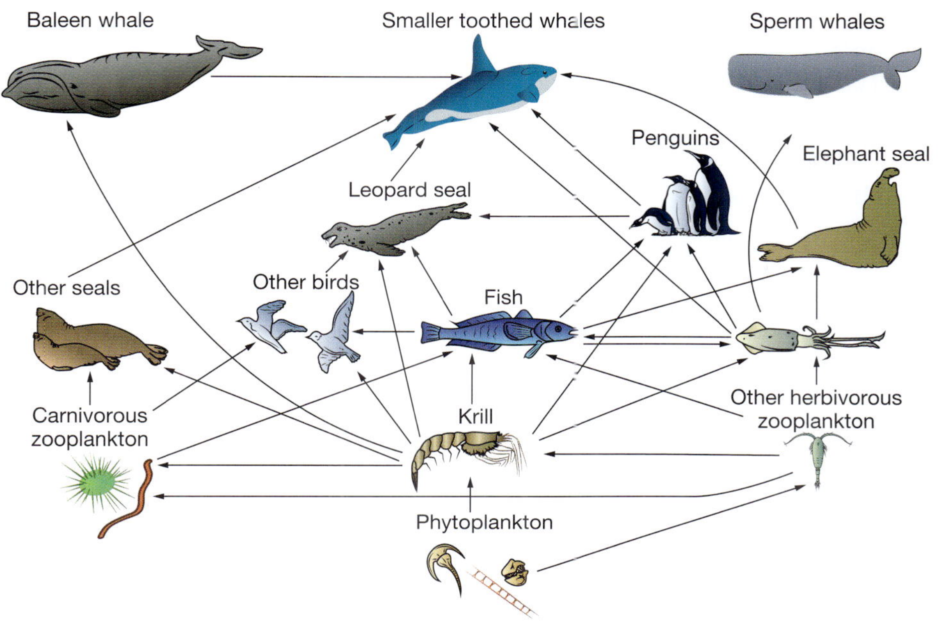

3.8.15 A simplified food web for Antarctica

The nutrient cycle

The nutrient cycle is the key biochemical process that enables the effective functioning of ecosystems. Like other biological systems, it is driven by solar energy. Photosynthesis is at the centre of the cycle. Energy from the Sun is used by plants to turn carbon dioxide and water into glucose and oxygen. In this way, the nutrient cycle interacts with other important biochemical cycles, such as the water cycle, the nitrogen cycle and the carbon cycle. Other organic chemicals, such as phosphorus, oxygen and sulphur, are also important in providing resources for the ecosphere and also operate in a cyclical fashion.

Although nitrogen makes up more than 75 per cent of Earth's atmosphere in a gaseous form, this is unusable to most organisms. However, through a series of biochemical processes, nitrogen is transformed into a form that is suitable for plants.

Figure 3.8.16 is a simplified model of the nutrient cycle. Plants grow by producing glucose. They combine it with minerals and other nutrients gained from soil to create their energy source. As they grow, plants produce organic material that becomes available for animals to eat. The energy stored as nutrients thus passes from one trophic level to the next. Animal dung is a by-product. This waste and any uneaten plant material returns to the soil to decay into humus. Decomposers, such as fungi and dung beetles, speed up this process by breaking down the material. Living plants then re-access these nutrients promoting new growth.

Nitrogen fixation involves blue-green algae and bacteria converting the nitrogen gas into compounds. The nitrogen is then assimilated into the tissue of the algae and bacteria. Animals consume these, so the nitrogen is assimilated into their own tissues. Certain plants, such as legumes (e.g. soya beans), have very high levels of bacteria that capture the nitrogen. Such plants are called nitrogen-fixing plants.

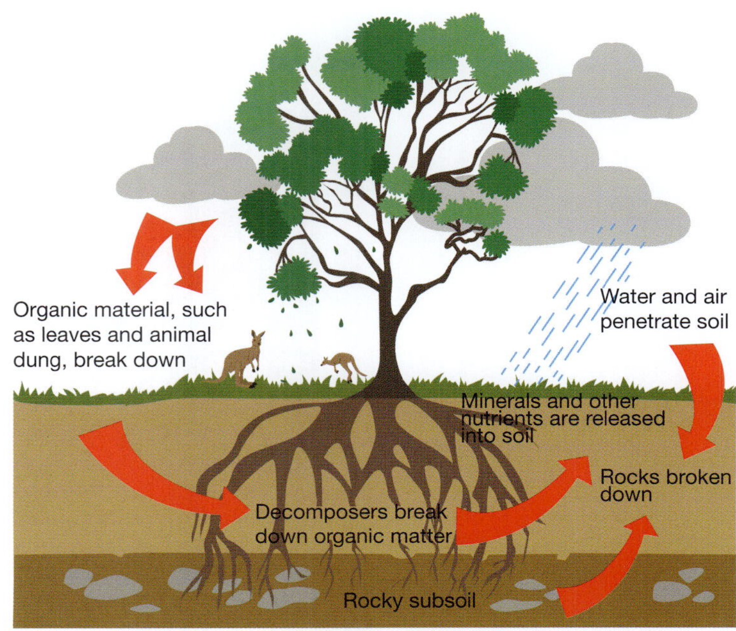

3.8.16 Simplified model of the nutrient cycle

When living organisms produce dung or die and begin to decay, ammonia (NH_3) and ammonium (NH_4) is created as a by-product. It gets stored in the soil. A special type of bacteria called a nitrifying bacteria converts this ammonia back into nitrates (NO_3). These are consumed by denitrifying bacteria. They create nitrogen gas as a by-product which then returns to the atmosphere. Thus, the whole process begins anew.

The carbon cycle

Carbon is the basic building block of the compounds necessary for life. Most plants in terrestrial ecosystems absorb the carbon they need through their leaves from the atmosphere in the form of carbon dioxide (CO_2). In aquatic ecosystems, microscopic plants called phytoplankton access the carbon dioxide dissolved in the water. Plants use carbon dioxide while making glucose through photosynthesis, creating oxygen as a by-product. When plants die, carbon is released back into the system. Animal respiration also produces carbon dioxide as a by-product.

Carbon is also stored deep underground in the form of oil, natural gas and coal. These sources of stored carbon are produced by decomposed plants and animals. This was compressed under enormous pressure millions of years ago. When humans access this stored carbon and burn it as fossil fuels, the carbon re-enters the carbon cycle as carbon dioxide. The amount of carbon dioxide released by fossil fuel use far exceeds the amount that can be absorbed and stored. This disrupts the natural carbon cycle. This is one of the leading causes of climate change. Thus, humans play a leading role in the carbon cycle.

Human impacts on the nutrient cycle

Human activities can disrupt how natural cycles operate, including those essential for ecosystems to effectively function. The nutrient cycle can be impacted by human actions, especially when they reduce the availability of nutrients for the ecosystem. Land clearing for agriculture and urban development often removes vegetation. This depletes the source of organic material to make humus, which replenishes soil nutrients. This can result in desertification, the loss of soil's ability to support life (see Figure 3.8.17).

3.8.17 Desertification can result from land clearing.

● **SPOTLIGHT**

Human impacts on the Gulf of Mexico

Humans can also have an impact on the nutrient cycle by adding in more nutrients than it can recycle. These often come from chemical fertilisers applied during intensive agriculture. Their impact is clear in the catchment areas of the Mississippi River and the Gulf of Mexico (see Figure 3.1.18). The Mississippi is one of the world's major river systems. Stretching 3800 kilometres, its vast catchment is almost 2 million square kilometres—the world's fourth largest. It includes 31 states in the USA and two Canadian provinces. Figure 3.8.18 reveals that this catchment has several major rivers acting as Mississippi tributaries, eventually flowing into the Gulf of Mexico.

3.8.18 The Mississippi River catchment area covers over 40% of continental USA.

This catchment contains some of the most intensely farmed regions in the USA and some large cities, such as Chicago, Memphis and St Louis, plus countless smaller urban centres. Farming and urban activities add vast quantities of extra nutrients into this river. Every year over 1.6 million tonnes of nitrogen enter the Gulf of Mexico. It comes from nitrogen-based fertilisers, sewage from urban areas and agriculture. Much of it arrives in spring when snowmelt significantly increases run-off and river flows.

The impact from such massive nitrogen loads can form huge dead zones. For example, it created the Gulf of Mexico Dead Zone, off the coasts of Texas and Louisiana, USA. In 2017 it measured a record-breaking 22 000 square kilometres (see Figure 3.8.19). Subsequent dead zones were smaller, but still exceeded 15 000 square kilometres. The nutrients promote algal growths that extract oxygen from the water. This leads to hypoxic water—water with very low oxygen levels—able to support very little life.

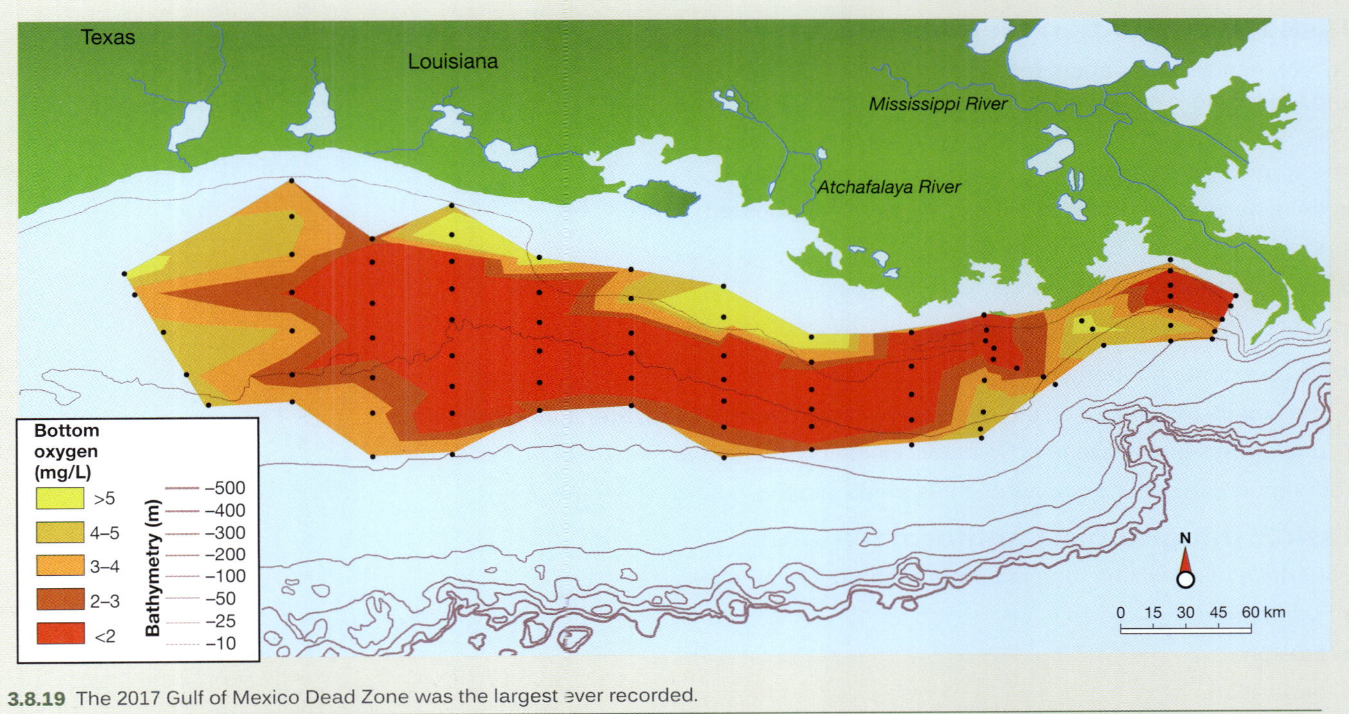

3.8.19 The 2017 Gulf of Mexico Dead Zone was the largest ever recorded.

Activities

Acquiring and processing geographical information

1 Define the term biomass.

2 Explain the importance of the Sun to ecosystem operations.

3 What is a trophic level?

4 Explain why plants are classified as producers in food chains.

5 Differentiate between primary, secondary and tertiary consumers.

6 Explain the difference between a food chain and a food web.

7 Explain why nutrient, carbon and nitrogen cycles are all referred to as cycles.

8 What is biological productivity?

9 Explain why different ecosystems have different rates of biological productivity.

Applying and communicating geographical understanding

10 Examine Figure 3.8.12. Describe the relationship between ecosystem productivity and size. What implications might this have for human use and managing different ecosystems?

11 Use Figure 3.8.14 to explain how energy is transferred through a food chain.

12 Study the food web in Figure 3.8.15.

- a Which species is the primary producer in this food web?
- b Which species would be considered tertiary consumers? Why?
- c Explain the consequences for this food web of a significant decline in krill.
- d If elephant seals were removed from the food web, what would be the possible consequences?

13 Create a flow chart to explain the nutrient cycle. Add annotations to:

- a explain why it is so critical to ecosystem functioning
- b describe how land clearing can disrupt the nutrient cycle and outline its consequences
- c outline the impacts of humans adding too many nutrients into the cycle.

APPLICATION AND CONSOLIDATION TASKS

Task 1: Presentation: Geomorphic events

Collect media reports about recent earthquakes and volcanic eruptions. Include video recordings, webpage screenshots and articles from newspapers or magazines. Write a synopsis of the material gathered and present the information to your class.

In your presentation include:

- the type of geomorphic event and where it was located
- the impact of the event
- the consequences and outcomes.

Task 2: Report writing: Ecosystems

Write an extended report on:

1 the factors that influence the operation of ecosystems—both terrestrial and aquatic

2 one ecosystem investigated in detail, including the factors that affect that ecosystem.

Your report can include visuals such as maps, graphs, diagrams and photos.

Task 3: Infographic: Landform features

Undertake research into one landform's features formed by:

- weathering
- erosion
- deposition.

Include detailed information on the processes and how the landform was created.

Continued from page 27.

Chapter glossary

groundwater water beneath the surface of the ground
hectopascal a unit of atmospheric pressure; 100 pascals
humus part of the soil produced by decaying organic (plant and animal) matter
infiltration the downward movement of water into the soil or rock
insolation the solar radiation that reaches Earth's surface
lapse rate the decline in temperature that occurs with increasing altitude throughout the troposphere
lithosphere the outer shell of Earth, consisting of solid rock (100–150 km thick), soil and geological formations
magma molten rock under Earth's crust, which is under immense pressure
mass movement the movement of soil or rock down a slope as a result of gravity
mesosphere the layer of the atmosphere between the stratosphere and the thermosphere, 50–80 km above the surface of Earth
moraine a deposit of soil or rock carried and left by a glacier
nutrient cycle the recycling of nutrients around an ecosystem to support the biosphere
oceanity the degree to which the weather in an area is impacted by the ocean
ozone layer the layer of gaseous ozone in the stratosphere that protects life on Earth by filtering out harmful ultraviolet (UV) radiation from the Sun
plate tectonics a theory that explains the movements of the lithosphere on top of the asthenosphere
percolation the act of liquid passing through porous layers, for example soil
precipitation the process by which water returns to Earth's surface from the atmosphere, most commonly as rain
run-off rain that flows across the land into streams
seasonality affected by seasons that result from changes in the tilt of Earth's axis
soil a mixture of organic materials and particles of rock and sand together with living microbes, air and water
stratosphere the layer of the atmosphere between the troposphere and the mesosphere, 20–50 km above Earth's surface
system any part of the universe that can be isolated from the rest of the universe to observe and measure changes
tectonic forces disturbances in Earth's crust that result from Earth's internal energy and create physical features, such as mountains, on the surface
tectonic plates sections of Earth's crust that move as distinct units on the asthenosphere on which they rest
terrestrial radiation stored heat emitted as long-wave radiation by Earth, including its land, oceans and atmosphere
thermosphere the layer of the atmosphere above the mesosphere, 80–100 km above Earth's surface, in which the temperature increases with altitude
transpiration the process by which leaves lose water into the atmosphere
trophic level a level in a food chain
troposphere the innermost layer of the atmosphere in which most of Earth's weather occurs
water budget the total amount of water available in an area, its inflows and outflows
water table the upper limit of the part of the ground saturated with water
weathering the process of change as a result of exposure to atmospheric conditions

CHAPTER

4 Land cover change

Land cover change describes human modifications made to the Earth's surface. For thousands of years, this was done to obtain food and other resources. However, the current rate and intensity are unprecedented, leading to large-scale changes in ecosystems and environmental processes on local, regional and global scales. These changes include the warming of the planet, biodiversity loss and the pollution of water, soils and air.

Mitigating the negative consequences of land cover change, while also sustaining the production of the resources needed to sustain our way of life, are among the greatest challenges facing humanity.

This chapter covers the nature, extent and outcomes of land cover change. It explores the natural cycles and processes in tandem with the role humans play in modifying global systems and how this contributes to land cover change. Deforestation, desertification and melting glaciers and ice sheets are investigated. Climatic and glacial cycles, geomorphic processes, and invasion and ecological succession of vegetation communities, are also further explored.

We abuse land because we regard it as a commodity belonging to us. When we see land as a community to which we belong, we may begin to use it with love and respect.

Aldo Leopold, American author, philosopher, scientist, ecologist, conservationist and environmentalist

4.0.1 Clearing rainforest for plantations

Chapter glossary

afforestation the planting of trees, or sowing of seeds, to create a forest in a barren land devoid of any trees

algal bloom a rapid increase or accumulation of algae in a waterway

anthropogenic caused by humans

bioaccumulation when substances, especially toxins, remain within an organism and increase with repeated exposure or ingestion

climax community the final stage of succession, in which the mix of species remains relatively stable until a disturbance such as fire occurs

eutrophication water pollution involving excess nutrients, often from fertilisers or sewage

evapotranspiration physical, chemical and biological changes that take place after a lake, estuary or slow-flowing stream receives inputs of nutrients and phosphates from natural erosion and run-off from the surrounding drainage basin

feedback loop occurs when an output of matter or energy is fed back into the system as an input and leads to changes in that system

geomorphological relating to landforms, their origin, evolution, form and distribution

glaciation the process or state of being covered by glaciers or ice sheets

habitat fragmentation the break-up of a habitat into smaller pieces, usually because of human activities

heat island an urban area that is significantly warmer than its surrounding rural areas due to human activities and the nature of the constructed environment

herbaceous plants plants without woody stems

hydrological drought drought that occurs when the surface flow (river flow) and lakes or reservoir levels decline below the long-term mean

land cover the physical and biological features of Earth's surface

periglacial the area marginal to a frozen or ice-covered region and which is subject to the seasonal effects of freezing and thawing

permafrost a subsurface layer of soil that remains frozen throughout the year, mostly found in polar regions

plant succession the change in the types of plant species occupying a given area over time

Pleistocene the geological epoch that lasted from about 2 580 000 to 11 700 years ago

primary succession ecological succession in an area without soil or where the soil is incapable of sustaining life (because of recent lava flows, newly formed sand dunes, or rocks left from a retreating glacier)

secondary succession ecological succession in an area in which natural vegetation has been removed or destroyed (e.g. by fire) but the soil has not been destroyed

shrub a woody plant, under 5 metres tall if it has a single main stem or 8 metres if it has multiple stems

This chapter's glossary words list continues on page 100.

UNIT 4.1

Nature and extent of land cover

Land cover refers to the physical and biological cover of Earth's surface. It includes natural features like water bodies (oceans, seas, lakes and rivers), areas of ice and snow, vegetation, and exposed rock and soil. It also covers constructed and managed environments (e.g. buildings, roadways, agricultural land).

Land cover types

Many land cover classification systems exist, with the most recent and widely used being The Global Land Cover-SHARE, developed by the UN's Food and Agriculture Organization's (FAO) Land and Water Division. It identifies 11 thematic land cover layers, which are presented in Figure 4.1.1.

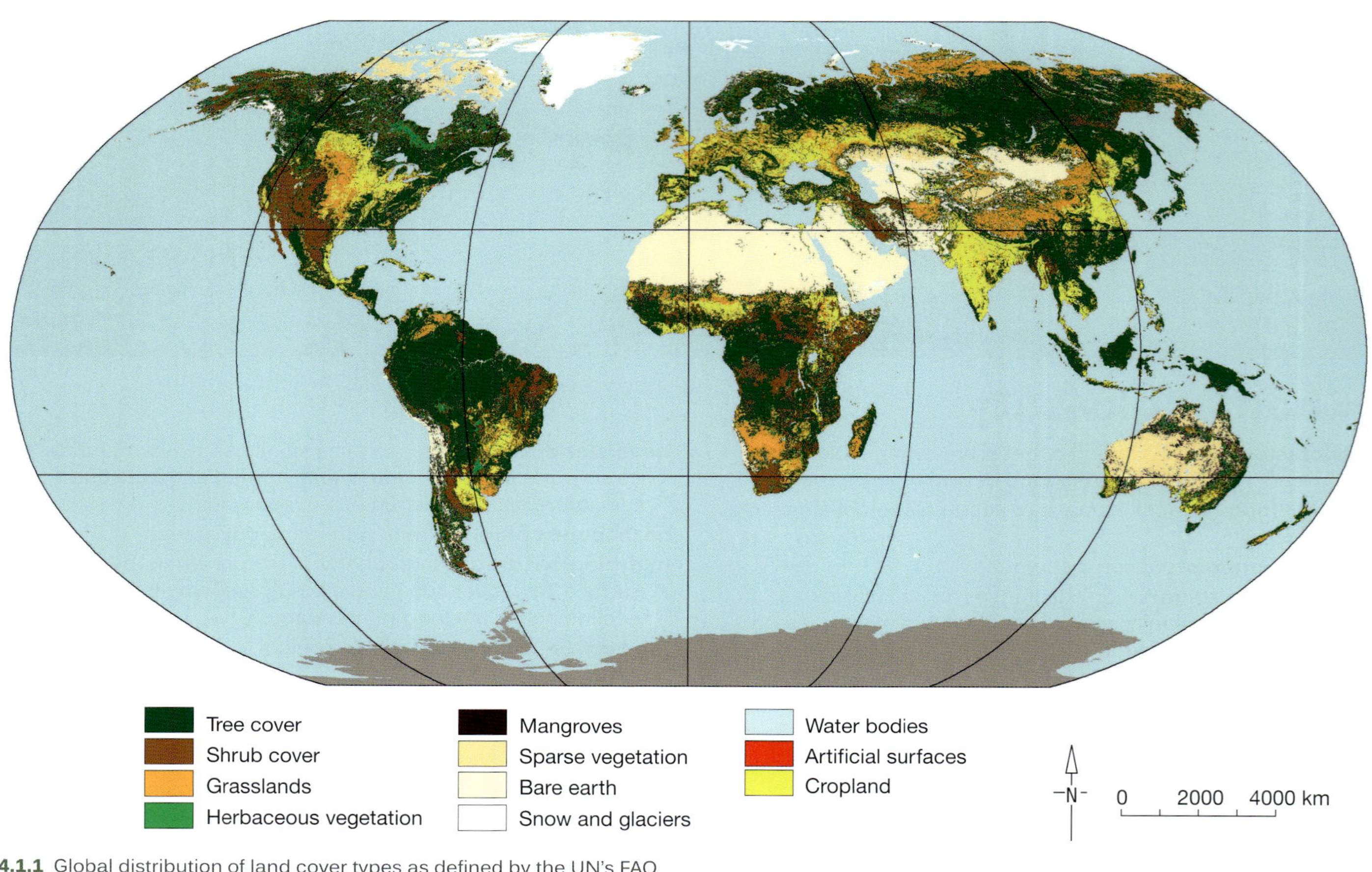

4.1.1 Global distribution of land cover types as defined by the UN's FAO

Tree cover

Any land with a natural tree cover of 10 per cent or more is considered to be a tree-covered area. This category includes areas planted as an **afforestation** project, areas seasonally or permanently flooded with fresh water, and coastal mangroves (see Figures 4.1.2 and 4.1.3).

4.1.2 Tree cover in a tropical rainforest

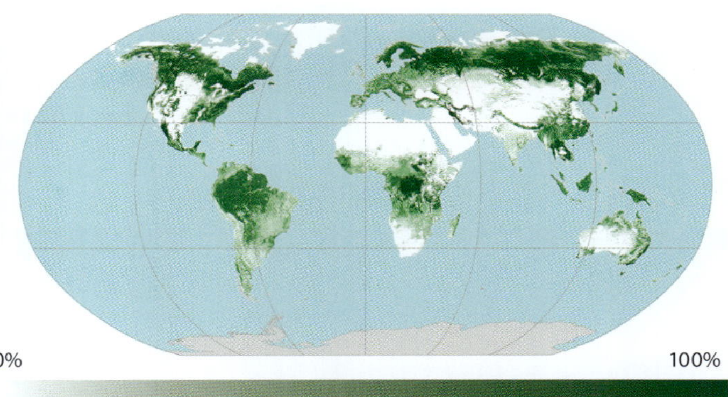

4.1.3 Global distribution of tree-covered areas is 27.7% of terrestrial land.

Shrub cover

Shrub-covered areas have a natural shrub cover of 10 per cent or more. It can include scattered trees, if they cover less than 10 per cent of this land, and **herbaceous plants**. This category includes areas flooded by inland fresh water, but not coastal salt or brackish water.

Shrubland is also called heathland or chaparral. Its diverse vegetation shares the characteristics of shrubs. Most shrubland is in warm temperate climates with mild, wet winters and hot, dry summers, including the Mediterranean; California, USA; Chile; South Africa; and southern Australia. Others occur in the semi-arid tropics and the Arctic, with smaller pockets found elsewhere. Australia has the greatest expanse and range of shrub cover due to its dry variable climates (see Figures 4.1.4 and 4.1.5).

4.1.4 Shrubland in King's Canyon, Northern Territory

4.1.5 The global distribution of shrubland covers 9.5% of terrestrial land.

Grasslands

Grasslands have natural herbaceous plants covering 13 per cent of land area, and include grasslands, prairies, steppes and savanna, irrespective of human or animal activities, such as grazing (See Figure 4.1.7). **Woody plants** (trees and shrubs) can be present if they cover less than 10 per cent.

The African savanna is dominated by grasses and dispersed trees and has a complex food web. It is hot, dry and prone to wildfires for half the year. A rainy season yields tall grasses for grazing herds such as zebras and wildebeests, preyed on by carnivores such as lions and cheetahs (See Figure 4.1.6).

The steppe grasslands stretch across Eurasia through to Mongolia into northeast China. Similar to savanna grasslands, they are typically drier and colder. Far from oceans with mountainous barriers, they lack humidity. Poor soil quality yields few other plants, so the main vegetation is herbs and seas of grasses up to 66 centimetres tall.

4.1.6 Savanna grasslands of Namibia, Africa

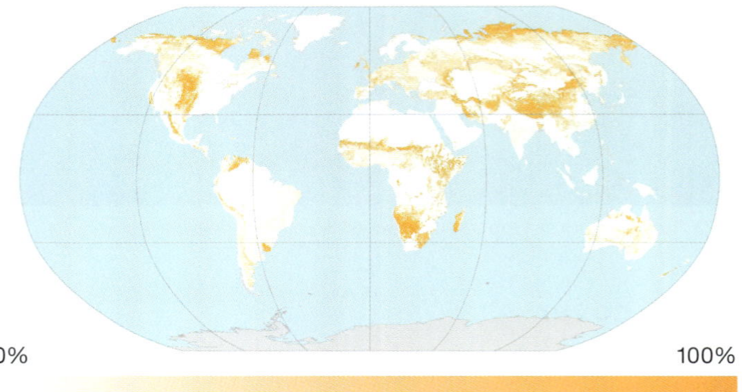

4.1.7 The global distribution of grasslands covers 13% of terrestrial land.

Herbaceous freshwater wetlands

Herbaceous vegetation is any area where vegetation has a cover of 10 per cent or more that is regularly flooded by fresh or brackish water, such as swamps and marshes, for at least two months of the year. Woody vegetation may be present if its coverage is under 10 per cent. Herbaceous plants are non-woody and **vascular**, and include grasses and grass-like plants. The Everglades of Florida, USA and the Okavango delta of Botswana are examples (see Figures 4.1.8 and 4.1.9). Kakadu National Park's floodplains in the Northern Territory have herbaceous swamp vegetation dominating areas submerged for six to nine months a year. Waterlilies are common, including the blue, yellow and white snowflake.

4.1.8 Herbaceous vegetation, Okavango Delta, Botswana

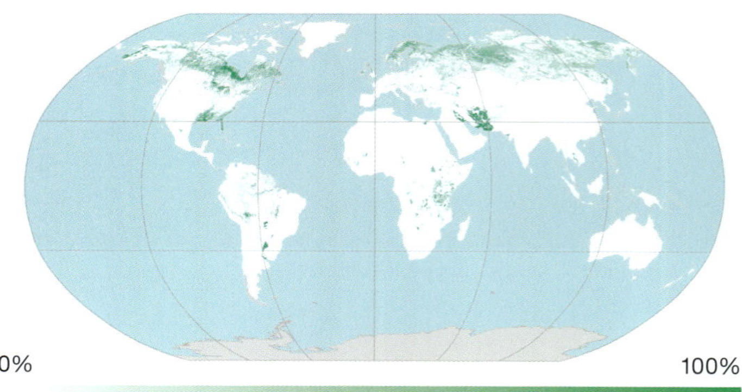

4.1.9 The global distribution of herbaceous vegetation covers 1.3% of terrestrial land.

Mangroves

Mangroves include any area with woody vegetation covering 10 per cent or more that is permanently or regularly flooded by salt and/or brackish water. They are found in coastal areas or river deltas and provide important environmental services including their role as a 'carbon sink' (see the box, Spotlight: Australia's mangroves as 'carbon sinks'; see Figures 4.1.10 and 4.1.11).

4.1.10 Mangrove forest, Krabi, Thailand

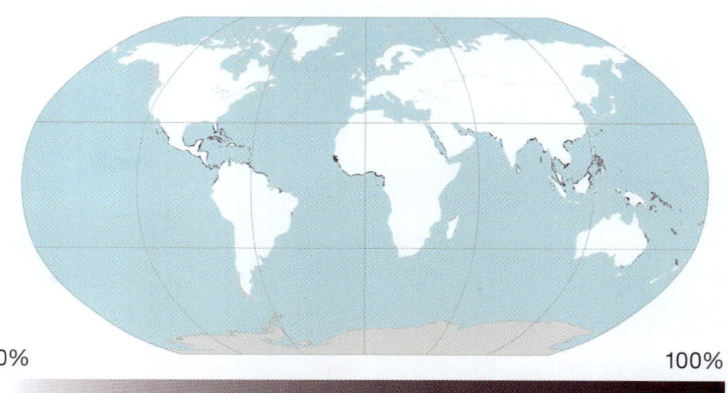

4.1.11 The global distribution of mangroves covers 0.1% of terrestrial land

● SPOTLIGHT

Australia's mangroves as 'carbon sinks'

Australia's mangrove forests, tidal marshes and seagrass meadows absorb some 20 million tonnes of carbon dioxide annually, making them an important 'carbon sink'. They can absorb carbon dioxide up to 40 times faster than forests and have five times the storage capacity of land trees. Mangroves store carbon in their soils as well as in the plants, which if undisturbed can be locked away for millennia. They are crucial in protecting coastlines from erosion, as fish nurseries, and in filtering water to maintain water quality.

When damaged by storms, heatwaves, dredging or land clearing, their stored carbon dioxide is released back into the atmosphere. In Australia, this means around 3 million tonnes of carbon dioxide is released into the atmosphere annually. Globally, these ecosystems are being lost twice as fast as tropical rainforests, despite being far smaller.

Australia's 25 760 kilometre coastline represents 5–11 per cent of the world's 'blue carbon' locked up in mangroves, seagrasses and tidal marshes. These coastal ecosystems are calculated to store 4000–6300 million metric tonnes of carbon dioxide. This is a significant amount, compared to the annual carbon dioxide emissions—501.5 million tonnes in 2021.

A record heatwave in 2015 along 1000 kilometres of the Gulf of Carpentaria's coastline induced mass mangrove dieback (see Figure 4.1.12). Recovery is slow with storm-damaged and dead mangroves stifling new growth. Significantly, these dead trees emit eight times more methane (a powerful greenhouse gas) as live mangroves. In 2019, a cascade of rising sea levels, heatwaves and back-to-back tropical cyclones killed or damaged a 400-kilometre stretch of mangroves. Protecting these ecosystems is a vital contribution to slowing climate change.

4.1.12 Mass mangrove dieback in the Gulf of Carpentaria in 2015

Sparse vegetation

Sparse vegetation is land with 2–10 per cent natural vegetation coverage. It includes permanently or regularly flooded areas and deserts, except the very driest. Desert plants adapt to the coarse, dry conditions with extensive root systems, small leaves, stems that store water and prickly spines that discourage animals. Cacti in the deserts of North and South America are desert plants. In Australian deserts, the trees include the silvery white ghost gum, mulga, sandalwood, northern cypress pine, sandhill wattle and western myall. Some shrubs, grasses and wildflowers are spinifex, saltbush, Mitchell grass, Sturt's desert pea and kangaroo paw (see Figures 4.1.13 and 4.1.14).

4.1.13 Spinifex grasses in the Pilbara, WA

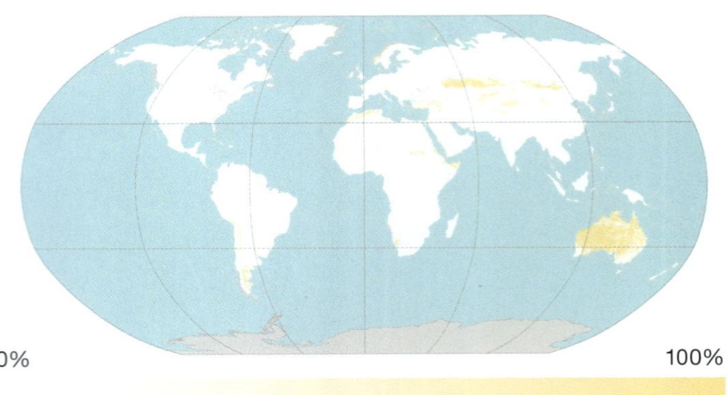

4.1.14 The global distribution of sparse vegetation covers 7.7% of terrestrial land.

Bare ground

Bare ground refers to land dominated by natural abiotic surfaces (e.g. bare soil, sand, rocks), with little to no natural vegetation cover (under 2%). This includes areas periodically flooded by inland water (e.g. lake shores, riverbanks, salt flats). It excludes coastal areas affected by tidal saltwater. Deserts are good examples, such as the Sahara (Africa), the Gobi (Asia) and the Simpson (Australia). Other examples are exposed rock surfaces, saltpans and areas covered with sand (see Figures 4.1.15 and 4.1.16).

4.1.15 The Gobi Desert of remote Mongolia, a landscape devoid of vegetation

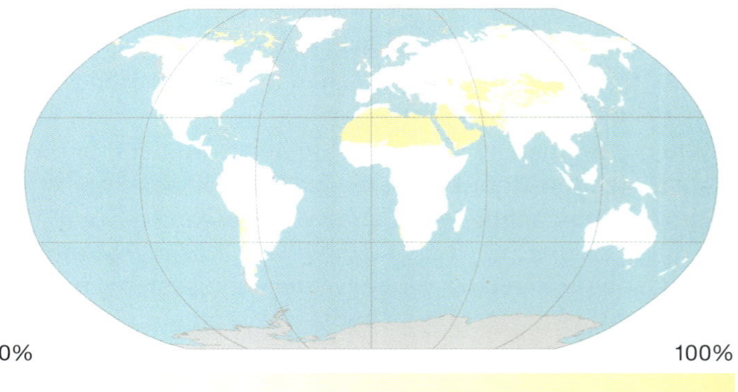

4.1.16 The global distribution of bare ground covers 15.2% of terrestrial land.

Snow and glaciers

Snow and glaciers include any area covered by snow or glaciers for at least 10 months of the year. The Antarctic ice sheet is Earth's largest single mass of ice, covering an area of almost 14 million square kilometres and containing 26.5 million cubic kilometres of ice. The ice sheet covers 98 per cent of the Antarctic continent. In Greenland, ice (see Figure 4.1.17 and 4.1.18) occupies about 82 per cent of the island's surface, and if melted would cause sea levels to rise by 7.2 metres. Sea ice (an expanse of frozen seawater) covering the Arctic Sea reached its annual minimum extent of 4.67 million square kilometres on 18 September 2022, the tenth lowest reading in 40 years of satellite data collection. September Arctic Sea ice extent is now shrinking at a rate of 12.6 per cent per decade, compared to its average extent from 1981 to 2010. Glaciers and ice sheets are discussed in detail in Chapter 6.

4.1.17 Elephant Foot Glacier and icefield, North Greenland

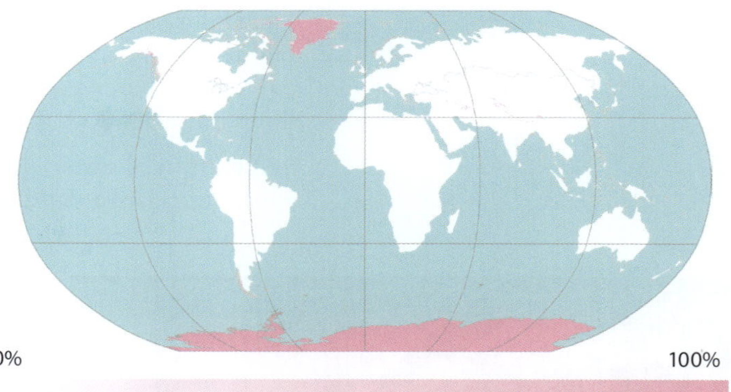

4.1.18 The global distribution of glacier and ice fields covers 9.7% of terrestrial land.

Water bodies

Any area covered by inland water for most of the year is called a water body (see Figure 4.1.20). Natural water bodies largely form by inland draining catchments, such as the Caspian Sea (the world's largest), North America's Great Lakes and Africa's Lake Victoria. Some water bodies can be frozen for up to 10 months of the year.

Artificial water bodies are engineered, mainly dams, which greatly impact freshwater ecosystems. Since the early 1980s around 180 000 square kilometres of land has been inundated by dam building. Dams fragment river systems, block animal migration routes and alter downstream flooding and sediment deposition patterns. Floodplains, riverbank zones and wetlands can be lost by building dams. Globally, dam construction is booming, most intensely in India, China and Brazil. Zimbabwe's Kariba Dam has the world's largest capacity (see Figure 4.1.19), forming Lake Kariba, 280 kilometres long and up to 32 kilometres wide. The storage capacity is 185 billion cubic metres and the surface area is 5580 square kilometres.

River diversions, wetland draining and excessive water extraction have led to surface water losses elsewhere. Any water surface imbalance is detrimental to the local biodiversity and ecosystems.

4.1.19 Kariba Dam lies in Kariba Gorge on the Zambezi River Basin between Zambia and Zimbabwe.

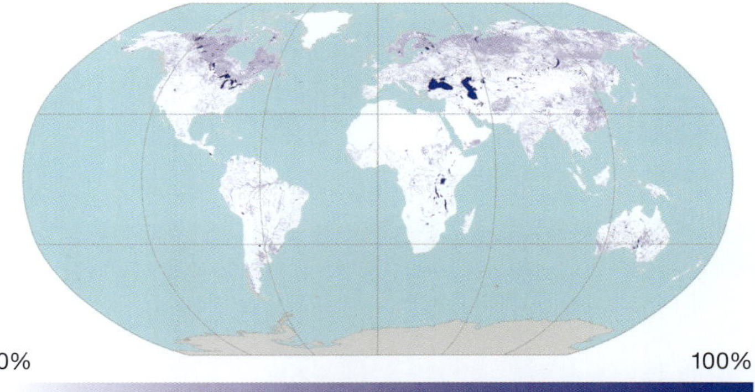

4.1.20 The global distribution of water bodies covers 2.6% of terrestrial land.

Artificial surfaces

Artificial surfaces are urban areas or features such as parklands, sports facilities, industrial areas, waste dumps and mining sites. Urbanisation drives much land-cover change. Buildings and artificial surfaces (e.g. roadways and parking lots) contribute to ecosystem loss and natural **habitat fragmentation**. They often replace local tree-covered areas, grasslands or shrublands.

Soil sealing also causes loss. It adds a hard, impervious layer of concrete or asphalt (see Figure 4.1.21) which degrades the soil and increases flood risk. The soil's moisture content, its biodiversity and the amount of organic matter all decline. The nutrient cycle slows down, as does the soil's ability as a diluter of pollutants.

Some artificial surfaces are built on existing cropland. Such conversions are especially apparent in Japan, Switzerland and the Netherlands (see Figure 4.1.22).

4.1.21 Motorway interchange in Osaka, Japan

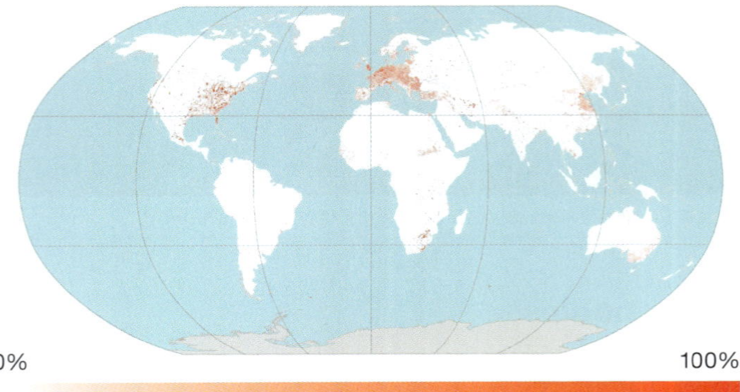

4.1.22 Global distribution of artificial surfaces covers 0.6% of terrestrial land.

Cropland

Cropland encompasses a wide range of crops, both herbaceous and woody. Herbaceous crops include wheat, rice, maize, soybean, barley, sorghum, cotton, sugar cane and hay plants (See Figure 4.1.23). Woody crop land cover features a layer of permanent crops. It includes orchards and plantations, such as fruit trees, coffee and tea plantations, oil palms, and rubber plantations. In some areas, woody and herbaceous crops are layered. This is seen in Mediterranean regions when wheat is grown in olive groves.

Agricultural expansion drives natural and semi-natural land losses worldwide (see Figure 4.1.24). Rapid population growth and rising living standards are increasing this expansion. Natural vegetation is replaced with far fewer plant types, lowering a community's plant diversity. Converting natural landscapes to croplands requires significant effort and expertise. Fertilisers and pesticides are added to maintain productivity, but overuse can degrade soils, reduce biodiversity and deplete nutrients. Sustainable agricultural practices are critical to offset the environmental impacts of cropland.

4.1.23 Sugar cane cropland, Sao Paulo state, Brazil

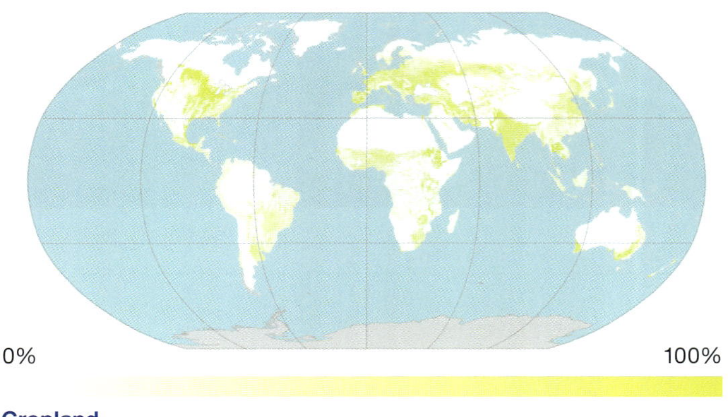

4.1.24 The global distribution of cropland covers 12.6% of terrestrial land.

Land use

Land use describes human activities such as agriculture, forestry and constructions that alter land-surface processes, including hydrology and biodiversity. It is also defined by the land's social and economic purposes, such as subsistence versus commercial agriculture or public versus privately owned land. Consider these differences between land cover and land use:

- Forests are classified as land use if they are used for selective logging or recreation (see Figures 4.1.25 and 4.1.26).
- A naturally occurring lake is land cover, but considered as land use when used for irrigation or recreation.
- Grasslands are a type of land cover, but when used for grazing sheep and cattle they are also considered a land use.
- The African savanna is land cover, but is classified as a land use if managed as a tourist safari park or wildlife refuge.

4.1.25 Pine forest is a land cover. If logged or used for recreation, it is a land use.

Land cover change

Changes in land cover can be traced back many millennia and are a direct result of people's need for resources, especially food. Initially this involved using fire to corral game, but it accelerated once farming practices developed, escalating deforestation and transforming vast natural areas into agricultural landscapes. Industrialisation concentrated human populations in urban areas, a process called urbanisation. This led to an intensification of agriculture on productive lands and the abandonment of marginal land. Rural populations gradually declined.

SPOTLIGHT

Observing land cover change

While changes in land cover extent can be assessed using field-based observations or remote sensing, local expertise is often needed to identify different activities in various parts of a landscape. For example, a satellite image of a vegetated area may be undisturbed old-growth forest, a selectively logged forest, secondary regrowth or a rubber tree plantation. Only the expertise of local land managers can provide such information.

From 1700 to 2020, the world's population increased from 650 million to 7.6 billion. Roughly a quarter of Earth's land area changed from trees and shrubs to crops and pasture:

- Forest and woodland declined from 41 to 31 per cent
- Grassland and similar vegetation declined from 24 to 13 per cent
- Shrubland declined from 6.5 to 2 per cent
- Cropland expanded from 2 to 11 per cent
- Pasture expanded from 4 to 23 per cent.

Land cover change is caused by both natural and human-related processes. While some parts of the world are marginally affected by land cover change, others have been transformed by human activity. In Australia's southern agricultural regions, up to 50 per cent of native forests and 65 per cent of native woodlands have been cleared or greatly modified. Even in areas where the type of land cover has not changed, it has been degraded by land use and management, such as using grasslands and shrublands for grazing livestock.

4.1.26 Rainforest being cleared in Borneo for oil palm plantations

Activities

Acquiring and processing geographical information

1. Distinguish between land cover and land use.
2. Outline the changes in land cover from a historical perspective.
3. Explain why expert knowledge is often required to supplement field-based observations and remote sensing when investigating the nature and extent of land cover change.
4. Outline the extent of land cover change in the 320 years from 1700 to 2020. To what extent does land clearing in Australia reflect this trend?

Applying and communicating geographical understanding

5. **a** Using the following data, create a pie graph of the distribution of terrestrial land cover.

   Tree cover: 27.7%
   Bare earth: 15.2%
   Grasslands: 13.0%
   Croplands: 12.6%
   Snow and ice: 9.7%
   Shrubs: 9.5%
   Sparse vegetation: 7.7%
   Water bodies: 2.6%
   Herbaceous vegetation: 1.3%
   Artificial surfaces: 0.6%
   Mangroves: 0.1%

   b Identify the four largest terrestrial land cover types. Which are the smallest land cover types?
6. Study the information that describes the various land cover types. Create a table to summarise it.
   - Label the columns: Land cover type, Location, Characteristics.
   - Label the rows: Tree cover, Shrub cover, Grassland, Herbaceous vegetation, Mangroves, Sparse vegetation, Bare earth, Snow and glaciers, Water bodies, Artificial surfaces, and Cropland.

UNIT 4.2

Land cover change: Natural causes

Changes in land cover can be initiated by a range of natural processes. These include non-**anthropogenic** climate change (see Chapter 13); geophysical processes, including continental drift, earthquakes, volcanic eruptions and landslides; **plant succession**; and fire and pests.

Natural climate change

Earth's climate has changed over time. The last time the climate was markedly different from the present was during the last **glaciation** (or ice age). It reached its peak about 24 000 years ago, when ice and snow cover was much greater than it is today. This period of glaciation was the last in a long succession of glaciations that characterised the **Pleistocene**, a geological epoch lasting from 2.6 million years ago to about 11 700 years ago.

During the last glaciation, the climate of the middle and high latitudes of the Northern Hemisphere cooled to such an extent that a vast ice sheet formed over central and eastern Canada and spread southward into what is now the USA, and westward towards the Rocky Mountains. It advanced at an average rate of 25 to 100 metres a year (see Figure 4.2.1).

At the same time, another vast ice sheet formed over the mountains of western Canada, northern Europe and north-west Asia. Large glacier systems developed in the European Alps, the Andes, the Himalayas and the Rockies. Smaller glaciers developed in all other ranges and mountain peaks at all latitudes, including Tasmania and Australia's Southern Alps.

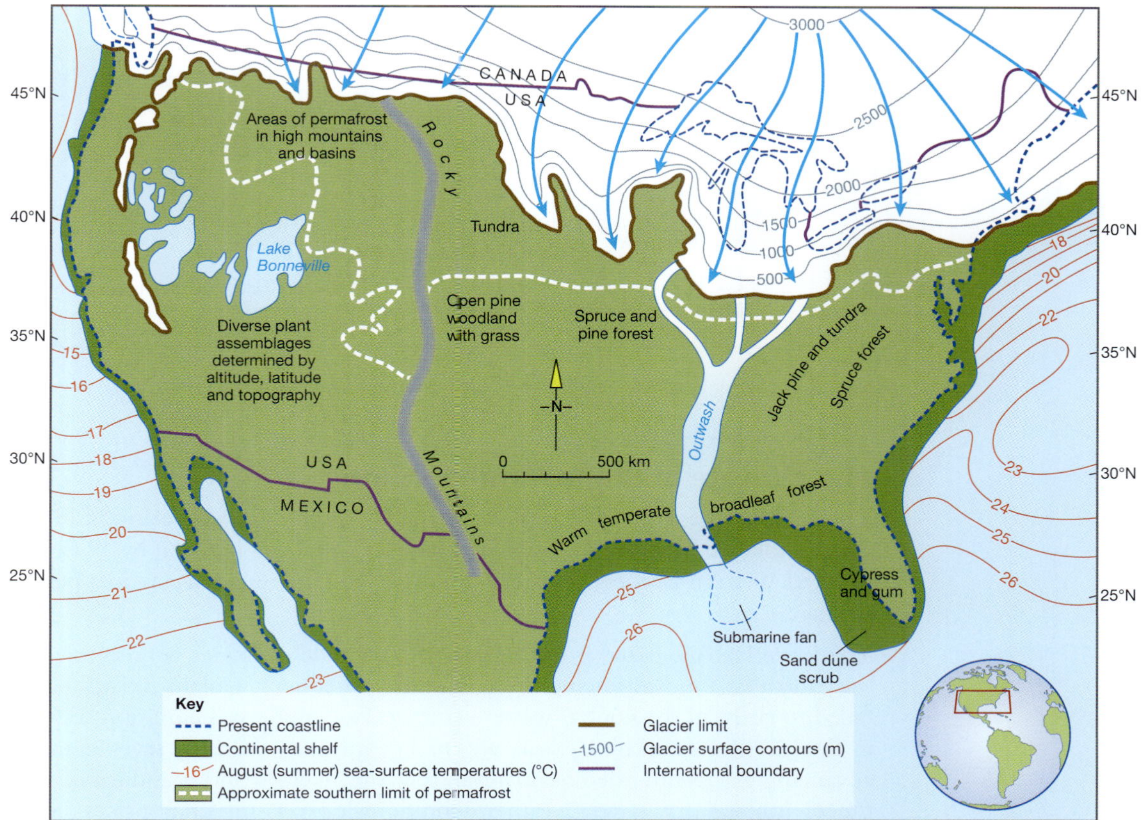

4.2.1 The geography of North America about 20 000 years ago, during the last glaciation. Sea levels were about 120 m lower, exposing large areas of the continental shelf. The southern margin of the ice sheet, shown here, was over 2.5 km deep in the northern parts of the continent.

Areas of sea ice also expanded during periods of glaciation, creating one vast northern glacial ice sheet that covered all of the Arctic and much of the sub-Arctic regions.

With the southward advance of the northern ice sheets, the **periglacial** zones spread to lower latitudes and lower altitudes. In Russia, **permafrost** extended 1000 kilometres south of the ice sheet's edge. In North America, the periglacial zone was largely limited to a small belt adjacent to the southern edge of the ice sheet. For the southern extent of permafrost in North America and Eurasia, see Figures 4.2.1 and 4.2.2.

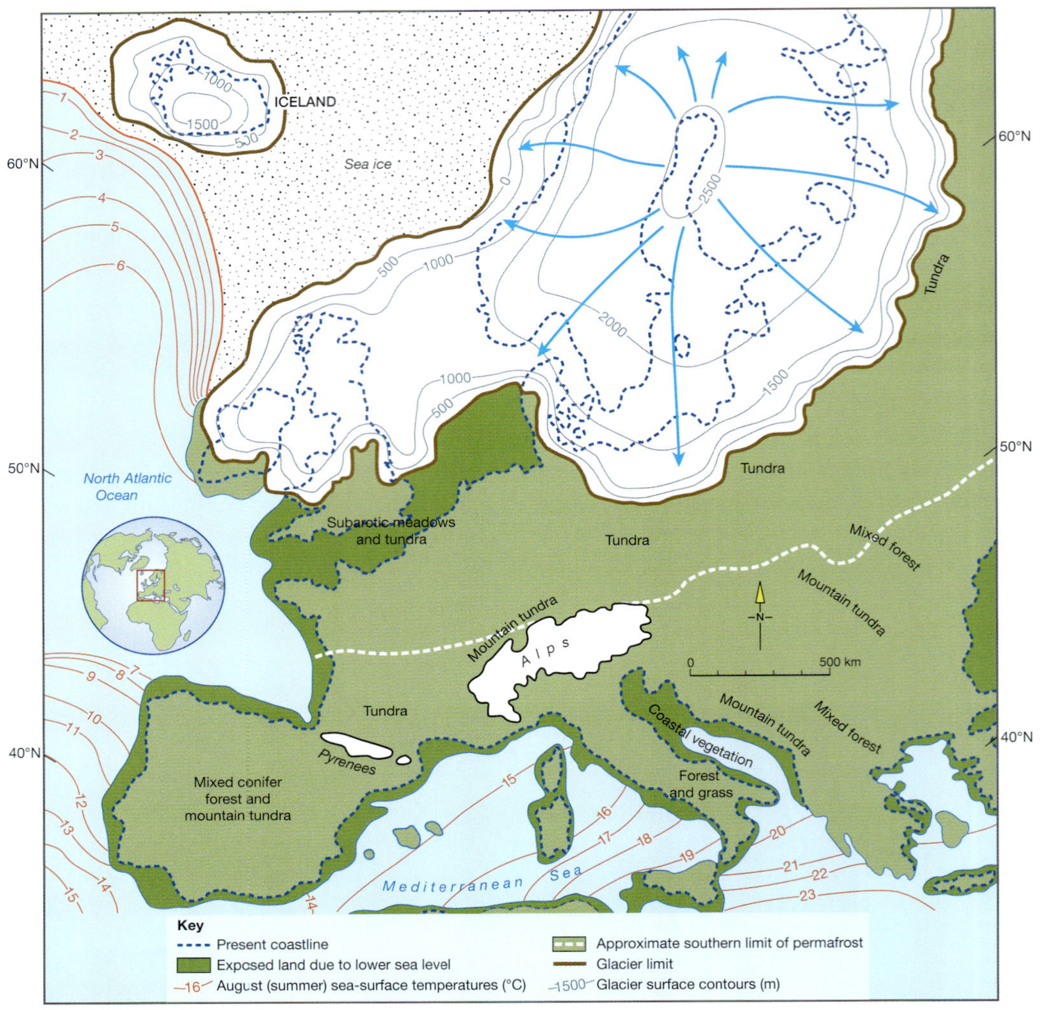

4.2.2 A vast periglacial zone separated the northern European ice sheet from the glacier-covered Alps.

The climate of the last millennium

The Middle Ages (the fifth to the fourteenth centuries) saw a period of relatively mild climate followed by a period when average temperatures reduced by 1° to 2°C. It lowered Western Europe's snowline by about 100 metres, ushering in a period of glacial advance.

The Little Ice Age lasted from 1300 to 1870. Colder, snowier winters saw sea ice in the North Atlantic advancing south, and more frequent, violent storms. Grain crops failed to ripen in the cooler, wetter summers, so famine was common. Life expectancy in England decreased by 10 years within a century. By the early seventeenth century, advancing glaciers invaded farmland in the Alps, Iceland and Scandinavia. By the nineteenth century, erratic weather conditions led to rising grain prices, epidemics and famines. This initiated large-scale migrations, notably to North America.

Mountain glaciers and North-Atlantic sea ice began to retreat with the general warming in the 1870s. It brought conditions favourable for crops in middle latitudes as the world's population expanded rapidly.

Today's anthropogenic warming is significant for its accelerating rate. Climate changed very slowly in the past, over centuries. Today's rate is over decades. The retreat of glaciers, ice sheets and sea ice, and the associated sea-level rise, is changing the planet's land cover. It can be observed and measured using spatial technologies.

● **SPOTLIGHT**

The Antarctic ice sheet

Antarctica is home to the larger of Earth's two polar ice caps. The ice sheet covers 98 per cent of the Antarctic continent—almost 14 million square kilometres—and contains 26.5 million cubic kilometres of ice. The ice sheet is up to four kilometres thick (see Figures 4.2.3 and 4.2.4).

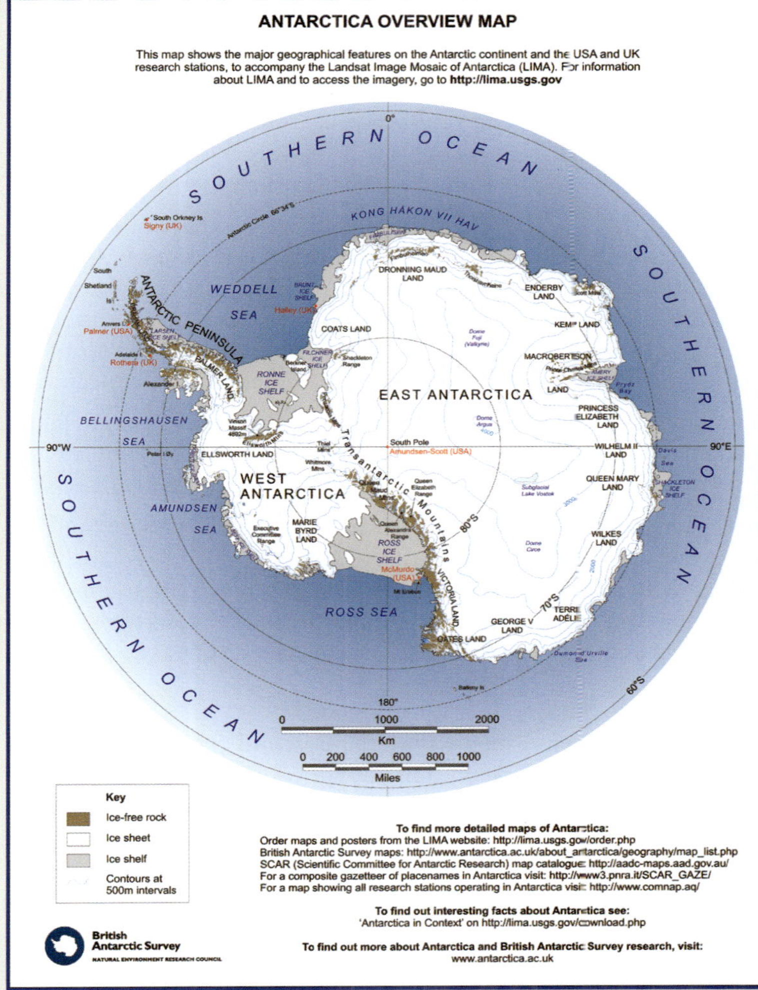

4.2.3 Antarctica can be divided into West Antarctica, East Antarctica and the Antarctic Peninsula. The peninsula is home to scientific stations operated by many nations and is a popular destination for cruise ship-based tourism. The numerous glaciers and floating ice shelves are changing rapidly here because this region is warming faster than the rest of Antarctica.

Roughly 61 per cent of Earth's freshwater is held in the Antarctic ice sheet.

This ice sheet is the focus of significant scientific attention. The melting Antarctic ice has added nearly 3 trillion tonnes of water to the oceans over the past 25 years, mostly from West Antarctica, where the rate of ice melting has tripled—from 58 billion to 175 billion tons per year. Antarctic ice losses have led to sharply rising sea levels over the last decade.

If the entire Antarctic ice sheet were to melt, sea levels would rise by about 58 metres. Even a 1-metre rise would degrade or destroy over a third of the world's coastal estuaries, wetlands, coral reefs and fertile river deltas. It would inundate low-lying countries such as Bangladesh and the island states of the Pacific and the Caribbean, and flood some of the world's largest cities, including Kolkata and Mumbai, India and Dhaka, Bangladesh.

4.2.4 The Antarctic ice sheet

4.2.5 The Australian tectonic plate moves north by about 7 cm a year, the fastest on Earth.

Geophysical and geomorphological processes

Geophysical and **geomorphological** processes such as **tectonics**, including earthquakes, volcanic activity and landslides, can initiate land cover change. Weathering, erosion and deposition also contribute.

Continental drift changed the distribution of the continents over tens of millions of years (detailed in Chapter 3). As landmasses traversed latitudinal zones they adjusted to new climatic conditions.

Australia separated from Antarctica when the supercontinent Gondwana broke up 90 to 30 million years ago. As Australia moved north (see Figure 4.2.5), the climate became warmer and drier, transitioning the vegetation land cover from diverse forest to today's scrub and eucalypt landscape.

During the Pleistocene epoch, significant changes in sea level linked Australia to Indonesia and New Guinea via land bridges. This allowed for new combinations of species and ecosystems as species dispersed and intermixed. Through isolation and time, Australia's flora and fauna evolved separately from other continents.

After it separated from Antarctica, southern Australia contained diverse forests; their relatives are still living in small pockets. These rainforests became less diverse and more fragmented over time, their distribution becoming scattered before human occupation.

The mainland's characteristic aridity developed from about 20 million years ago. Recent cycles of increasing and decreasing aridity were part of the glacial cycles of the last million years or so.

There is evidence of regular fires in south-eastern Australia dating back 25 million years. Eucalyptus trees were absent from the forests, appearing on the east coast about 20–25 million years ago. While evidence that suggests eucalypts appeared in response to more frequent fires, they were not common until 50 000–200 000 years ago. They responded positively to First Nations people's fire management practices.

Volcanic activity

Volcanic activity can cause land cover change within the blast and deposition zone, but a volcano's influence reaches far beyond its location. When widely distributed gases, dust and ash reach the upper atmosphere they upset its circulation patterns. Eruptions in the tropics can affect climate in both hemispheres, while those at mid or high latitudes are usually limited to their own hemisphere. Massive eruptions throw gases and dust particles into the atmosphere, which can block incoming solar radiation and cool the planet for months or even years. Sustained episodes can increase ice and snow cover extent, causing vegetation pattern changes.

4.2.6 Monument Valley, USA, is shaped by weathering, erosion, transportation and deposition.

Weathering, erosion, transportation and deposition

The geomorphological process of weathering, erosion and deposition also impacts land cover (see Figure 4.2.6). Typically, this impact is felt at a more localised scale than geophysical processes such as volcanic eruptions. These processes are detailed in Chapter 3.

● SPOTLIGHT

Mt St Helens

The 1980 eruption of Mt St Helens, USA, was the first such event observed in scientific detail. Satellite imagery revealed land cover changes and the rate of plant succession over the following decades. An earthquake followed, collapsing the northern face, causing the largest recorded landslide. Hot rocks, ash, gas and steam exploded up and out, spreading volcanic debris (grey in images) over 600 square kilometres. Debris obliterated forest cover up to 27 kilometres away. The avalanche buried 23 kilometres of the North Fork Toutle River in rock and debris to an average of 46 metres deep. Volcanic mudflows (lahars) poured down rivers and gullies of the southern half of the mountain. Mt St Helens was reduced from 2950 metres to about 2550 metres. Eventually, the river carved a shallow, braided path through the buried valley.

Heat and noxious gases sterilised the surface, which became buried under ash, mud and rock. Nearly every living creature in the area perished, but traces of life survived—seeds, spores and fungi. Plant succession processes began and land cover was slowly re-established (see Figures 4.2.7 to 4.2.10).

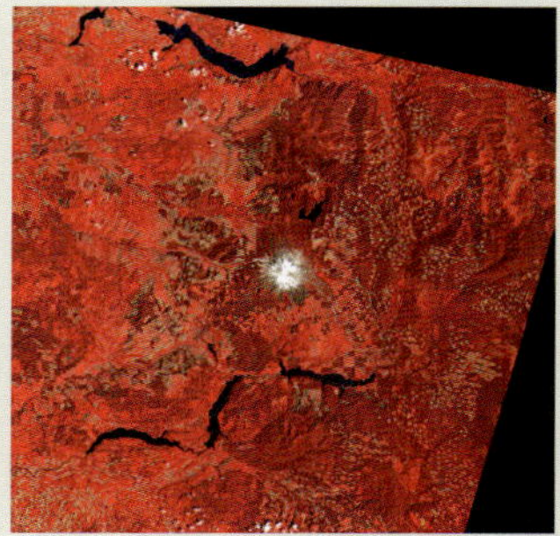

4.2.7 Mt St Helen's in 1979, before the eruption. Areas that appear red are vegetation.

4.2.8 Mt St Helens in 1980, immediately after the eruption

4.2.9 Mt St Helens in 2000

4.2.10 Mt St Helens in 2016

Invasions and ecological succession

An invasive plant is a non-native species that spreads, causing habitat damage and threatening biological diversity. Such invasions typically occur when some, often human, intervention results in a loss of natural controls or the introduction of an exotic species, such as blackberry.

Plant succession is the change in the types of plant species occupying a given area over time. It typically involves the processes of pioneering, establishing, sustaining and reproducing. It can also be described in terms of primary, secondary and climax plant communities. Beginning with a few pioneering species, a community increasingly diversifies, then stabilises and self-perpetuates.

Plant succession may be initiated in two ways. Forming a new, unoccupied habitat, uninfluenced by pre-existing communities is called a **primary succession** (e.g. from a lava flow or landslide, or a disturbance such as fire, strip mining or logging). **Secondary succession** follows the disruption of a pre-existing community. When succession results in a stable, self-perpetuating plant community it is called a **climax community**.

Fires and pests

Fire is a major threat to forests, shrublands and grasslands, as it changes the structure of habitats and the mix and diversity of species present. It has an impact on soil properties such as texture, porosity, organic matter, nutrient availability and biota. This then has an impact on plant life. Drought, disease, insect infestation, overgrazing or a combination of these can intensify this impact. Uncontrolled animal populations such as rabbits can lay waste to communities such as grasslands.

Pests threaten forests, shrublands and grasslands. Insect infestations can cause 'dieback' in forests, the progressive death of trees within a year or two after symptoms appear.

4.2.11 Clearing fungal-diseased common ash trees in the UK

Symptoms are often subtle, slow to develop and usually uniform throughout the tree's crown.

Trees and shrubs are also subject to diseases, extremely high or low temperatures, and fluctuations in soil moisture during long-term weather cycles. Weakened trees and shrubs are more susceptible to insect attack and fungal invasion. These include borer-type insects, canker diseases and fungi that cause root rot (see Figure 4.2.11).

In Europe, landscapes are transformed by a fungus (*Hymenoscyphus fraxineus*) that causes widespread dieback and has a mortality rate of up to 85 per cent in ash tree forests. Most attempts to control its spread have failed, even removing trees in infected areas, as the fungus grows on forest leaf litter (see Figure 4.2.11).

Tree root systems are especially vulnerable to soil environment changes. These include **soil compaction**, drainage pattern changes, excessive moisture, lack of water, removing or adding soil over the root system, and excess pesticide or fertiliser use.

Invasive animal species, such as feral cats, foxes and cane toads, can damage habitats. Without controls, rabbits and feral pigs can strip the vegetation exposing the soil to erosion. Some can kill native animal species and carp can increase the **turbidity** of waterways.

Activities

Acquiring and processing geographical information

1. Explain the relationship between climate and land cover.
2. When was Earth's climate markedly different from now? What was it like?
3. Identify the defining climatic event of the Pleistocene.
4. Outline the effects of the cooling associated with the last glaciation event. What was the impact on sea levels?
5. Identify the geophysical processes initiating land cover change.
6. Outline the role of fire in shaping land cover in Australia over time.
7. Outline the impacts of the Mt St Helens eruption on the surrounding area. What was the role of plant succession in repairing the damage?
8. Define plant succession. Outline the processes involved. Distinguish between primary and secondary succession. What is meant by the term climax community?
9. Outline the impacts of fire and out-of-control animal species on habitats.
10. Explain what dieback is. What are the likely causes?

Applying and communicating geographical understanding

11. Construct a flow diagram describing the nature of the world's climate over the past 1000 years. Use the following headings in your diagram:
    - a Middle Ages
    - b Little Ice Age (1300–1870)
    - c Post–1870s
12. Study Figure 4.2.1. Using information from the map, write a brief report describing the nature and extent of the North American ice sheet and permafrost about 20 000 years ago.
13. Study Figure 4.2.2. Using information from the map, write a brief report describing the nature and extent of the European ice sheet and permafrost about 20 000 years ago.
14. Study Figures 4.2.3 and 4.2.4 and the related text.
    - a Write one or two paragraphs describing the physical geography of the Antarctic ice sheet.
    - b Outline the impacts of climate change on the ice sheet. State why any significant melting of the ice sheet would prove problematic.
15. Study Figure 4.2.5.
    - a Describe Australia's progress northwards over the last 90 million years.
    - b Briefly describe the relationship between continental drift and land cover change.
16. Study Figure 4.2.6. Working in groups, identify and describe the role of weathering and erosion in the formation of this landscape. Sketch the photo and annotate the features, describing what agents of erosion appear to be primarily responsible for the formation of this landscape.

UNIT 4.3
Land cover change: Human causes

The human causes of land cover change include a rapidly growing population, along with improvements in material standards of living and technological advances that have greatly increased human capacity to transform landscapes and anthropogenic climate change (detailed in Chapter 13).

Population growth

The environment's ability to cope is threatened with becoming overwhelmed by population growth and a rising demand for consumer goods, particularly from the developing world's rapidly growing middle-class. An increasing number of consumers and increasing demand for food challenge the limits of growth. Already, the adverse impacts of production are evident and numerous. Demand for energy parallels production. It is largely sourced by using fossil fuels, which have increased greenhouse-gas emissions. This has fuelled climate change, which threatens the stability of entire ecosystems.

Developed nations and a handful of rapidly developing nations currently account for just 18 per cent of the global population. The rest of the world's population (i.e. the remaining 82 per cent) aspire to the developed world's living standards. This is most apparent in the heavily populated parts of South and East Asia, especially China and India. The demands this will place on food and industrial production will deplete Earth's natural resources, especially soil and water resources. The biosphere will be seriously threatened. Species extinctions, largely as a result of deforestation, will accelerate.

Technology

Advances in technology have greatly enhanced the ability of humans to transform Earth's terrestrial landscapes. Large-scale earth-moving equipment can change the topography and clear large areas of forest. Genetically modified crops have been developed to grow in areas once deemed marginal or unsuitable. Humans can build large dams and related water-distribution infrastructure to supply water to once-arid landscapes.

Examples of major dam infrastructure are:

Snowy Mountains Scheme, Australia: water is diverted from the Snowy River into the Murray and Murrumbidgee rivers to expand irrigated crop production in an area once dominated by grazing.

State Water Project (SWP), USA: constructed in the 1960s and 1970s, it supplies water to over 27 million people and 750 000 acres of farmland. It is one of the world's most extensive systems of dams, reservoirs, power plants, pumping plants and aqueducts, and is key to California's economy (see Figure 4.3.1).

4.3.1 California's aqueducts demonstrate human capacity to transform land cover.

Anthropogenic climate change

After nearly 1000 years of relatively stable climatic conditions with near-surface atmospheric temperatures remaining fairly steady, global temperatures began to rise. Current climate changes since around 1975, along with projected changes, are many times faster than natural changes, which take place over hundreds or thousands of years.

Significantly, the world's leading scientific bodies, including the UK's Royal Society, the US National Academy of Sciences, the US National Oceanic and Atmospheric Administration (NOAA), the US National Aeronautics and Space Administration (NASA), and Australia's CSIRO and Academy of Sciences, together with 97 per cent of the world's climate scientists, agree that:

- the world's climate is warming at an accelerating rate and is primarily caused by the burning of carbon-rich fossil fuels, which add carbon dioxide to the atmosphere. A secondary cause is the clearing of forests, which take up carbon dioxide from the atmosphere
- climate change will continue to accelerate unless we act to reduce emissions. Such actions are possible and affordable, and will result in significant improvements in people's health and the environment more generally. They will also boost economic activity. The sooner we act to reduce emissions, the lower the economic and environmental costs will be.

The consequences of global warming are mounting. Additionally:

- atmospheric concentrations of carbon dioxide and other greenhouse gases that warm the atmosphere are rising rapidly. After remaining below 300 parts per million (ppm) for more than 400 years, atmospheric concentrations of carbon dioxide rose from 318 parts per million in 1960 to 421 parts per million in May 2022
- in the Arctic, the extent of summer sea ice has been retreating since 1979. In 15 years (2007 to 2021) the Arctic recorded the lowest 15 minimum extents in sea ice in the 43-year satellite record
- almost all mountain, valley and piedmont glaciers, and most tidewater glaciers are retreating. Glacier National Park in the USA, was home to 150 glaciers in 1910. It has since decreased to under 30, and most of those remaining have shrunk by two-thirds. Within 30 years most, if not all, the remaining glaciers will disappear
- in Alaska, USA and Siberia, Russia, frozen ground (**permafrost**) is melting, releasing masses of carbon that have been locked away in the frozen soil for thousands of years. It also emits methane, another of the greenhouse gases. On entering the atmosphere, methane and carbon dioxide accelerate the rate of climate change, even as humans try to reduce their reliance on fossil fuels
- sea levels are rising at an accelerating rate. This rise is driven by the expansion of the ocean water as its temperature increases and increasing run-off from land-based ice (glaciers and ice sheets). Sea levels may rise by up to 1.3 metres by the end of the century.

Chapter 13 covers more on climate change.

Did you know

The decade 2011–2020 was the hottest ever recorded on Earth. It had eight of the 10 hottest years on record. The other years in the top 10 were 2005 and 1998.

Earth is now about 1.2°C hotter than it was at the beginning of the industrial age in the mid-to-late-1800s. This number is important because in 2015 global leaders adopted a goal of preventing 1.5°C of warming since the rise of big industry.

SPOTLIGHT

Australia burns

The 2019–20 Australian bushfire season, arguably the most devastating in Australian history, started with a series of uncontrolled fires in June 2019. Given the prolonged drought affecting NSW and much of Queensland at the time, the blazes proved difficult to control. Many burned well into February 2020. Record high temperatures and strong winds were all that were needed to create catastrophic fire conditions.

By the time the fire crisis peaked, an estimated 18.6 million hectares (186 000 square kilometres) had been burnt, over 5900 buildings (including 2779 homes) had been destroyed, and at least 33 people had been killed. The extent of habitat destruction was alarming. An estimated 1 billion animals were killed, and some endangered species were driven to the brink of extinction.

These bushfires emitted 400 megatonnes of carbon dioxide into the atmosphere, as much as Australia's average annual carbon dioxide emissions, in three months. These contributed to global warming, which heightens the likelihood of recurring megafires that will release yet more emissions. Scientists call this a climate **feedback loop**.

The fires became the centre of a public debate about the relationship between climate change and the growing incidence and severity of fire and drought in Australia.

The prolonged period of drought was followed by well-above-average rainfall caused by the occurrence of two major weather patterns—La Niña in the Pacific and the negative Indian Ocean Dipole—along with warmer than average waters around northern Australia. Flooding was widespread in south-eastern Australia and South East Queensland in early and late 2022. The Bureau of Meteorology links Australia's wild temperature and rainfall variability to global warming caused by human activities.

Chapter 14 covers more on bushfires.

Activities

Acquiring and processing geographical information

1 Briefly outline how humans have impacted the nature and rate of land cover change.

2 Describe how population changes impact food and energy demands. How will growing standards of living in developing countries impact Earth and its resources?

3 Outline the key elements of the scientific consensus about the nature and causes of climate change.

4 Outline the evidence to support the claims made about anthropogenic climate change.

Applying and communicating geographical understanding

5 As a class, brainstorm examples of the ways in which humans, via advances in technology, have been able to transform terrestrial land cover.

6 Study the box, Spotlight: Australia burns, then answer the following questions:

a What factors fuelled the devastating bushfires?

b What was the impact of the fires?

c How did it impact the public debate surrounding the link between climate change and the incidence and intensity of bushfires?

UNIT 4.4

Impacts of land cover change

Land cover change is amplified by land clearing. Marginal land is often cleared for agriculture and food production, while fertile land and valuable habitat are lost to urban expansion, industry and transport infrastructure. As the world population grows, the land-clearing rate escalates.

Impacts on ecosystem services

The clearing of natural vegetation causes a range of environmental impacts:

- **Change in radiation balance:** cleared land has a higher albedo than vegetated land.
- **Decline in soil water-holding capacity:** soil porosity and its water-holding capacity are reduced by clearing natural vegetation for agriculture. Soil compaction by grazing livestock often results. This raises the risk of **hydrological drought,** especially during dry seasons. Exposed soils are susceptible to erosion during heavy rainfall. Soil erosion greatly impacts agriculture, local economies and ecosystems.
- **Decline in precipitation:** the cloud formation rate declines when forests are cleared for agriculture because the rate of **evapotranspiration** declines.

Anthropogenic changes in the landscape contribute to regional and global climate change. While not fully understood, converting forested land to agricultural land changes local climates by changing the solar radiation and water balance. Changes in precipitation and temperature patterns eventually harm the sustainability of agricultural systems.

● SPOTLIGHT

Land cover change and a warming planet

Land cover change releases greenhouse gases into the atmosphere, contributing to climate change (see Figure 4.4.1). Deforestation, especially combined with tillage-based agriculture, releases carbon dioxide.

Altered surface hydrology increases methane emissions (e.g. through wetland drainage and the flooding of rice paddy fields), as does cattle farming. Agricultural products increase nitrous oxide emissions (e.g. in organic nitrogen fertilisers, irrigation and biomass combustion).

While the complex relationship between greenhouse gas emissions and land cover changes is not fully understood, it is known that land cover change and land use account for about 12.5 per cent of human carbon emissions. An added uncertainty is how much sulphur dioxide and particulates (produced by biomass combustion) might contribute to regional and global cooling. As airborne particulates, they reflect incoming sunlight and this might affect cloud cover.

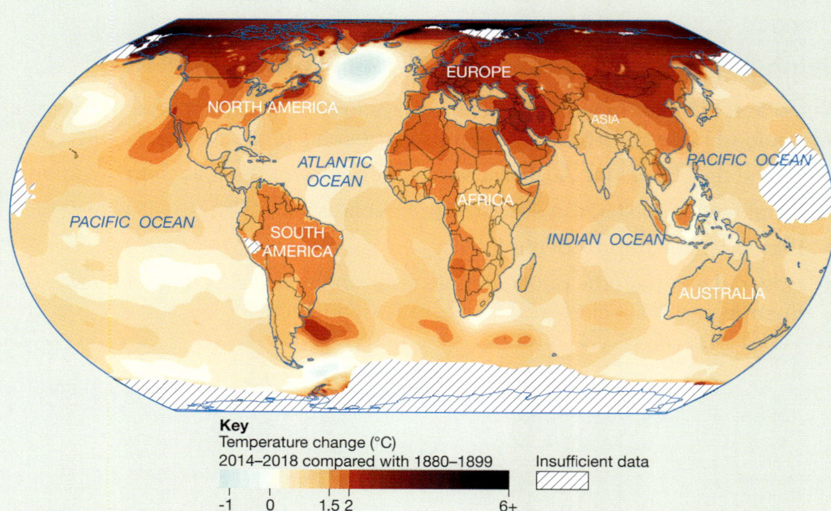

4.4.1 Temperature change 2014–18 compared to 1880–1899. The Arctic region has warmed more markedly than elsewhere, driving the decline in sea ice and retreat in terrestrial ice sheets and glaciers.

Land cover change and biodiversity

Clearing forests and grasslands for agriculture reduces biodiversity. Species loss is immediate and often complete; even partial forest habitat loss has an impact. When existing habitats are fragmented (see Figure 4.4.2) the forest's edges become exposed to external influences. This reduces the extent and environmental integrity of core habitat areas. Smaller habitat areas typically support fewer species. For any species needing undisturbed core habitat, habitat fragmentation can result in their extinction. By clearing land, an area becomes exposed to invasion by exotic (non-native) plants, animals and diseases, especially in remnants near human populations.

4.4.2 Fragmented boreal forests in Ontario, Canada

Land cover change and pollution

Removing land cover vegetation can cause water, soil and air pollution. Soil becomes exposed to water and wind erosion, which ultimately reduces its fertility, especially when fire or large-scale machinery is used. Erosion releases phosphorus, nitrogen and sediments into streams and other aquatic ecosystems, increasing sedimentation, turbidity and **eutrophication**. The latter can trigger aquatic plant growth, which depletes dissolved oxygen.

Modern agricultural practices can result in waterway and groundwater pollution (e.g. nitrogen and phosphorus fertilisers and concentrated livestock feedlots). This can cause **algal bloom** outbreaks in waterways that deplete oxygen levels. Some of these rapidly accumulating algae contain toxins that kill fish and harm livestock and native fauna. (See Unit 3.8.)

Modern agricultural practices can cause chemicals such as herbicides and pesticides to accumulate in ground and surface-water bodies. They can accumulate in living things, be passed up the food chain and result in **bioaccumulation**.

Burning vegetation to clear land for agriculture is a major contributor to air pollution. Mining has a greater impact and when poorly managed can pollute waterways and groundwater with toxic metals.

Land cover change in urban areas

Urban expansion radically transforms land cover. Natural vegetation and farmland are replaced with built environments such as housing, factories and roads. This creates urban **heat islands**, making cities typically hotter (see Figure 4.4.3). This can be caused by dark surfaces, such as asphalt and roof tiles, absorbing and storing more solar energy than grass and trees; industrial processes and transportation generating heat, and eliminating shade and transpiration by removing vegetation.

On average, cities are 1.5°–2.0°C warmer than surrounding vegetated lands. Planting street trees and creating parklands mitigates this because parks are cooler than their surrounding area. Shady trees prevent ground surfaces from overheating and evapotranspiration pleasantly cools the air beneath the canopy. Plants transpire water into the atmosphere, which evaporates to cool the atmosphere around its aerial parts. The effect is more pronounced at night. The park's size and plant community dictate the intensity and impact of this phenomenon. Figure 4.4.3 illustrates the heat island of London, UK.

Urban run-off is a source of environmental pollution in aquatic environments. It is often heavily contaminated with oils and other pollutants from streets and cars.

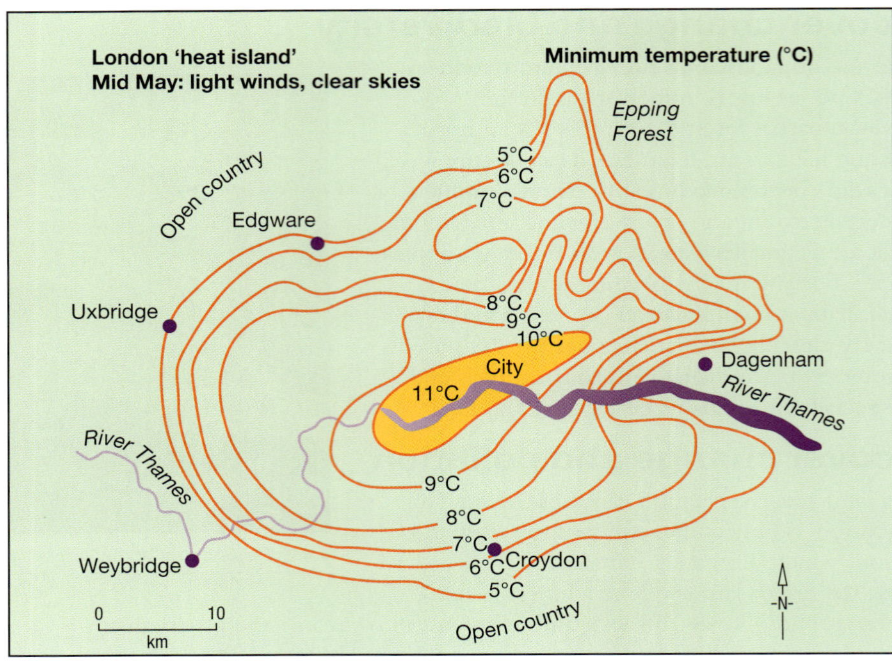

4.4.3 London, UK's heat island effect

Activities

Acquiring and processing geographical information

1 Outline the impact land cover change has on:
   a radiation
   b the soil's water-holding capacity
   c precipitation
   d climate change
2 What is the percentage of anthropogenic carbon emissions linked to land cover change?
3 Explain how the release of sulphur dioxide and particulates complicates the analysis of the effects of anthropogenic climate change.
4 Outline the impact of land cover change on biodiversity. Why is habitat fragmentation considered such a problem?
5 Identify the types of pollution associated with land cover change.
6 Outline the causes and impacts of algal blooms.

Applying and communicating geographical understanding

7 Study Figure 4.4.1. Write a paragraph outlining the global pattern of warming.
8 Write a paragraph or two explaining why urban areas are measurably warmer than their surroundings.
9 Study Figure 4.4.3. Use data from the map to describe the heat island effect associated with London.
10 As a class, brainstorm the strategies authorities could initiate to mitigate the effects of the heat island effect.

Continued from page 79.

Chapter glossary

soil compaction the increased density of soil when it is compressed by the trampling of livestock. The compression forces air and water from the soil

tectonic (processes) disturbances in Earth's crust that result from Earth's internal energy and create physical features, such as mountains, on Earth's surface

turbidity the degree to which light will pass through liquid, usually a measure of the amount of mud or other particles suspended in the liquid

vascular related to, or having, vessels or ducts that convey fluids like blood or sap

woody plants plants that have a hard stem or trunk of wood

APPLICATION AND CONSOLIDATION TASK

Task 1: Interpreting memes

A meme is an image, video or piece of text, generally humorous, that is copied and spread rapidly by internet users. Memes are often used by both private and public organisations to spread messages. Memes have been used successfully during political campaigns to discredit ideas or cast them into doubt. Being able to read or interpret a meme is an important life skill. It is also important to identify the issues being addressed, the techniques being used and where the meme originated.

Use the internet to find a meme to study. Choose a meme with a political or environmental message.

- Where did you see the meme?
- Who sent you the meme or where did it originate?
- What is the meme about? Why?
- What can you see in the meme? What words, if any, are used?
- What events provided the background or context in which the meme was created?
- Has the creator used symbolism, irony, analogy or exaggeration to help communicate their message?
- Can you identify the perspective or point of view of the creator?
- Is the meme persuasive? If so, why?
- What alternative perspectives or points of view are there about this issue?

CHAPTER

5 Deforestation

Forests are one of Earth's greatest natural assets. They are home to more than half of all species, a rich variety of life that keeps many of our most vital natural systems operational—from keeping our climate stable by absorbing carbon dioxide and releasing oxygen to regulating our water supply and improving its quality.

More than a billion people live in, or adjacent to, forests, and depend on them for fuel, food and medicines. Forests provide timber-based resources for homes, paper and furniture. Humans have already destroyed around 40 per cent of the world's forests. Slowing down the rate of deforestation, and protecting and sustainably managing our existing forests is essential.

This chapter investigates deforestation as an example of land cover change and the nature, extent and rate of deforestation in selected places. It explores how deforestation interrupts natural systems; how deforestation impacts at a range of scales, including climate change; and the strategies used to protect forests and promote the sustainable use of forest resources.

I see a future in which nature gives us a helping hand. Instead of destroying the natural world, why can't we use it to solve the kinds of problems that we are facing?

Frances Arnold, American scientist and Nobel Laureate

5.0.1 Rainforest is cleared and burned for the palm oil industry.

Chapter glossary

carbon cycle the sequence of processes whereby carbon compounds are interconverted in the environment, involving the incorporation of carbon dioxide into living tissue by photosynthesis and its return to the atmosphere through respiration, the decay of dead organisms and the use of fossil fuels

clear-fell to cut down every tree in an area

deforestation the clearing of a forest or a stand of trees

environmental services the qualitative functions of natural assets of land, water and air

forest degradation entails a reduction or loss of the biological or economic productivity and complexity of forest ecosystems, resulting in the long-term reduction of the overall supply of benefits from the forest, which includes wood, biodiversity and other products or services

forest fragmentation the division of continuous habitat into smaller and more isolated fragments

habitat the natural home or environment of an animal, plant or other organism

plantation forests managed forests consisting of one or two species of trees

primary forest naturally regenerated forests of native tree species where there are no clearly visible indications of human activity and the ecological processes are not significantly disturbed. Sometimes referred to as old-growth forests

reforestation the re-establishment of forest formations after human-induced or natural disruption

secondary-growth forests a stand of trees that result from secondary ecological succession. This occurs in an area where trees were removed by human activities

UNIT 5.1

Nature and spatial distribution of the world's forests

Forests are critical to life on Earth because they provide a range of important **environmental services**, including **habitat** for plants and animals. They also supply oxygen and protect Earth's watersheds. A forest can inspire wonder, providing a place to interact with the natural world.

Did you know?

Visits to forest environments can have positive impacts on human physical and mental health. Many people have a deep spiritual relationship with forests.

Types of forests

Globally, forests cover 4.06 billion hectares (30.8 per cent of land), but their distribution is uneven. Just five countries—Brazil, Canada, China, Russia and the USA—contain over half of the world's forests (see Figures 5.1.1 and 5.1.2). Of these, nearly half are intact, with **primary forest** making up a third. They are dominated by native tree species and there is no visible signs of human activity, leaving their ecological processes largely undisturbed. Nine per cent of global forest cover exists as unconnected fragments. Primary, or old-growth forests, are irreplaceable for their biodiversity, carbon storage and ecosystem services, such as cultural and heritage values. **Secondary-growth forests** result from secondary ecological succession after disturbance by human activity, including **clear-felling** or by natural forces such as cyclones. **Plantation forests** are commercially managed forests that account for about seven per cent of the world's forested area.

Worldwide, forest diversity varies with latitude, soil, rainfall and prevailing temperatures (see Figure 5.1.3). The three principal types of forest are coniferous (or boreal or taiga), temperate, subtropical and tropical rainforests (see Figure 5.1.4).

Coniferous forests: covering 33 per cent of the land surface, are dominated by cone-bearing trees, such as pines and firs (see Figure 5.1.7). They thrive in northern latitudes in areas with warm summers and cool to cold winters. They span eight countries: Canada, China, Finland, Japan, Norway, Russia, Sweden and the USA. Boreal forests are those evolving under short growing seasons and severe winters with several months of snow cover.

Temperate forests: covering 25 per cent of the land surface, are found in Europe, East Asia, North America and in some parts of South America (see Figure 5.1.6). Typically, they include

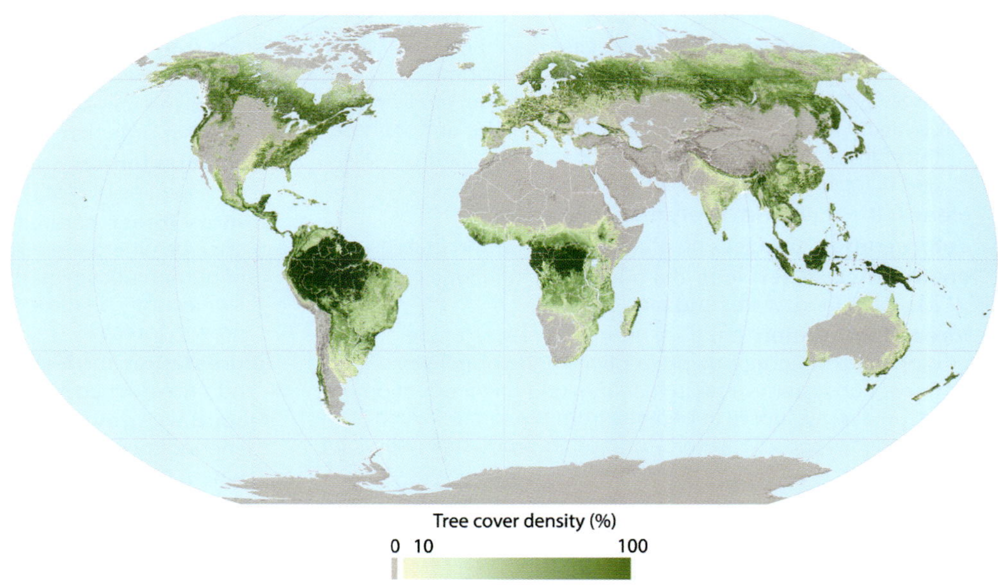

5.1.1 World forest density by percentage

coniferous and broad-leafed trees, such as oaks and elms. Many broadleaf trees are deciduous, with leaves that turn orange, yellow and red in autumn before falling off. They are found in the mid-latitudes with a temperate climate—moderate rainfall evenly spread throughout the year, mild to warm summers and cool to cold winters.

Tropical and sub-tropical rainforests: these are the most biologically diverse and complex forests (see Figure 5.1.5). Tropical forests need abundant rainfall (average over 60 millimetres a month) and constant warmth. They are found in equatorial regions of South America, Africa and South-East Asia. Subtropical forests are found on the east coasts of North America and Asia, and some mountainous semi-tropical regions of Australia and New Zealand. They also need hot summers with abundant rainfall, but tolerate colder winters with moderate rainfall. Both types act as 'carbon sinks', soaking up greenhouse gases.

Collectively, forests account for over 80 per cent of the world's terrestrial biodiversity and over 50 per cent of all biodiversity. They are home to thousands of species of plants, animals, fungi and bacteria. Rainforest ecosystems are especially complex and vulnerable owing to their high level of diversity in relatively small areas. In contrast, savanna grasslands have lower biodiversity spread over large areas, which reduces vulnerability.

Naturally regenerating forests account for 93 per cent of the world's forested area, the other 7 per cent being planted forests.

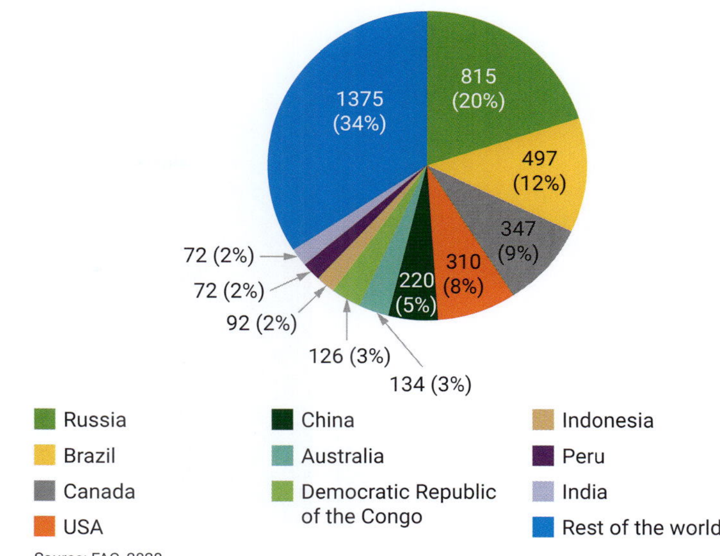

5.1.2 Global distribution of forest showing the 10 countries with the largest forest area by million hectares and percentage of the world's forests.

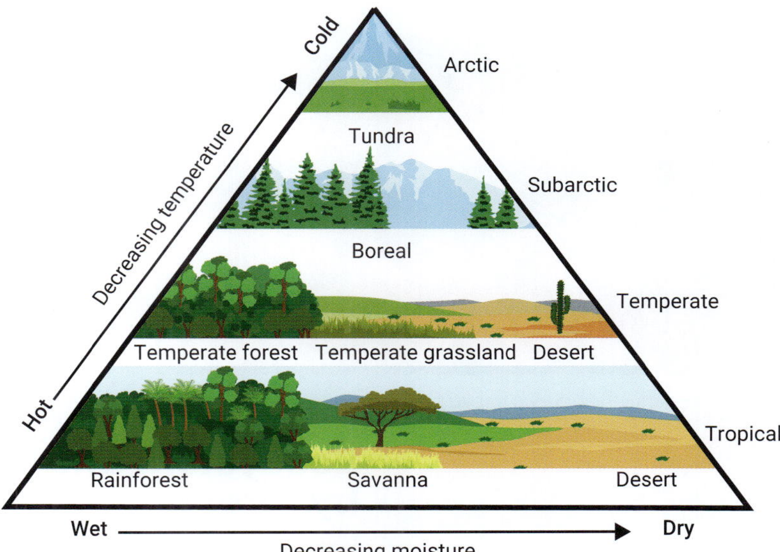

5.1.3 The relationship between temperature, moisture and forest type

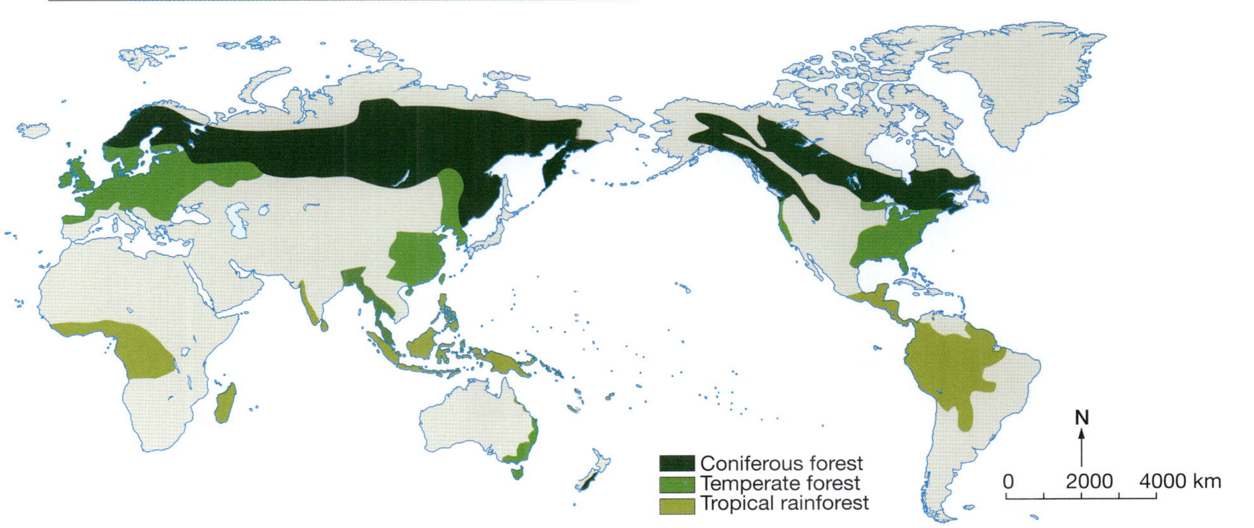

5.1.4 Distribution of the world's forests by type

5.1.5 Tropical rainforest, Khao Yai National Park, Thailand

5.1.6 Temperate deciduous forest, Laurentian Mountains, Canada

5.1.7 Coniferous (boreal) forest, Carpathian Mountains, Ukraine

Ecosystem and economic services provided by forests

Forests provide a range of ecosystem and economic services. These are shown in Figure 5.1.8.

Ecosystem services

Store atmospheric carbon

Influence local and regional climate

Supports energy flow and chemical (nutrient) cycling, including photosynthesis and oxygen production

Wildlife habitats—biodiversity

Absorb and release water

Purify water and air

Soil erosion retention

Economic services

Fuel wood

Food

Timber for construction and furniture

Pulp for paper and cardboard

Livestock grazing

Amenity, recreation and spiritual enrichment

Tourism

Employment

Education and research

5.1.8 Ecosystem and economic services provided by forests

Through photosynthesis, forests remove carbon dioxide from the atmosphere and store it as organic compounds as part of Earth's global **carbon cycle**. In doing so, forests help stabilise average atmospheric temperatures and moderate Earth's climate. Forests also produce oxygen, purify water and reduce run-off and flooding by storing water and then releasing it slowly back into catchments.

Forests support biodiversity, providing habitats for two-thirds of terrestrial species. They support the livelihood of around 300 million people directly and 1 billion people indirectly.

Forests also provide a range of raw materials, including timber. More than half of the wood harvested is used as biofuel for cooking and heating. The rest is mainly used to make timber and paper.

Forests also play a role in maintaining human health and traditional medicines (used by 80 per cent of the population). Many modern medicines are derived from forest species. Certain chemical compounds in forest plants are blueprints for prescription drugs. Forests also remove air pollutants, including some human carbon dioxide emissions.

Did you know?

Globally, 18 per cent of the world's forest area, or more than 700 million hectares, falls within legally established protected areas such as national parks, conservation areas and game reserves.

Natural causes of deforestation

The natural causes of **deforestation** include forest fires, disease, parasites and extreme weather events like cyclones and floods. However, these impacts are not entirely independent of human activity. The intensity of naturally occurring forest fires, for example, is affected by human land management practices. Disease and parasite attacks often result from humans upsetting the ecological balance. The frequency and intensity of extreme weather events are increased by anthropogenic climate change.

Forest fires can be ignited by lightning strikes. If it happens frequently, the regeneration rate is affected and the nature of the land cover can change permanently. Parasitic insects can attack and even kill trees. Bark beetles bore through the bark of trees to access nutrients in the inner layers. If they eat all the way around the tree, they will kill it because the tree cannot send nutrients up and down the trunk. Defoliating insects eat the green needles and leaves. Mistletoe-like plants (clumpy, parasitic growths) rob nutrients from the host tree, weakening it and letting root diseases attack the underground root systems. Extreme weather events can physically damage trees and prolonged flooding can cause death.

Activities

Acquiring and processing geographical information

1. Explain why forests are critical to life on Earth.
2. Distinguish between primary and secondary forests. What is a plantation?
3. Identify the three principal types of forest. Describe the characteristic features of each.
4. Explain why tropical and subtropical rainforests are so important.
5. Explain the relationship between the photosynthesis of forest trees and plants and global climate.
6. Outline the natural causes of deforestation.

Applying and communicating geographical understanding

7. Study Figure 5.1.1. With the aid of a world map, identify the parts of the world with the densest tree cover.
8. Study Figure 5.1.2. Using data from the graph, describe the global distribution of the world's forests.
9. Study Figure 5.1.3. Outline the relationship between temperature, moisture and forest type.
10. Study Figure 5.1.4. Write three sentences describing the global distribution of tropical and subtropical rainforests, temperate forests and coniferous (boreal) forests.
11. Study Figures 5.1.5, 5.1.6 and 5.1.7. Describe the nature of the landscape shown in each photograph. Note the level of diversity evident and any information that indicates the climate present in the region.

UNIT 5.2

Deforestation and its impacts

Deforestation involves the conversion of the existing tree-based land cover into non-forest land uses, including subsistence farming, commercial agriculture, animal grazing, mining, timber harvesting and urban expansion (see Figure 5.2.1). In the five years from 2015 to 2020, the annual extent of deforestation was an estimated 10 million hectares, down from 12 million hectares in 2010–2015 and 16 million hectares in the 1990s. An estimated 420 million hectares of forest have been lost worldwide through deforestation since 1990. But not all forest resource exploitation is damaging. Appropriately managed, selective logging is sustainable, but clear-felling can destroy whole ecosystems and the habitat on which species depend.

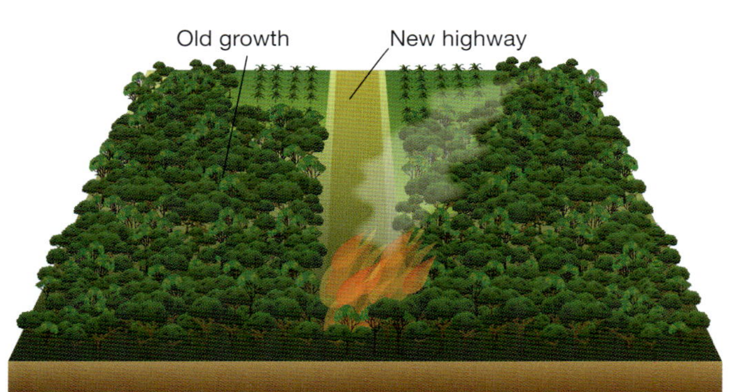

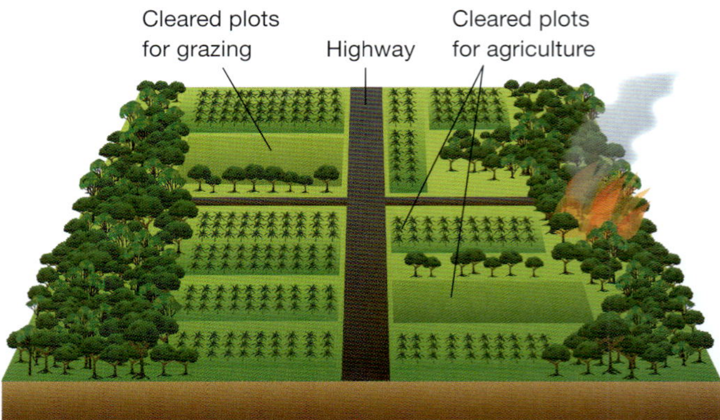

5.2.1 Building roads into previously inaccessible forests is often the first step in the harvesting of forest resources. It paves the way to forest fragmentation, destruction and degradation of forest ecosystems.

Forest fragmentation is the division of forests into smaller and more isolated fragments. Examples include clearing forests for agriculture, transport infrastructure, power and pipeline corridors, and urban subdivisions. Over time, the edges of each remnant become degraded and may become too small and too isolated from other fragments to support viable forest ecosystems, ultimately resulting in deforestation. Forest fragmentation and the clear-felling of forests have profoundly altered the characteristics and connectivity of forests and caused severe biodiversity losses.

While some forests are cleared, others are expanding (typically on land previously used for agriculture), a process referred to as afforestation. So, while the global loss of forest is around 10 million hectares per year, the net annual loss is less. The net change in forest cover is calculated by deducting the increase in forested land from the area that has been deforested.

In 2010–2020 the average annual net loss was 4.7 million hectares (down from 7.8 million hectares per year in 1990–2000 and 5.2 million hectares a year in 2000–2010). Africa had the largest annual rate of net forest loss in 2010–2020, at 3.9 million hectares, followed by South America, at 2.6 million hectares. Asia had the highest net gain of forest area in 2010–2020 at 1.2 million hectares, followed by Oceania and Europe. Significantly, both Europe and Asia recorded substantially lower rates of net gain in 2010–2020 than in 2000–2010. Despite this decline in the rate of deforestation, more than 80 million hectares of primary forest were lost from 1990 to 2021. See Figures 5.2.2 to 5.2.4.

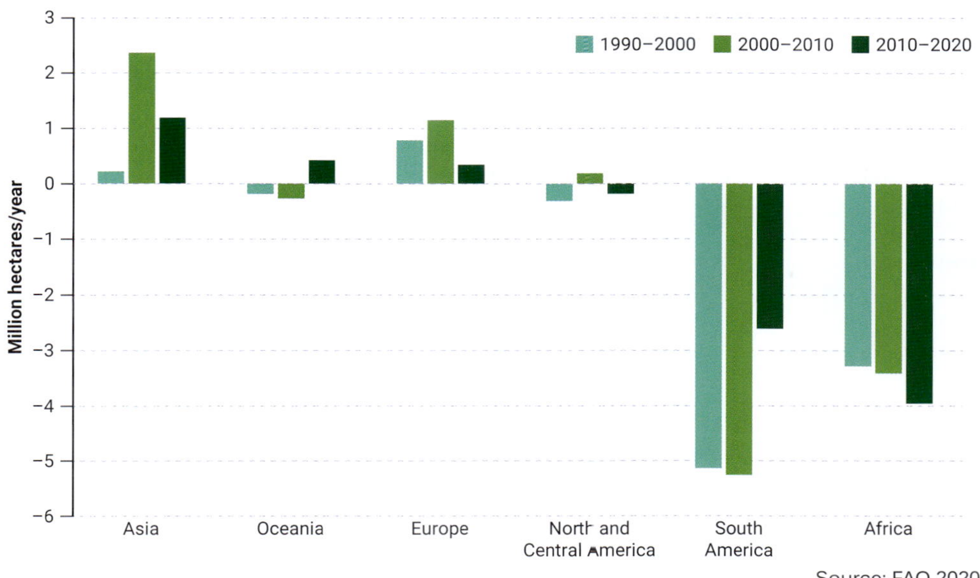

5.2.2 Net forest area change by region, 1990–2020 (million ha per year)

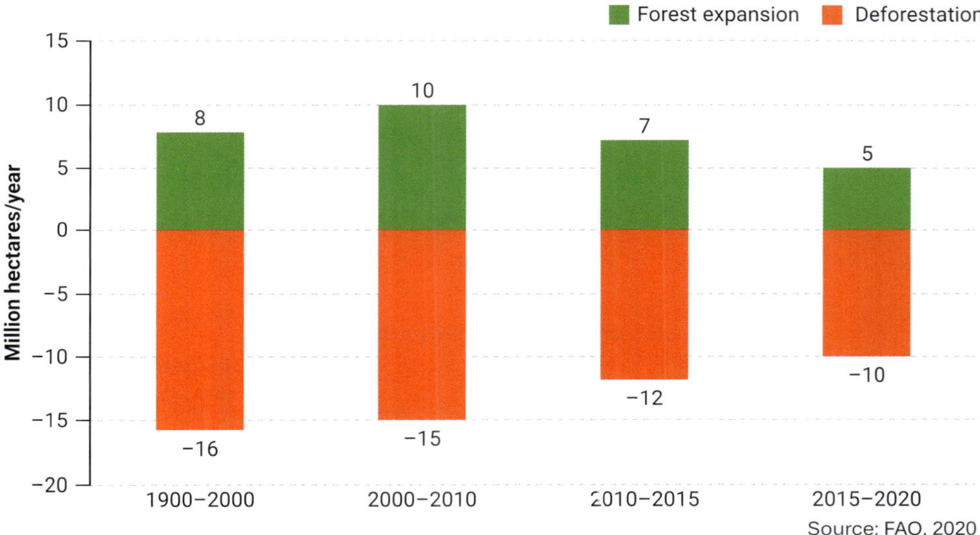

5.2.3 Global forest expansion and deforestation, 1990–2020 (million ha per year)

Causes of deforestation

Agricultural expansion is the principal driver of deforestation and forest fragmentation and associated biodiversity forest loss. Agriculture accounted for 73 per cent of tropical forest loss between 2000 and 2010.

Deforestation and **forest degradation** are driven by many political and socioeconomic forces interacting at global to local levels. Population growth, demographic trends and economic development are the principal drivers of environmental change, including deforestation. Since the 1970s, the global population has doubled and the global economy has almost quadrupled. It lifted billions of people out of poverty, but significantly altered the environment through habitat change, loss and degradation, unsustainable agricultural practices, invasive species, and resource overexploitation, including illegal logging and wildlife trade.

Global market pressures, dietary preferences and waste throughout agricultural supply chains drive demand for products, which drives deforestation and degradation. In Africa, population pressure and poverty are the principal threats. Survival needs drive people to convert forests to cropland and to harvest fuel wood at unsustainable levels. Elsewhere, affluence drives deforestation through rising consumption patterns.

The UN Food and Agricultural Organisation (FAO) found that large-scale commercial agriculture (primarily cattle ranching and soya bean and oil palm cultivation) is the most significant driver of deforestation, at 40 per cent from 2000 to 2010. Local subsistence agriculture accounts for roughly 33 per cent, urban expansion (10%), infrastructure (10%) and mining (7%). Sometimes land-use change was preceded by forest degradation, such as unsustainable or illegal wood removal. Specific drivers differ significantly between regions (see Figure 5.2.4).

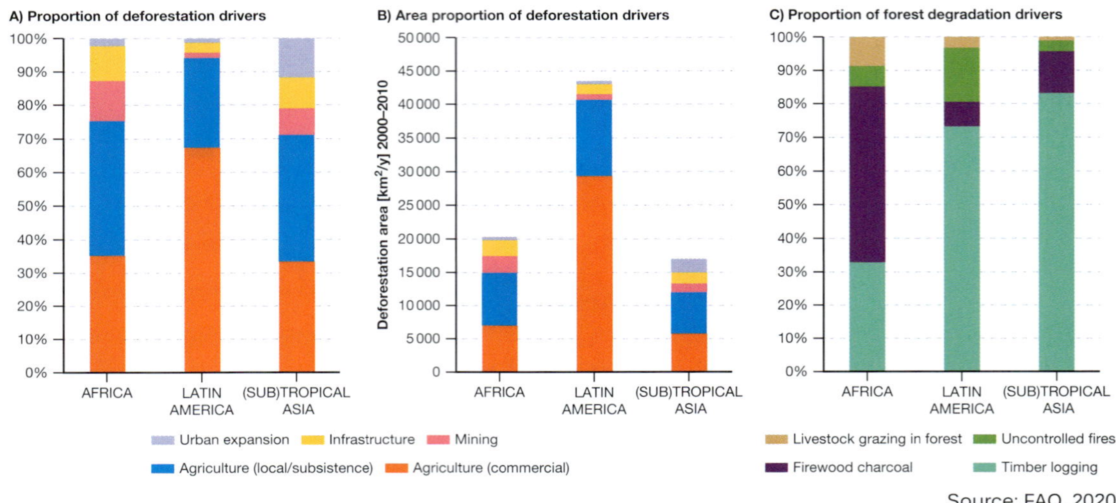

Source: FAO, 2020

5.2.4 Drivers of deforestation and forest degradation by region, 2000–2010

State of the world's tropical rainforests, 2020

Figure 5.2.5 summarises the state of the world's tropical rainforests in 2020.

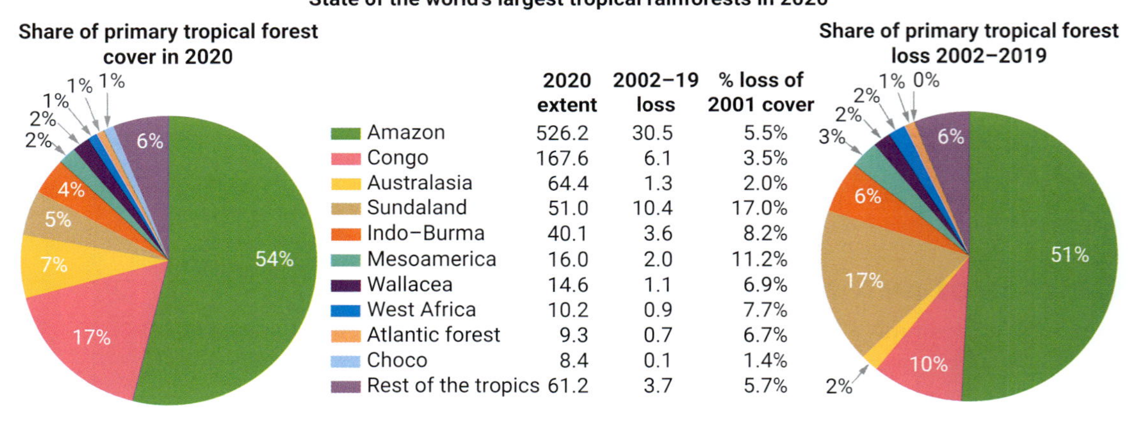

	2020 extent	2002–19 loss	% loss of 2001 cover
Amazon	526.2	30.5	5.5%
Congo	167.6	6.1	3.5%
Australasia	64.4	1.3	2.0%
Sundaland	51.0	10.4	17.0%
Indo–Burma	40.1	3.6	8.2%
Mesoamerica	16.0	2.0	11.2%
Wallacea	14.6	1.1	6.9%
West Africa	10.2	0.9	7.7%
Atlantic forest	9.3	0.7	6.7%
Choco	8.4	0.1	1.4%
Rest of the tropics	61.2	3.7	5.7%

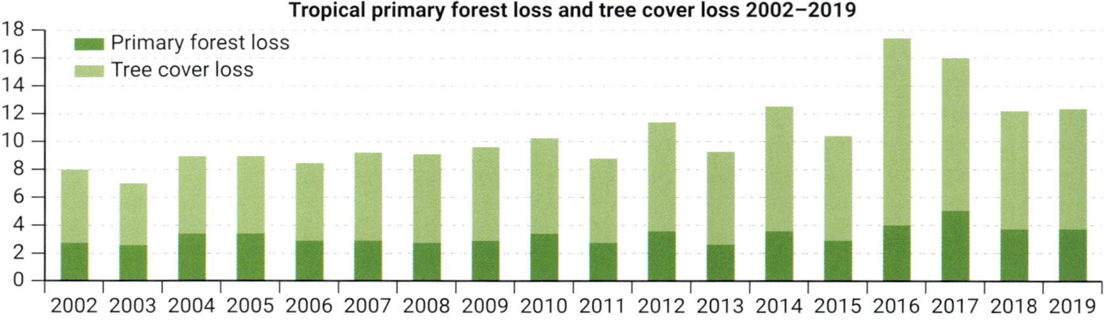

Source: FAO, 2020

5.2.5 The state of the world's tropical rainforests, 2020

Deforestation in Australia

Around 44 per cent of Australian forests and woodlands have been cleared since 1788. This reflects the more recent history of land clearing and agricultural land uses, plus the legacy of over 65 000 years of Indigenous land management practices.

The least-modified vegetation cover is in the north and centre of Australia, along with the eastern and south-western ranges of the mainland and the eastern ranges and south-western Tasmania (see Figure 5.2.6). In these places, about 80 per cent of vegetation is classified as residual or modified. Conversely, the most-modified or replaced vegetation is in the intensive-use zones of the eastern and southern mainland, and the midlands and north of Tasmania. In these zones, an average of only 40 per cent (range 15–69%) of vegetation is classified as residual or modified. In both instances, the vegetation's regenerative capacity remains intact.

5.2.6 Clear-felling of native eucalypt forest in Tasmania's Derwent Valley

Land clearing and fragmentation pose the greatest threat to biodiversity. Of the 1250 plant and 390 terrestrial animal species listed as threatened, 964 plants and 286 animals have deforestation and resulting habitat fragmentation or degradation identified as a threat.

Land clearing rates are broadly stable or decreasing everywhere except Queensland. There, relaxing the tree-clearing legislation led to an increase in clearing rates of both remnant and non-remnant vegetation. In the period 2010–2018, the total area of land cleared in Queensland was 2 446 600 hectares. Over 90 per cent of cleared Queensland forest was replaced by pasture from 2016 to 2019 (see Figure 5.2.7). From 2001 to 2021, Australia lost 8.73 million hectares of tree cover, equivalent to a 21 per cent decrease in tree cover since 2000. Fortunately, the rate of loss is slowing and natural regrowth means there has been a net increase in vegetation over recent years.

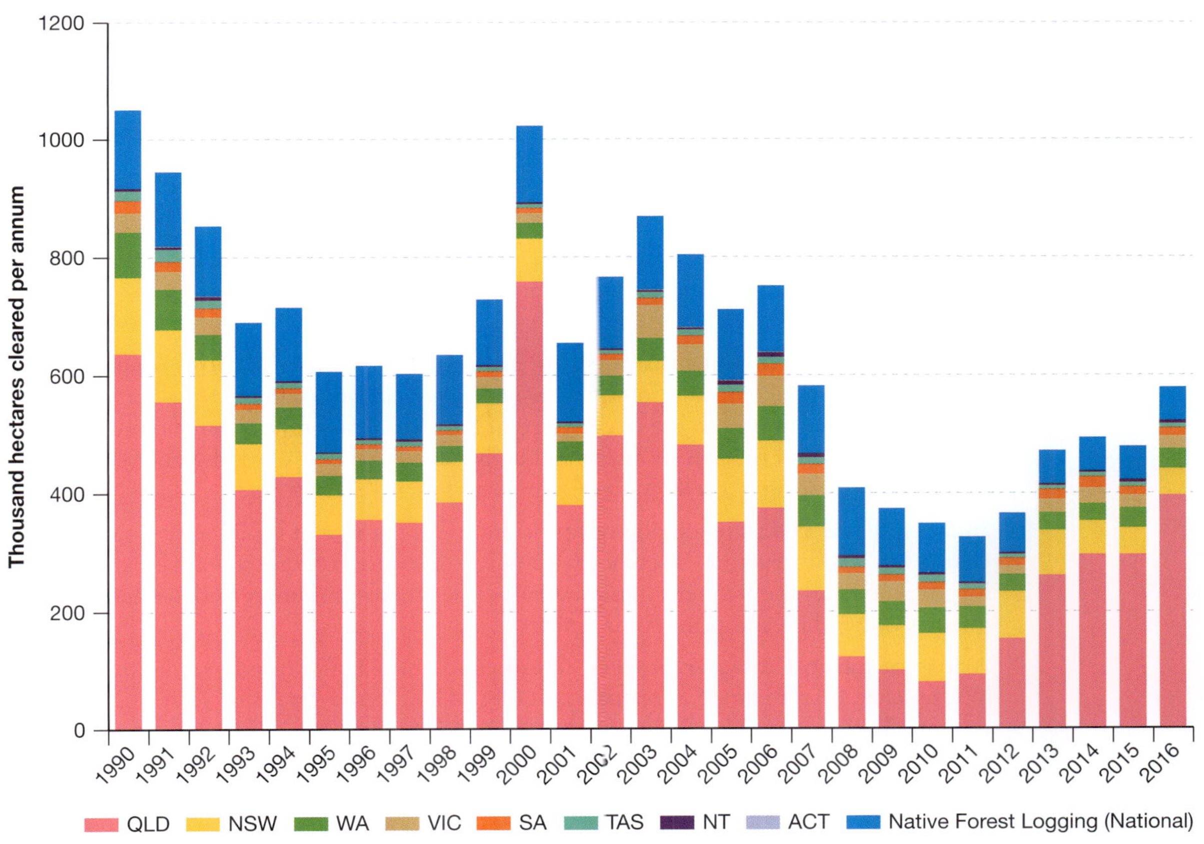

5.2.7 Deforestation and land clearing in Australia, 1990–2016

Impacts of deforestation and land clearing in Australia

Deforestation and land clearing in Australia have a range of impacts.

Animal endangerment and extinction: tens of millions of animals are killed every year, and species are being driven to extinction, by deforestation and land clearing. Amphibians and fish are also affected as rivers and wetlands are polluted and damaged.

Wiping out plant communities: whole plant communities are threatened by deforestation and land clearing. Almost all of Australia's threatened plant communities are found in areas of high current or historic deforestation and land clearing.

Climate change: deforestation and land clearing impacts climate change in two ways. Firstly, the removal of vegetation results in a decline in the drawdown of carbon dioxide from the atmosphere. Secondly, when burned, forests release large amounts of stored carbon and other greenhouse gases into the atmosphere. The World Wildlife Fund for Nature (WWF) estimates that the greenhouse gas emissions from deforestation and land clearing in Australia are equal to about half of all the country's coal-fired power station emissions.

Dryland salinity: when vegetation is removed, water that was once absorbed by the root systems of trees penetrates deeper into the soil profile where it accumulates. As it does this, it raises the water table. As the water rises, it brings dissolved salt towards the surface. If the accumulation is large enough, salt scars develop and most vegetation can't grow. Across Australia, millions of hectares of land, and tens of thousands of farms, have been affected by salinity. By 2050, it's projected that more than 17 million hectares of land will have been degraded by salinity.

Soil erosion: when vegetation is removed, soil is exposed to erosion by wind and running water. This impacts the fertility of soils and the prospects of forest cover restoration.

River system pollution: the erosion of soil and increase in run-off negatively impacts water quality, temperature, sediment and nutrient loads of waterways, lakes and wetlands. In doing so, it changes the physical properties of water bodies and the habitat they provide, threatening the species that depend on clean water.

Pollution of the Great Barrier Reef and other marine environments: there is a direct link between deforestation and the quality of water flowing into marine environments. Sediment, pesticides and nutrient load all impact fragile ecosystems like the Great Barrier Reef, where they promote algae growth that adversely affects corals. Pesticides can leach into river systems and then into the sea, where it affects marine life.

Invasive species incursion: the clearing of native vegetation can result in weed invasions, and the fragmentation of habitat encourages the spread of animal predators such as cats and foxes.

Increased frequency and duration of droughts: forests, as with other types of vegetation cover, take up and transpire moisture into the atmosphere, together with heat. This process contributes to the occurrence of localised cloud formation and rainfall. The clearing of the land reduces moisture and increases albedo (see Units 3.3 and 4.4). The net result of clearing is a reduction of rainfall and increased heating.

Increased fire risk: the logging of forests increases the rate of localised drying and changes in forest structure. Together with heightened fire risk due to climate change, logging increases the frequency and severity of destructive bushfires with its devastating impacts on wildlife and plant communities.

A loss of natural disaster buffers: forests can act as a buffer to natural disasters. Within river catchments, for example, the loss of land cover results in more and quicker run-off. The incidence and severity of flooding increases as a result.

Diminished sources of food and medicine: Australia's native plants are a rich source of indigenous medicine and food. These plants are also studied in some contemporary medical research. Deforestation depletes these resources and opportunities for future medical discovery.

Loss of amenity: deforestation robs people of places to engage with nature. Forests can inspire people, and, in the case of those Aboriginal and Torres Strait peoples whose Country is located in forest regions, can also form an important source of spiritual enrichment and identity. They are also important recreational spaces.

Loss of cultural heritage: Aboriginal and Torres Strait Islander peoples have strong and enduring personal and cultural connections to Country. This includes Australia's forests, the loss of which can destroy sites of cultural significance.

SPOTLIGHT

Deforestation in Borneo's rainforests

Borneo's rainforests are estimated to be at least 130 million years old. Of great ecological importance, it is home to 15 000 species of plants (including 3000 trees,) 420 birds and 222 mammals. Many are only found in Borneo, including Borneo pygmy elephants, proboscis monkeys, black shrews and Bornean orangutans. It is estimated that only 50 000 Bornean orangutans live in the wild, making it one of the most critically endangered species on Earth (see Figure 5.2.8).

5.2.8 Bornean orangutan mother with her infant

Deforestation in Borneo is widespread and largely involves the use of fire (see Figure 5.2.9). Rainforest fires have become an annual event. In the Indonesian state of Kalimantan, illegal fires that raged during 2019 spread acrid smoke as far as West Malaysia and Singapore. The use of fire is driven by a growing demand for lucrative palm oil, which is used in products ranging from chocolates to shampoo. The money that farmers earn from palm oil is four times what they would earn from growing rice or producing rubber. See Figure 5.2.10.

Land clearing by fire also destroys the peatlands that lie beneath it, one of the world's largest natural terrestrial 'carbon sinks'. Natural forest fires in Borneo are rare. Peat contains partially decayed plant material that soaks up excess water during the monsoon season and stays damp during the dry season, preventing fires.

Clearing forests by fire devastates local biodiversity, so the number of threatened plants and animals is rising, threatening the island's biodiversity heritage.

Though most are deliberatively lit, global warming has increased the frequency and intensity of all fires. Losing forest cover diminishes its capacity to sequester, or store, carbon and produce oxygen, and reduces the flow of moisture into the atmosphere.

Continued on next page

Island of Borneo

Map location

Kota Kinabalu

Brunei

South China Sea

Sulawesi Sea

MALAYSIA

Kuching

Pontianak

INDONESIA

Samarinda

Sulawesi

Banjarmasin

Java Sea

N

Forest cover loss on Borneo

- Up to 1900
- 1900–2000
- 2000–2020
- Forest cover left in 2020 (projection)
- International border

0 100 200 300 km

Source: WWF

5.2.9 Borneo rainforest loss 1900–2020

5.2.10 A palm oil plantation on land cleared of rainforest

Activities

Acquiring and processing geographical information

1. Define deforestation. What does it involve?
2. What was the amount of forest lost in the period 2015–2020? What trend is evident?
3. Explain the process of forest fragmentation. How does it contribute to the loss of biodiversity?
4. Explain what afforestation is. To what extent does it offset forest loss?
5. Outline the principal causes of deforestation.

Applying and communicating geographical understanding

6. Study Figure 5.2.1. Explain how the construction of roads in forest areas often results in forest fragmentation, destruction and degradation.
7. Study Figure 5.2.2. Using data from the graph, outline the net forest loss being experienced by different regions. Where, for example, is the rate of deforestation increasing and decreasing?
8. Study Figure 5.2.3. Using data from the graph, describe the global trends in forest expansion and deforestation.
9. Using data provided in the text under the heading, 'Causes of deforestation,' construct a pie graph showing the relative contributions of commercial agriculture, urban expansion, infrastructure and mining to deforestation.
10. Study Figure 5.2.4 and then answer the following questions:
    - a In which regions does commercial agriculture make its largest and smallest contributions to deforestation?
    - b In which region does the collection of fuel wood make its greatest contribution to forest degradation?
    - c In which region does timber logging make its largest proportional contribution to forest degradation?
11. Study Figure 5.2.5. Drawing on the data presented, write a report outlining the state of the world's tropical rainforests in 2020.
12. Study the text under the heading, 'Deforestation in Australia', and complete the following activities.
    - a List the factors that have determined the current distribution and state of Australia's forested lands.
    - b Study Figure 5.2.7. Write an extended response summarising the impacts of deforestation in Australia.
13. Study the box, Spotlight: Deforestation in Borneo's rainforests and complete the following activities.
    - a Explain what is special about the rainforests of Borneo.
    - b Outline the factors driving deforestation in Borneo.
    - c Explain why the burning of Borneo's peatlands is so environmentally damaging.
    - d Describe the impact of climate change on the rainforests of Borneo.
    - e Study Figure 5.2.9. Compare the cover in 2020 against that of 2000 and 1900.

UNIT 5.3

Sustainable management of the world's forests

> **Did you know?**
> Sustainability is an approach to environmental management that focuses on meeting the needs of the present without compromising the ability of future generations to meet their needs. It is both a goal and a way of thinking. Progress towards environmental sustainability depends on the maintenance or restoration of both economic and social environmental functions that sustain all life and human wellbeing.

Efforts to combat deforestation have gathered pace over the past few decades, primarily because of a growing appreciation that forest loss and the use of fire to clear land is negatively impacting the global carbon cycle. Slowing, if not reversing, the rate of deforestation is seen as critical to limiting the warming of the planet to well below 2°C, and preferably to 1.5°C, compared to pre-industrial levels.

Applying the principle of sustainable development is critical to the successful management of the world's forest resources. An integrated approach to land use planning, within a sustainable framework, provides a strategy for balancing competing land uses. The development of such strategies requires meaningful participation from stakeholders. This ensures that strategies are accepted, and that stakeholders will implement and monitor the plans.

Actions designed to address the issue of deforestation

The establishment of protected forest areas

The global network of protected areas, such as national parks, nature reserves, conservation reserves and game reserves, has expanded rapidly in recent decades, reaching almost 240 000 designated protected areas. Collectively, these areas now protect just over 2 billion hectares, equivalent to 15 per cent of Earth's land surface. Thousands of these protected areas are specifically designed to protect forests.

Globally, 18 per cent of the world's forest area, or more than 700 million hectares, is found within legally established protected areas. The largest share of forest in protected areas is found in South America (31%) and the lowest in Europe (5%). See Figures 5.3.1 and 5.3.2.

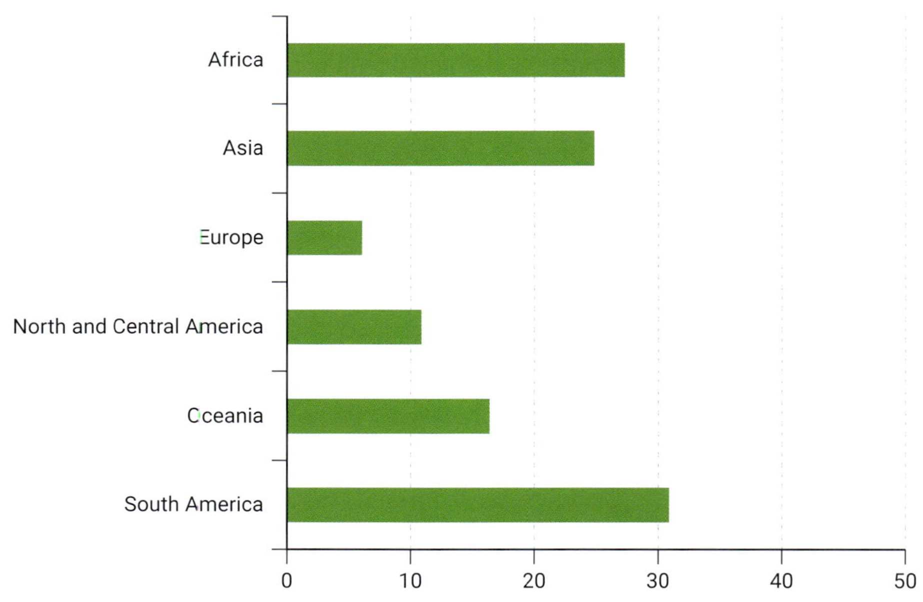

Note: Data for Europe include Russia. If Russia is excluded, 18 per cent of Europe's forest area is in protected areas.

Source: FAO, 2020

5.3.1 The percentage of forests in legally protected areas

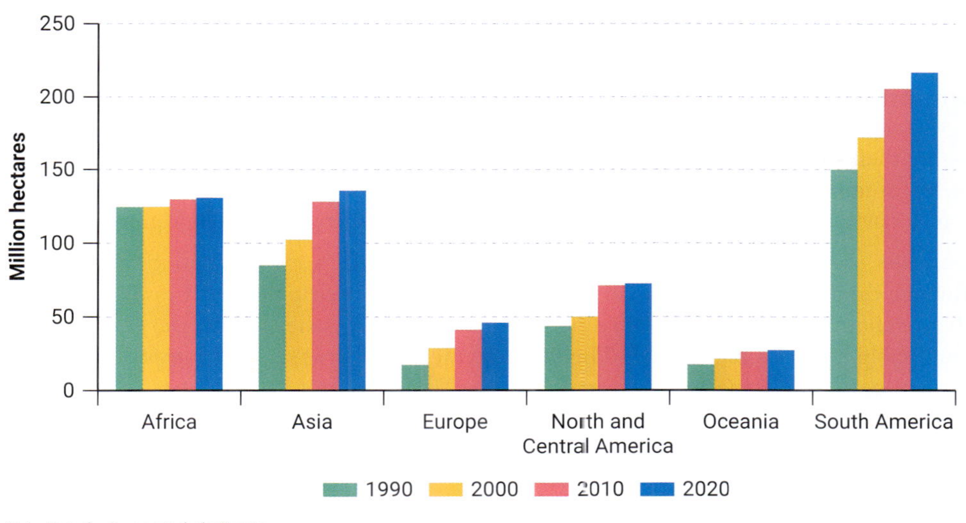

Source: FAO, 2020

5.3.2 Areas of forest within protected areas by region, 1990–2020

Did you know?

Sydney's 151-square-kilometre Royal National Park, the second oldest national park in the world, gained protection in 1879. The first designated forest reserve, the Marakele Forest Reserve in Sri Lanka, was established in 1875.

To be effective, protected areas need to be large enough to allow ecological processes to be sustained indefinitely without being negatively affected by activities in surrounding areas. They must be managed in ways that meet both the interests of local communities and the demands of biodiversity conservation, which can be conflicting. For example, where the main drivers of deforestation are subsistence agriculture and fuel wood harvesting, the development of alternative forest-based livelihoods is often necessary. This might involve the development of sustainably produced forest products and services (see Figure 5.3.3).

5.3.3 In protected forest areas, local communities need to be supplied with alternative sources of energy if their impact on the forest is to be minimised.

Another strategy to maintain biodiversity is the retention of biological (or habitat) corridors. A habitat corridor is an area of habitat that connects wildlife populations separated by human activities. These, often narrow, forest corridors allow for an exchange of individuals between populations, which may help prevent the negative effects of inbreeding and reduced genetic diversity that often occur within isolated wildlife populations. Corridors may also help facilitate the re-establishment of populations that have been reduced or eliminated due to events such as fires or disease outbreaks.

Non-government organisations

Non-government organisations (NGOs) have been central to efforts to save the world's forests. NGOs serve as watchdogs and agents of change. They hold governments and businesses to account by facilitating collective action and community engagement. They have become skilled at capacity-building, networking and partnerships. Some of the best known and successful forest-related NGOs are the Global Forest Coalition, Rainforest Action Network, World Wildlife Fund (WWF), Rainforest Trust and Conservation International.

By bringing the plight of the world's forests to the public's attention, often through the skillful use of the media, NGOs have been successful in shaping public discourse and debate (see Figure 5.3.4). This has forced governments and businesses to act.

5.3.4 A protest in Brazil to denounce indiscriminate burning in the Amazon

Consumer and shareholder activism

Environmental groups and other NGOs have worked to 'name and shame' corporations that profit from deforestation and land clearing. By pushing for consumer boycotts and putting pressure on shareholders, they have helped modify corporate behaviour. Certification schemes have also identified products that meet certain environmental and human rights standards.

Consumer-based approaches encourage customers and shareholders to boycott products from illegally cleared land. NGOs also pressure importing firms and manufacturers, which, in turn, pressure their suppliers to prove their products do not come from deforested land. They may develop or join an internationally recognised forest certification standard. Such initiatives advance environmental standards, labour rights, workplace safety and human rights.

Consumer pressure on large beef importers, like McDonald's, encouraged them to influence the Brazilian beef ranching industry to change their practices. Importers may enhance their reputation by purchasing sustainable beef via schemes like the Global Roundtable for Sustainable Beef. The Chinese meat industry is a major importer of Brazilian beef and China has partnered with the WWF to protect the Amazon forests.

Government responsibilities

Governments are forest regulators and often large-scale forest owners. Their major role is to create conditions that ensure all forests are sustainably managed. Success means establishing clear processes for land-use planning and decision-making that involve community and corporate stakeholders.

Governments need to enforce forest-related laws and regulations. Illicit exploitation and trade in timber is a global problem with serious implications for biodiversity conservation and ecosystem services. Illegal activities include harvesting, transport, processing, purchasing or selling forest products. The driving forces are complex, but include inadequate legal protection, absent law enforcement, corrupt local officials and a limited capacity to develop and implement effective land use plans. Countries importing timber products without ensuring they are legally sourced are encouraging illegal land clearing.

Demand for timber commonly drives illegal logging, especially high-value species such as mahogany. However, it can also result from clearing land for plantation commodities like oil palm and soya beans.

Efforts to combat illegal logging have been spearheaded by trade regulations in consumer countries. Importers must be able to demonstrate that their timber was harvested legally. Australia's illegal logging laws, similar to the European Union and the USA, were designed to combat the trade in illegally sourced timber. Its framework includes the *Illegal Logging Prohibition Act 2012* (the Act) and the *Illegal Logging Prohibition Regulation 2012* (the Regulation). The Act makes it a criminal offence to import illegally sourced wood, pulp and paper products intentionally, knowingly or recklessly into Australia. The Regulation outlines the due diligence process that businesses must undertake. It requires importers of regulated timber products and Australian processors of raw logs to minimise the risk that the wood has been illegally logged.

International initiatives

There are several notable international initiatives designed to slow the rate of deforestation.

The Forest Carbon Partnership Facility (FCPF): commonly referred to as REDD+ is an incentive-based scheme that helps developing countries to access funding, principally from the Green Climate Fund, for verified results in:

- reduced emissions from deforestation and forest degradation
- the sustainable management of forests
- the conservation and enhancement of carbon stocks.

To date, 50 countries have submitted a baseline of emissions against which they can measure their progress in reducing emissions from deforestation and forest degradation. These countries represent more than 30 per cent of the global forest area and more than 70 per cent of the global loss of forests. As of January 2020, nine countries have reported reductions of 8.82 billion tonnes of emissions due to reduced rates of deforestation and forest degradation. The UN-REDD Programme, an initiative of FAO, the United Nations Development Programme and the United Nations Environment Programme, supports the nationally-led REDD+

processes (see Figure 5.3.5). It promotes the engagement of all stakeholders, including indigenous peoples and other forest-dependent communities, implementing REDD+.

The New York Declaration on Forests (2014) is a voluntary (non-binding) international agreement to halt global deforestation. It has been endorsed by more than 200 bodies, including governments, transnational corporations, groups representing indigenous communities and NGOs. Importantly, it specifically includes commitments by the private sector to eliminate deforestation from the supply chains of major agricultural commodities.

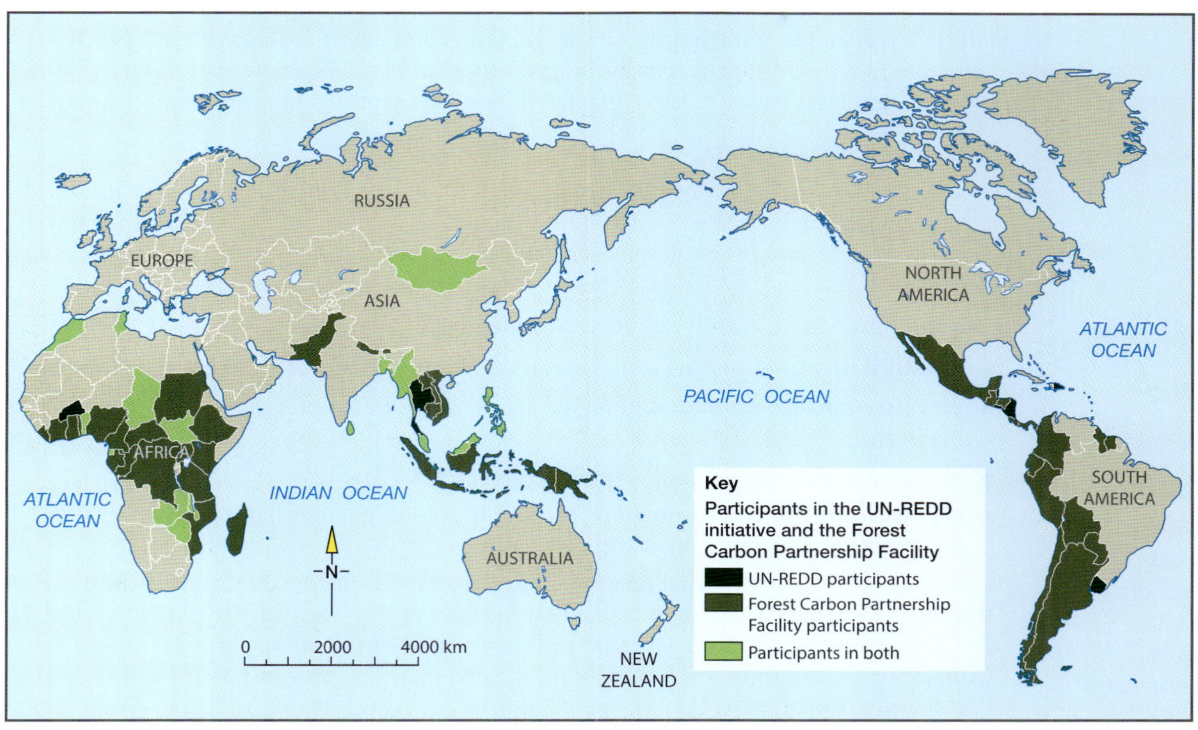

5.3.5 Countries participating in the UN-REDD initiative and the FCPF (REDD+)

● **SPOTLIGHT**

Cocoa: Towards a deforestation-free commodity chain

Almost 70 per cent of the world's supply of cocoa originates from West Africa where smallholder farmers rely on cocoa as their major cash earner (see Figure 5.3.6). Historically, cocoa has been an important driver and direct cause of deforestation. The expansion of cocoa growing into forests is often driven by declining cocoa yields from established plantations. The soils of freshly cleared forested land are often more fertile.

Governments in the region, together with private operators, have made a range of commitments to end deforestation in cocoa supply chains. They aim to safeguard biodiversity and ecosystem services while avoiding revenue loss and impacts on local livelihoods. Public–private initiatives, such as the Cocoa & Forests Initiative of Ghana and Côte d'Ivoire and the Green Cocoa Landscape Programme in Cameroon, aim to support the sustainable intensification and climate resilience of cocoa production, the prevention of further deforestation and the restoration of degraded forests. These initiatives are aligned with national REDD+ policies and plans.

5.3.6 Women preparing cocoa beans to be dried in Côte d'Ivoire.

The role of agribusiness

Growing international pressure has encouraged an increasing number of agribusinesses (agriculture-based businesses) to commit to zero deforestation. Commodity investors are also encouraged to adopt business models that are environmentally and socially responsible, and that involve and benefit local producers, distributors and other contributors to the commodity value chain. This can be achieved through the joint design of sustainable land use plans on corporate land.

In places where large-scale commercial agriculture is the principal cause of deforestation, effective regulation, with appropriate social and environmental safeguards, is needed. Initiatives, such as voluntary certification schemes and commitments to zero deforestation, also have a positive impact. These are examined in more detail below.

Forest restoration

The UN's *Sustainable Development Goals Report 2019* found that 20 per cent of Earth's surface was in a degraded state. To address the issue, the UN declared 2021 to 2030 the Decade on Ecosystem Restoration. The aim is to prevent, halt and reverse ecosystem degradation and raise awareness of the importance of ecosystem restoration.

Forest restoration can involve a range of strategies. They include:

- **rehabilitation:** restoration of the species mix and structure of an existing ecosystem
- **reconstruction:** restoration of native plants on land used previously for other purposes
- **reclamation:** restoration of severely degraded land devoid of vegetation
- **replacement:** the replacement of existing species with a new species better able to cope with a changing climate.

Did you know?

The State of the World's Forest report 2020, noted that there are 1.7 billion to 1.8 billion hectares of potential forest land in areas that have been previously degraded, dominated by sparse vegetation, grasslands and degraded bare soils.

Land restoration initiatives include the Latin American Initiative 20×20 (launched in 2014), which aims to restore 50 million hectares of degraded land by 2030. As of 2023, there were more than 100 land restoration and forest conservation projects in Latin America and the Caribbean. Another example is the African Forest Landscape Restoration Initiative (AFR100), launched in 2015. AFR100 aims to bring 100 million hectares of degraded land under restoration by 2030.

Natural regeneration

The natural regeneration of forests can be assisted and managed by humans. It includes forest restoration, **reforestation** and afforestation (see Figures 5.3.7 and 5.3.8). Assisted natural regeneration (ANR) involves any set of interventions aimed at enhancing and accelerating the natural regeneration of native forests.

5.3.7 Forestry worker and volunteer plant trees as part of reforestation efforts, Okayama, Japan

ANR is a relatively simple and inexpensive approach to forest regeneration. It involves removing or reducing barriers to natural succession. It often involves the re-establishment of trees from seeds adapted to local soil and climate conditions. It may also involve the encouragement of pollinators, herbivores and seed-dispersal agents of colonising tree species.

Human interventions may also include protecting an area from disturbances such as fire, stray domestic animals and human activities such as trail bike riding, and reducing competition from grasses and other vegetation forms that hinder the growth of naturally regenerating trees.

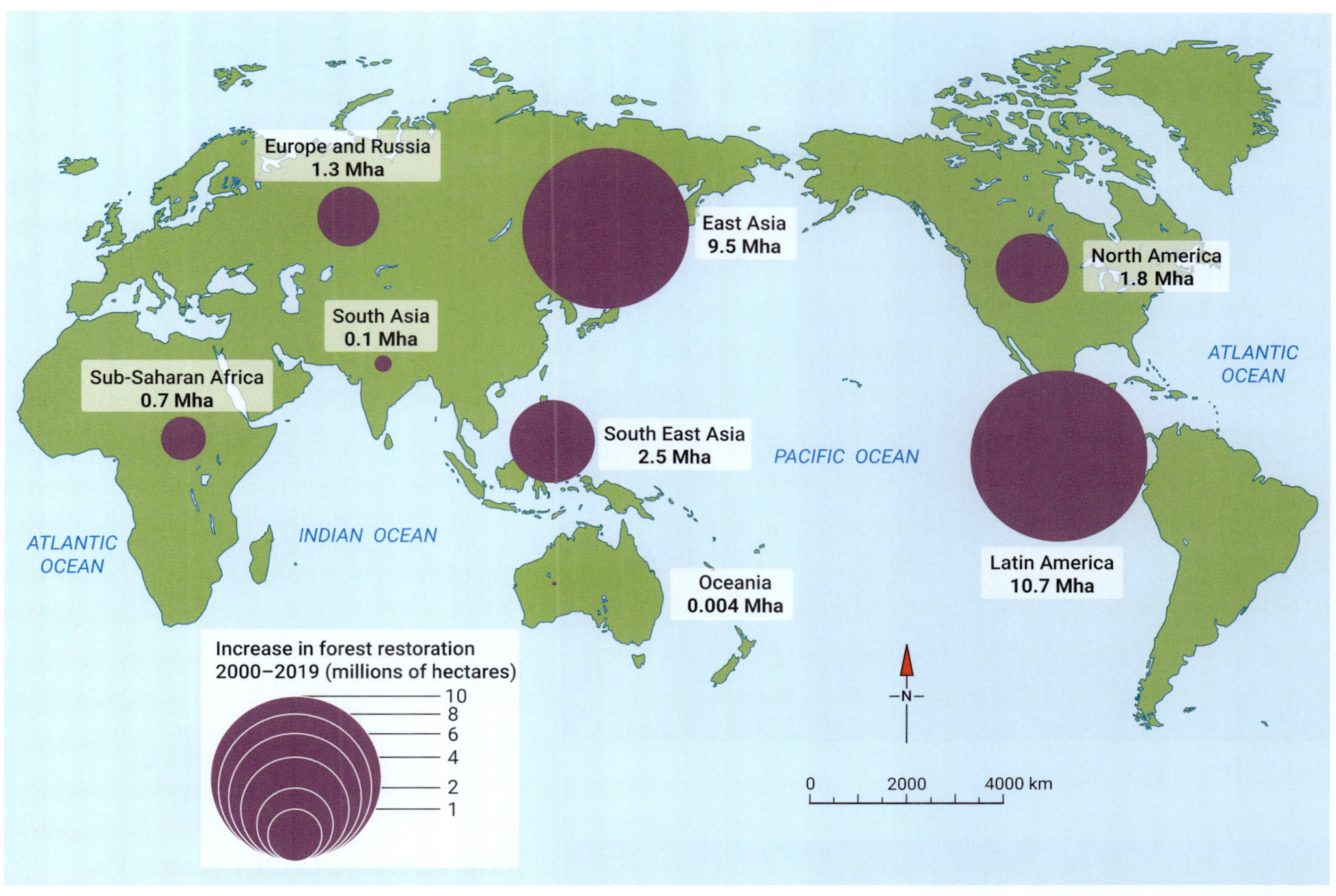

Source: FAO, 2020

5.3.8 Increase in forest area through forest restoration, reforestation and afforestation, 2000–2019

Activities

Acquiring and processing geographical information

1. Explain what is driving the effort to slow down the rate of deforestation.
2. Name the criteria which must be met for a protected area to be effective in maintaining biodiversity.
3. Explain why biological (habitat) corridors are important.
4. Outline the role and characteristics of NGOs in terms of their efforts to protect the world's forests.
5. Explain how consumer and shareholder activism has proved to be an effective way of battling deforestation.
6. Outline the role that governments play in protecting the world's forests.
7. Explain how agribusiness can contribute to a slowing rate of deforestation.
8. Outline the ways in which the natural regeneration of forests can be assisted by humans.

Applying and communicating geographical understanding

9. Explain the concept of sustainability. Outline what is required to achieve sustainable development.
10. Study Figure 5.3.1. Using data from the graph, describe the global distribution of protected forest areas.
11. Study Figure 5.3.2. Identify the regions which have seen the greatest rate of increase in protecting forest areas since 1990. Which have shown the lowest rate of increase?
12. Study the box, Spotlight: Cocoa: Towards a deforestation-free commodity chain. Outline how a deforestation-free commodity chain has been achieved within the cocoa industry.
13. Study Figure 5.3.8. Identify the regions of the world that have the best record of forest restoration, reforestation and afforestation.

UNIT 5.4

Deforestation in the Amazon Basin

5.4.1 Amazon rainforest, Brazil

5.4.2 Amazon jaguar is just one of the 427 mammal species known to live in the Amazon forest.

The Amazon Basin of South America is vast (see Figure 5.4.1). It covers about one-third of South America, spans eight countries and features a mosaic of intersecting and overlapping ecosystems. The Basin covers at least 6 million square kilometres (an area nearly twice the size of India) and is home to Earth's largest rainforest (See Figure 5.4.3). The forest, which covers about 80 per cent of the Basin, supports one-fifth of the world's terrestrial species, including many found nowhere else on Earth (see Figure 5.4.2). It is also home to more than 30 million people, including hundreds of indigenous groups. The Amazon is drained by the world's largest river, as measured by the volume of flow and the size of its drainage basin.

Despite its vast size, and its importance to the planet, there is still much to learn about the complexity of its contribution to Earth's ecosystem services. Given that it is largely surrounded by mountainous plateaus, much of the basin is remote and difficult to access.

The Amazon rainforest's environmental services

The importance of the Amazon rainforest and the environmental services it provides extend well beyond the borders of Brazil. The rainforest is an enormous 'carbon sink'. It draws down carbon from the atmosphere and sequesters it in the soil and plants of the forest. Additionally, the plants, transform un-sequestered carbon dioxide into oxygen, which is pumped back into the atmosphere. Up to 20 per cent of Earth's oxygen originates from the Amazon forests. Without this sink and oxygen-producing capacity, global atmospheric carbon dioxide concentration would increase more rapidly than it is, resulting in even higher temperatures. When the forest burns, the stored carbon dioxide is released into the atmosphere, thereby adding to the warming of the planet.

The Amazon rainforest also pumps vast quantities of moisture into the air by transpiration. About half of this moisture falls as rain within the basin. The rest travels to other parts of South America via 'atmospheric rivers' and contributes to precipitation in these areas. As the forest is lost or fragmented, this transfer of moisture is disrupted, impacting rainfall patterns across the continent.

● **SPOTLIGHT**

Land cover change in the Amazon

Over the past four decades, an extraordinary change has taken place in the Amazon Basin. Remote sensing reveals widespread deforestation associated with the expansion of agriculture and grazing. The scale and scope of this transformation are unmatched anywhere else.

Figure 5.4.3, a false-colour satellite image of the Amazon and adjacent regions, highlights key differences in vegetation, moisture levels and other surface features. The darker green areas show where forest areas—mostly tropical humid rainforests—thrive and have not been severely changed or degraded by human activity.

5.4.3 False-colour image of South America showing the nature and extent of land cover change

The lighter green areas in Venezuela, Guyana, Suriname, French Guiana and southern and eastern Brazil are generally tropical savanna. These woodland–grassland ecosystems often have trees, but they are spaced far enough apart that the canopy does not appear fully closed. While tropical savannas receive plenty of rain during the wet season, they typically have vegetation that can withstand the region's lengthy dry season. Rivers and reservoirs appear as navy blue in the image. The brown areas are seasonally flooded wetlands.

Areas affected by human activity stand out. Forest areas that were converted to pasture generally appear yellow. Savanna converted to cropland is generally pink, especially if fields are fallow or have exposed soil.

Causes of forest lost in the Amazon

The main causes of forest loss in the Amazon are clearing land for cattle grazing (63 per cent), small-scale farming (12 per cent), commercial crops (9 per cent), fire (6 per cent), selective logging (4 per cent) and tree plantations (including palm oil) (1.5 per cent).

Deforestation, often achieved through the use of fire, represents the single greatest threat to the Amazon rainforest. Figure 5.4.4 shows the trend in deforestation since 1988, compared to 1970. Almost 20 per cent of the pre-1970 forest cover had been lost by the end of 2021—down from 4.1 million square kilometres to 3.3 million square kilometres. The rate of loss peaked in 1995 when 29 059 square kilometres was lost. The average annual rate of loss in the period 2010–2021 was a more modest 7184 square kilometres. This reduction in the rate of loss can be attributed to a range of initiatives, some of which have been introduced by international bodies like the European Union.

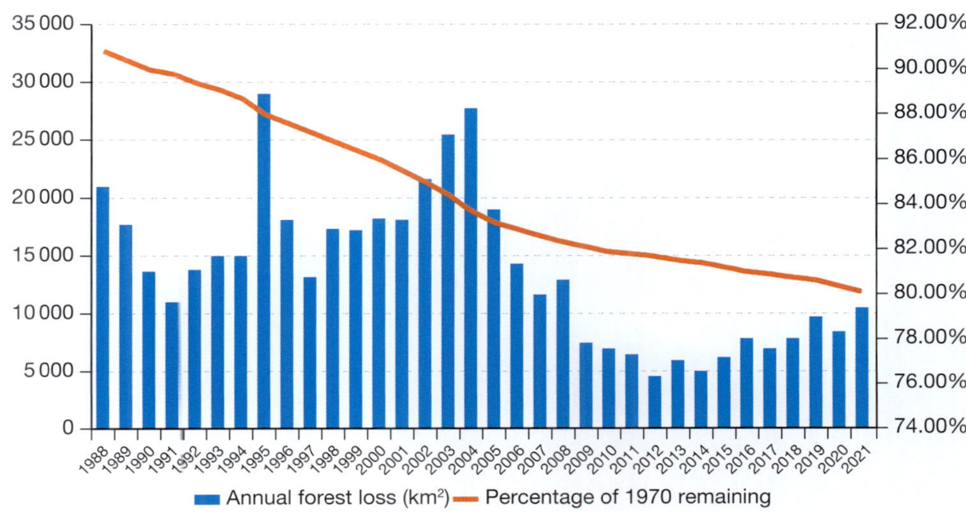

5.4.4 The trend in annual forest loss (1988–2021) and natural forest cover compared to 1970

The clearing of the rainforest is typically done using the slash-and-burn technique (see Figure 5.4.6). The forest is first cleared using bulldozers during the wetter months of the year (November–June). The piles of felled trees are set alight as the dry season arrives (July–October). Often these fires spread into the adjacent, uncleared rainforest where they are difficult to extinguish. Much slash-and-burn activity is done illegally. While most countries in the Amazon Basin have laws against deforestation, they are rarely enforced.

The inauguration of president Jair Bolsonaro, in January 2019, resulted in a relaxation of environmental protection measures. As a result, Brazil had more than 70 000 fire outbreaks in 2019, an 84 per cent increase on the same period in 2018. More than half of these fires were in the Amazon rainforest. Higher temperatures and a dryer than normal summer, both attributed to global climate change, increased the intensity of the fires.

Farmers took advantage of weaker enforcement of environmental regulations. After Bolsonaro became president, the Brazilian environment agency issued fewer penalties, and ministers sided with those responsible for land clearing rather than the indigenous groups who live in the Amazon. In the 2022 presidential election, Bolsonaro was defeated by Luiz Inácio Lula da Silva. Lula, as he is known, made climate a cornerstone of his presidential campaign. He pledged to end deforestation in the Amazon.

Satellite imagery is used to track the incidence of fire in the Amazon. While naturally occurring fires do happen, 99 per cent of fires are the result of human intervention, either on purpose or accidentally. Satellite imagery highlights a relationship between the points of fire ignition and roads and existing agricultural areas. Very few fires start in remote areas of the Amazon Basin.

Did you know?

Scientists fear that the continued destruction of the Amazon rainforest could push climate change towards a tipping point, after which the region would enter a feedback loop of decline. The result would be rainforests converting into savanna, releasing 200 billion tonnes of CO_2 into the atmosphere. Temperatures would increase further, and the environmental services of the Amazon rainforest would be greatly diminished to the detriment of the entire planet.

The Amazon fires of 2019

In 2019, fires raged across the Amazon Basin (see Figure 5.4.5). The destruction was so great that it was represented as one of the greatest environmental disasters of the decade. The impact of the fires prompted a wave of global concern and focused attention on the valuable environmental services provided by the Amazon rainforests.

Importantly, there was not just one large fire, but thousands of fires spread all over the Brazilian Amazon—in particular, along the southern edges of the rainforest where the encroachment of agriculture and grazing has resulted in the greatest rate of land cover change. The vast majority of fires were lit by cattle ranchers, loggers and speculators who want to clear and use the land.

5.4.5 A photo of an Amazon blaze taken with a drone

5.4.6 Slash-and-burn land clearing

Saving the forests of the Amazon

Brazil faces an enormous challenge: how to balance economic growth with the preservation of the Amazon rainforest. Many NGOs are campaigning to protect what remains of this amazing natural wonder. Mongabay, a website that publishes news focusing on tropical rainforests, identifies several strategies for protecting the rainforests of the Amazon.

Forest rehabilitation

The restoration of entire ecosystems is possible in regions where remnants of the original forest cover remains intact and population pressure is not too great. Small clearings surrounded by forest recover quite quickly. Large areas of cleared land may recover over time, especially if there is human intervention in the reforestation process. After several years, once–barren fields can support vegetation in the form of pioneer species and secondary growth. While secondary forests are generally low in diversity, the forest cover will be substantial enough to allow some species to return. Additionally, newly forested areas can be used for the sustainable harvest of forest products and low-intensity logging.

Increased productivity of formerly forested lands

Increasing the productivity of already cleared rainforest lands using improved agricultural technologies can generate higher-yielding crops and reduce the need to clear additional areas of rainforest.

Expansion of protected areas

Expanding the protection of critically important habitats is seen as essential to maximising the survival of the Amazon's biodiversity. The key to such efforts include:

- identifying biological hotspots and prioritising their protection
- providing the funding needed by enforcement agencies
- sufficient funding to maintain protected areas
- making sure local communities are involved in decision-making processes
- improving the living conditions of local communities. Local communities need to be both partners in and beneficiaries of conservation efforts.

Development based on the sustainable use of existing rainforests

Applying the principle of sustainability is potentially one of the best ways of protecting rainforests. In agriculture, this could involve the adoption of the techniques used by the indigenous peoples of the basin. These techniques could be used to increase the productivity of degraded forest lands and promote sustainable use of forest resources. Additionally, the application of agroforestry techniques (the growing of trees or shrubs around or among crops or pastureland) and polyculture practices (the practice of growing more than one crop at the same place and time) can increase productivity and minimise the need to clear an extensive area of rainforest. Ending subsidies for large landowners would also slow down the rate of deforestation.

Traditional, sustainable tree–harvesting practices have the potential to sustain forests as functional ecological systems, while providing the economic benefits associated with the resource.

Related strategies include:

- restricting the trade of specific rainforest tree species, for example, mahogany
- ending subsidies for sawmills and road construction will reduce the financial viability of forest clearing
- adopting best-practice logging procedures to reduce the impact on surrounding vegetation, for example, directional tree felling
- establishing stream buffer zones to protect water quality
- minimising soil disturbance

Did you know?

In 2021, Amazon rainforest destruction rose 24% from 2020, to 10 476 km^2. This was a 39% increase from the 7536 km^2 deforested in 2018. Scientists agree that protecting the Amazon is critical to stop catastrophic climate change because it absorbs vast amounts of CO_2.

5.4.7 Law enforcement officers on patrol in the Amazon

- reducing wood waste
- limiting the gradient of roads to prevent excess erosion
- establishing plantations on already degraded lands to minimise the clear-felling of forests.

Land policy reform

Under Brazilian law, the Amazon rainforest is defined as an 'open access resource', meaning that it is freely available for exploitation. As a result, there is little incentive for squatters, farmers or developers to use forests or resources sustainably. Many simply clear another area of the forest when the land being used is no longer viable, because of declining soil fertility. Developers gain the right to unoccupied land just by using it for at least one year and a day. This 'use' could simply involve grazing a few cattle on the land or burning an area of forest. Laws restricting such practices are urgently needed. Alternatively, existing laws could be enforced. For example, the 1996 law forbidding landowners from cutting down more than 20 per cent of the forest on their land.

Law enforcement

While Brazil has several laws aimed at reducing the rate of deforestation and encouraging the sustainable use of forest resources, the agency responsible for their enforcement (the Environmental Protection Agency) is so underfunded that it is largely ineffective (see Figure 5.4.7). Corruption further undermines attempts at law enforcement. The agency estimates that as much as 80 per cent of all logging in the Amazon is illegal.

International attention

In November 2021, more than 100 countries with around 85 per cent of the world's forests agreed in the COP26 climate summit's first major agreement to end deforestation by 2030. This was an improvement on a similar 2014 agreement, because it now included Brazil. Those signing the 2014 agreement, the New York Declaration on Forests, pledged to halve the rate of deforestation by 2020 and end it by 2030. But in the 2014–2021 period, deforestation increased.

Activities

Acquiring and processing geographical information

1 Identify the key features of the Amazon Basin's geography.

2 Outline the environmental services provided by the Amazon's rainforests.

3 Outline the nature and extent of land clearing in the Amazon.

4 Explain how political factors have impacted on deforestation in Brazil.

5 Outline the causes of the Amazon fires of 2019. What was their impact?

6 Summarise the various strategies that could save the rainforests of the Amazon Basin.

Applying and communicating geographical understanding

7 Study Figure 5.4.3. Identify the areas on the false-colour satellite image that are covered by tropical humid rainforests, tropical savanna and woodland grassland ecosystems. Identify those areas affected by the activities of humans.

8 Draw a pie graph illustrating the causes of deforestation in the Amazon. Include an 'others' category.

9 Study Figure 5.4.4, then complete the following tasks.

a Using data from the graph, write a short report describing the trends in annual forest loss between 1988 and 2018.

b Using data from the graph, describe the trend in the percentage of 1970 forest cover remaining.

c Describe the trend in the estimates of natural forest cover 1970–2021.

APPLICATION AND CONSOLIDATION TASKS

Task 1: Infographic: Ecosystems and economic services

Using Figure 5.1.8, create an infographic outlining the ecosystem and economic services provided by forests.

When you are outlining the services provided by forests, ensure you include the following information in your infographic:

- specific examples for each service provided
- both the direct and indirect benefits of the services provided by forests

Task 2: Report writing: State of the World's Forests

Write an extended report on the State of the World's Forests.

Access the most recent *State of the World's Forests* report from the FAO and analyse the progress made in protecting the world's forests since 2020. In your report:

- describe the main findings of the report
- outline and describe changes and updates
- analyse the initiatives discussed in the report
- add visuals (maps, graphs, diagrams, photos).

Task 3: Presentation: Rainforest investigation

Investigate an example of how an international initiative has proved effective in protecting forests in a specific country and/or region. You can either select one of the examples in this chapter or another (in consultation with your teacher).

In your presentation:

- outline the international initiative
- explain whether or not the situation has improved since 2020 or since it started
- explain whether the initiative has been successful or if changes have had to be made
- outline the future prognosis
- add visuals (maps, graphs, diagrams, photos).

Task 4: Class debate: Protecting forests

Conduct a class debate on the topic 'Humanity has a moral and ethical responsibility to protect the world's forests'.

To prepare for the debate:

- write down all the arguments for and against the topic
- re-read this chapter (and undertake further research) to find data to support your argument.

CHAPTER

6 Glaciers and ice sheets

Glacial ice moving across the surface of the land has created some spectacular landscapes and landforms. Currently, 10 per cent of Earth's land area is under glacial ice. During the last ice age, glacial ice covered 32 per cent of Earth's surface, drastically reshaping and changing the land beneath.

Historically, such transformations were slow. Today, the extent and rate of glacier and ice sheet retreat has increased due to climate change. Understanding glaciers and ice sheets means understanding that they are open systems interacting with other global systems.

This chapter investigates the power of glacial ice to transform the land surface. Though mainly found in remote places, these frozen landscapes are greatly impacted by human interruptions of natural systems. The alarming rate at which glaciers and ice sheets are melting poses significant consequences for the immediate environment. At scale, large glacial ice losses will have a global impact.

Where the glacier meets the sky, the land ceases to be earthly, and Earth becomes one with the heavens; no sorrow lives there anymore, and therefore joy is not necessary; beauty alone reigns there, beyond all demands.

Halldór Laxness, Icelandic poet

6.0.1 Ice falls off a glacier into the sea in Alaska, USA.

Chapter glossary

ablation the process of a glacier or ice sheet losing more ice mass than it accumulates

benthic invertebrates organisms without a backbone that live in the sediment at the bottom of a water body

calving the process of chunks of ice breaking away from a glacier to become icebergs

cirque a deep, rounded hollow with steep sides found in a mountain valley head that has been eroded by a glacier

cryoconite dark dust or airborne sediment composed of dust, pulverised rock particles from volcanic eruptions, soot from fires and particulates from diesel engines and coal-fired power stations

cryospheric relating to the cryosphere, the parts of Earth's surface where water is frozen in solid form, including glaciers, ice sheets, ice caps, sea ice, lake ice, river ice, snow cover, and frozen ground

glacial abrasion where the glacial ice and the rocks it carries grind and scrape against a valley's sides or the bedrock underneath

glacial accumulation the zone of a glacier in which the amount of snow deposited exceeds the ice lost

glacial lake a body of water filling a void or depression previously occupied by glacial ice, within a cirque or dammed by the terminal moraine of a retreating glacier

glacial lake outburst flood a catastrophic discharge of water where all the water from a glacial lake rushes down a mountain when the terminal moraine holding it back is breached

glacial plucking where the moving ice of a glacier lifts out and removes angular blocks of rock, which are then embedded in the ice and move with it

glacial trough a long, U-shaped valley that was carved out by a valley glacier that has since disappeared; a trough tends to have high, steep, straight sides and a flat valley floor

heavy fuel oil the residual fuel oil remaining after crude oil has been stripped of its more valuable components in the oil refining process

moraine a deposit of soil or rock carried and left by a glacier

moulin a vertical shaft or hole worn in a glacier by meltwater falling through a crack in the ice

mountain glacier glaciers that originate in high alpine areas and form in cirques; also called alpine or cirque glaciers

peak water the tipping point of glacier melt supply when run-off in glacier-fed rivers reaches its maximum due to accelerated melting; after peak water, water flow will diminish as glacial storage has been depleted

permafrost a subsurface layer of soil that remains frozen throughout the year, mostly found in polar regions

retreat when a glacier or ice sheet reduces in size because ice melts or ablates more quickly than it accumulates

terminal moraine a rubbly heap of debris dropped at the snout or lowest end of a glacier

tidewater glacier valley glaciers that flow far enough to reach out into the sea

valley glacier glaciers that originate in the snowfields of high mountain ranges and flow through steep walled valleys

UNIT 6.1

The power of glacial ice

A glacier is a system of flowing ice that moves under the force of gravity. It originates on land through the **accumulation** and recrystallisation of snow. Glacial ice is a potent agent of erosion and drastically alters the land it covers. Glaciers carry immense weight, scouring soil and rock material in their path and transporting debris, depositing it far from its original site.

The source of glacial ice: The snowfields

Glaciers form in the world's coldest areas in high altitudes or high latitudes, where more snow accumulates each year than melts. Newly fallen snowflakes are light and loose, but as the snow accumulates it compacts and has fewer air spaces. Eventually, under the weight of more than 24 metres of snow, compaction leads to glacial ice forming (see Figure 6.1.1).

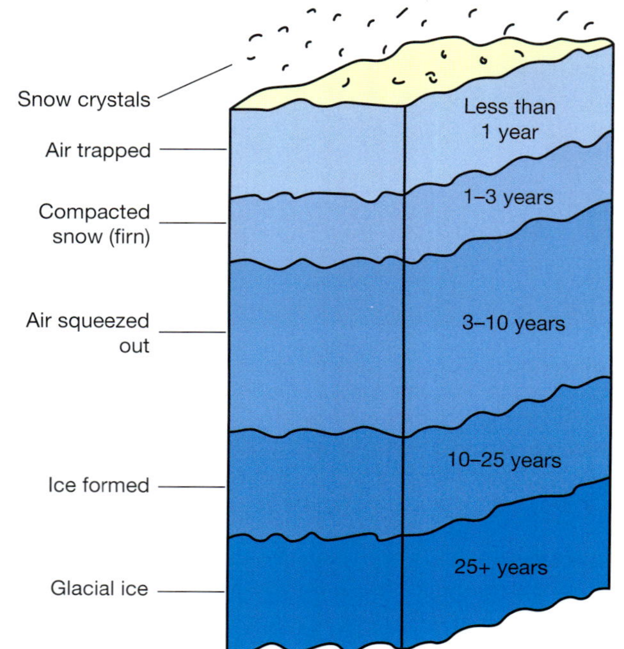

6.1.1 The transformation of snow crystals into glacial ice

Over time, the ice mass builds until it is thick enough to flow under its own weight, pulled by gravity. The weight of accumulated ice deforms ice crystals, making them slippery. Most glaciers creep just 25 centimetres daily, but in Antarctica, some move less than a metre a year. A few surge at speed, including Greenland's Jakobshavn Glacier, generally considered the fastest glacier in the world, able to cover up to 45 metres per day.

Types of glaciers

The principal types of glaciers are **mountain glaciers**, **valley glaciers**, **tidewater glaciers** and ice sheets (or ice caps and icefields).

Mountain glaciers

Mountain glaciers develop in high mountainous regions, often flowing out of icefields and spanning several peaks or entire mountain ranges (see Figure 6.1.2). They form in small bowls with steep sides called **cirques**. They are also known as cirque glaciers. They terminate before they reach the sea. The largest mountain glaciers are found in Arctic Canada, Alaska, the Andes and the Himalayas.

6.1.2 The Gorner Glacier, Zermatt, Switzerland

Valley glaciers

Valley glaciers originate in the snowfields of high mountain ranges. They are typically long and narrow, occupying previously formed valleys. Such glaciers carry the ice to lower elevations where warmer temperatures melt it. They terminate on land. Their distinctive landform features are visible in Figure 6.1.3. Mountain and valley glaciers are **retreating** at faster rates as a result of global warming.

Tidewater glaciers

Tidewater glaciers are valley glaciers that flow far enough to reach into the sea (see Figure 6.1.4). Tidewater glaciers are responsible for **calving** icebergs. Due to variations in precipitation, they undergo periods of advance and retreat. The general trend now is retreat, due to rising global temperatures.

Ice sheets

Ice sheets form in the polar regions where prevailing temperatures are low enough for ice to accumulate. Over thousands of years, layers of snow build up, forming a mound of ice that thickens

6.1.3 Yosemite National Park, California, USA, has glacial valley features from melted glaciers.

6.1.4 Three glaciers flowing into Prince William Sound, Alaska, USA.

to thousands of metres deep. Under such immense weight, ice deforms and then flows outwards in all directions from a central accumulation zone. These continent-sized ice sheets overwhelm the land, even causing Earth's crust to subside under them. During the Pleistocene ice ages, ice sheets estimated to be one to two kilometres thick covered much of the Northern Hemisphere. They exerted approximately 1000 tonnes of pressure on each square metre of the crust underneath causing it to sag under the weight.

Glacial landforms

The landform features associated with mountain, valley and tidewater glaciers are illustrated in Figures 6.1.5 and 6.1.6.

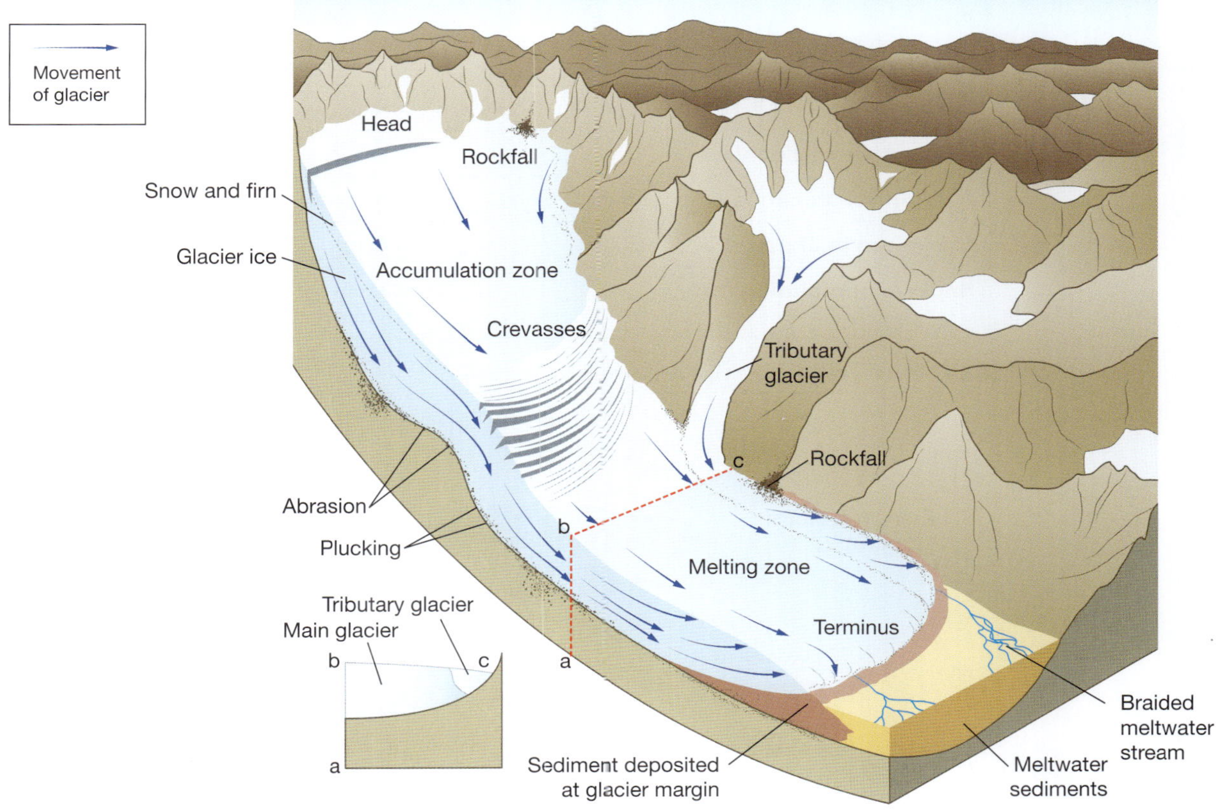

6.1.5 Landforms of an active valley glacier

Did you know?

The oldest glacier in Antarctica may be 1 million years old, while the oldest glacier in Greenland is thought to be 100 000 years old.

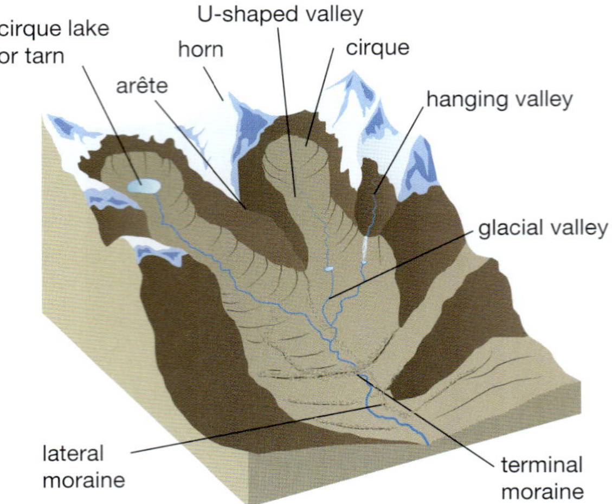

6.1.6 Distinctive landforms left by a valley glacier

Antarctic and Greenland ice sheets

Today, Earth has just two major ice sheets, one on Antarctica and one on Greenland (see Figure 6.1.7). The Antarctic ice sheet is the largest single mass of ice on Earth, covering 98 per cent of the continent. As illustrated in Figures 6.1.8 and 6.1.9, much of the ice sheet is over three kilometres thick and its extreme weight has depressed large parts of the continent's surface, leaving it below sea level. There are mountains near the edges of the continent, where the highest ranges and peaks emerge above the icy surface. These mountains funnel glacial ice towards the sea. The Greenland ice sheet covers 80 per cent of the island. It is more than three kilometres thick in its centre, where the ice is at its thickest, but then thins towards its edges.

Features	Greenland	Antarctica (east & west ice sheet)
Area	1.71 million km^2	12.37 million km^2
Volume of ice	4 million km^3	25.71 million km^3
Ice cover	approximately 80%	98%
Percentage of global ice	10%	90% of global ice
Average ice thickness	1.6 km average	2.16 km average
Maximum thickness	3200 m	4776 m
Snow accumulation rate		303 mm per year in the west and 118 mm in the east

6.1.7 Features of the Greenland and Antarctic ice sheets

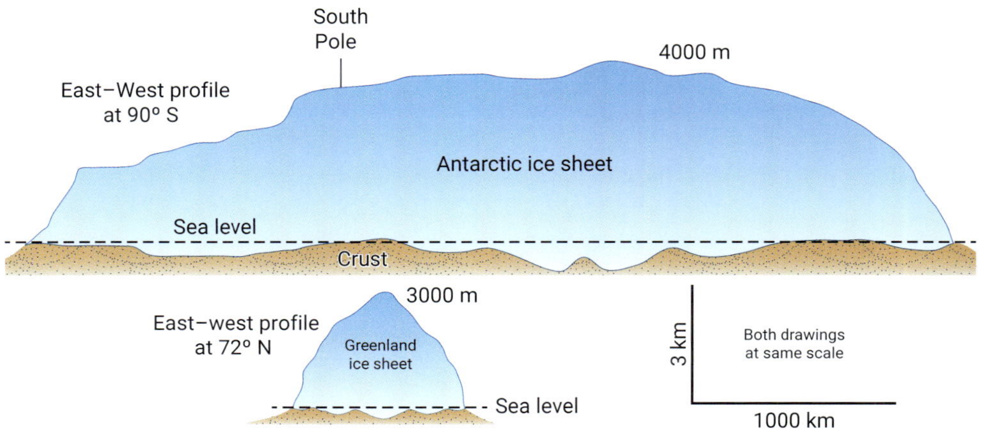

6.1.8 Cross-sections of Antarctica and Greenland, highlighting their vastly different sizes

Ice sheets take on a dome-like shape at the centre and flow outwards. The ice is at its thickest in the centre. The ice is pushed outwards, at a rate of only a few centimetres a year, until it reaches the ocean, where ice speeds can reach hundreds of metres or even several kilometres a year along fast-flowing outlet glaciers. For both Antarctica and Greenland, much of their ice flow terminates in the surrounding ocean. As glaciers flow into water, calving is common as ice chunks break off the end of the glacier and float away as icebergs. Antarctica is surrounded by ice shelves where permanent floating ice extends out over the ocean (see Figure 6.1.9).

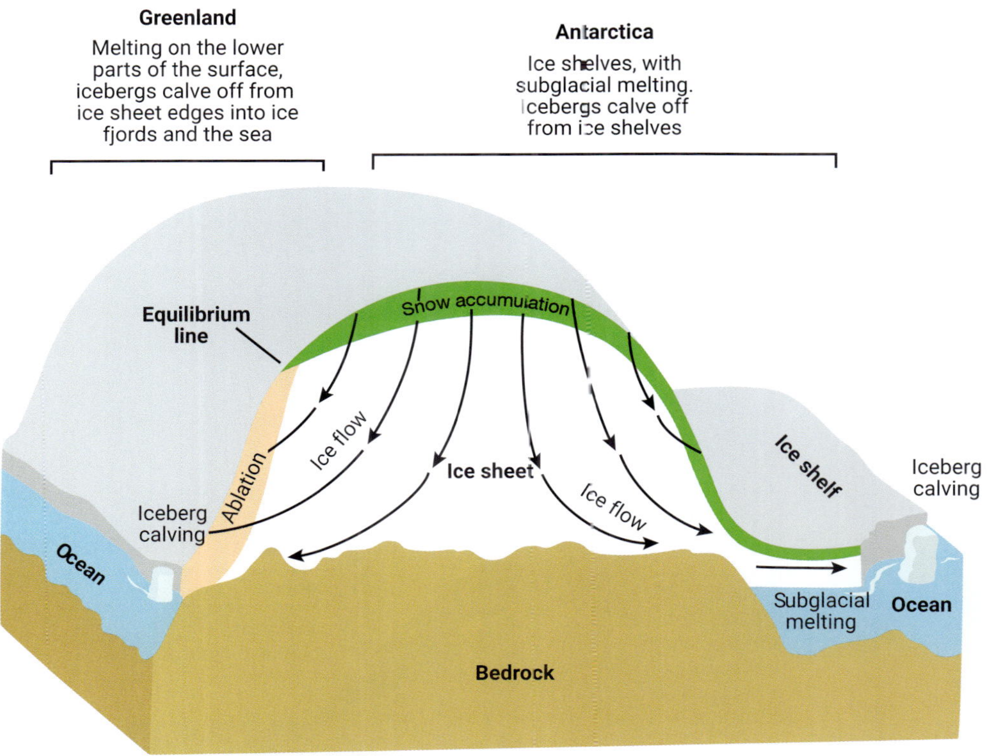

6.1.9 Ice flow on the Greenland and Antarctic ice sheets

Ice shelves receive ice in several ways: the flow of ice from the ice sheet, surface accumulation (snowfall) and the freezing of marine ice to their undersides. Ice shelves lose ice by melting from below (from relatively warm ocean currents), melting above (from warm air temperatures) and from calving icebergs. These are normal parts of their **ablation**. Ice shelves can be up to 2000 metres thick, with a cliff edge up to 100 metres high.

Activities

Acquiring and processing geographical information

1 Distinguish between mountain, valley and tidewater glaciers.

2 Identify the parts of the world where glaciers form.

3 Explain how freshly fallen snow becomes glacial ice.

4 Explain what enables glacial ice to flow.

5 Demonstrate the differing speeds at which glaciers move, with reference to examples.

6 Explain how an ice sheet deforms the land underneath it.

7 Contrast the direction of flow between a valley glacier and an ice sheet.

8 Explain how an ice shelf gains and loses ice.

9 Describe an iceberg and identify where they originate in both the Northern and Southern Hemispheres.

Applying and communicating geographical understanding

10 Study Figures 6.1.2, 6.1.3, 6.1.5 and 6.1.6. Explain how a valley glacier transforms the mountain landscape. Identify and describe the main landform features created by the erosive power of the glacier.

11 Study Figures 6.1.7, 6.1.8 and 6.1.9. Write a brief report comparing the Antarctic and Greenland ice sheets.

UNIT 6.2

Glaciers and ice sheets as open systems

Glaciers and ice sheets are open systems and have similarities with other systems like rivers, where flows occur due to gravitational forces. They have a series of linked inputs, transfers, stores and outputs, through which both energy and material are cycled (see Figure 6.2.1). The stores in this system are the atmosphere where the snow comes from, the glacial ice and the underlying bedrock. The ice moves over it and geomorphological processes—erosion, transportation and deposition—transfer the debris with the moving ice. Meltwater is a system output. So too are the distinctive landforms, such as **moraines**, **glacial troughs**, outwash plains and icebergs.

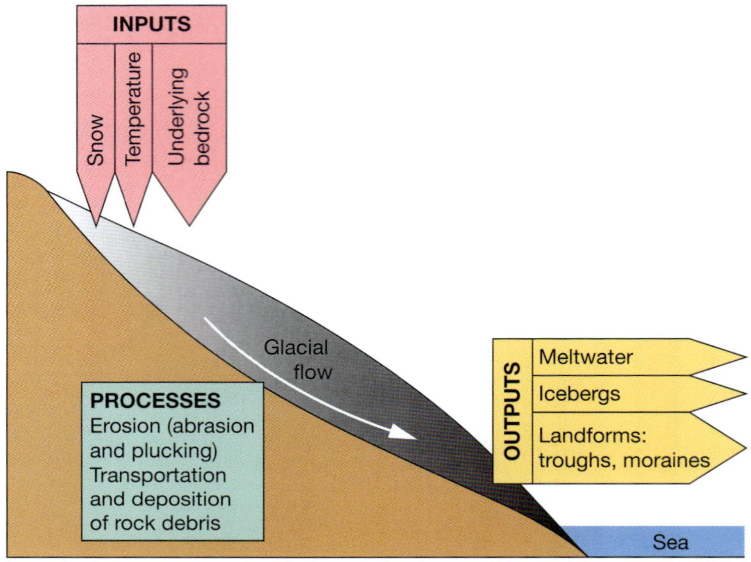

6.2.1 A glacier operating as an open system

Like other natural systems, glaciers respond to changes in inputs. They alter the rates of ice flow and determine outputs. When an increase in temperature causes an increase in melting that exceeds the input of new ice, a glacier loses mass until it reaches a new balance between accumulation and ablation. Lower temperatures and higher snowfall have the opposite effect, causing a glacier to gain mass. This relationship between total accumulation and total ablation at a point in time is called mass balance. Averaging this over a year determines a glacier's annual net balance. This determines whether it is positive or negative, indicating if a glacier's size has increased or decreased. Most glacial systems have had a negative balance since at least the 1970s.

Ice in the glacial system

Ice is the most significant input into a glacial system. Snow enters as precipitation and is compressed into ice. It builds up in the accumulation zone from where it is transferred downslope in a valley glacier or outward toward the margins of an ice sheet. Ice leaves the system in the ablation zone. Here, it calves and floats away as icebergs, melts or evaporates. The balance between the rate a glacier or ice sheet accumulates ice and the rate at which ice ablates determines the system's size. Temperature is the controlling factor.

The surface of a glacier or ice sheet reflects its operation as a system. The snowline generally shows the boundary between the accumulation and ablation zones. The surface appears smooth and white above the snowline, as fresh snow covers any irregularities caused by melting. Below the snowline, the surface is broken by crevasses, as gaps are left where snow melts or evaporates (see Figures 6.2.2 and 6.2.3).

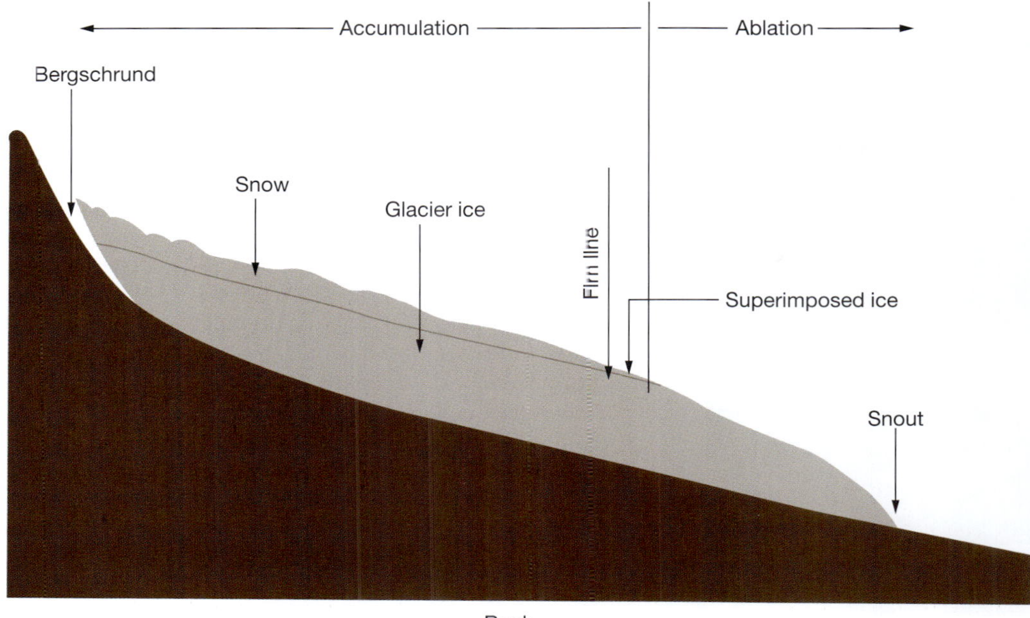

6.2.2 The accumulation and ablation zones on a glacier

Rocks and sediment in the glacial system

Rock is another important input into the system. Rocks embedded into the base of glaciers and ice sheets act as powerful agents of erosion, with the transported rock material being deposited near the margins of ablation zones. The main processes are **glacial plucking** and **abrasion**. Plucking is where the ice lifts out and removes angular blocks of rock, which move with the ice to then grind and scrape the valley sides or bedrock underneath, in a process known as abrasion.

Unlike rivers, which sort and smooth the rocks and sediment they carry, glaciers carry large rocks and small fragments side by side. Being embedded in the solid ice they remain angular and are deposited when the ice melts and the glacier retreats. This builds up a heap of debris, known as a **terminal moraine,** at the glacier's snout. Meltwater forms streams that exit the snout, carrying finer sediment to the outwash plains. Many of the largest rivers in the world, including the Ganges, the Yangtze and the Indus, have their sources in elevated, glaciated areas.

The meltwater that emptied from the continental Pleistocene ice sheets (2.58 million to 11 700 years ago) created extensive outwash plains that covered hundreds of square kilometres. Sand and gravel were dropped closest to the margins of the ice sheets, with the finer clay and silt carried the furthest. These finer sediments were pulverised rock particles generated by the grinding of the bedrock under the power and weight of the ice sheets. Due to their rich mineral content, they contributed to the fertility of the US Midwest's deep soils, where some of the world's most productive agricultural land is found. Glacial rock flour is the finest sediment produced. Being very light, it remains suspended in lake water for a long time. Sunlight reflecting

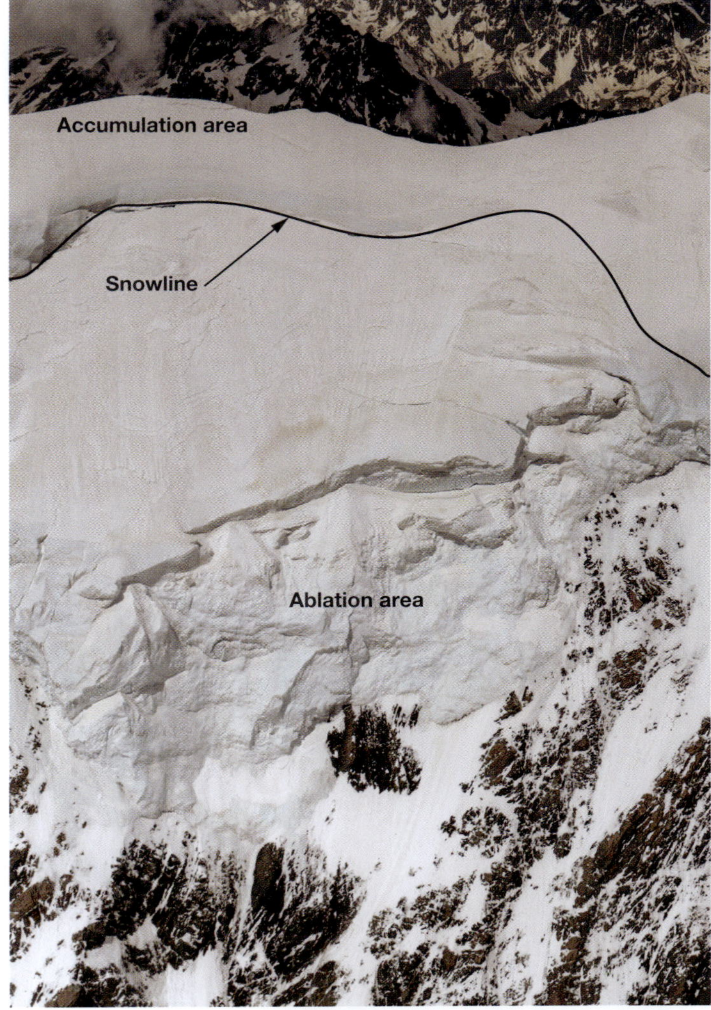

6.2.3 The surface of a glacier is different above and below the snowline.

6.2.4 The distinctive colour of Peyto Lake, Canada is evidence of its glacial origin.

off these particles gives the lakes a distinctive turquoise green (see Figure 6.2.4).

Part of an assemblage of open systems

The state of the atmosphere and temperature (which is largely a product of latitude and altitude) determines the input of snow and regulates the amount of energy entering the system. Temperature drives the growth or retreat of a glacier or ice sheet, thereby establishing the size and velocity of flow. It also sets the ice thickness and the rate at which it melts, the quantity and spatial extent of the distribution of deposited rock fragments and debris. This is why there is concern now about the extent and rate of glacial and ice sheet retreat related to climate change. Key to addressing the problem is understanding that glaciers and ice sheets are open systems, interacting with Earth's other systems.

Activities

Acquiring and processing geographical information

1 Identify the inputs, processes and outputs of a glacial system.
2 Describe how the mass balance of glaciers is determined.
3 Explain what determines the size of a glacial system.
4 Distinguish between the accumulation zone and the ablation zone of a glacier.
5 Contrast the ways in which glaciers and rivers transport rocks and sediment.
6 Distinguish between plucking and abrasion of a glacier as it erodes the valley sides or bedrock underneath.
7 Describe a terminal moraine and how it is formed.
8 Assess the importance of the Pleistocene ice sheets to the agricultural productivity of the Northern Hemisphere.

Applying and communicating geographical understanding

9 Study Figure 6.2.1. Draw two diagrams to show the response of the glacial system to:
   a warmer temperatures
   b cooler temperatures.
10 Study Figures 6.2.2 and 6.2.3. Demonstrate how the surface of a glacier moving downhill reflects its operation as a system.

UNIT 6.3

Land use change and glacier and ice sheet retreat

Land provides for human livelihood and wellbeing, including the supply of food, fresh water and multiple other resources. Tracing changes in land use over time demonstrates the extent to which humans have increasingly interrupted Earth's natural systems. Despite their remoteness in high latitudes or altitudes, glaciers and ice sheets are sensitive to alterations in natural systems occurring elsewhere in the world.

Changes in land use over time

While early agricultural practices interrupted natural systems, they were mostly localised. Much of the natural world, notably forests, was intact, maintaining a balance in the natural cycles. The Industrial Revolution began industrialising and urbanising agricultural societies from 1760, and land transformation began accelerating. Fossil fuel use increased dramatically, upsetting Earth's natural balance.

● SPOTLIGHT

A brief history of land use

Humans have a long history of altering terrestrial ecosystems by hunting, foraging, land clearing, agriculture and other activities. For example, Aboriginal and Torres Strait Islander peoples have been modifying Australia's environment for at least 60 000 years. Around 8000 years ago, agricultural land uses spread throughout Mesopotamia and the Fertile Crescent areas of South-West Asia, followed by China, India and Europe. Intensive land use patterns developed in India, China, Africa and South America. By around 6000 years ago, agricultural expansion had spread across most continents, leading to native vegetation clearing and culling, and the domestication of herbivores. Native flora and fauna were replaced with intensive crop and livestock management practices, as human populations grew in size and density.

Hotter oceans also expand and melt ice, causing sea levels to rise. The measured sea levels of the decade 2010–2020 were the highest, based on records dating back to 1900. Scientists expect about 1 metre of sea level rise by 2100, potentially displacing 150 million people worldwide.

Warmer oceans melt floating ice

The ocean absorbs heat from the atmosphere and the warmer ocean water melts floating ice from below, increasing its vulnerability and speeding up iceberg calving. If warmer water reaches under the ice shelf to where the ice sheet or glacier is attached to the bedrock (known as the grounding line), it can weaken it. This makes it even less stable and it loses more ice over time (see Figures 6.3.2 and 6.3.3). As these ice shelves thin, the ice held in glaciers and ice sheets flows more rapidly into the ocean.

● SPOTLIGHT

Ocean temperatures hit a record high

The heat in the world's oceans set a new record for the fifth consecutive year in 2022, showing irrefutable and accelerating heating of the planet. The world's oceans are the clearest measure of the climate emergency because they absorb more than 90 per cent of the heat trapped by the greenhouse gases emitted by fossil fuel burning, forest destruction and other human activities (see Figure 6.3.1). The amount of heat being added to the oceans is equivalent to every person on the planet running 100 microwave ovens all day and all night.

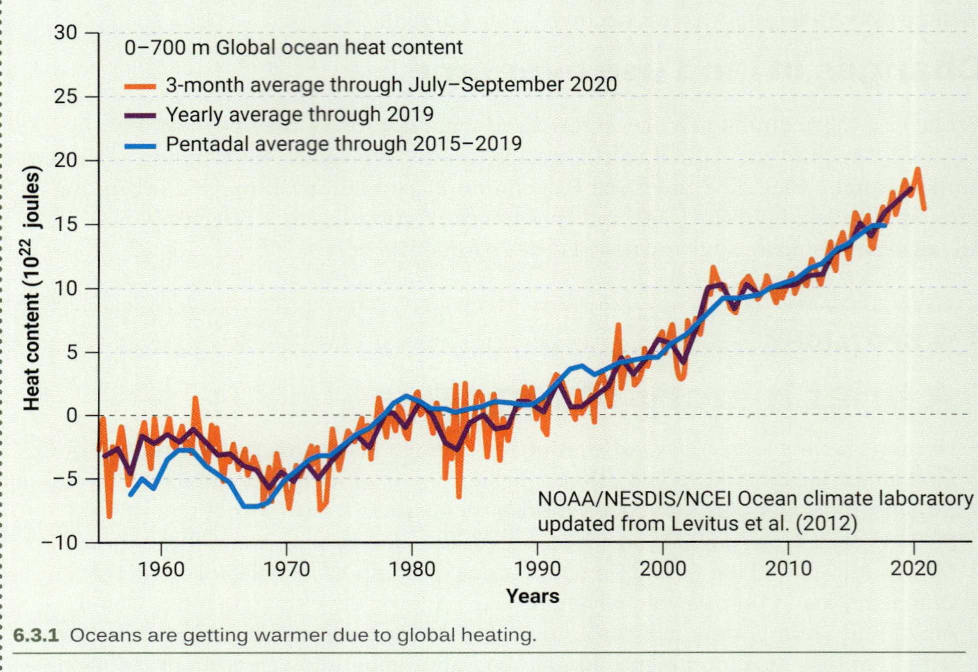

6.3.1 Oceans are getting warmer due to global heating.

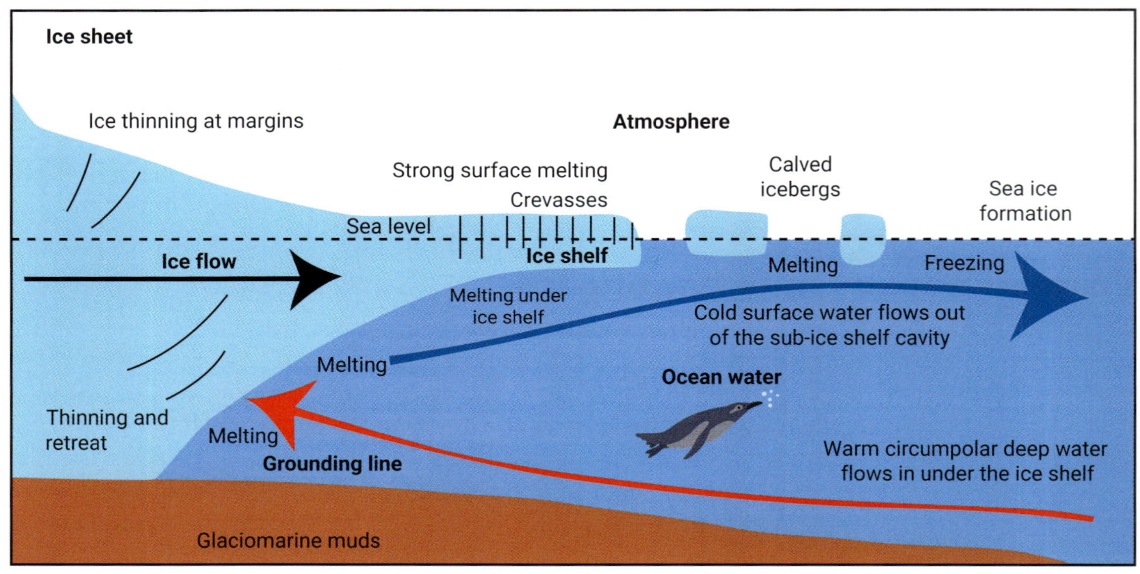

6.3.2 Warm water flowing under the ice shelf melts it at the grounding line, making it more unstable, which then speeds up ice flow.

Dust and soot lower snow and ice albedo

6.3.3 The Ross Ice Shelf, the largest Antarctic ice shelf

Land uses in places well away from glaciers and ice sheets are also having an impact on melting. From the snow-capped peaks of New Zealand to Himalayan glaciers and Greenland's ice sheets, layers of dust and soot darken the surface of glaciers and ice sheets.

Windborne dust, soot and pollution that settle on the surface of glaciers and ice sheets discolour them and lower their albedo. Snow and ice have a high albedo as the white colour of the surface is highly reflective, bouncing incoming radiation back into space. Pristine white snow reflects incoming solar radiation back into the atmosphere. This has always been a natural defence against melting. However, as humans have changed land cover, intensified agriculture and become increasingly urbanised and industrialised, soot and dust have become mobilised and are found settling on the surface of remote ice sheets and glaciers. When snow becomes blanketed by darker-coloured particles of dust and soot, this coating

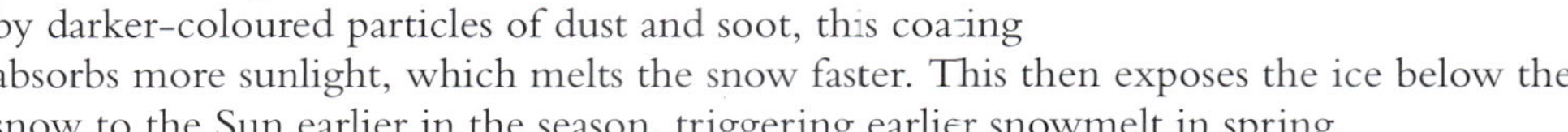

absorbs more sunlight, which melts the snow faster. This then exposes the ice below the snow to the Sun earlier in the season, triggering earlier snowmelt in spring.

Dust-related melting in Greenland could add at least 2 centimetres more to the globally averaged sea-level rise by 2100 than current estimates suggest.

Dust

Dirt darkens snow surfaces as windstorms pick up and carry dust for thousands of kilometres before settling. In 2019, dust stained the Southern Alps of New Zealand orange-brown after drought-loosened iron oxide-rich soil particles were blown 2000 kilometres across the Tasman Sea by north-westerly winds (see Figure 6.3.4). Dust from the Sahara impacts the albedo of the European Alps (see Figure 6.3.5) and dust from the arid lands of the western and southwestern USA impacts the snowpack of North America. Dust from north-western China impacts Himalayan glaciers.

Dust is not only generated by poor agricultural practices and drought; mining also causes problems. In Chile, there are 24 114 glaciers and some of the largest ice sheets outside the polar regions. They are at risk from mining-generated dust. Because of this, environmentalists are pushing for more protective measures against mining dust to be put in place.

Did you know?

Pure snow generally has a visible albedo of 0.95. It reflects more than 95% of the visible light that hits it. Soot reflects less than 10% of incoming light (albedo 0.1). Tiny particles of car tyres and brakes are blown from densely populated areas of the Northern Hemisphere onto the Arctic, reducing its albedo.

6.3.4 Dusty glaciers in New Zealand, November 2019

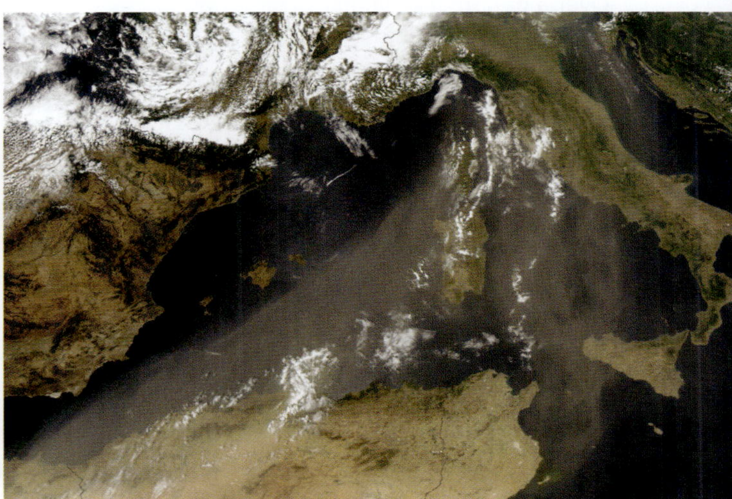

6.3.5 Saharan dust heading towards the Italian Alps

6.3.6 Surface water entering a moulin on the Ferpècle Glacier in the Swiss Alps.

6.3.7 A large moulin on Langjökull Glacier, Iceland

Soot

Soot is a black powdery substance, consisting mainly of particulate carbon, and is known as black carbon. Because it is darker than dust, it absorbs more sunlight and so is much more potent at warming and melting ice than dust. As well as settling on the surface of glacial ice, soot particles can also form the nuclei of snow crystals. Then, with their additional heat, they are responsible for changing a feathery and highly reflective white snow crystal into a darker more compact grain which then absorbs more heat (see Figures 6.3.6 and 6.3.7).

Historically, increased levels of soot being carried in the atmosphere began with the Industrial Revolution. It is an unwanted by-product of the incomplete burning of fossil fuels. Research has confirmed that the reason the Little Ice Age (1300–1870) ended in Europe was the coating of soot deposited on the Alps by the Industrial Revolution.

Soot is also produced by the incomplete burning of organic matter. Forest fires and wood-burning stoves are common sources of soot (see Figure 6.3.8). As development has pushed the agricultural frontier in the Amazon Basin, smoke plumes have billowed from extensive forest fires and travelled with the wind to darken the Andes' glacial snow with soot.

The use of biomass fuels such as wood, crop waste and dried animal dung in open fires and cooking stoves also results in high levels of airborne particulate matter. Many people living in poorer regions of China, India and Tibet depend on such fuels as their major energy source. The impact is evident in the Tibetan Plateau of the Himalayas, as soot darkens snow and ice in the region.

Did you know?

Over a period of 20 years, each molecule of black carbon traps 3200 times more heat than CO_2.

6.3.8 A traditional cooking stove in northern India releases soot into the atmosphere.

● SPOTLIGHT

Historic coal soot in the Himalayas

New research published by the Proceedings of the National Academy of Sciences has found that soot dating back to the beginning of the Industrial Revolution made its way across Europe to settle on the top of the Himalayas, over 10 000 kilometres away. Soot has been found in an ice core extracted from the Dasuopu glacier, at an elevation of over 7900 metres, on Shishapangma mountain (see Figure 6.3.9).

The research began in 1997 when scientists travelled to Dasuopu to drill ice cores from the glacier. The ice cores have provided a trove of data, including snowfall records, atmospheric circulation and other environmental changes over time.

Geographers studying an ice core they believe formed between 1499 and 1992, found higher-than-natural levels of several toxic metals, including cadmium, chromium, nickel and zinc, from around 1780. All these metal trace elements are by-products of burning coal, and were probably transported by winter winds, which travel around the globe from west to east.

6.3.9 By-products of coal were found in an ice core from Shishapangma in the Himalayas.

Activities

Acquiring and processing geographical information

1 Identify the greatest dislocation to Earth's natural balances and when it started.

2 Explain how warmer atmospheric and ocean temperatures are accelerating the melting of the ice sheets of Antarctica and Greenland.

3 Describe how the main windborne pollutants from human activities are being carried to and settling on remote ice sheets and glaciers.

4 Define soot and explain why it is more potent than other pollutants in melting glacial ice.

Applying and communicating geographical understanding

5 Study Figure 6.3.1. Using data from this graph, answer the following questions.

   a Identify the year when oceans first started getting warmer.

   b Compare the rate of warming between last century and this century.

UNIT 6.4

The extent and rate of glacial retreat

Glacial ice is especially sensitive to climate change. The edges of ice caps are shrinking and glaciers are retreating. When there is warmer air over a glacier or ice cap, the ice on the surface melts, which directly reduces the mass of the glacier. Once meltwater penetrates cracks on the surface of the ice, it can lead to further melting and accelerate ice flow. As these cracks open, they allow more water to drain down to the bed and spread out across its base lubricating it.

Evidence of glaciers and ice caps melting

In the past decade, Earth scientists have documented record-high average annual surface temperatures and have been observing changes in the distribution of ice. Since the early 1900s, with few exceptions, glaciers around the world have been rapidly melting and retreating at unprecedented rates. Glaciers are melting, calving into the sea and retreating on land, and ice caps are shrinking. During the last century, several glaciers, ice caps and ice shelves disappeared altogether, with many more shrinking so quickly that they may vanish within a matter of decades.

The World Glacier Monitoring Service (WGMS) has tracked changes in more than 100 glaciers worldwide. There are 37 that have records dating back 30 years and are accepted as climate-reference glaciers. Figures 6.4.1 and 6.4.2 reflect how these reference glaciers have been losing their mass balance over recent decades as melting occurs. The total mass loss over time appears to be accelerating as the downward curve steepens.

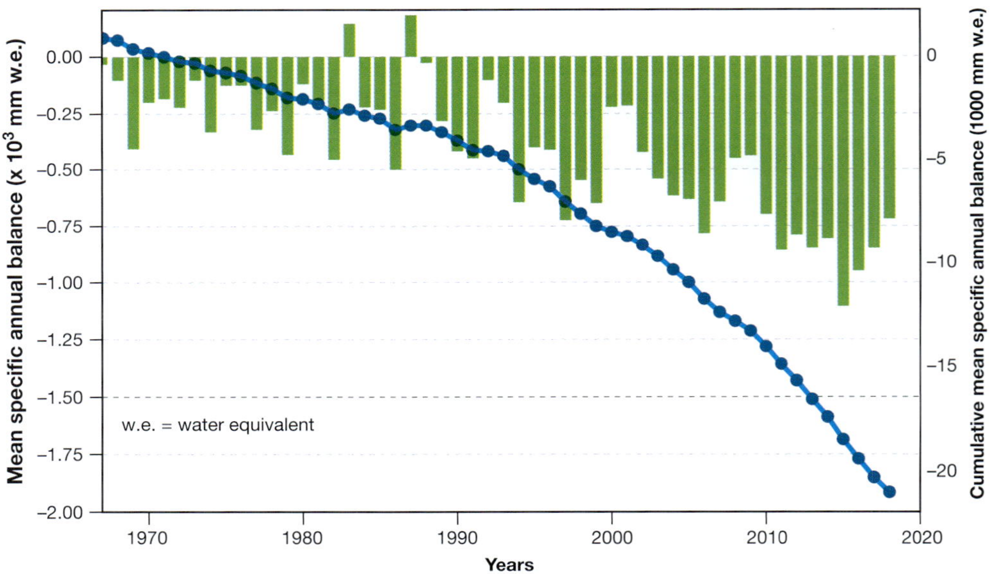

Source: American Meteorological Society 100, Si-S306

6.4.1 Mass balance of the WGMS 37 reference glaciers 1950–2022

Decade	Loss per year in metres
1980s	–0.228
1990s	–0.443
2000s	–0.676
2010s	–0.921

Source: NOAA

6.4.2 Glacier loss from the 1980s to 2010s

● **SPOTLIGHT**

Measuring changes in Earth's gravitational pull

As NASA Gravity Recovery and Climate Experiment (GRACE) twin satellites orbit Earth, they continuously measure the slightest changes in Earth's gravitational pull. Changes in the distribution of water are the most significant cause of gravity changes, so scientists can measure the weight of water and track its shift from ice caps and glaciers to the oceans. See Figures 6.4.3 and 6.4.4.

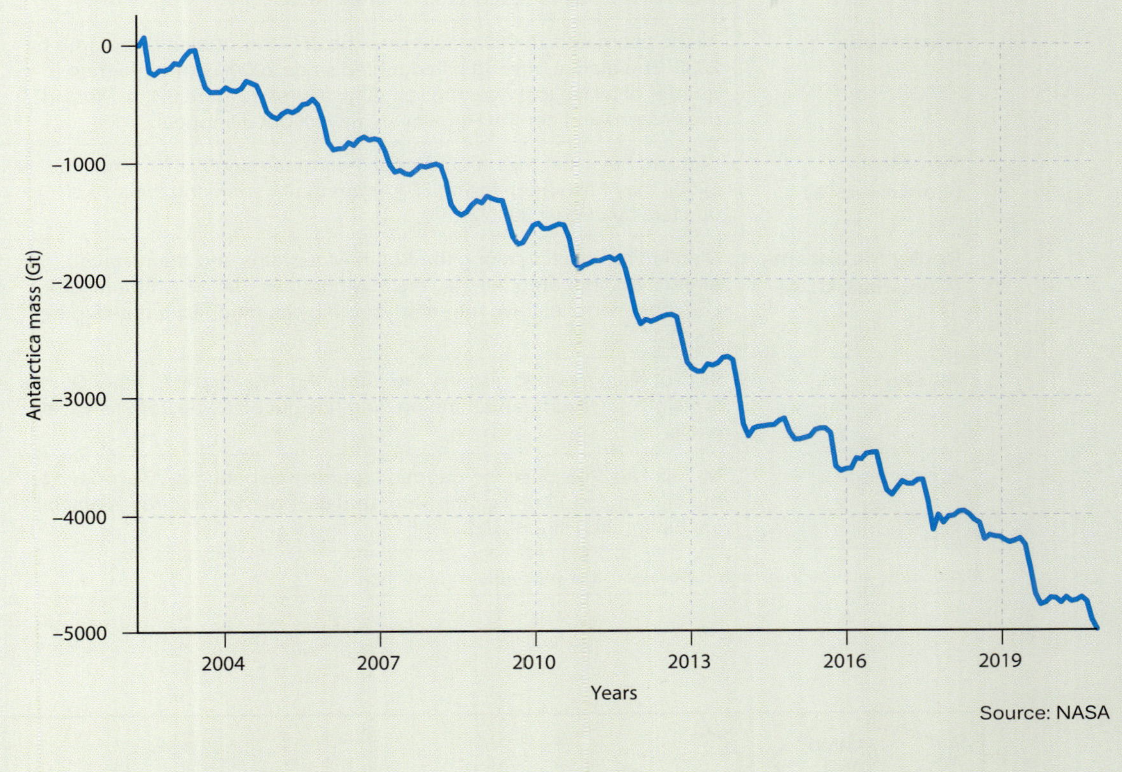

6.4.3 Antarctica mass variation since 2002. GRACE ice mass measurements of the Antarctic ice sheet

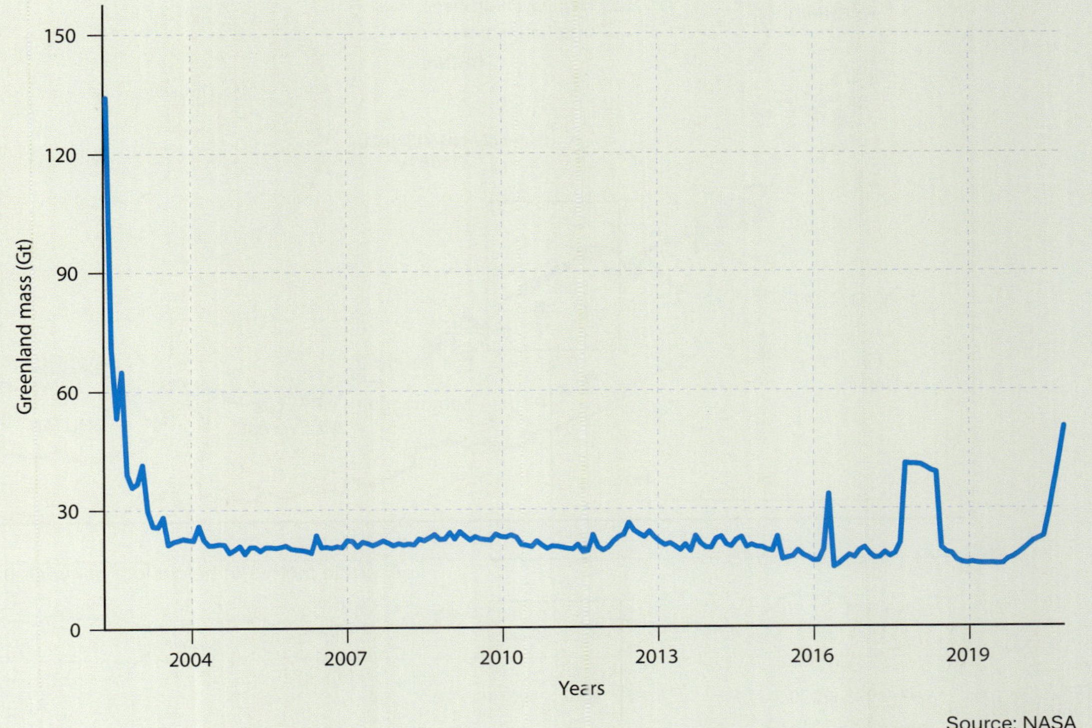

6.4.4 Greenland mass variation since 2002. GRACE ice mass measurements of the Greenland ice sheet

Alpine summits and glaciers losing ice cover

Since 1970, several hundred glaciers across the world have been losing 0.5–1.3 metres of thickness each year; this is at least twice the average loss for the twentieth century. It is predicted that by 2100, alpine summits may have lost 90 per cent of the ice that covered them in the 1920s. Figures 6.4.5 to 6.4.7 highlight just some of the loss.

European Alps	Glaciers lost half their volume between 1850 and 1975. Since 1970, 40% of their remaining glacial ice has melted. The Argentière Glacier on Mont Blanc has receded 1150 m since 1850.
Himalayas	All the estimated 15 000 glaciers will be 35–75% smaller by volume by 2035. The glacial melting has doubled since 2000, with more than a quarter of all ice lost over the last 40 years (see Figure 6.4.5). Most in the eastern and central Himalayas may all but disappear.
Tien Shan (Central Asia)	Glaciers have lost over a quarter of their mass since 1970. Since the 1960s they have been losing 60 km^2 annually, shrinking the size of glaciers by almost 3000 km^2.
Rocky Mountains (Canada & USA)	Glaciers are melting along the Rocky Mountains and many have already disappeared. All glaciers in Montana's Glacier National Park, USA, for example, have retreated, some by as much 85% (see Figure 6.4.6).
Alaska	99% of Alaska's 2000 glaciers are retreating. The Grand Plateau Glacier is visibly narrowing and thinning, and has pulled away from its valley walls.
Africa	Mt Kilimanjaro's once ice-capped summit has been partly exposed for the first time in 10 000 years. More that 80% of the mountain's glacial ice has melted since 1912.

6.4.5 Extent of ice loss on the world's major glaciated mountains.

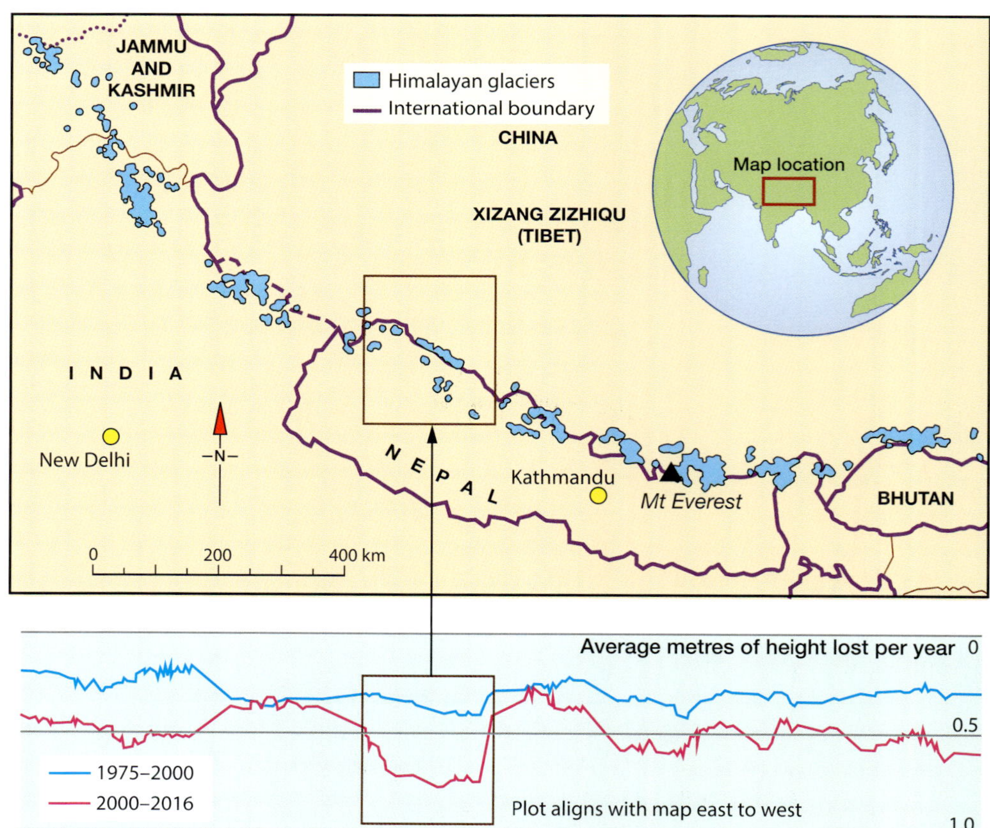

6.4.6 Eight billion tonnes of ice are lost every year from the Himalayas. Not replaced by snow, the lower-level glaciers are shrinking 5 m annually.

6.4.7 The Grinnell Glacier in Glacier National Park in Montana's Rocky Mountains has retreated to the mostly shaded, upper confines of its basin.

● **SPOTLIGHT**

The Andes Meltdown

The glaciers in the Andes Mountains, South America, are rapidly shrinking due to warming temperatures. A staggering 98 per cent of glaciers have shrunk and some mountain glaciers have retreated 9 kilometres over the last century. Many sit in low altitudes in the southern Andes, which makes them vulnerable to rising air temperatures. The Patagonian icefields at the southern end account for 83 per cent of all ice loss in South America (see Figure 6.4.8).

Tropical glaciers in the Peruvian Andes are thinning and losing mass, shrinking at their fastest rate in the past 300 years. They have lost almost half their surface area since the 1970s.

Over 70 per cent of the world's tropical glaciers are found in Peru in high mountain ranges around the Equator, many concentrated in the Cordillera Blanca, or White Range. They are acutely sensitive to rising temperatures. Small icy regions in Columbia, Venezuela and Bolivia are also rapidly melting. Glaciers in the western Andes of Bolivia have lost more than two-thirds of their mass since the 1980s. Those at lower altitudes are the most vulnerable, losing 1.35 metres of glacial ice annually, double the rate of high-altitude glaciers. Usually under 40 metres deep, they will probably disappear within decades.

6.4.8 Olivares Alpha Glacier, Chile, has lost 66% of its mass since 1953.

Shrinking of polar ice sheets

Satellite surveys show that polar ice sheets are melting six times faster than in the 1990s, when Greenland and Antarctica combined lost 81 billion tonnes of ice a year. The loss increased to 475 billion tonnes in the 2010s. From 1992 to 2017, the combined ice loss from both sheets was 6.4 trillion tonnes, with 60 per cent from Greenland.

Increasing sea temperatures drive the melting Antarctic ice sheet. The ocean melts the outlet glaciers, speeding up their flow. In Greenland, ice loss has been triggered by a combination of atmospheric and sea temperature rises, exacerbated by complex feedback loops that are speeding up surface melting (see Figures 6.4.9 and 6.4.10).

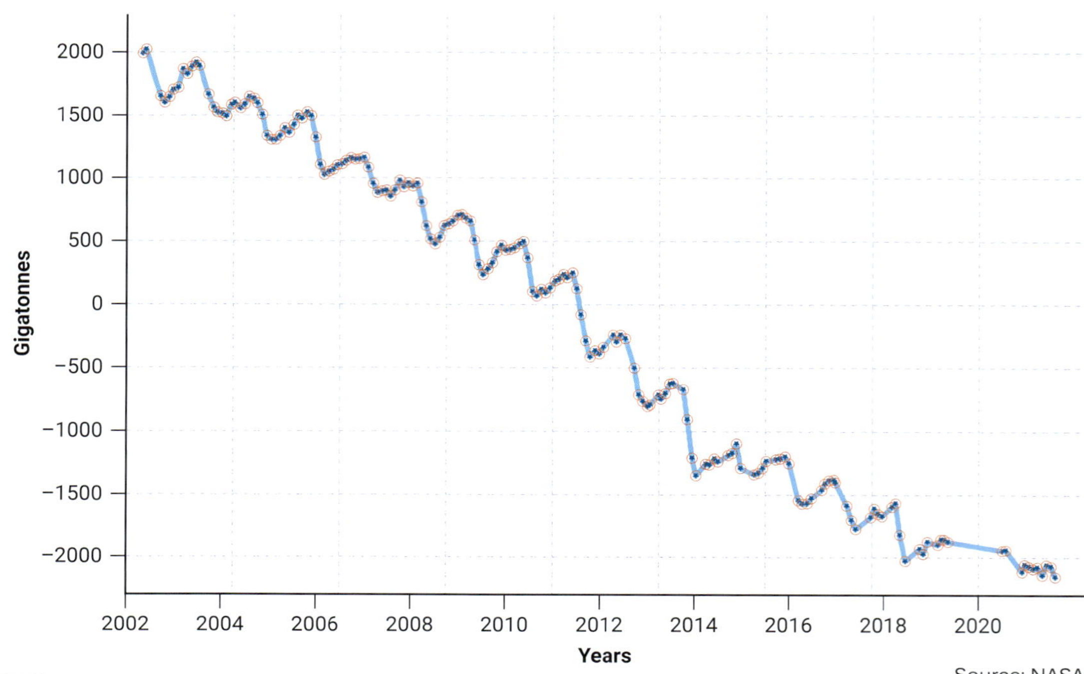

6.4.9 Estimated total mass change in Greenland's ice sheet 2002–2019 from GRACE

6.4.10 The edge of Greenland's ice sheet south of Ilulissat

Did you know?

Australian scientists have used NASA satellite data to calculate the weight of the Greenland ice sheet. During the summer of 2019, the loss of ice has been calculated to be enough to cover all of Tasmania in water almost 5 metres deep.

Greenland is losing glacial ice seven times faster than in the 1990s. This could create another 7 centimetre rise in sea levels by 2100. Glacial ice is reacting to the rapid Arctic warming, with temperatures rising 0.75°C in the decade 2010–20. Most ice loss comes from surface melting where water runs off into the ocean. Melting is most apparent on the edges of the ice sheet where the glaciers carrying ice from the interior calve and crumble into the sea. The speed the ice moves towards the sea has also more than doubled. Seawater thaws the ice from below and this is thought to be a growing driver of ice loss.

Temperatures are rising across the Arctic at about twice the global average. This causes surface melting of the ice sheet, particularly its edges. Summer temperatures are now among the highest on record, often over 20°C near where the Jakobshavn Glacier enters the ocean. Across Greenland, 600 billion tonnes of ice are now lost during summer. If sustained, the melt could raise global sea levels by 2.2 millimetres a year. Recent persistent high pressure systems across the Arctic have brought warm air to northern parts of the ice sheet. This has helped to maintain clear skies leading to a lack of snowfall. If this becomes a regular summer weather pattern, future melting could double. A Danish meteorological team estimated that over 30 billion tonnes of ice melted in three days in the summer of 2019.

As warming melts the edges of the ice sheet, rock that has been crushed by it is exposed. As it dries it releases mineral dust. This blows back over the retreating ice sheet, settling on its surface, darkening it, reducing its albedo and accelerating melting. This is a significant feedback loop as melting causes even more melting. Much dust is generated elsewhere in the Arctic as there is a longer snow-free season, leaving soil exposed. Soot also blows in, which absorbs heat more effectively. Large wildfires in northern Russia and Alaska produce most of the black particulates settling on Greenland's ice sheet. Together with the dust, they form **cryoconite**, which accelerates melting.

With the surface of the ice sheet becoming increasingly wet in the warmer summers, microbes and algae thrive on the cryoconite. The algae also produce a dark pigment as a sunscreen, which further boosts the ice's absorption of solar energy. This is another feedback loop, as the absorption of heat by the dark cryoconite deepens the holes where the algae lives. This stops the water from freezing, creating a suitable environment for more to grow. **Moulins** develop and this fuels more melting (see Figure 6.4.11).

6.4.11 Cryoconite matter is made up of microscopic mineral and organic particles carried by the winds to fall on the ice.

Activities

Acquiring and processing geographical information

1. Describe what has been happening to ice sheets and glaciers since the early 1990s. Explain how this has created a feedback loop.
2. Assess the extent of ice cover from the world's major glaciated mountains this century.
3. Identify which mountain range in the world has lost the most glacial ice relative to its size and account for its vulnerability and the speed with which glaciers are melting.
4. Describe how ice cover in the Arctic and Antarctica has been lost and explain what the main drivers of melting are in both the Antarctic and Greenland.
5. Explain why the rate of melting of the Greenland ice sheet exceeds that of the Antarctic ice sheet.
6. Describe and account for the extent of melting of the Greenland ice sheet in 2019.
7. Explain how the melting of the edges of the Greenland ice sheet creates a feedback loop.
8. Explain how a combination of airborne pollutants combine to form cryoconite, which can exacerbate melting and form moulins, and how these accelerate the speed with which a glacier flows.

Applying and communicating geographical understanding

9. Study Figure 6.4.1. Using data from this graph, answer the following questions.
   a. Identify the two years in which the reference glaciers gained mass.
   b. Calculate how many years the reference glaciers lost mass, as a percentage of the time frame represented.
   c. What has happened to the rate of melting over the period shown? Validate your answer.
10. Study Figures 6.4.3 and 6.3.4. Compare the rate of change of the ice masses in Antarctica and Greenland. Account for the difference.
11. Study Figure 6.4.5. Complete the following tasks.
    a. Assess the severity and extent of ice loss on the world's major glaciated mountains.
    b. Investigate recent statistics of ice melts in these mountains to establish the current situation and whether ice loss has accelerated or abated.
12. Study Figure 6.4.6. Compare the rate of melting in the Himalayas moving from west to east.
13. Study Figure 6.4.9. Calculate the percentage change in the mass of the Greenland ice sheet from 2002 to 2019.

UNIT 6.5

The impact of glacier and ice sheet retreat

The retreat of glaciers and ice sheets can have significant and even devastating effects on the immediate environment, such is the scale of the loss of glacial ice.

Rising sea levels

> **Did you know?**
> If all the ice in Greenland melted, sea levels would rise by 7 metres.

The loss of land ice directly causes the sea level to rise, putting at risk low-lying islands and coastal communities throughout the world. Rising sea levels loom as one of the most damaging long-term impacts of the climate crisis. Figure 6.5.1 shows how the sea level has risen since 1900. The accelerated increase reflects the ice sheet mass loss from both Antarctica and Greenland. The Greenland ice sheet has become the largest contributor, currently raising the level by more than 1 millimetre each year.

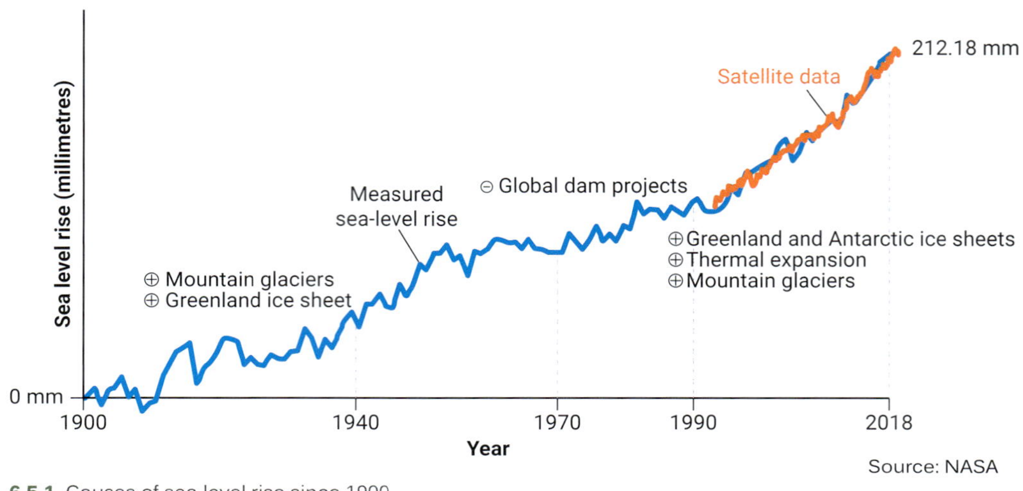

6.5.1 Causes of sea level rise since 1900

> **Did you know?**
> Greenland and especially Antarctica were quite stable at the start of the 1990s. It took until the 2010s for the ice sheets to react to decades of a warming climate.

6.5.2 Invasion of a low-lying coastal area by the sea

The Intergovernmental Panel on Climate Change's (IPCC) most recent mid-range prediction for global sea level rise in 2100 is 53 centimetres. But the new analysis of satellite data suggests that if current trends continue, seas will rise by an additional 17 centimetres. It is estimated that every centimetre of sea level rise leads to coastal flooding and coastal erosion, disrupting people's lives around the planet. For example, an extra 17 centimetres would mean the number of people exposed to coastal flooding each year would rise from 360 million to 400 million (see Figure 6.5.2).

Disruption of ocean currents

Fresh water pouring off the Greenland Ice Sheet is altering the salinity of seawater and disrupting the ocean conveyor belt in the Atlantic Ocean. This global process, the Atlantic Meridional Overturning Circulation (AMOC) ensures heat and energy are distributed globally and ocean waters are continually mixed. Warm water moves northward and colder, denser, saltier water moves south, moving the current. Adding fresh water reduces the water's salinity, making it harder to drive the current southward.

If the AMOC continues to weaken it will affect the weather patterns of countries bordering the Atlantic Ocean. Europe may experience both hotter summers and colder winters, and more destructive storms could hit the US coast. Rainfall patterns in the tropics could also be altered.

Long-term decline in water resources

Nearly one-sixth of humanity lives near streams and rivers that source their water from glacial meltwater originating in the headwaters of drainage basins. Some are already finding that run-off is decreasing, while others face an uncertain future.

In response to accelerated melting, decreasing glacial mass will deliver increasing water in the basin with more run-off until the point of **peak water** is reached. Beyond that, water flow will diminish as glacial storage is depleted. The timing of reaching peak water depends on geographical location. Research indicates that it has already passed for almost half of the basins in the Alps, Andes and Rocky Mountains, where flow is now diminishing. The smaller the glacier and the closer it is to the Equator, the quicker peak water is reached. Many of the smaller tropical glaciers in the Andes have already reached peak water and may disappear altogether. Being further away from Equator, larger glaciers in Alaska are not expected to reach peak water until near the end of this century. While the Himalayas sit at a low latitude, their high altitude tempers melting and they are expected to reach peak water in the middle of the century.

Potential water crisis in the Hindu Kush

The Hindu Kush–Himalayan (HKH) region is the source of the waters flowing in the Ganges, Brahmaputra, Indus, Yangtze and Yellow rivers. This water sustains close to 2 billion people. These rivers drain the Himalayas, delivering sediment-laden water that supports intensive agriculture on the region's fertile soils. Any change in the flow of water from melting glaciers will severely impact farming, the availability of drinking water and the generation of hydro-electricity. An accelerated spring melt is expected to increase water flow out of the mountains until 2050, when peak water is reached, before river flows start to plummet. Food security in the region is threatened.

Andes water shortages

With ever-increasing climate stress, the availability of water is becoming critical, particularly in semi-arid and arid regions, where other sources of water are limited. Meltwater from glaciers in the Andes ensures that many rivers continue to flow in the dry season and during droughts. In Peru, for example, the source of the Rio Santa is in the Cordillera Blanca. The river supplies drinking water and irrigation for large-scale agriculture schemes along the dry Atacama coast. Further glacial retreat in the upper reaches of the Rio Santa is expected to reduce the flow of the river in the dry season by 30 per cent.

For many large cities in the region, glacial melt from the Andes is a critical natural resource. Santiago (population 6.8 million) in Chile, sources as much as two-thirds of its water from river systems fed by glaciers high up in the Andes. La Paz (population 2.7 million) in landlocked Bolivia, relies on glaciers in the surrounding Andean mountains for its water supply. The city has three main dams, but they have almost run dry as they can no longer rely on glacial melt water (see Figure 6.5.3). Water rationing has been introduced and a state of emergency declared.

6.5.3 Low water levels in Milluni Zongo reservoir near La Paz, Bolivia

Significant changes to flow regimes in glacierised catchments are expected as glacier melt diminishes. Stream and river flow will become more dependent on unpredictable rain events. There are many challenges facing those seeking water security, and similar struggles across the world could lead to conflict between neighbouring countries in the future, as water resources become scarce.

Hydro-electricity

Glacial run-off contributes significantly to the generation of hydro-electric power in many countries: 91 per cent in Iceland; 70 per cent in Austria and 25 per cent in Switzerland. The Mauvoisin Dam in Switzerland collects run-off from nine glaciers. Its power plant output is expected to increase until 2050 as run-off increases, but then fall as the glacier melt drops towards the end of the century.

Did you know?

In Alberta, Canada, agriculture relies on meltwater from glaciers in the Canadian Rockies, which are expected to lose 90% of their current volume by 2100.

Did you know?

A warming climate may see a longer growing season in Greenland and farming may return to a similar state to what it was a thousand years ago. Dairy cattle have recently been brought in and there are now locally-grown potatoes and broccoli.

Glacier tourism numbers follow retreating ice

Glacial mountain landscapes rank as some of the world's most popular tourist destinations. In 2019, more than 4 million people visited Banff National Park in Canada and 672 000 visited Glacier Bay National Park in Alaska. In New Zealand, glacier tourism contributes more than US$85 million to the economy annually.

Clearly, the loss of mountain snow and ice will mean such sites may lose their appeal. Initially, it may spur people to want to see glaciers before they are lost, but accessing some glaciers is becoming increasingly difficult. Hiking up onto New Zealand's two iconic glaciers, the Fox and Franz Josef glaciers was stopped in 2015 because it is too dangerous. The rapid melting of glaciers has left the steep sides of the glacial trough exposed and vulnerable to rockfalls. The only way onto these shrinking glaciers now is by helicopter (see Figure 6.5.4).

6.5.4 Franz Josef glacier, New Zealand, from a helicopter cockpit

Too many visitors may also degrade the glacial surface and threaten the ecological integrity of the region. In 2016, the Xinjiang government banned tourism on glaciers in the Tien Shan Range in the far northwest of China, ensuring that the glaciers be observed from a distance rather than having people step onto them.

Glacial lake outburst floods

Melting glaciers can also trigger catastrophic discharges of water where the contents of a whole **glacial lake** rushes down a mountain at once, known as **glacial lake outburst floods**. As the run-off from glaciers has increased, there has been a significant increase in the formation of glacial lakes. As glaciers retreat up a mountain they leave behind a large void or depression that was formerly occupied by glacial ice. Meltwater forms a glacial lake. On the downslope side, the water is held back by the terminal moraine of the glacier, which sits as a pile of unconsolidated, rocky rubble that had been pushed forward by the glacial snout. As melting continues, the lake increases in size. These lakes become deadly if the moraine is breached allowing the water to escape. This will happen if the loose rocks shift when ice at the core of the moraine melts. Even a small crack is quickly enlarged by surging water being pulled downslope by gravity. This leads to the glacial lake emptying without warning, sending a destructive flood wave down the valley below.

6.5.5 Creek flowing after a flood emptied the Tulsequah Glacier glacial lake, Canada

Glacial lake outburst flood waves comprise water mixed with sediment, rock debris and chunks of ice which makes them deadly, as illustrated in Figure 6.5.5. They have significant societal impacts causing a loss of life as communities are engulfed, as well as damaging land, property and infrastructure in remote regions. They have killed over 12 000 people worldwide—in South America, the European Alps, Central Asia and Iceland. They also negatively impact the ecology of systems downstream.

The number of glacial lakes in the world has increased. Research using over 250 000 Landsat images showed that between 1990 and 2018 the number and size of glacial lakes increased by more than 50 per cent from roughly 9400 to over 14 300. The water volume in the lakes also increased by about 50 per cent. Lakes at high latitudes exhibited the fastest growth. Some countries have begun mitigation strategies to reduce flooding risks by glacial lakes. In Nepal, for example, officials lowered the water level in Imja Lake, a glacial lake near Mt Everest, by more than 3.5 metres.

Loss of biodiversity

The rapid shrinking of glaciers is having cascading impacts on downstream systems. The continual melt from glaciers keeps rivers flowing during summer and drought periods when other sources are depleted. This creates a perennial riverine habitat, which is especially important in arid and semi-arid regions. The inevitable reduction in such river flow once glaciers have passed peak water will place vulnerable ecosystems at risk. The high Andean wetlands are particularly sensitive to glacial melting and are considered ecosystem sentinels for the impact of climate change. Many species depend on the slow-growing mossy pillows of cushion plants of the wetlands that are struggling to keep up with the pace of melting. Some specialist organisms are starting to disappear, including the meltwater stonefly, as its larvae depend on the cold, glacial streams.

Diverse ecological communities have evolved in the harsh environmental conditions of glacier-fed streams and rivers. Many of the aquatic species that have evolved in mountainous environments require cold temperatures to survive. **Benthic invertebrates**, or organisms that live in the sediments at the bottom of rivers, drive gross primary production. They are critical to energy flows and nutrient cycling within the ecosystem, from the lower trophic levels of algae and detritus feeders to upper trophic levels of fish, amphibians, mammals and birds. Many of these temperature-sensitive benthic organisms become extinct when water temperatures rise, disrupting the basis of the aquatic food chain. Such changes to glacier-fed stream habitat may similarly have an adverse effect on cold water fish such as trout and salmon.

Did you know?

The only bird known to nest on a glacier is the white-winged diuca finch, also known as the glacier bird, of the Andes. As glaciers retreat, it will have to move further up the mountains to find ice to build its nest on. Eventually, it may run out ice completely.

Did you know?

The continued thawing of the Greenland ice sheet could release radioactive waste. It was buried by the US Army in the 1950s, believing it would be buried forever.

Water contamination

An unexpected impact of the glaciers melting is the release of contaminants that have been locked up in the ice for a long time. Many originated as emissions from industrial activity, often thousands of kilometres away, with particulates and compounds carried by wind and settling on the ice, subsequently becoming part of it. Black carbon, mercury, pesticides and other pollutants have been historically deposited in the snowpack. The freeing of these harmful substances could significantly reduce water quality downstream.

Loss of spiritual significance

Glacial mountain peaks have spiritual significance for indigenous peoples throughout the world. Their sophisticated understanding and connection to the natural world are ingrained in their cultures. They recognise the mountains as the source of the streams and rivers on which their very existence depends. In India, the Gangotri Glacier is considered a sacred place and receives thousands of pilgrims each year.

Some indigenous peoples consider glaciers gods. In Peru, the loss of ice and snow from mountain peaks is thus associated with the god's departure and the end of the world.

Permafrost is melting

Permafrost sits beneath the surface across vast swathes of the higher latitudes in the Northern Hemisphere. It is permanently frozen ground that can be more than a kilometre thick. With Arctic temperatures increasing at more than double the global rate, ice within the permafrost is melting, causing the ground to collapse (see Figure 6.5.6).

6.5.6 The colour of melting permafrost indicates its high organic content.

Sinkholes and hollows pockmark the surface. Once the organic material that was held within the permafrost thaws, it decomposes and releases both carbon dioxide and methane into the atmosphere. The permafrost has long served as a 'carbon sink' and it is believed to hold twice as much carbon as the atmosphere. By releasing greenhouse gases, it is creating yet another climate change feedback loop with the potential to accelerate warming.

● **SPOTLIGHT**

A framework integrating the effects on ecosystem services

The impacts of glacial and ice sheet retreat on human activities are far-reaching. As Figure 6.5.7 predicts, there may be some initial increase in the value of some activities up to when peak water is reached, but beyond the peak, they will all decrease as the glacial storage is depleted. Figure 6.5.8, by the National Academy of Sciences of the USA, highlights the complexity of adapting to this change. The outer ring highlights the broad groups of ecosystem services provided by glacier-fed watersheds. The inner ring highlights specific services potentially altered by glacier retreat and disappearance. The centre highlights the complexity of interactions among the various services, which will necessitate trade-offs as society adapts to **cryospheric** change.

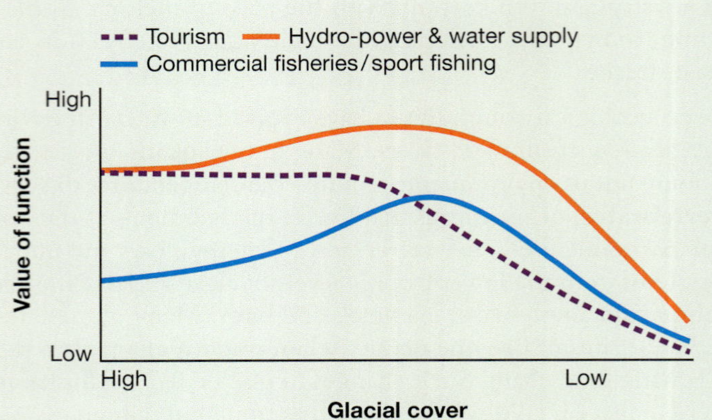

6.5.7 Projected impact of decreasing glacial cover on human activities

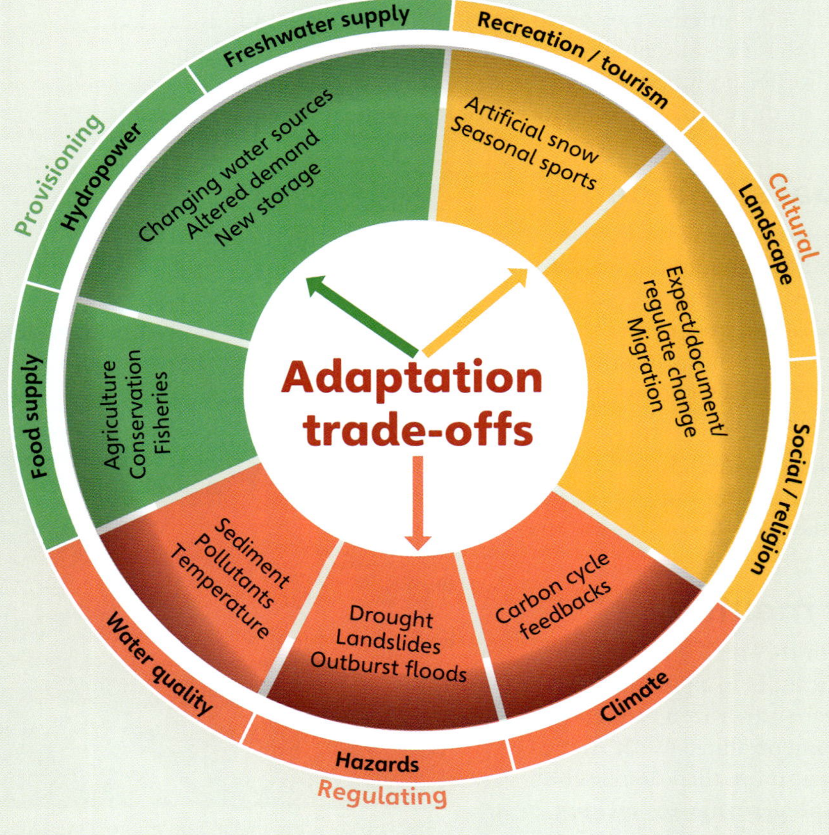

6.5.8 The impact of glacial retreat on human activities

Activities

Acquiring and processing geographical information

1 Identify the two contributing causes of sea level rise and analyse their respective impacts.

2 Assess the impact of sea level rise on coastal communities.

3 Outline the predictions of the Intergovernmental Panel on Climate Change on global warming.

4 Explain how fresh water pouring off Greenland could disrupt ocean currents and what this would mean for global weather patterns.

5 Describe how water supplies will be impacted by the melting of glaciers and what will happen once peak water is reached. Provide illustrative examples to determine the likely impact on water resources for many communities relying on glacial meltwater.

6 Describe a glacial lake outburst flood. Explain why they are dangerous to those living downstream.

7 Explain how glacial meltwater becomes contaminated.

8 Outline the spiritual significance of glacierised mountain peaks and what the loss of their ice and snow signifies to those that have a spiritual connection with them.

9 Explain why holes are appearing in the ground above the permafrost in the Arctic and how this may trigger a feedback loop that would accelerate climate change.

Applying and communicating geographical understanding

10 Study Figure 6.5.1 and, using data from this graph, answer the following questions.

- a By how much did the sea level rise from 1900 to 2018?
- b What contributed to this rise from 1900 until 1990?
- c What contributed to this rise from 1990 until 2018?
- d What evidence from the graph is there that the rate of rising sea levels has accelerated from the 1990s?
- e What additional data has been available since the 1990s?

11 Study the information explaining the various impacts of glacier and ice sheet retreat. Draw up a table summarising the information provided. Use the following as a guide.

	Cause	Effect	Examples
Rising sea levels			
Disruption of ocean currents			
Long-term decline in water resources			
Declining tourism			
Glacial lake outburst floods			
Loss of biodiversity			
Contamination of water			
Loss of spiritual significance			

12 Study the box, Spotlight: A framework integrating the effects on ecosystem services, and complete the following tasks.

- a Describe the impact of the loss of glacial cover on human activities and how this changes over time.
- b Write a paragraph or two explaining the complexity of addressing the issues arising from glacial retreat and disappearance as society adapts to the change.

13 As a class, brainstorm the strategies that could be undertaken at a local, regional and global level to slow the rate of melting of the Greenland ice sheet.

UNIT 6.6

Sustainable management: Banning heavy fuel use in Arctic shipping

The complexity of the interconnections between the natural systems of ice sheets and glaciers has become evident, as have the feedback loops that increase the rate of warming. It is by understanding the damage that can be done by one of these feedback loops, namely the settling of black carbon particulates on glacial ice, that is spurring a campaign to ban the **heavy fuel oil** (HFO) responsible.

Why heavy fuel oil is a threat to the Arctic

HFO is the dirtiest marine fuel. It is the residual oil left after crude oil has been stripped of valuable components during refining. Being cheap it's widely used for long-haul shipping. A spill from a ship does not disperse like regular oil. It leaves a highly viscous thick, tar-like sludge that emulsifies in seawater forming a paste. This is difficult to capture and gets trapped under the ice. Sensitive marine environments are at risk as it persists in the environment far longer than other fuels.

These heavy fuels burn slowly, emitting a high concentration of black carbon particles that contribute to the Arctic's more rapid warming. Reducing soot emissions is the fastest and most economical means to slow down that rate. HFO spills:

- are 50 times more toxic to fish compared to medium and light oil spills
- pose a severe risk to Arctic residents, many of the 4 million inhabitants depend on the sea for survival
- can cause hypothermia and death in seabirds and marine mammals
- produce harmful pollutants such as black carbon, sulphur and nitrogen oxides.

The growing use of heavy fuel oil in the Arctic

Dwindling polar ice makes traversing the Arctic Ocean easier for ships, attracting many countries and corporations to this route. HFO is the most commonly used fuel by bulk carriers, tankers and cargo vessels in the Arctic (see Figure 6.6.1). Russia is the biggest user as it needs to deliver supplies to remote coastal communities that lack road or rail connections (see Figures 6.6.2 and 6.6.3). With sea ice vanishing, Russia sees the Northern Sea Route (NSR) as a Suez Canal alternative. Shipments between Europe and Asia could possibly use the NSR as a shortcut for 10 months of the year. During the coldest months, sea ice means ships needed an icebreaker to traverse the passage.

There has already been explosive growth in shipping on the NSR, with 31.5 million tonnes of goods moved in 2019, up more than 60 per cent from 2018. In 2020,

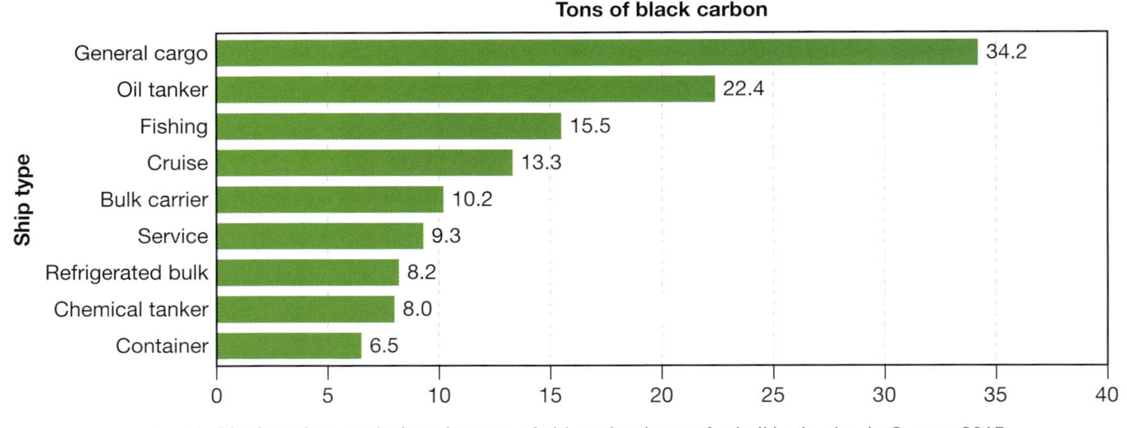

6.6.1 Black carbon emissions by type of ship using heavy fuel oil in the Arctic Ocean, 2015

commercial shipping began in May, earlier than ever before. A Russian carrier loaded with liquefied natural gas crossed the NSR to China without an icebreaker for the first time. Even the most icy and difficult parts of the route provided safe passage. The ease of early-season navigation will expedite the growth of cargo traffic in the Arctic.

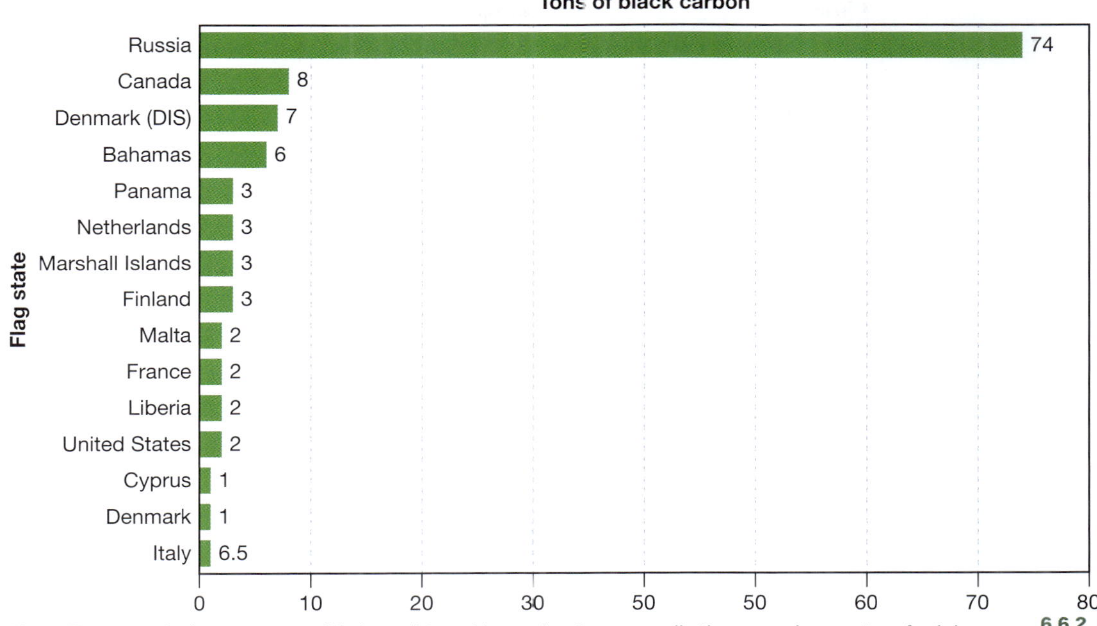

Note: Flag state is the country a ship is registered in, and not necessarily the owner's country of origin. Panama, Liberia, Marshall Islands, Bahamas and Malta register a large number of foreign vessels. DIS is the Danish International Shipping Register.

6.6.2 Black carbon emissions by ship's flag state, from ships using heavy fuel oil in the Arctic Ocean, 2015

6.6.3 A cargo ship with containers at Anadyr, Chukotka in far east Russia

Action to ban the use of heavy fuel oil in the Arctic

With the growth of shipping in the Arctic, the use of HFO will rise and with it more accelerated melting of glacial ice. Robust measures are needed to reduce the risks arising from the use and carriage of HFO. Concerns have long been held about the reliance on this dirty fuel in such a delicate and remote environment. It led to the International Maritime Organization (IMO) prohibiting its use in Antarctica in 2011. Efforts to include the Arctic region in that ban fell short. The IMO has been working for a decade to have heavy fuels banned in the Arctic. They have only secured this for some of the waters of the Norwegian archipelago of Svalbard.

Considerable pressure has been placed on the IMO by environmental non-profit organisations. At the forefront has been the Clean Arctic Alliance, which led a multi-year campaign to ban HFO in the Arctic. They gained increasing support from many of the member states of the IMO. In 2020, the IMO agreed on a draft regulation to phase out the use and carriage of HFO, recommending a ban on Arctic shipping in the circumpolar north, starting from July 2024. This is the first-ever agreement to limit greenhouse gas emissions from shipping. Although Russia was initially hesitant, it eventually supported the measure, along with the other countries bordering the Arctic Ocean. The designated boundary of the zone where the bans apply is shown in Figure 6.6.4. There were some key exceptions:

- ships engaged in search and rescue or oil spill preparedness and response
- nation-states whose coastlines border the Arctic waters can waive the ban if their ships are flying their own flag and they are operating in their sovereign waters where they have jurisdiction. This was integral to securing Russia's support as they rely heavily on domestic journeys servicing coastal communities.

6.6.4 The International Maritime Organization's Arctic boundary designation

Other operators in the Arctic move to cleaner fuels

Apart from regulatory efforts, several shipping operators, including those operating in Arctic waters, have begun to phase out HFO or have even committed to transitioning to carbon-neutral fleets over the coming decades.

Norway's Hurtigruten implemented a voluntary ban on HFO a decade ago and is now moving to even cleaner vessels. The company took ownership of the first hybrid-powered, ice-strengthened expedition cruise ship in June 2018. The vessels can operate for short stints on silent, emission-free fuel cells. Ponant, a French cruise-ship operator with frequent voyages to the Arctic, no longer uses HFO in its fleet of five cruise ships.

Maersk, the world's largest shipping operator, and first company to sail a container ship through the Arctic in September 2018, is committed to operating a carbon-neutral fleet by 2050. It has already achieved a 46 per cent reduction in emissions compared to the 2007 baseline. The company says that the next five to ten years will be crucial for developing and investing in clean fleet technology.

Activities

Acquiring and processing geographical information

1 Describe heavy fuel oil (HFO).
2 Outline the impacts that HFO can have on the Arctic environment.
3 Account for the increasing use of HFO in the Arctic.
4 Describe the work that has been undertaken by the IMO in addressing the threat of HFO.
5 Outline the draft regulation that was agreed on in 2020. Explain what exemptions were needed to gain the support of Russia and other countries bordering the Arctic Ocean.
6 Investigate the current situation regarding regulations banning the use of HFO in the Arctic.
7 Describe how other operators in the Arctic Ocean are moving towards cleaner fuels. What may be motivating them to do so?

Applying and communicating geographical understanding

8 Study Figure 6.6.1. Using data from this graph, answer the following questions.
   a Calculate the total tons of black carbon emitted by all ships using HFO in the Arctic.
   b Calculate the percentage of combined emissions from general cargo ships and oil tankers.
9 Study Figure 6.6.2. Using data from this graph, calculate the percentage of Russia's contribution to black carbon emitted in the Arctic. Suggest why it dominates.

APPLICATION AND CONSOLIDATION TASKS

Task 1: Class debate

Conduct a class debate on the topic 'Atmospheric carbon levels are likely to be reduced in the future and the balance in glacial systems will be restored'.

To prepare for the debate:

- write down all the arguments for and against the topic
- re-read this chapter (and undertake further research) to find data to support your argument.

Task 2: Report writing

Write a short report assessing how land use changes have interrupted the operation of glacial systems and contributed to the accelerated melting of ice sheets and glaciers; and the impacts of the melting ice sheets and glaciers.

In your report, consider the framework that integrates the effects on ecosystem services (see Figure 6.5.7).

Task 3: Create an infographic

Prepare an infographic promoting the benefits of cruise ships bringing tourists to the Arctic using cleaner fuels. Your task is to encourage tourists to select a tourism operator that only uses clean fuels and has sustainable practices.

Your infographic should include the following information:

- what HFO is
- why HFO is a problem
- what the tourism operator is doing
- what the benefits are of using cleaner fuels
- reasons to take a cruise to the Arctic.

SECTION

2 People, patterns and processes

The footprint of Earth's 8 billion people is evident everywhere we look. Not even the most remote polar outpost, the most inhospitable desert, the most distant island, the deepest ocean trench or the world's highest mountain peak is free of human impacts. The agricultural systems that supply the food and fibres we need, the industrial complexes that produce the goods we consume, the urban centres that house more than half the world's population, and the transport networks vital to the movement of people, goods and services, have had far-reaching effects on the global environment. To this, we add the world's cultural diversity—its mosaic of customs and traditions, religions, languages, social organisation, forms of government and economic systems.

The world's cultural landscapes arose from the complex interactions between cultures and the natural environment. Each culture has its own practices, preferences, values and aspirations. The outcome of all such interactions is a 'humanised' version of this planet's landscapes.

Content focus

In this section of the text, students have the opportunity to examine evidence of human diversity across Earth's surface. In doing so, they develop an understanding of the spatial patterns and extent of the human footprint, and the human transformations shaping those patterns. They investigate how systems of production, distribution and consumption shape places through the study of population change, resource consumption and international integration. To consolidate this knowledge, and better understand the human processes that shape the unique character of places, students have the opportunity to focus on either place and cultural change, political power and contested spaces, or human resilience in diverse environments.

In this section

CHAPTER

7 Human activity on Earth

Human activity has impacted Earth's surface and its atmosphere. Nearly 95 per cent of Earth's surface shows some form of human modification, with 85 per cent bearing evidence of multiple forms of human impact. The anthropogenic stressors that dominate the more highly modified biomes include dense human settlements, agricultural land uses, networks of infrastructure and industrial activities.

Only 5 per cent of the world's lands are largely unaffected by humans, while 44 per cent are described as having a low modification. The remainder has a moderate to a high degree of modification—34 per cent (moderate); 13 per cent (high); and 4 per cent classified as (very high).

This chapter studies Earth's surface from a human perspective, focusing on the diversity and extent of the human activity and the spatial patterns related to settlement, infrastructure, industrial production and services. Also examined are the spatial patterns of the world's indigenous peoples and their languages.

It is our collective and individual responsibility ... To preserve and tend to the world in which we all live.

Dalai Lama, Tibetan spiritual leader

7.0.1 The busy Shinjuku district of Tokyo, Japan

Chapter glossary

capital wealth available to a business, in the form of money or assets

deindustrialisation the relative decline (and in extreme cases, absolute decline) in industrial employment in core industrial regions of the developed world

developed country an industrialised country

developing country a country in the process of becoming industrialised

economic activity a process that leads to the manufacture of a good or the provision of a service

economic restructuring the significant and enduring changes in the nature and structure of the economy, brought about primarily by the emergence of the global economy

environmental degradation any deterioration evident in the state of an environment through depletion of resources such as air, water and soil, habitat destruction, the extinction of wildlife, and pollution; any change or disturbance perceived to be damaging

green revolution the development and use of high-yielding varieties of rice and other staple crops

infrastructure the public and private physical structures required for the functioning of an economy and society

international division of labour the specialisation of particular countries in distinct branches of production, whether this be in certain products, or selected parts of the production process

megacity a very large city, typically one with a population of over 10 million people

service sector the provision of services (intangibles) rather than the production of products or goods. It involves providing services to other businesses, as well as final consumers

settlement a place in which people live

urbanisation an increase in the proportion of people living in towns and cities

UNIT 7.1

The diversity and extent of human activity on Earth

Humans are the world's dominant species. People have power, evident in the technologies used to sustain, add to, or degrade the planet's natural endowment, and the resources necessary to sustain life and economies. Humans can determine which forests are to be cleared or protected and redirect the flow of river systems. They pollute waterways and drain aquifers. Use of fossil fuels warms the climate and increases ocean acidity. Habitat destruction contributes to species extinction.

At the same time, the quality of life most people experience has greatly improved—a process driven by economic growth, scientific research, interventions of governments, and the creative endeavours and grassroots activism of people.

SPOTLIGHT

Ecological footprint

The simplest way to define the concept of 'ecological footprint' is to think of it as the area of biologically productive land and water required to produce the goods consumed and to assimilate the wastes they generate. More simply, it is the amount of the environment necessary to produce the goods and services necessary to support a particular lifestyle.

Figure 7.1.1 shows the ancient footprints of Acahualinca. They belonged to 15 Paleo-Indians from around 6000 years ago. While few of our footprints will last as long, our ecological footprints are likely to have environmental impacts many times greater than those left in the mud of Managua.

7.1.1 Footprints preserved in volcanic ash and mud in Managua, Nicaragua

Human achievement

In improving material wellbeing, humans have developed an amazing range of useful (and some not-so-useful) materials and products. These include high-yielding crops and sophisticated food supply chains that greatly improve the variety and quality of consumed food. Humans learnt how to cure diseases and prolong life, split the atom and send satellites into space. High-tech building products, such as carbon fibres that allow aircraft to fly further using less fuel, have been developed. There are new sources of energy (geothermal, hydrogen, solar and wind), which are gradually reducing reliance on fossil fuels. Artificial environments in the form of buildings and cities have been created. Computers enhance problem-solving and robots perform work in highly automated factories. Electronic networks enable instantaneous global communication, transforming the ways in which people interact, and providing access to vast amounts of information. New forms of **economic activity** have become possible. Many workplaces are safer and the work performed is often more rewarding.

The air and water are getting cleaner in many parts of the world. People are better protected from the toxic chemicals that once polluted our bodies and caused health problems. Humanity has successfully protected some endangered species and ecosystems and restored some wetlands. Forests are growing back in some areas that were cleared.

Globally, life expectancy is increasing, infant mortality is declining, more children have access to education and the population growth rate has slowed down.

These improvements in life and environmental quality, underpinned as they are by scientific research and technological advances, are possible through the wealth generated by economic activity. Humans are now globally connected with growing access to information.

Challenges remain

These advances are still not universally shared. While great strides have been made in alleviating global poverty, too many people still live in circumstances most would find unacceptable. More than 1.2 billion people have been lifted out of poverty since 1990, but 9.2 per cent of the world's population (almost 720 million people) continue to live in extreme poverty (down from 36 per cent in 1990).

Despite advances in wellbeing, human activity still strains Earth's natural functions that enable global ecosystem services to sustain future generations. People are living unsustainably by wasting, depleting and degrading Earth's life-sustaining capital—a process geographers call **environmental degradation**.

Human activities directly affect an estimated 83 per cent of Earth's land surface (excluding Antarctica). These lands are used to grow crops, graze livestock and produce energy. They also accommodate urban areas, industry, mining, forestry, transport **infrastructure,** and a range of commercial and recreational activities.

In many places, forests and grasslands are shrinking, deserts are expanding and topsoil is eroding. The atmosphere is warming, ice sheets and glaciers are melting at an accelerating rate, sea levels are rising, and ocean acidity is increasing. The frequency and intensity of extreme weather events are increasing in many areas. Rivers and underground aquifers are running dry as the rate of extraction exceeds the speed at which the water cycle can replenish them. The world's fisheries are being depleted. People litter the land and oceans with wastes faster than they can be recycled by natural processes. Twenty per cent of the world's coral species have been lost and many others are threatened.

Did you know?

Species loss is occurring at a rate at least 100 times faster than pre-humanity. This rate is expected to accelerate and result in anthropogenic mass extinction.

Spatial patterns of settlement, infrastructure and economic activity

The most distinctive ecological footprints, and those that have had the greatest impact in terms of land cover change, are **settlements**, infrastructure and **economic activity**. A settlement is any place in which people live. They can range in size from an isolated homestead to the world's largest **megacities**. Infrastructure includes all those structures associated with utilities (e.g. water, electricity, gas and telecommunications distribution) and transport (by road, rail, air and water). Economic activity is any activity of providing, making, buying or selling commodities or services by people to satisfy their needs and wants. It includes manufacturing, distributing or utilising products or services. The changing spatial patterns created by settlements, infrastructure and economic activity are of particular interest to geographers.

Activities

Acquiring and processing geographical information

1. In your own words, outline the power of humans to impact the planet's natural endowment.
2. Explain what is meant by the concept of an ecological footprint.
3. State the ways in which the material wellbeing of humanity has been improved by the ingenuity of people.
4. Distinguish between settlements, infrastructure and economic activity.

Applying and communicating geographical understanding

5. As a class, brainstorm the ways in which people have modified the planet's natural endowment, beginning with your immediate surroundings.

UNIT 7.2
Patterns of settlement

The spatial distribution of the world's largest urban centres is shown in Figure 7.2.1. For the first time in human history, more than half of the world's population lives in urban places (towns and cities). The urban population stands at 3.7 billion and is projected to almost double by 2050.

This 'big shift' results from the process of **urbanisation**. Much of the shift from rural to urban living is occurring in **developing countries**. This global transformation has vast implications for broad issues, including food, water and energy consumption. Moving towards greater urban concentrations makes city life a reality for more of the world's population.

The big city is a relatively recent development. In 1900 there were only 12 cities with a population of over 1 million people, including London, New York, Paris, Beijing and Tokyo. By 1950, this number had risen to 83, and to 512 by 2020. Projections indicate that by 2030, 662 cities will have over 1 million residents.

Megacities and small cities

Cities with more than 10 million inhabitants are often called megacities. In 2020, there were 33 megacities, projected to rise to 41 by 2030. See Figures 7.2.1 and 7.2.2.

Of the cities predicted to rise above 5 million inhabitants, 15 are in Asia and 10 in Africa. In 2030, 63 cities are projected to have 5–10 million inhabitants. Most of the world's cities have fewer than 5 million inhabitants.

	0.5–1 million	1–5 million	5–10 million	>10 million (megacities)
2020	598	467	48	33
2030 (projected)	710	597	63	41

7.2.1 The number of cities by population in 2020 and the estimated number by 2030

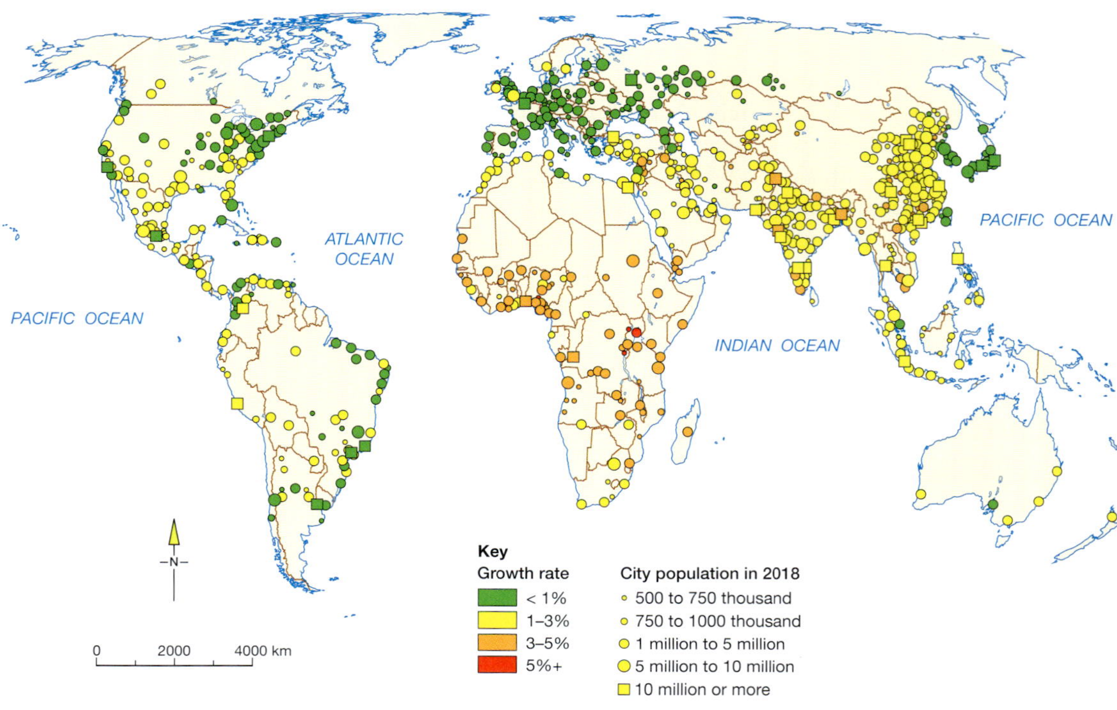

7.2.2 Urban centres by population and growth rate, 2018–2030

The environmental footprint of cities

The concentration of the world's population in urban settlements has both positive and negative environmental consequences. Having people live in just two per cent of the global land area allows for wildlife conservation and biodiversity hotspots to remain relatively undisturbed. However, urbanisation concentrates pollutants that disproportionately impact the physical setting of urban concentrations.

There is one qualification. Urban sprawl, a term initially used to describe the process of suburbanisation of land-rich **developed countries**, such as the USA, Canada and Australia, is now occurring in cities globally. The physical extent of urban areas is growing much faster than their populations, consuming more land for urban development. The expansion of cities has vital implications for energy consumption, greenhouse gas emissions, climate change and environmental degradation. Urban sprawl, in the absence of additional public transport, increases peoples' reliance on energy-demanding private transport. In response, governments seek new technologies (e.g. electric vehicles and building new public transport infrastructure) to reduce the carbon footprints in cities.

● **SPOTLIGHT**

Evolution of human settlements

Humans have lived together in settlements for thousands of years. They first appeared when advances in agriculture enabled surplus food to be produced. From this, emerged a division of labour and trade. Farmers could trade surplus food for goods produced by people with specialised (non-agricultural) skills (e.g. blacksmiths, potters, weavers). Markets—the space in which trade took place—became a central land-use feature of the earliest cities.

By the fourth and third millennium BCE, relatively sophisticated civilisations, along with the first cities, had developed in the river valleys of Mesopotamia. This stretched across parts of modern-day Iraq, Syria, Iran, Turkey, India, China and Egypt.

Historically, city dwellers represented a small proportion of humanity. This is no longer the case.

Did you know?

The ancient city of Uruk on the banks of the Euphrates River in Mesopotamia is considered the world's oldest city. First settled around 4500 BCE, at its height (circa 2000 BCE) Uruk had 50 000–80 000 residents living in a 6 km^2 walled area.

Activities

Acquiring and processing geographical information

1 State the proportion of the world's population living in urban places. What process is responsible for this development?

2 Describe the distribution of the world's megacities.

3 Outline the positive and negative environmental consequences of urbanisation and urban sprawl.

4 Outline the evolution of urban settlements.

Applying and communicating geographical understanding

5 Study Figure 7.2.2. With the aid of an atlas, describe the spatial distribution of cities that have a growth rate of greater than three per cent.

UNIT 7.3

Patterns of infrastructure

The infrastructure our way of life depends on is all around us. Infrastructure includes all public and privately owned structures associated with utilities. These include the distribution of water, electricity, gas and telecommunications, waste removal and treatment (such as sewage, household refuse and recyclables) and transport (road, rail, air and water). Add to this, facilities provided for the public good, including schools, hospitals, parks and other recreational facilities (see Figures 7.3.1 and 7.3.2).

In the past, the world's best infrastructure was found in developed countries. This is now not always the case. The infrastructure in many developed countries is ageing and is starved of funding by governments that prioritise lower taxes above providing public goods and services. In the USA, for example, infrastructure has been decaying for decades and is seriously lacking investment. Meanwhile, some developing countries, notably China, invested heavily in very modern and advanced infrastructure. In the world's least developed countries, there is an ongoing deficit in the infrastructure needed to meet the needs of people.

7.3.1 Power grids distribute the energy needed to power the economy and homes.

7.3.2 Transport infrastructure, like the interchange of Los Angeles and Century freeways, keeps large cities functioning.

SPOTLIGHT

America's decaying infrastructure

In 2020 it was estimated the USA needs to spend around US$4.5 trillion by 2025 to fix the country's roads, bridges, dams and other infrastructure. The American Society of Civil Engineers has given the country's crumbling infrastructure a D+ grade. Roads and bridges, airports and schools need urgent renewal (see Figure 7.3.3). Newer networks critical to modern life, like broadband internet, have not yet reached many parts of the country. In 2021, US Congress approved a US$1 trillion infrastructure bill to rebuild the nation's deteriorating roads and bridges, and fund new climate resilience and broadband initiatives.

7.3.3 The iconic Brooklyn Bridge, New York City, USA, is in urgent need of repair.

Infrastructure building

China

Infrastructure is seen as the key to China's future economic success and central to a commitment to lift millions out of poverty. The country has invested heavily in constructing new transport infrastructure, including highways, railway systems, port facilities and airports (see Figure 7.3.4).

As part of its 2020 post-COVID-19 relief package, China ramped up plans to construct new digital infrastructure nationwide—including building 5G networks, artificial intelligence (AI) and the Internet of Things. Vast amounts will go towards intercity high-speed rail, ultra-high voltage power transmission and electric vehicle charging stations. The new infrastructure investment is expected to be US$1.43 trillion to US$2.51 trillion over 2020–2025.

7.3.4 The world's largest single-building airport terminal, Beijing's futuristic Daxing international airport opened in 2019 and cost US$11.4 billion.

● **SPOTLIGHT**

Shanghai's metro system

In just 27 years (1993–2020) Shanghai built the world's biggest metro system. Its extensive network has 700 kilometres of track with 408 stations on 18 lines. Second in the world by annual ridership, with 3.88 billion rides taken in 2019. Only Tokyo's metro is busier. Over 10 million people use the system on an average workday.

Expansion plans will extend the network to 25 lines over more than 1000 kilometres of track by 2025. By then, every location in central Shanghai will be within 600 metres of a metro station. In 2020, Sydney had only one 36-kilometre-long metro line with 13 stations from Tallawong to Chatswood.

7.3.5 Shanghai Metro ticket machine

Did you know?

China has just under 38 000 km of high-speed railway passenger track—about two-thirds of the world's commercial high-speed railway tracks. It plans to extend this to 70 000 km by 2035. It also has the world's most extensively used railway service, with 2.29 billion high-speed train trips taken in 2019.

Over 1000 sets of trains operate at speeds of 250 km/h to 350 km/h with the fastest trains capable of a maximum test speed of 487.3 km/h.

Sydney

After decades of a focus on reducing taxes and government debt, in 2008 the NSW Government initiated the greatest infrastructure spending in the state's history. Much of it was financed by borrowing money and selling off state assets, such as electricity generation and distribution networks, and motorways.

Major construction projects include the City and Southwest Metro to extend the existing Metro Northwest to a 31-station, 66-kilometre standalone metro line, due for completion in 2024. The Metro West line will connect Parramatta to the city. Road projects include the West Connex motorway, a 23-kilometre link connecting the M4 with the M5 and M8. Eventually, it will link to the planned Western Harbour Tunnel and Beaches Link and M6 Motorway heading to the city's south. The NorthConnex link between the M2 and the M1 has been completed. The city's biggest infrastructure project is the new Western Sydney Airport at Badgery's Creek.

The evolution of infrastructure

This infrastructure evolution is driven by technological advances. The world's railway infrastructure developed in response to steam power, the key technological advance driving the Industrial Revolution. It led to a period of decline in the system of canals built to transport raw materials and finished products. Notable exceptions are the Suez and Panama canals, created to accommodate large ocean-going vessels, central to world trade development. Internal combustion engines led to road-based infrastructure, particularly the rapid expansion of private car ownership after World War I. Airport infrastructure developed in direct response to aviation technology advances. Communications infrastructure developed in response to technological advances in the information sector.

● SPOTLIGHT

Changing technology of shipping and ports

Sail has been used to power vessels for up to 8000 years. The first significant change in shipping technology occurred with sail-powered clipper ships in the 1840s and 1850s. They were built for speed with streamlined iron hulls and large sail areas designed to catch the slightest breeze, representing the peak of sailing ship technology.

By the 1870s clipper ships were replaced by steam-powered vessels that could maintain greater average speeds. Some vessels initially combined both sail and steam, but eventually steam triumphed. Steamships were replaced by diesel-driven vessels mid-twentieth century. These vessels grew bigger and became highly specialised in cargo carrying.

The next generation of vessels is likely to be powered by liquefied natural gas (LNG). The relatively low emissions and cost advantages of LNG are major incentives.

The technology for handling cargo on the waterfront has also changed. Before containerisation was introduced in the 1950s, goods were mostly carried as general cargo. They were delivered to the dock by horse-drawn dray, rail, or later, motorised vehicles, and stored in warehouses before the ship arrived. Once the vessel arrived, the stored cargo was taken to the wharf and then loaded onboard using cranes. Unloading reversed this operation (see Figure 7.3.6). Sydney's remaining heritage finger wharves are a legacy of this mode of cargo handling.

By the 1960s the increasing size of ships and containerisation heralded a dramatic transformation in ports worldwide. Their existing infrastructure became increasingly obsolete as new container-handling facilities were built near waters deep enough to accommodate the larger vessels (see Figure 7.3.7). Port facilities that had been serving maritime trade for over a century were abandoned. Such land has often undergone large-scale urban renewal. Sydney's Barangaroo (see Figures 7.3.8 and 7.3.9) and London's Docklands are notable examples.

By 2020 the world's shipping fleet had grown to almost 52 000 commercial vessels over 1000 tonnes. These ships carried more than 11 billion tons of cargo. In terms of tonnage, 29 per cent of ships were tankers, 43 per cent bulk carriers, 13 per cent container ships and 15 per cent other types of vessels. The technological advances in shipping and cargo handling were critical factors in the expansion of world trade and a key driver in global economic integration.

7.3.6 Ships being unloaded at Millers Point, Sydney, in the 1920s

7.3.7 Goods are handled in highly automated container terminals, Port Botany, Sydney.

7.3.8 Barangaroo, 1937. The port infrastructure reflects the technology of shipping and cargo handling in place at the time.

7.3.9 Bangaroo, 2022

Activities

Acquiring and processing geographical information

1. Name the types of infrastructure addressed in this section of the text.
2. Explain why some of the most advanced infrastructure is now found in developing, rather than developed countries.
3. Outline the evolution of infrastructure.

Applying and communicating geographical understanding

4. Study Figures 7.3.1 and 7.3.2. Identify the types of infrastructure evident in each image. How do these images inform our understanding of the type of structures our lifestyle depends on?
5. What benefits will arise from funds being spent on addressing America's decaying infrastructure?
6. Why is the Chinese government so focused on developing the country's infrastructure? What are China's infrastructure priorities?
7. Contrast the emphasis given to public transport infrastructure in Shanghai and Sydney.
8. Outline changes in the technology of shipping and cargo handling since the age of sail. What have been the implications of these changes for the world's ports?
9. Study Figures 7.3.8 and 7.3.9. Describe the changes observed in the port-related infrastructure shown in the two photographs.

UNIT 7.4

Patterns of economic activity: Agriculture

Agriculture is the process of cultivating crops and raising livestock for subsistence and/or profit. Over millennia, people developed sophisticated ways of transforming land into food and fibre-producing systems, reflecting the nature of their relationship with their environment.

In studying the spatial pattern of agriculture it's vital to understand the diverse ways in which humans have learnt to modify the biophysical environment to feed themselves, their families and the global population (see Figure 7.4.1).

There is no definitive point in time at which agriculture originated. Before people developed crop-growing and livestock-grazing skills, they obtained food by hunting and gathering.

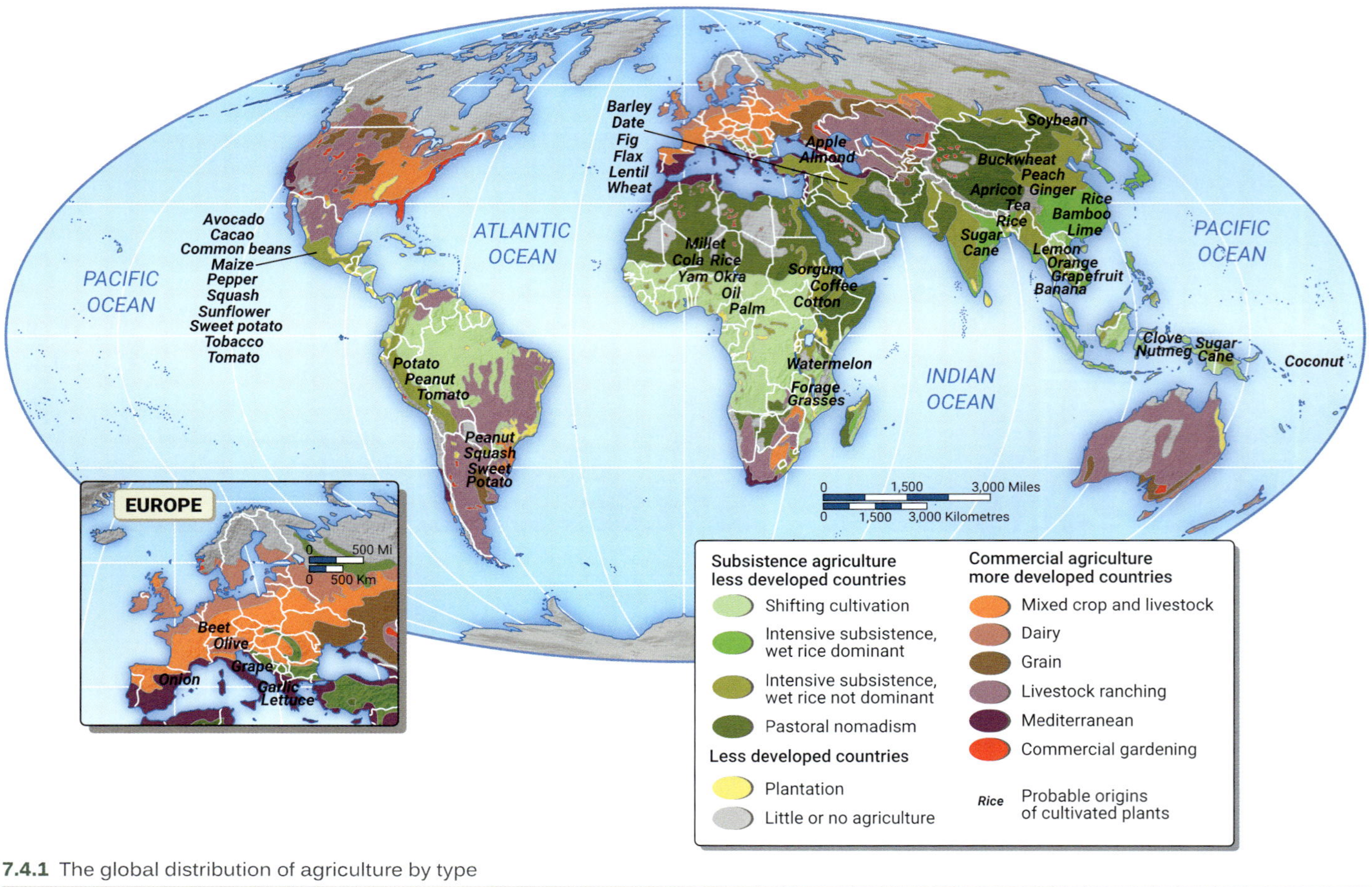

7.4.1 The global distribution of agriculture by type

Subsistence agriculture

Subsistence agriculture slowly replaced hunting and gathering when people developed the ability to domesticate animals and plant species. This advance allowed them to settle in one place. Subsistence agriculture is where a farmer and their family consume most of what they produce. It has declined, but is still widely practised by people living in developing countries. There are usually three kinds of subsistence activities.

Shifting cultivation focuses on maintaining the fertility of the soil by rotating cultivated fields. Plots are cultivated until the soil shows signs of exhaustion. They are abandoned to allow the native vegetation to regrow while cultivators move on to another plot.

This style is now largely confined to developing countries in tropical regions, especially rainforests in Central and West Africa, and parts of South-East Asia, where climate, rainfall and vegetation combine to produce soils that lack nutrients. Slash-and-burn clearing is typically the first stage of the farming cycle. Larger trees are felled and their remains burnt. Crops are planted in the ash-enriched soils. The clearing is used for two to three years until soil fertility declines. It is abandoned and the forest is allowed to regenerate.

Intensive subsistence agriculture focuses on the effective and efficient use of small areas of land to maximise crop yields. Large inputs of labour and fertiliser are required. Such practices can support large population densities. It is mostly found in Asia, especially in India, China and South-East Asia. These regions generally have high population growth rates and limited land, as well as a range of environmental restraints. This has led people to reshape the landscape. Terraced hillsides are common, and the fields are planted year after year, as adding fertiliser helps maintain soil fertility. Rice production dominates, especially in wetter areas (see Figure 7.4.2).

7.4.2 Women practise intensive agriculture by planting rice in a paddy field in Lombok, Indonesia.

Pastoralism involves traditional practices around managing domesticated livestock such as cattle, goats, camels and reindeer. Animals are bred and herded to provide food, clothing and shelter. It is typically found in regions that are too cold or dry to support subsistence agriculture. These include savanna grasslands, deserts and steppes (lightly wooded grassy plains found in cooler continental areas of south-eastern Europe or Siberia). Today, pastoralism is largely confined to parts of North Africa (particularly the Sahel, see Figure 7.4.3), the savannas of central and southern Africa, the Middle East and Central Asia.

7.4.3 A Fulani man herding cattle, practising pastoralism in the Sahel

Commercial agriculture

Commercial agriculture sees farmers producing crops and livestock primarily for sale. It can be practised on either an extensive or intensive scale.

Extensive commercial agriculture uses relatively small inputs of labour, fertilisers and **capital**, relative to the land area being farmed. Examples include cattle and sheep grazing and grain growing, such as wheat, oats and millet (see Figures 7.4.4 and 7.4.5).

Extensive farming is found in the mid-latitudes of most continents, and in semi-arid regions where water for cropping is not available. Its nature means it requires less rainfall than intensive farming.

Did you know?

One-third of the food produced globally is either lost or wasted. Reducing this waste is critical to improving food and nutrition security, meeting climate goals and reducing environmental stress.

7.4.4 and 7.4.5 Extensive agriculture examples include wheat harvesting in Canada and sheep grazing in New Zealand

7.4.6 A cattle feedlot near Hereford, Texas, USA

Intensive or industrial agriculture requires capital-intensive agricultural inputs, such as fertiliser, pesticides and large-scale machinery, to maximise output per unit of agricultural land. It is widespread in developed nations and increasingly common worldwide. It involves agribusinesses (many of which are transnational corporations) and industrial-scale infrastructure. Cattle feedlots (see Figure 7.4.6), large climate-controlled greenhouses and poultry battery cages are examples. Much of the meat, dairy, eggs, fruit and vegetables sold in supermarkets is produced this way.

All forms of commercial agriculture have become increasingly reliant on selective breeding, genetic engineering and new technology to increase yields. This involves farmers analysing growing conditions (weather, soil, water, weeds and pests) using technology such as drones and sensors to help them manage the land most effectively.

Did you know?

Agriculture is an important driver of economic growth and accounts for around 4 per cent of global GDP (gross domestic product). In some developing countries, this can be over 25 per cent of GDP.

Extensive agriculture has a range of advantages:

- Less reliance on inputs of labour per unit of land, especially since costly and intrusive alterations to the land (e.g. terracing) are not required.
- Large-scale machinery can be used more efficiently over large, flat areas of land.
- Labour efficiencies result in generally lower product prices and higher returns.
- Animal welfare is less of an issue compared with, for example, high-capacity feedlots and battery cages.
- Animals typically graze on pastures native to the area, rather than relying on introduced species.
- Soil management is easier, there is less use of fertilisers and chemicals.
- Large amounts of grain and livestock can be produced relatively cheaply.

Intensive agriculture has some advantages:

- Yields are much higher, at least in the short term.
- The area of land required is smaller, given the intensive, capital-intensive nature of the enterprise.

● **SPOTLIGHT**

Plantations

Plantations are a form of commercial agriculture. They are large-scale estates specialising in crops such as cotton, coffee, tea, cocoa, sisal, oil palms, fruits and rubber trees. They are typically monocultures—the same plant species is grown across a given area (see Figure 7.4.7).

Their environmental impact is often focused on the site where the plantation is established. Natural forest is cleared, reducing biodiversity and habitat.

Crops are typically destined for a global market, playing an important role in integrating the economies of developing countries into the global economy.

7.4.7 Tea plantation pickers, Sri Lanka

● SPOTLIGHT

Mediterranean agriculture

Six major regions of the world have a Mediterranean type of climate. These are the countries of the Mediterranean Basin (France, Spain, Italy (see Figure 7.4.8), Algeria, Morocco, Turkey, Tunisia, Israel and the Northern Nile valley of Egypt), California in the USA, Central Chile, the southern part of South Africa, the lower Murray–Darling Basin of South Australia and south-west Western Australia.

Regions that host Mediterranean agriculture feature similar biophysical environment characteristics. These include erratic rainfall, mild temperatures, irregular topography and proximity to large water bodies. The resulting land uses are intensive, highly specialised and varied. Subsistence agriculture often occurs side-by-side with commercial farming. Crops such as wheat, barley and vegetables are often grown for domestic consumption, while others, such as citrus fruits, olives and grapes, are mainly for sale.

7.4.8 Vineyards and olive groves, Tuscany, Italy

Today, food and fibre production are big businesses. The food industry is especially complex. It includes the diverse range of businesses that participate in the supply chain and provide much of the food energy consumed globally. These businesses involve farmers and ranchers; those involved in research and development; food processors; the transport and logistics industry; farm machinery suppliers; and food distributors, marketers, wholesalers and retailers. Only subsistence farmers—those who survive on what they grow—fall outside the scope of the modern food industry.

Fibre production includes commonly used products such as cotton, hemp, jute, flax and sisal. Plant fibres are used in the manufacture of textiles and paper.

Did you know?

Agriculture, forestry and land-use change are responsible for about 25% of greenhouse gas emissions. Agricultural mitigation must be part of any response to climate change.

The evolution of agriculture

Agriculture evolved via a series of distinct phases or revolutions. Significantly, they did not occur everywhere simultaneously. Therefore, some parts of the world remain largely unaffected by such changes. Generally, the world has progressed from largely subsistence-based agricultural systems to predominantly capital-intensive, commercial systems.

The first agricultural revolution was founded on the development of seed-based agriculture using ploughs and draft animals (see Figure 7.4.9). It originated when crops were domesticated. Animal-based practices began when humans domesticated animals, including sheep and goats.

Seed-based agriculture originated in several regions almost simultaneously, each with a distinctive cluster of plants drawn from the local flora:

- Mesoamerica (Mexico and Guatemala): corn, beans, squash, papaya, tomatoes, chilli, peppers (capsicum)
- The Fertile Crescent (the Middle East from the Nile Valley to the Tigris and Euphrates rivers): wheat, barley, grapes, apples, figs, melons, lentils, dates
- North China (Yellow River Valley): rice, soybeans
- Africa: sorghum, cowpeas, yams, oil palm
- South America: potatoes, sweet potatoes, cassava, peanuts, pineapples

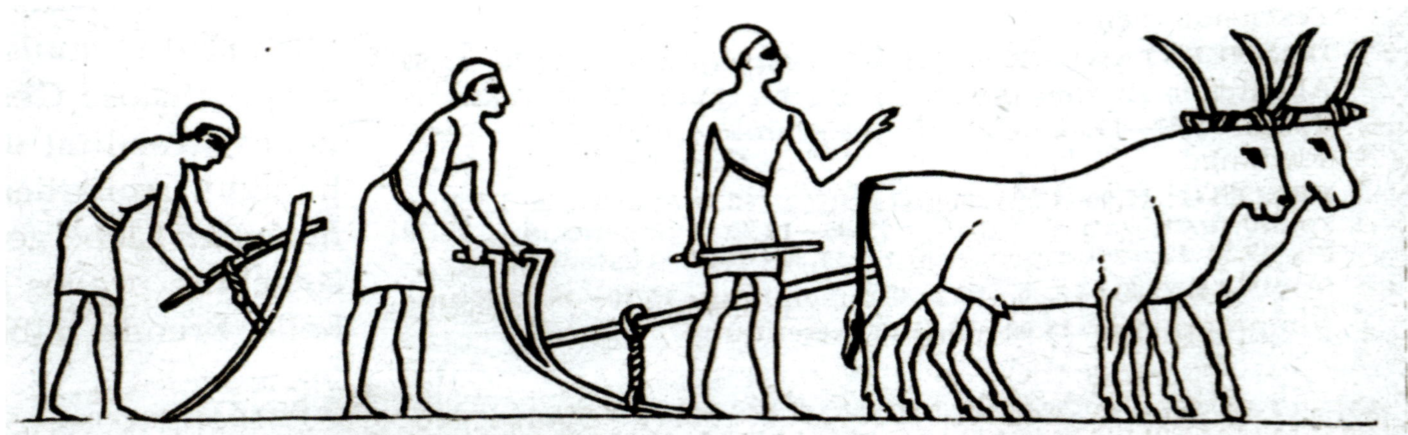
7.4.9 An engraving of people ploughing fields with oxen, Ancient Egypt

Domesticating plants and animals led to a settled way of life. Villages emerged bringing new types of social, economic and political organisation. On the fertile floodplains of the Tigris, Euphrates and Nile rivers, complex civilisations developed. Their new agricultural practices spread to other parts of the world.

7.4.10 Engraving of a hay harvest by horse-drawn machinery in Australia, circa 1880

The second agricultural revolution had several defining phases. The first included dramatic improvements in crop yields, developments such as fertilisers and field drainage systems, and improvements to the horse-drawn tools used in agriculture. This phase coincided with the social changes that stemmed from the Industrial Revolution in the late eighteenth century. Consequently, subsistence-based agricultural systems began to break down. The system of peasant farmers dependent on a land-owning elite was gradually replaced by private property ownership, and communal lands were replaced by enclosed, individually owned land.

Many other defining developments then were in response to far-reaching social and economic changes accompanying the Industrial Revolution. The most vital was the creation of a commercial market for food, the inevitable outcome of an emerging urban industrial workforce. The associated technological innovations helped transform agriculture. New types of horse-drawn machinery helped to increase yields (see Figure 7.4.10). Improvements in transport technologies aided food distribution.

7.4.11 Mechanical harvesting of grapes in a vineyard in France

The third agricultural revolution began in the late nineteenth century and extended into the early twenty-first century. Advances in farm mechanisation and artificial fertilisers, along with the emergence of global food supply chains, transformed agriculture. Mechanisation resulted in the gradual replacement of most human labour with machines, including tractors, combine harvesters, reapers and pickers (see Figure 7.4.11).

Advances in chemical-based farming applications have also played a role in increasing yields. Synthetic fertilisers promote plant growth, and herbicides, fungicides and pesticides are used to reduce diseases and pests. Biological innovation is also important.

Advances in food processing, cold storage technologies and sophisticated packaging have greatly increased the shelf-life of food and enabled it to be transported over long distances. These developments expanded the relationship between farms and manufacturing firms, a key factor in the industrialisation of agriculture (see Figure 7.4.12).

7.4.12 Beef burgers on the production line in a meat processing and packaging plant

Agricultural industrialisation resulted in farms becoming part of a vertically integrated industrial process that includes production, processing, distribution, marketing and retailing, rather than distinct production units.

The **green revolution** saw high-yielding varieties of rice and other staples being developed and used, largely in Asia and Mexico. Combined with using synthetic fertilisers and small-scale mechanisation, this enabled many developing countries to meet the needs of rapidly growing populations.

A relatively recent development is the increasing focus on non-traditional agricultural exports, especially in developing countries. More governments are moving from a focus on self-sufficiency to commercial crops that are internationally competitive and earn foreign exchange Examples include fruit, vegetables and flowers (see Figure 7.4.13).

7.4.13 A worker in a large greenhouse sprays roses, grown for European markets, in Lusaka, Zambia.

Activities

Acquiring and processing geographical information

1 Identify the drivers of the changing global patterns of economic activity.
2 Define agriculture.
3 Distinguish between shifting agriculture, intensive subsistence agriculture and pastoralism. Give examples of each.
4 Distinguish between extensive commercial agriculture and intensive industrial agriculture. Give examples of each.
5 Outline the advantages of extensive agriculture. What are the advantages of intensive commercial agriculture?

Applying and communicating geographical understanding

6 Study Figure 7.4.1. With the aid of an atlas, write a report describing the global distribution of at least two forms of subsistence agriculture and two types of commercial agriculture.
7 What are the distinguishing characteristics of plantations? What is their environmental impact?
8 Describe the distinguishing characteristics and spatial distribution of Mediterranean agriculture.

Did you know?

Poor diets are the leading cause of death worldwide. In 2020, nearly 690 million people—8.9 per cent of the population—went hungry. The cost of a healthy diet is unaffordable for more than 3 billion people.

UNIT 7.5

Patterns of economic activity: Industrial production

Industrial production refers to the output of the manufacturing, mining and utilities sectors.

Manufacturing is producing goods for use or sale. It involves raw material processing or the assembly of component parts into finished goods through human labour, tool use, machinery and chemical processing. Modern, large-scale manufacturing enables the mass production of goods using assembly line processes and advanced technologies.

The manufacturing sector is a key part of the global economy, accounting for nearly 14.4 per cent of the global GDP in 2021. China is now the world's manufacturing powerhouse, overtaking the USA in 2010. It now accounts for 28.4 per cent of the global manufacturing output and the USA, 16.6 per cent (see Figure 7.5.1).

Rank	Country	Millions of US$
1	China	4 865 827.26
2	USA	2 497 131.58
3	Japan	995 308.97
4	Germany	803 213.14
5	South Korea	461 104.31
6	India	443 911.66
7	Italy	314 095.61
8	UK	274 872.28
9	France	262 642.86
10	Russia	256 958.27
11	Mexico	230 067.71
12	Indonesia	228 324.76
13	Türkiye	181 887.68
14	Ireland	174 522.85
15	Canada	170 222.21
16	Spain	164 651.64
17	Brazil	155 191.76
18	Switzerland	144 042.17
19	Thailand	136 650.26
20	Poland	113 608.85
Australia is ranked 26 (US$86 154.94 million)		

Source: World Bank

7.5.1 World's Top 20 manufacturing countries in US dollars, 2021

The changing spatial pattern

Manufacturing's global spatial pattern has changed significantly over the past 50 years. A key feature is **economic restructuring** and the emergence of a new **international division of labour**.

Global economic integration has been accompanied by the process of **deindustrialisation**—the relative decline (in extreme cases, absolute decline) in industrial employment in core industrial regions of the developed world. The workers most affected were those engaged in labour-intensive manufacturing enterprises with limited profitability. The net effect was a shift in manufacturing jobs to developing countries, resulting in a new international division of labour. The manufacturing jobs retained in developing countries were those in high-value-added manufacturing dependent on a skilled workforce. The expanding **service sector** helped cushion the blow of economic restructuring in developed countries.

The process of deindustrialisation was most apparent in the older industrial regions of the USA, Western Europe and Australia. Manufacturing employment in Western Europe decreased by a third to half in countries such as France, the Netherlands, Norway and Sweden. The most pronounced decline occurred in the UK. The Manufacturing Belt of the US north-east was also seriously affected.

Major manufacturing concentrations

Today there are three principal manufacturing regions—North America, Western and Eastern Europe and South and East Asia—and several smaller, emerging regions.

North America

North America has several major manufacturing concentrations. Most significant is the long-established US Manufacturing Belt, now known as the Rust Belt due to the impacts of economic change on labour-intensive manufacturing. This region encompasses much of the north-eastern USA, including Illinois, Indiana, Ohio and Michigan, and the very southern part of Ontario, Canada.

Other manufacturing concentrations in the USA include California, Texas, North Carolina, Georgia and Florida. Texas has the advantage of no personal or corporate tax and cheap energy.

Eastern and Western Europe

A major manufacturing region, it extends from Eastern Russia and Eastern Europe, through the Rhine-Ruhr Valley of Germany to the UK. Its major exports are oil, synthetic plastic, machinery, wood and raw metals.

South and East Asia

This major manufacturing region is responsible for most of the world's textile, clothing, automobile and electronics production. It is concentrated around ports that access the Pacific Ocean. These include regions around northern Beijing, Shanghai and Nanjing, and the Guangdong region around Hong Kong. Smaller regions are found in India, South Korea, Japan, Vietnam and Bangladesh.

Evolution of manufacturing

The manufacturing of simple tools and weapons dates from early human history. It became more sophisticated as early artisan skills improved. In the pre-industrial world, these skills were passed down through generations often within a family. Most manufacturing occurred in rural areas, where home-based manufacturing supplemented subsistence agriculture.

The factory system accompanied the earliest stages of urbanisation. It developed in the UK from the Industrial Revolution, later spreading around the world. Its defining characteristic was machinery use, initially powered by water, but later by steam and electricity (see Figure 7.5.6 on page 180).

In the USA, an increased reliance on economies of scale ensued. Factories became centralised into industrial centres and interchangeable components were standardised. Combined with a growing population and rising standards of living, America emerged as a major industrial power.

● SPOTLIGHT

The decline of the US industrial north-west

Once the industrial heartland of the USA, the US Manufacturing Belt has experienced deindustrialisation since the 1980s. The region once specialised in manufacturing large-scale, finished industrial and consumer goods, as well as transporting and processing raw materials for heavy industry. The transfer of labour-intensive manufacturing overseas, increased automation, and the decline of the US steel and coal industries powered the decline.

The belt stretched in a wide arc across nine states (see Figure 7.5.2). It skirted the southern shoreline of the Great Lakes and the St Lawrence Seaway, a vital waterway for ships sailing from the Atlantic Ocean to the Great Lakes. It was distinct from the surrounding rural lands of the Corn Belt, known as 'America's breadbasket'.

7.5.2 The US Rust Belt

The region's once-powerful industrial sector centred on industries like steelmaking, automobile manufacturing and coalmining. A general economic decline led to population loss and widespread urban decay. The population of Detroit, Michigan, once North America's car-making capital, declined by almost 30 per cent between 2000 and 2020 (see Figure 7.5.3).

Some cities and towns adapted by diversifying and/or transforming their economies. Others experienced severe economic distress and a range of adverse social consequences. The rise of right-wing political activism, exploited by former president Donald Trump, is attributed to the alienation experienced by those adversely affected by deindustrialisation.

New England also faced industrial decline, but transformed its economy by basing it on services, advanced manufacturing and high-tech industries.

7.5.3 The abandoned Packard Automotive Plant in Detroit, Michigan. It opened in 1903, and once employed 40 000 people.

● SPOTLIGHT

Zhengzhou, China

Zhengzhou, the capital and largest city of Henan Province in China, is a major manufacturing hub. Industrial regions like this are representative of the restructured global economy (see Figures 7.5.4 and 7.5.5).

It boasts a vast network of industrial zones and technology parks, producing a wide range of products. Large corporations are engaged in information technology (IT), biomedicine, aviation and e-commerce.

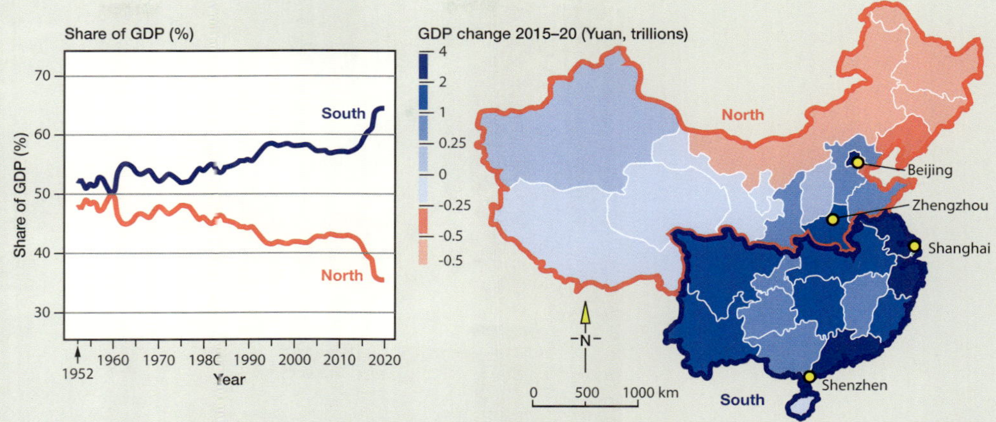

7.5.4 Within China, economic growth is not evenly spread spatially. Economic growth (largely focused on manufacturing) has been stronger in the south-east of the country than in the north.

Foxconn's factory complex 'Apple City' produces 80 per cent of the world's iPhones—over 100 million a year, and employs 300 000 people. Another 14 manufacturers supply 12.5 per cent of the world's smartphones.

Zhengzhou is a major centre of automotive production, with Yutong Bus, Shaolin Bus, Zhengzhou Nissan and Haima Motor all having a major presence. Yutong Bus, China's largest bus and coach manufacturer, produced over 67 000 electric buses in 2020, which were exported to over 130 countries. Zhengzhou Nissan, a joint venture between China and Japan, manufactures 240 000 SUVs and pickup trucks, while Haima Motor produces 300 000 cars annually.

7.5.5 Modern residential towers, Zhengzhou

The city also supplies two-thirds of China's frozen food. Half of the country's 10 largest frozen food producers are based here. Sanquan Foods operates a 5000-worker factory manufacturing over 100 000 frozen dumplings every hour. The Synear Group commands 30 per cent of China's frozen food market.

Some other key Zhengzhou industries are textiles, carbonated beverages, cigarettes, fertiliser and processed meats.

Henry Ford revolutionised the automobile industry and the concept of manufacturing worldwide. He introduced the moving assembly line of the Model T Ford, with workers performing specific steps of the production process. This innovation increased output from one car every 12 hours to one car every 93 minutes (see Figure 7.5.7). By increasing efficiency and reducing costs, it ultimately became the preferred model of manufacturing worldwide. Mass production of consumer goods made them far more affordable and accessible.

Automation is gradually reducing its reliance on human operators, a trend accelerated by modern advances in computer-based technologies and robotics (see Figure 7.5.8).

7.5.6 Wason railroad car manufacturing company facilities in Springfield, Massachusetts, USA, 1872

7.5.7 Henry Ford's Model T motor car assembly line in 1913

7.5.8 Robots weld car parts on an assembly line at an auto plant.

Activities

Acquiring and processing geographical information

1 Define manufacturing. What contribution does it make to the global economy?

2 Define deindustrialisation. Where has it been most apparent? What have been its effects?

3 Locate the world's principal manufacturing regions.

4 Outline the evolution of manufacturing.

Applying and communicating geographical understanding

5 Study Figure 7.5.2. Identify the principal population concentrations of the USA.

  a Where is the US Manufacturing Belt located?

  b What manufacturing activities were carried out in the Rust Belt prior to its decline?

  c What factors contributed to the decline in the region?

  d What have been the impacts of industrial decline in the region?

6 Study the box, Spotlight: Zhengzhou, China and complete the following tasks.

  a Locate Zhengzhou. What types of manufacturing take place in the region?

  b Outline the ways in which Zhengzhou reflects the changes taking place in the global distribution of manufacturing.

  c Using the data presented in Figure 7.5.4, write a report outlining the spatial distribution of economic growth in China and how this has changed over time.

UNIT 7.6

Spatial pattern of the world's indigenous peoples

Indigenous peoples, often called 'first peoples', are culturally distinct groups descended from the earliest known inhabitants of a particular place.

First Nations Australians

First Nations Australians, or Aboriginal and Torres Strait Islander peoples, are descendants of Australia's original inhabitants. With their history spanning over 65 000 years, they are one of the world's oldest continuous cultures. They have strong connections to family, land, languages and cultures. Their cultures form the foundation for social, economic, and individual wellbeing, and are increasingly recognised as a crucial part of Australia's national identity.

The world's indigenous peoples

Worldwide, there are anywhere from 250 to 600 million indigenous people, from about 5000 nations (see Figure 7.6.1). Official designations and terminologies of who is considered to be indigenous vary from country to country. Indigenous status is applied differently around the world. In the Americas, Australia and New Zealand, indigenous status generally applies to groups directly descended from the peoples who lived there before the arrival of Europeans. In Asia and Africa, home to most indigenous peoples, numbers are difficult to determine due to differences in the ways indigenous status is determined.

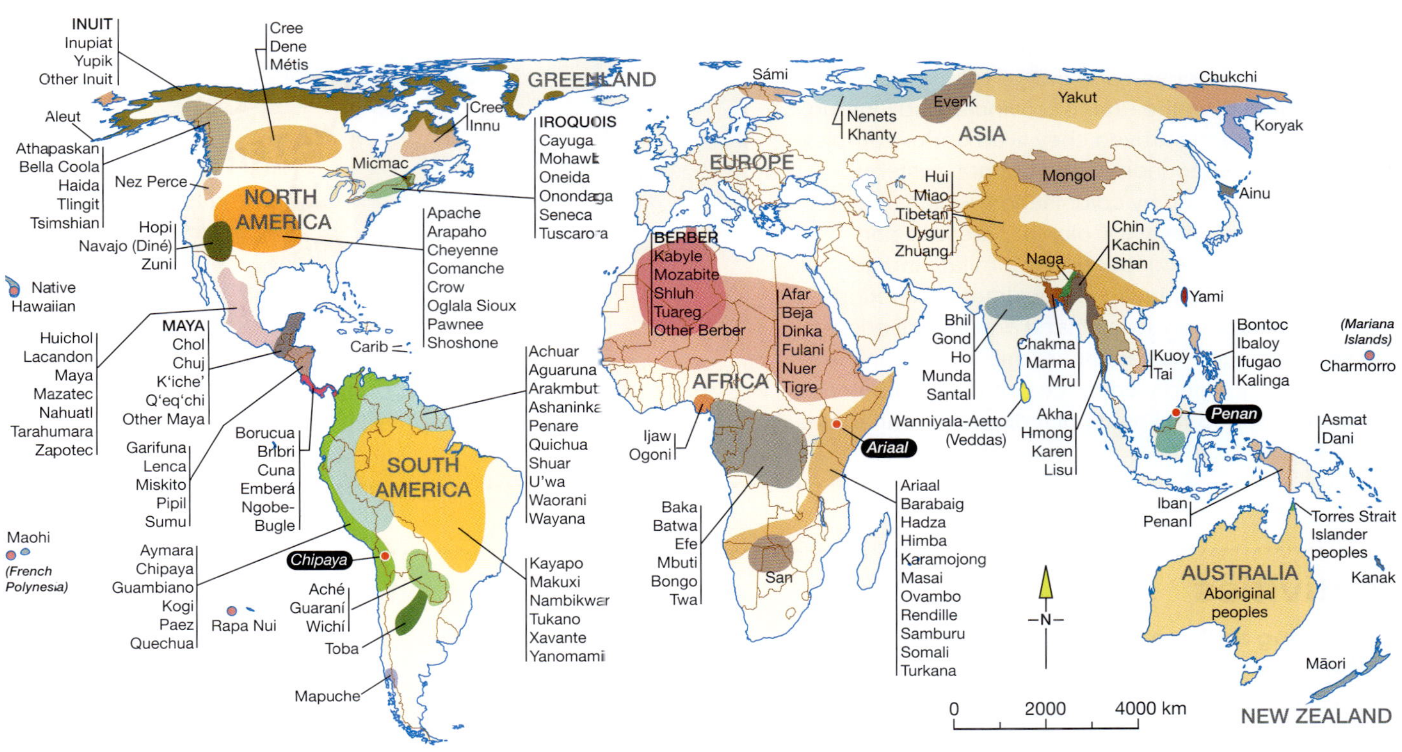

7.6.1 World map of indigenous peoples

Colonisation and dispossession

The lands these people once occupied have usually been colonised by external powers with different worldviews, often resulting in dispossession and disadvantage. This typically involved notions of racial superiority and the imposition of new land uses and economic systems. Many were dispossessed of the lands and resources their cultures depended on.

They may still be disadvantaged across a range of social and wellbeing measures. Languages, ways of knowing and key elements of indigenous cultures are threatened or have been destroyed.

Indigenous peoples typically keep their cultural heritage alive by passing on their knowledge, speaking and teaching their languages, and protecting sacred sites and objects. For many of the world's indigenous people, and specifically Aboriginal and Torres Strait Islander peoples, land is at the core of all spirituality and the relationship to, and the spirit of, Country remains central to their worldview.

Recent decades have seen a growing appreciation of the value of indigenous cultures and their contribution to nations.

● SPOTLIGHT

Indigenous Canadians

The ancestors of Canada's First Nations peoples (The Inuit and Métis) can trace their cultural heritage back 20 000 years. There are now just over 1.6 million people in Canada identified as indigenous, making up 4.9 per cent of the population. The Inuit primarily inhabit the northern regions. Their homeland, Inuit Nunangat, includes much of the land, water and ice contained in the Arctic region. Métis peoples are of mixed European and indigenous ancestry and live mostly in the Prairie provinces and Ontario. Before Europeans arrived, they lived under complex social, political, economic and cultural systems. This was destroyed by colonisation, which altered their traditional lifestyle. Colonial practices and policies sought to control and assimilate indigenous peoples, and continue to impact later generations. Acts of segregation, loss of land and lack of access to services, such as healthcare and education, have had devastating consequences on the health and socio-economic wellbeing of Indigenous Canadians (see Figure 7.6.2).

7.6.2 Canada's Coast Salish peoples march against the Trans Mountain pipeline expansion, to protect their spiritual land.

Activities

Acquiring and processing geographical information

1. Explain what an indigenous person is. What do indigenous people have in common?
2. Outline the impact that colonisers have had on indigenous peoples.
3. How is indigenous culture transmitted from one generation to the next?

Applying and communicating geographical understanding

4. Study the box, Spotlight: Indigenous Canadians. In what ways has the experience of the indigenous peoples of Canada resembled that of Aboriginal and Torres Strait Islander peoples?

UNIT 7.7

Spatial pattern of the world's languages

Language is a set of speech norms associated with a particular group of people. They are also a part of the larger cultural heritage of the community that speaks the language.

There are 7151 identified languages in the world. Some, such as English, are spoken worldwide. Others are spoken by just a handful of people in remote places. The global pattern of languages is shown in Figure 7.7.1. The spread of European languages, including English, Spanish and Portuguese, was closely associated with the age of exploration and colonisation (see Figures 7.7.2 and 7.7.3).

Language is a principal way to pass cultural understanding from one generation to the next. It is also how people interact across cultures. English is the principal means of intercultural communication, a trend driven partially by popular culture (especially music and film) and commerce.

Did you know?

Nearly 80 per cent of the world's population speaks only one per cent of its languages. When the last speaker of a language dies, the world loses the knowledge that was contained in that language.

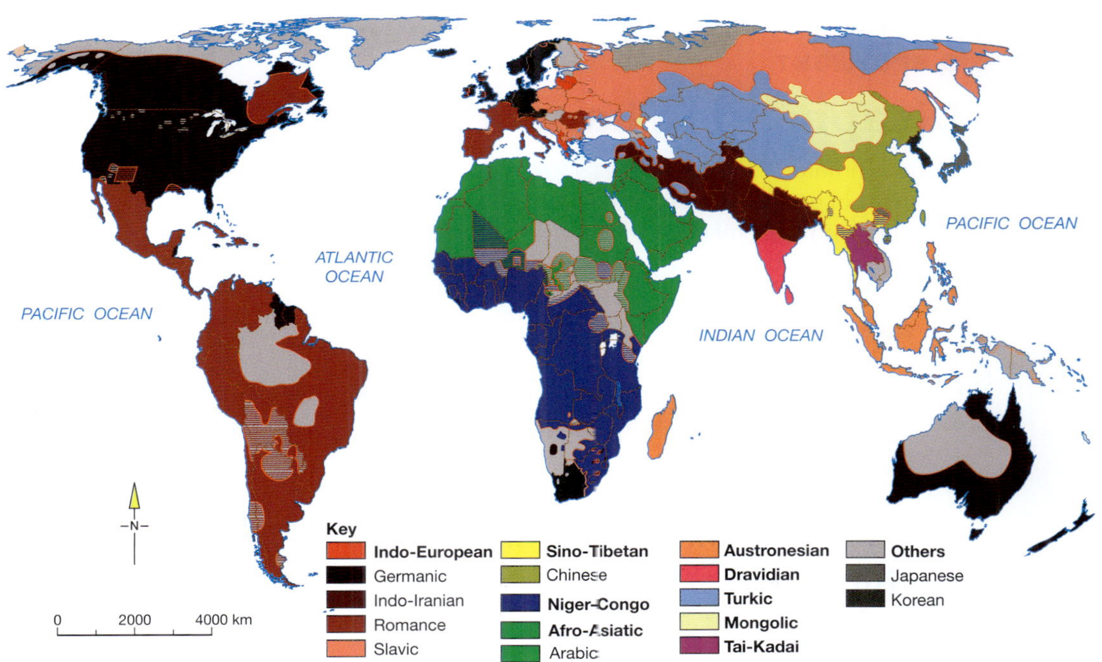

7.7.1 World map of languages

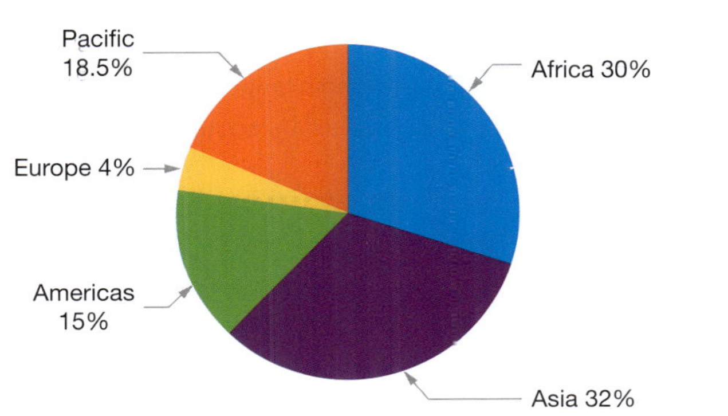

7.7.2 Percentage of the world's languages by region

Did you know?

Asia has the most indigenous languages, closely followed by Africa. Combined, they account for nearly two-thirds of the world's languages. Myriad factors—terrain, cultural history, the spread of ancient civilisations—play into how many languages originated within a certain area.

Did you know?

There are nearly 850 languages spoken in Papua New Guinea, making it the world's most linguistically diverse country.

Did you know?

Of the 250 Aboriginal and Torres Strait Islander languages that were spoken at the time of European colonisation, around 120 are still spoken in Australia today.

Language	Number of speakers	Language family
English	1.5 billion	Indo-European
Mandarin Chinese	1.1 billion	Sino-Tibetan
Hindi	602.2 million	Indo-European
Spanish	548.3 million	Indo-European
French	274.1 million	Indo-European
Standard Arabic	274.0 million	Afro-Asiatic
Bengali	272.7 million	Indo-European
Russian	258.2 million	Indo-European
Portuguese	257.7 million	Indo-European
Urdu	231.3 million	Indo-European

Source: Ethnologue, 2022

7.7.3 Ten most widely spoken languages

● **SPOTLIGHT**

Disappearing languages

Of the estimated 7151 languages spoken in the world today, nearly half (3045) are in danger of extinction this century. Some endangered languages vanish when the sole surviving speaker dies. Others are lost gradually. Many indigenous languages are overwhelmed by the dominant language spoken in school, media, shops and the workplace.

The world regions where languages are disappearing the fastest are Northern Australia, Central South America, North America's upper Pacific coastal zone, Eastern Siberia, and Oklahoma and Southwestern USA. All are occupied by indigenous people speaking diverse languages.

Activities

Acquiring and processing geographical information

1 Use the information and data provided in this unit to write a report outlining the spatial distribution of the world's languages.

2 Study Figure 7.7.1 and complete the following tasks:

- **a** Using an atlas, outline the spatial distribution of the Indo-European language family.
- **b** Outline the distribution of the Germanic language family.

3 Study Figure 7.7.2. What does this data tell us about the spatial distribution of languages?

APPLICATION AND CONSOLIDATION TASKS

Task 1: Infographic: The future of agriculture

Create an infographic outlining the future of agriculture. Your infographic should include:

- a summary of the key stages in the evolution of agriculture
- the impact of the green revolution
- the future of agriculture.

Task 2: Presentation: Manufacturing

Investigate the changing global distribution of car manufacturing. Your presentation needs to include information about how car types have evolved over time, for example, the future of electric cars and how this has and will impact car manufacturing.

Your presentation can include visuals (maps, graphs, diagrams, photos).

Task 3: Indigenous people

Study the world map showing the distribution of indigenous peoples (Figure 7.7.1). Select an indigenous people of interest to you and undertake research. In doing so:

- Locate the place where the people are found. For how long have they occupied these lands?
- Outline the people's spiritual and cultural links to the natural world.
- Describe how the indigenous cultural heritage of the people is passed from one generation to the next.
- Describe the impacts of colonisation on the people.
- Outline the relative wellbeing of the indigenous people compared to the non-indigenous population of the place.
- Describe the extent to which indigenous cultural heritage contributes to the national identity of the country or countries in which the people are found.

CHAPTER

8 Our increasingly integrated world

It was once thought that the processes of cultural and economic integration (often simply referred to as globalisation) would bring the people of the world closer together and impose a sense of sameness on places. While there is evidence of the latter occurring, the opposite appears to be happening in geopolitics. While global economies continue to converge and people stand ready to adopt and adapt elements of other cultures, there is mounting evidence of cultural divergence that is leading us into a period of heightened **nationalism**, social conflict, growing inequality and increased segregation.

Old national identities have proven to be surprisingly durable, and, in some instances, have been radicalised. Divisions have also emerged in the democracies of the developed world. These divisions were especially apparent between those disadvantaged by economic integration and those who benefited from the process.

This chapter examines the increasingly integrated nature of the world, the factors responsible for the nature and pace of integration, the processes of cultural diffusion, adoption and adaptation, and how cultural integration impacts on sense of place. The drivers of integration examined include technological change, the role of world trade and transnational corporations, world cities, migration and tourism.

COVID-19 plunged the whole world into a period of profound economic disruption. Travel was halted. Borders were shut. Hundreds of millions spent weeks in lockdown. The speed at which the virus and its consequences spread demonstrated just how integrated the world had become.

Grant Kleeman, author

8.0.1 People from all parts of the world are more connected than ever.

Chapter glossary

assimilation the process through which individuals and groups of different cultural backgrounds acquire the basic habits, attitudes, and modes of a dominant culture

cultural adaptation the modification of a culture to incorporate aspects of another culture

cultural adoption the acceptance and integration of different cultural elements into one's own

cultural diffusion the dispersion, or spread, of different cultural elements between countries and peoples

cultural integration the circumstances that arise when people from one culture adopt elements of another culture while maintaining their own culture

free market an economic system of supply and demand

global culture a set of shared experiences, norms, symbols and ideas that unite people at the global level

globalisation the process by which businesses or other organisations develop international influence or start operating on an international scale

high culture when elements considered to be of lasting value, such as art, literature, theatre, ballet, opera and classical music are incorporated; some critics consider its content to be 'highbrow' or 'intellectual' when compared with popular culture

hybrid technologies technology-based items that combine elements of different technologies

imperialism a policy of extending a country's power and influence through colonisation, use of military force, or other means

intellectual capital using ideas, knowledge or inventions as a means of gaining material wealth through a business enterprise; specialised knowledge of how a product works creates jobs in areas such as information technology (IT) support

melting pot a place where a variety of peoples, cultures or individuals assimilate into a cohesive whole

multiculturalism a public endorsement and recognition of cultural diversity

nationalism asserting the interests of a nation, viewed as separate from the interests of other nations; extreme and fanatical devotion to one's own nation

neoconservatism a political ideology characterised by an emphasis on free-market capitalism and interventionist foreign policy

popular culture considered to be more mainstream than high culture; it is associated with light forms of entertainment, such as sporting events, television, comics and rock music; also called pop culture

sovereignty the supreme, unrestricted power to govern a state

supply chain the sequence of processes involved in sourcing raw materials, their transformation of resources into usable inputs, the production of manufactured components, and the assembly of finished products from source to the point of consumption

tariff taxes or duties on imports

tax haven a country or territory where resident individuals or resident companies pay little or no tax

trade liberalisation the removal or reduction of restrictions or barriers on the free exchange of goods between nations

world cities the centres of global economic and cultural authority

UNIT 8.1

The processes of economic and cultural integration

Economic integration is the growing interdependence of national economies. This interdependence is driven by a reduction in, and ultimately the removal of, **tariff** and non-tariff barriers to the free movement of goods or services, and production factors between countries.

The global economy

The long boom from the end of World War II to 1970 was characterised by high economic growth rates, based on the production of consumer goods by multi-plant, multi-product firms. In the 1970s, firms began diversifying and expanding by increasing their market share. This resulted in mergers and corporate takeovers. By the 1990s, firms focused on core activities, leaving non-core activities to specialist providers.

Such changes coincided with increasing international competition. The global economy emerged, characterised by a system of production, marketing and finance that uses international trade and communication to move goods, money, information and people between countries. The global economy:

- operates globally. Money has no borders, worldwide US$5.3 trillion in foreign currencies is traded daily
- uses networks. Buy shoes from Italy online, get an illness diagnosed by a Filipino doctor in Edinburgh
- values information. In today's knowledge-based economy, **intellectual capital** drives the value of products. More people are employed in service industries, particularly those involving IT and finance
- decentralises power. The internet enables anyone to publish, email, create or download information anywhere in the world
- rewards openness. The fact that the internet is a censorship-free zone, where information and ideas can no longer be repressed with ease, can be both positive and negative
- promotes specialisation. Goods are ordered directly online, cutting out wholesalers and retailers. Customers can customise products
- evades government regulation. Laws in place to protect consumers in one country may not exist in another country
- manipulates markets. Share prices can drop dramatically via market rumours on social media.

These changes and their impacts result from a set of processes known as economic restructuring—significant and enduring changes in the economy's nature and structure brought about primarily by the global economy's emergence.

Economic restructuring

The impacts of economic restructuring occur at varying geographical scales, from the global to the local. It has led to money and other resources rapidly moving in and out of industries and sectors. It has often involved the development of new ways of carrying out and rewarding work.

Rapid technological change and ongoing economic restructuring threaten to increase inequality and contribute to the development of a two-tier society. Workers who lack the necessary skills run the risk of being left behind. This will open the way for the increased social disruption caused by retrenchments, factory closures and changes to the way production is organised.

The structural changes occurring, mainly in the developed world, are called deindustrialisation. It is sometimes associated with a process known as jobless growth, where production increases without an increase in employment. A characteristic of the post-industrial society is the increased demand for services, resulting in the service sector becoming the dominant form of employment.

Cultural integration

Cultural integration is the blending of two or more cultures. It is a form of cultural exchange in which one cultural group either adopts or adapts the beliefs, practices and rituals of another group without necessarily sacrificing the characteristics of its own culture.

Cultural integration generally enriches cultures. People are allowed to blend their beliefs and ideas. They do not have to give up their culture.

The concept of cultural integration leads us to the notion of a multicultural society. **Multiculturalism** reflects cultural integration at work. So does the idea of the 'global village', where, through technology and trade, a seemingly borderless world is created. Cultural integration is also linked with the adoption of a mass consumer culture where everything from fashion to sport, and from music to TV, becomes integrated into a country's culture. Cultural integration is viewed by some as a negative—a threat to national **sovereignty** and cultural diversity. Integration is especially negative when it's forced. Forced integration leads to a culture that is watered down, not full. The process requires time to help people develop the skills and comfort necessary to move forward. Additionally, people also may need to feel comfortable to blend their cultures.

Defining culture

Culture can be defined as the way of life of a group of people. It includes those elements of human existence, such as the traditions, customs, languages, belief systems, art, architecture, music, food and institutions, that are passed down from one generation to the next. It includes the material goods the group creates and uses, and the skills it has developed. It also includes territorial affiliation and shared history, and shared political and education systems.

Culture can also be described as the set of shared meanings that is lived through the material and symbolic practices of everyday life. This set of shared meanings can include values, beliefs, practices, and ideas about childhood, race, gender and sexuality.

Culture can also be considered in different contexts. In a classical sense, **high culture** can be defined in terms of classical standards and aesthetic excellence in, for example, literature, music and dance. Alternatively, we can think of culture as the activities and characteristics of a particular group, such as a working-class culture, corporate culture or teenage culture. It can also be conceptualised as a national culture (e.g. Australian culture) or the culture of a particular ethnic group.

Cultural change

The values, beliefs, ideas and practices that help define culture are routinely subject to re-evaluation and redefinition. They can be transformed from both within and outside a particular group. Cultures are dynamic and reflect the complex interplay of social, political, economic and historical factors. Importantly, technologies, products and people move from place to place. When cultures come into contact through migration, trade, or the latest telecommunications technologies, they impact on each other. Some people fear that such interactions will threaten cultural diversity. Others see this as an ongoing process enriched by the interacting cultures. Advanced communications make it inevitable that all cultures will be connected.

Cultures have evolved through contact with others for millennia. Historically, distant cultural influences spread slowly, limited by distance. The internet and rapidly evolving telecommunications mean cultural influences can spread worldwide at the click of a mouse. Other influences include the media, world trade, and the ease (and increased affordability) of long-distance travel.

● SPOTLIGHT

Marriage equality in Australia

In 2017, the wording of the *Marriage Act 1961* changed, replacing the words 'man and woman' with 'two people' in the Act's definition of 'marriage'. Over 60 per cent of eligible Australians voted yes when asked 'Should the law be changed to allow same-sex couples to marry?' Legalising same-sex marriage reflected a significant change in cultural attitudes 56 years after the Act was proclaimed (see Figure 8.1.1). While some still oppose same-sex marriage, it is now legally recognised in most Western liberal democracies.

8.1.1 Rally in favour of same-sex marriage in Brisbane, 2015

8.1.2 The influence of mass consumer culture is evident in Delhi, India.

Critiques of cultural integration

Some fear that cultural cloning is an inevitable outcome of the cultural influence of global brands and the English language. Critics argue that Western influences will eventually erase every cultural difference, creating one big 'McWorld' (see Figure 8.1.2). Everywhere will look like everywhere else, offering the same fast-food outlets, television shows, Hollywood movies and the same brands in cloned shopping malls. Events on the other side of the world will create a sense of immediacy, as if happening locally.

The logical outcome of such a proposition is that national differences and regional cultures will disappear as a global marketplace brings uniform dispersion of people, tastes and ideas. However, such an outcome is highly unlikely. Places and regions will inevitably change in a new global context. Even in the information age, geography still matters; consider transport costs, variations in resource endowments, territorial impulses, the resilience of local cultures and past legacy.

● SPOTLIGHT

Is there a global culture?

The answer is, no! There is no definitive indication we have achieved that level of integration. People worldwide do share an increasing awareness and familiarity with a common set of consumer goods and elements of **popular culture** (see Figure 8.1.3). But these are configured in different ways in different places and are not indicative of a single **global culture**. The local scale interacts with the global, often producing hybrid cultures. Sometimes traditional local cultures become the subject of global forces; sometimes it is the other way around.

8.1.3 Warriors from the Samburu tribe, Kenya, using a laptop. While people of different cultures become increasingly familiar with consumer goods, they retain key aspects of their own culture.

● SPOTLIGHT

COVID-19 disaster strikes Italian football game

On 19 February 2020, around 44 000 soccer fans headed from Bergamo into Milan (both in the Lombardy region), to watch Atalanta's greatest victory, against Spain's Valencia. Social distancing was yet to come. A crammed stadium let COVID-19 spread through untold numbers of cheering, hugging, asymptomatic carriers. It was a contagion disaster (see Figure 8.1.4).

Medical experts believe their team's 4–1 win was the catalyst that soon turned the Lombardy region into one of the world's deadliest places. By late 2022, Italy had had over 23.4 million COVID-19 cases, with 3.76 million in this region. Of Italy's 179 000 deaths, 43 012 were from Lombardy.

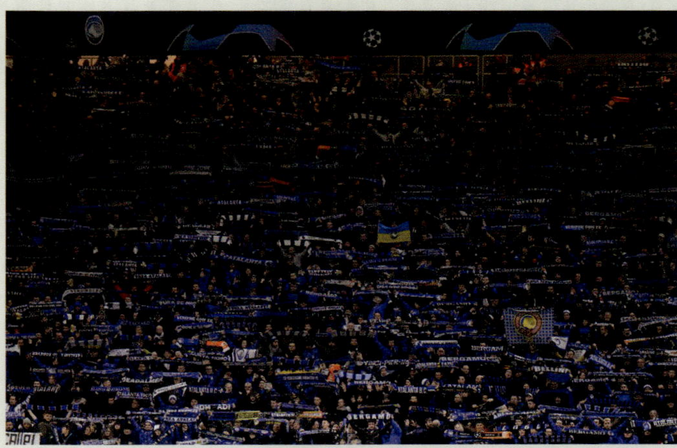

8.1.4 The crammed San Siro Stadium in Milan, Italy, 19 February 2020

International integration: Drivers and impacts

The nature and rate of international integration have changed over time. Responsible factors include advances in transport and telecommunications technologies, the emergence of transnational corporations (TNCs), world trade and global commodity chain growth, cultural **imperialism**, the dissemination of film and media, and government actions.

The impacts involve cultural diffusion, adoption and adaptation. **Cultural diffusion** is the dispersion, or spread, of different cultural elements between countries. When confronted with new cultural influences, people can reject them outright, adopt the change, or adapt. **Cultural adoption** involves accepting and integrating different cultural elements into one's own. **Cultural adaptation** involves modifying one culture to incorporate aspects of another.

Activities

Acquiring and processing geographical information

1. Define economic integration. What has driven this process?
2. Draw a flow chart to explain the nature of the long boom of the period immediately following World War II. Include the nature of economic activity during the 1970s, 1980s and 1990s.
3. Explain the link between economic restructuring and deindustrialisation.
4. Explain what is meant by the term cultural integration. What are its benefits? What are the negative outcomes of the process?
5. Explain the process of cultural change. What are the factors that can initiate changes? What are the consequences of these processes?
6. Define cultural diffusion and distinguish between cultural adoption and adaptation.

Applying and communicating geographical understanding

7. Study the section of text dealing with culture. Construct a diagram illustrating the elements of culture.
8. As a class, brainstorm examples of cultural and social change that reflected a rejection of long-term cultural norms in Australia.
9. Rather than result in the emergence of a global culture what has been the outcome of global integration?
10. Study the box, Spotlight: COVID-19 disaster strikes Italian football game. Outline the role the Milan football match had on the spread of the COVID-19 pandemic in Italy.
11. Working in groups, brainstorm the elements of Australian culture that have their origins in other parts of the world. Compare your group's list with other groups. Compile a class list. What conclusions can you draw from this activity?

UNIT 8.2

Drivers of international integration: Transport and telecommunications technologies

8.2.1 The Boeing Dreamliner's light weight saves fuel; its airframe is nearly half carbon fibre, reinforced plastic and other composites.

Advances in transport and communications technologies have transformed the nature and rate of economic and cultural integration.

Developments in aviation technologies—especially the Boeing 747 in 1969 and new fuel-efficient aircraft such as the Boeing 787 Dreamliner in 2011—helped lower transport costs and increase international tourism and trade between countries (see Figure 8.2.1). Although less freight is carried by air than by sea, air-freighted goods are more valuable. Airfreight is also more attractive where high-tech products are manufactured near airports. Transport by air is swift, and often vital in complex **supply chains**.

Shipping and cargo handling technologies

Technological developments in shipping and cargo handling have been central to international trade expansion. As ships have increased in size, transport costs have sharply declined. Specialised bulk carriers, oil tankers and container ships lower costs and reduce how long ships are in port. Containers can be carried by road, rail or ship. This means goods can be packed into a container at the point of origin, travel by multiple modes of transport, and be unpacked at the final destination. This innovation has transformed cargo handling.

Land transport

Higher investments in rail and road infrastructure have made land transport faster and more competitive over longer distances. High-speed rail systems and new cargo handling equipment have cut delivery times and increased efficiency.

Communication-based technologies

Australians have access to a sophisticated range of communications technologies. These include an extensive network of submarine cables and satellites that connect Australia with the rest of the world. This infrastructure facilitates the transmission of vast volumes of data at high speed. This has made global communications faster, more efficient and much less costly.

Satellite-based technologies include mobile phones, geographical positioning systems (GPS) and satellite television services. These communications systems have the advantage of being wireless and are more flexible than traditional cable-based technologies.

Hybrid technologies are evolving rapidly. Digital platforms have revolutionised the way people live their lives by enabling them to use their smart televisions and smartphones to undertake many of the functions available via the internet. Subscription services such as Netflix allow programs to be viewed on a phone or a tablet, not just on a television.

Did you know?

In 2017, tests on Volvo's self-driving technology struggled to identify kangaroos on the road. Their jumping confused the computer and it couldn't determine how close the animal was.

● **SPOTLIGHT**

China's 'One Belt, One Road' initiative

Announced in 2013, China's US$900 billion 'One Belt, One Road' initiative aims to create new trade corridors to countries and regions to China's west. Notably, these include Central Asia, the Middle East and Europe (see Figure 8.2.2). The goal is to improve trade relationships primarily through investing in transport-related infrastructure (road and rail networks and ports). China envisages a new era of **globalisation** that will benefit the world economically and cement its status as a major economic and political power.

China claims this will boost poorer countries' economies, as well as its less-developed border regions and those in its west. Chinese manufacturing would benefit from infrastructure investments, especially enterprises that supply steel and heavy machinery. Markets for Chinese manufacturing will open, as it transitions to higher value-added industrial goods. China may lend as much as US$8 trillion for infrastructure in 68 countries, which account for 65 per cent of the global population and a third of global GDP.

The initiative is not without its critics. China's human rights violations and the initiative's environmental impacts are sources of criticism. Some see it as a debt trap for developing countries and an example of economic imperialism.

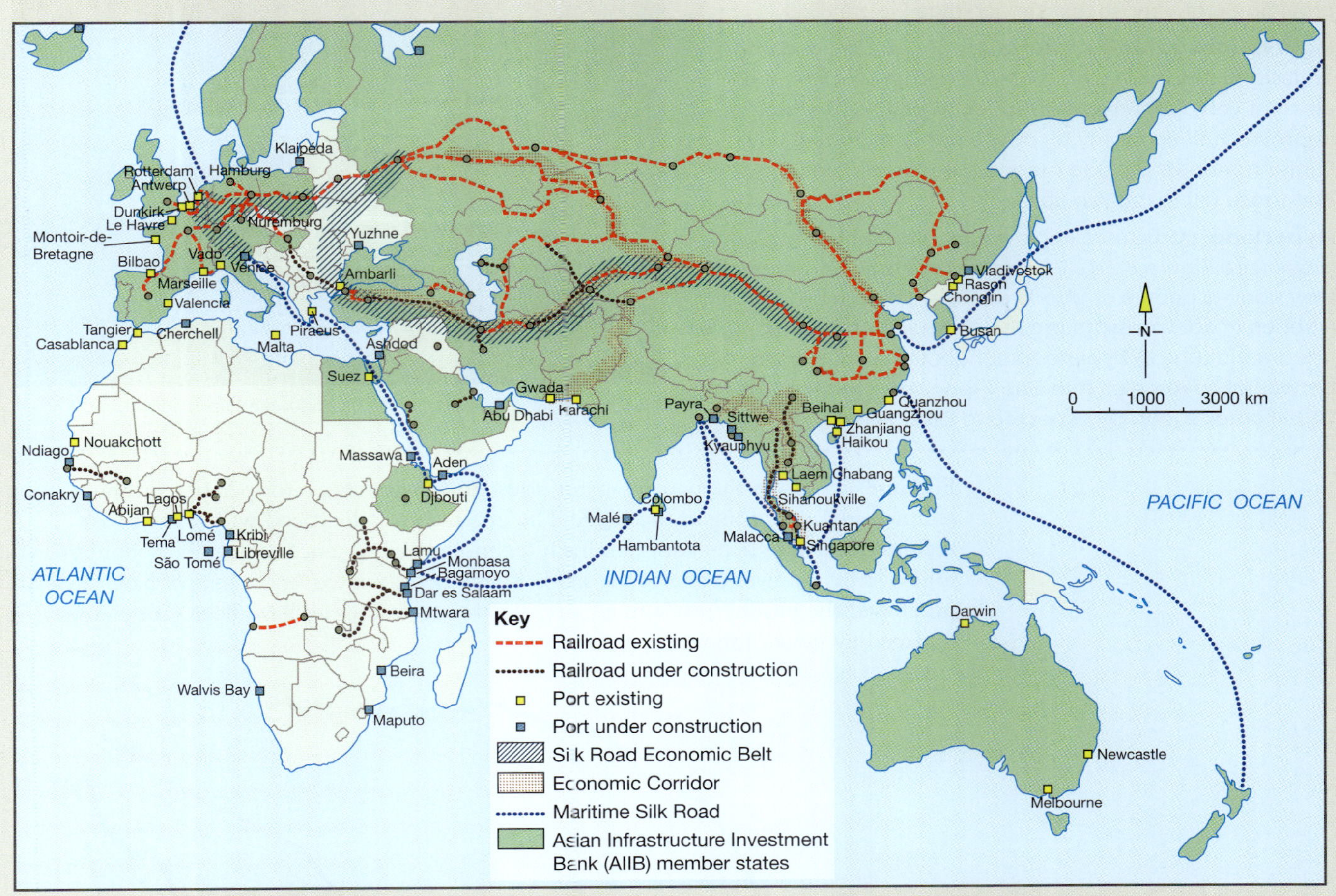

8.2.2 China's 'One Belt, One Road' initiative includes land routes (the 'Belt') and maritime routes (the 'Road').

● SPOTLIGHT

What of the future?

Transport technologies are on the cusp of a revolution. New efficiencies in existing transport modes are poised to reshape how people and goods move. Four key technologies are at the forefront.

The Internet of Things: a system of interrelated computing devices that can transfer data over a network without people needing to interact. These vast connected networks will impact many aspects of our lives.

Autonomous vehicles: continued research and development of autonomous car technology is likely to create a safer alternative to human-controlled vehicles. Google and Tesla have already created autonomous vehicles. San Francisco, USA, has two companies providing a driverless taxi service (see Figure 8.2.3).

Lightweight vehicle materials: concern about carbon emissions has increased pressure to manufacture fuel-efficient vehicles. A weight reduction of just 10 per cent improves fuel economy by over 6 per cent. Potentially lighter materials include magnesium-aluminium alloys and those using carbon fibre.

Hyperloop: pods travelling through a system of sealed tubes (see Figure 8.2.4). Free of air resistance or friction, they convey people or goods at super-fast speeds. A series of compressors propels the pods through a pneumatic tube. A hyperloop connection is proposed between Los Angeles and San Francisco, with the 560-kilometre journey expected to take 35 minutes.

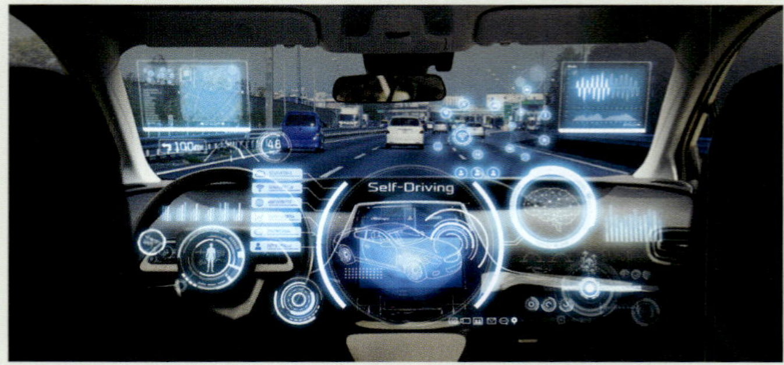

8.2.3 The view from an autonomous vehicle

8.2.4 Is the hyperloop the future of land-based transportation?

Figure 8.2.5 shows the percentage of households with access to specific communications technologies over time in the USA. The graph shows the speed with which new technologies have been adopted. Each technology experiences a period of rapid growth followed by a slower phase as the market becomes saturated (most people have already purchased the product). In some cases, existing technologies have been completely replaced by newer innovations.

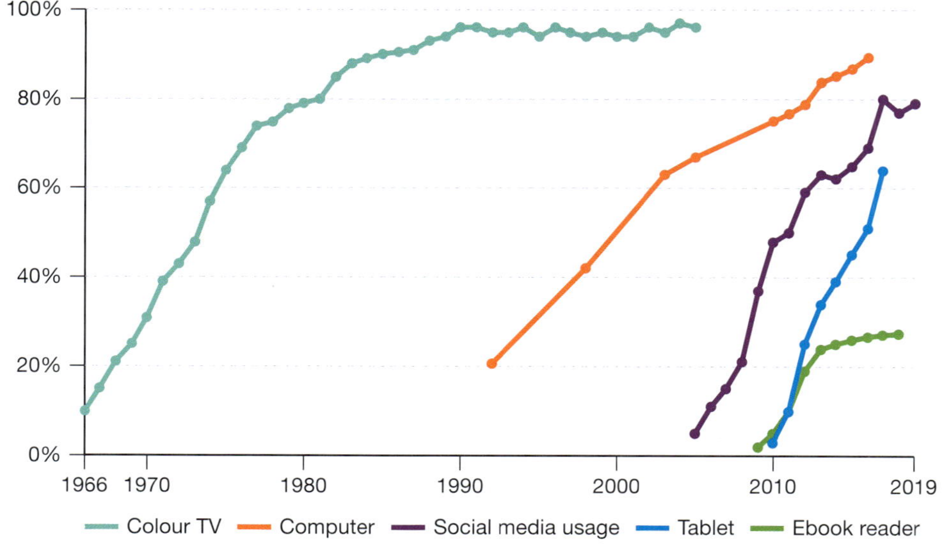

8.2.5 The take-up of selected technologies by US households

In Australia, the development of the National Broadband Network (NBN) has significantly improved access to the internet for many individuals and businesses. Costing $51 billion, the network is the largest nation-building infrastructure project in Australian history. The network aims to connect 90 per cent of all Australian homes, schools and workplaces to broadband services with speeds of up to 100 megabits per second—100 times faster than those used by many households and businesses in the past. The other 10 per cent of premises are linked by next-generation wireless and satellite technologies that will deliver broadband speeds of 12 megabits per second. Eighty-six per cent of Australian households have access to the internet.

The development of advanced communications and transport technologies has brought great benefits, especially to those living in the developed world and the emerging economies of South and East Asia. Many isolated and/or poor regions of the globe have limited access to these technologies. These regions have, until recently, effectively been shut out of the growing global networks in trade, transport, finance and communications. Even today they remain underrepresented in terms of internet access. Africa, with a population of more than 1341 billion (17.2 per cent of the world's population), accounts for just 11.5 per cent of the world's internet users. In countries such as Eritrea, Somalia, Guinea-Bissau and the Central African Republic, less than 10 per cent of the population have access to the internet. Even in cities, internet access is often very slow due to the age and limited capacity of the infrastructure.

Even within developed countries, there are inequalities in access to information and communication-based technologies. The aged and the poor have a lower level of access than the young and the more affluent.

Activities

Acquiring and processing geographical information

1. List the factors that determine the nature and rate of international integration over time.
2. Outline the role played by developments in shipping and cargo handling technologies in the growth of international trade.
3. Summarise the principal developments in land transportation.
4. Explain what hybrid technologies are and how might these revolutionise the way we live our lives.
5. Explain why the NBN has been such an important innovation in the Australian context.
6. Outline the impact of advanced telecommunications technologies on those living in developing countries.

Applying and communicating geographical understanding

7. Study the box, Spotlight: China's 'One Belt, One Road' initiative and answer the questions below.
   - a What is the principal objective of China's giant infrastructure initiative?
   - b What are the main elements of the initiative?
   - c What are the benefits of the initiative for China's neighbours and China?
8. Study Figure 8.2.5. Using data from the graph, write a report outlining the rate at which the technologies shown have been adopted by US households.

UNIT 8.3

Drivers of international integration: World trade and the role of TNCs

Advances in transport and communications-based technologies have facilitated the emergence of a new global pattern of world trade. The rapid expansion of world trade, a key feature of the integration of national economies into a global economic system, has been a highly significant development of the last century. Figure 8.3.1 shows the growth in the volume of world trade, while Figure 8.3.2 shows the increase in the value of world trade.

World trade volumes today are roughly 40 times the level recorded in 1950 and the value of this trade has increased by over 300 times. As of 2019, world trade volumes have expanded 4.2 per cent annually on average since 1995. For developing countries, the figure is 6.3 per cent; for developed countries, it is 3.1 per cent.

The COVID-19 pandemic caused a major disruption to world trade. Supply chains were disrupted, putting upward pressure on prices. Added to this was the Russian invasion of Ukraine, which greatly increased the cost of key energy sources (natural gas, coal and petroleum). The recovery in world trade gained momentum in 2022–23.

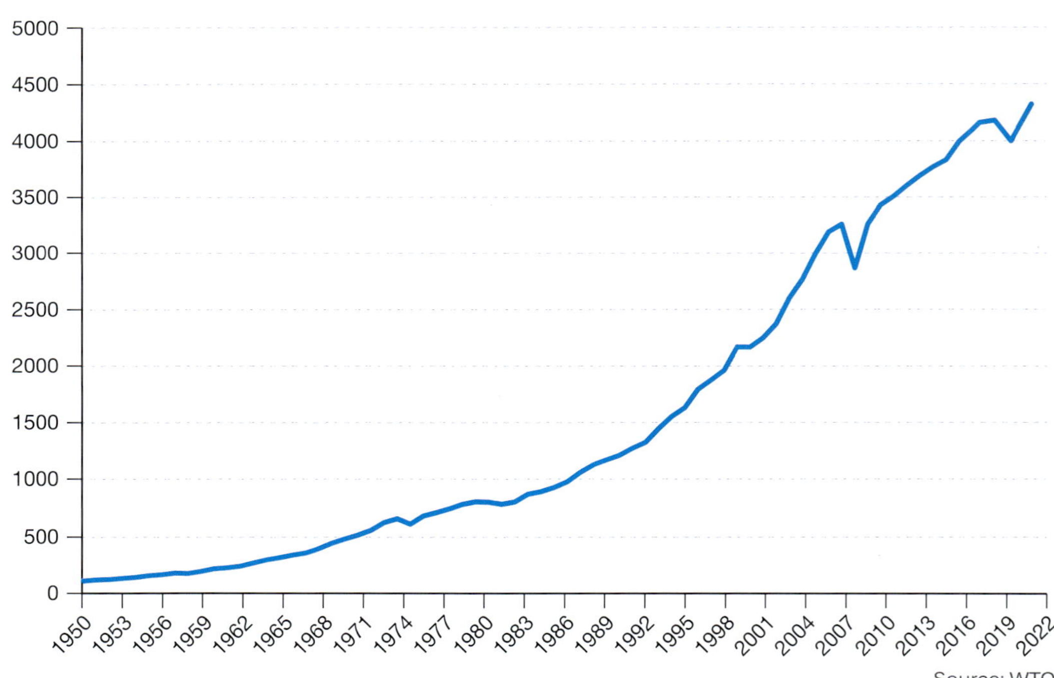

8.3.1 Growth in the volume of world trade, 1950–2022.

Global commodity chains

Today, countries exchange not only final products but also intermediate inputs. This creates an intricate global network of economic interactions. TNCs are key players and depend on global **supply chains** operating smoothly. It also involves the logistics involved in the transport, storage and distribution of each element of the supply chain. Significantly, automation is rapidly changing supply chains. It can be applied to most operational processes to increase efficiency, speed, reliability and accuracy. In sophisticated supply chain systems, used products may re-enter the supply chain at any point that they can be recycled.

The stability and resilience of supply chains are now critical to the global economy's functioning. The ability to identify and quickly respond to disruptions within supply

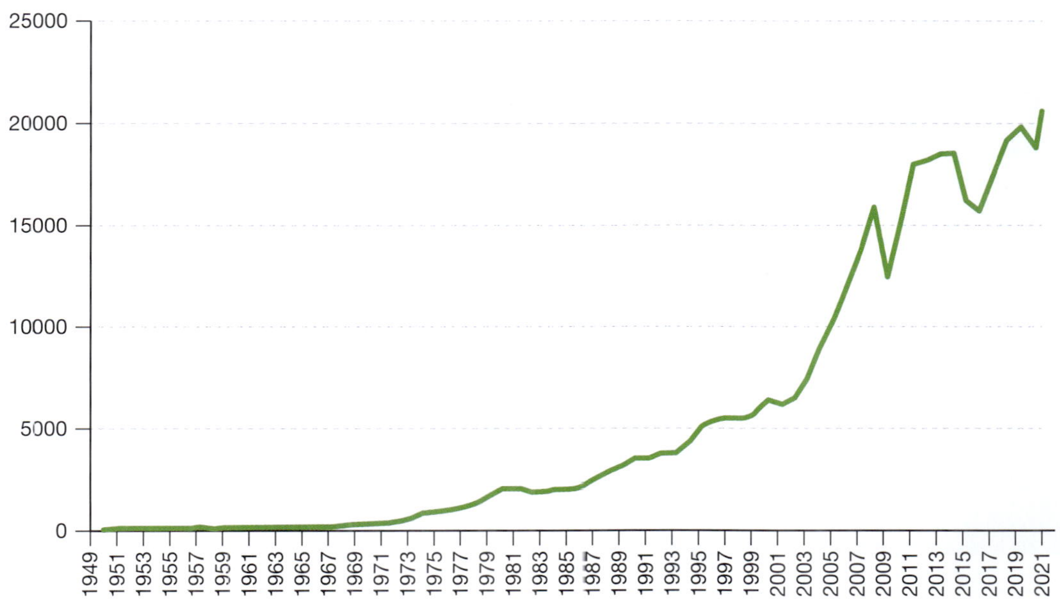

8.3.2 Increase in the value of world trade, 1950–2021.

chains is critical to TNC operations. The COVID-19 pandemic showed how fragile the global supply chain can be. In early 2020, the closure of Chinese factories (imposed to slow the spread of the virus) resulted in shortages in components and consumer goods.

8.3.3 Containerisation led to efficiencies (lower costs) in transporting goods.

The emergence of TNCs and sophisticated global supply chains, combined with rapidly expanding world trade, was aided by technological developments in shipping and cargo handling (see Figure 8.3.3).

The development of complex networks of economic interaction has been driven by the quest for increased efficiency. The efficiency gains from trade led to a major labour redistribution accompanying the relocation of labour-intensive manufacturing to the low-cost countries of East and South Asia. Reshuffling workers from less to more efficient producers resulted in job losses in the developed world's older manufacturing regions, as discussed in Unit 7.5. For some groups of people, trade hurts their wages and employment opportunities. However, it also has a positive effect by lowing consumer prices and increasing product availability.

Only some households gain a positive net effect. Workers who lose their jobs can be affected for long periods. Any positive effect of lower prices is inadequate to compensate for lost earnings.

These structural economic changes also disrupted political and social stability. Those people and regions disadvantaged by economic restructuring have in some instances been alienated from the political processes that facilitated the restructuring. In the 2016 US presidential election, Donald Trump was successful in regions negatively impacted by economic restructuring. His slogan 'Make America Great Again' represented his promise to bring manufacturing jobs back to US regions where they had been lost. A record number of people voted in the 2020 election. The election, in which Trump was denied a second term, can be viewed as a battle between those disadvantaged by global economic and cultural integration (principally the white, working-class without college degrees) and those keen to embrace it (the young, minorities, women, especially those based in US cities and suburbs).

● SPOTLIGHT

Transnational corporations

The emergence of TNCs has been central to the development of economic and cultural integration. These are businesses that operate internationally. The production and distribution of goods and services is increasingly dominated by these corporations. Most have headquarters in the developed world (see Figure 8.3.5).

The earliest TNCs date from the mid-nineteenth century, but it was not until the 1950s that their numbers started to increase significantly. Their economic dominance increased rapidly. Most TNCs are very large—of the world's 100 largest economic entities (defined by revenue), 71 are corporations. Ten of the largest TNCs rank in the top 25 of the largest entities in the world (see Figure 8.3.4).

They are powerful. Sales of the world's 10 largest companies exceed the combined gross national product of the world's 100 economically least developed countries, including all African countries. The 300 largest TNCs own or control at least one-quarter of the entire world's productive assets.

Rank	Entity	Revenue (US$ billions)
1	USA	3363
2	China	2465
3	Japan	1696
4	Germany	1507
5	France	1288
6	United Kingdom	996
7	Italy	843
8	Brazil	632
9	Canada	595
10	Walmart (USA)	482
11	Spain	461
12	Australia	421
13	State Grid (China)	330
14	Netherlands	323
15	South Korea	304
16	China National Petroleum	299
17	Sinopec Group, (China)	294
18	Royal Dutch Shell	272
19	Sweden	248
20	Exxon Mobil (USA)	246
21	Volkswagen (Germany)	237
22	Toyota Motor Corp. (Japan)	237
23	Apple (USA)	234
24	Belgium	232
25	BP (UK)	226
26	Mexico	224
27	Switzerland	224
28	Berkshire Hathaway (USA)	211
29	India	220
30	Norway	200

Source: Forbes

8.3.4 The world's top 30 economic entities (countries and corporations), 2019

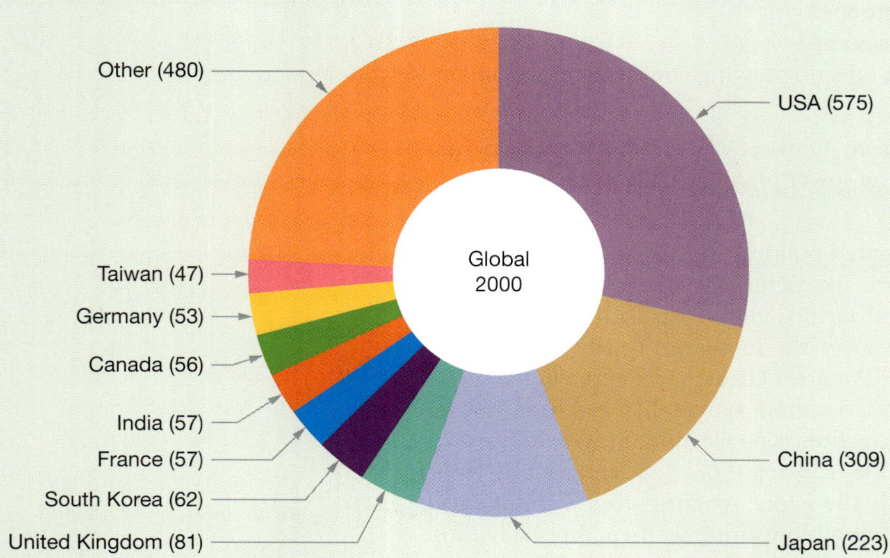

8.3.5 Where the world's top 2000 TNCs are based, 2019

Growth of transnational corporations

Most TNCs started as small enterprises producing goods for sale in their domestic markets. Initially, they used their profits to expand locally, often taking over other businesses within their industry. Before long, some firms outgrew their domestic market and sought to expand overseas. Usually, this involved establishing an overseas subsidiary through direct investment. During the 1960s and 1970s, many developing countries that had been targets of direct TNC investment tried to exercise some control over TNC activities by insisting on a degree of local equity (ownership and control) in such investments.

Partly in response to these restrictions, TNCs developed systems of corporate control, which, while not requiring direct ownership of capital (such as factories), still enable them to exercise control over the production, marketing and distribution of products. To achieve this, they retain ownership of a product, usually by a **patent**. Most TNCs now operate through a combination of joint ventures, licensing and franchise agreements, production agreements and subcontracting.

Method of operation

Like any business, the principal aim of the TNC is to generate a profit, which can then be reinvested to accumulate additional capital. By minimising production costs and by creating a worldwide demand for their products through global marketing campaigns, TNCs are often able to generate massive profits and minimise taxes paid. These profits are then returned to the parent company based in a developed country or 'parked' in a low-tax jurisdiction such as the Irish Republic. Little profit ever reaches the poorly paid, low-skilled workers of the developing world.

The proliferation of TNCs and the global economy's growth has, to some extent, diminished the sovereignty of the individual nation-state. Decisions made by TNCs are usually for the good of the company worldwide and may not be in the best interests of the individual countries in which TNCs operate (see Figure 8.3.6). Decisions made in the boardrooms of New York, London or Tokyo, for example, can affect employment opportunities throughout the world. Individual national governments may be unaware of such decisions until after they are made. General Motors Corporation's decision to abandon the Australian automobile market was made in Detroit. It did not notify the Australian Government of its intentions, despite receiving billions of dollars in government assistance over the preceding two decades.

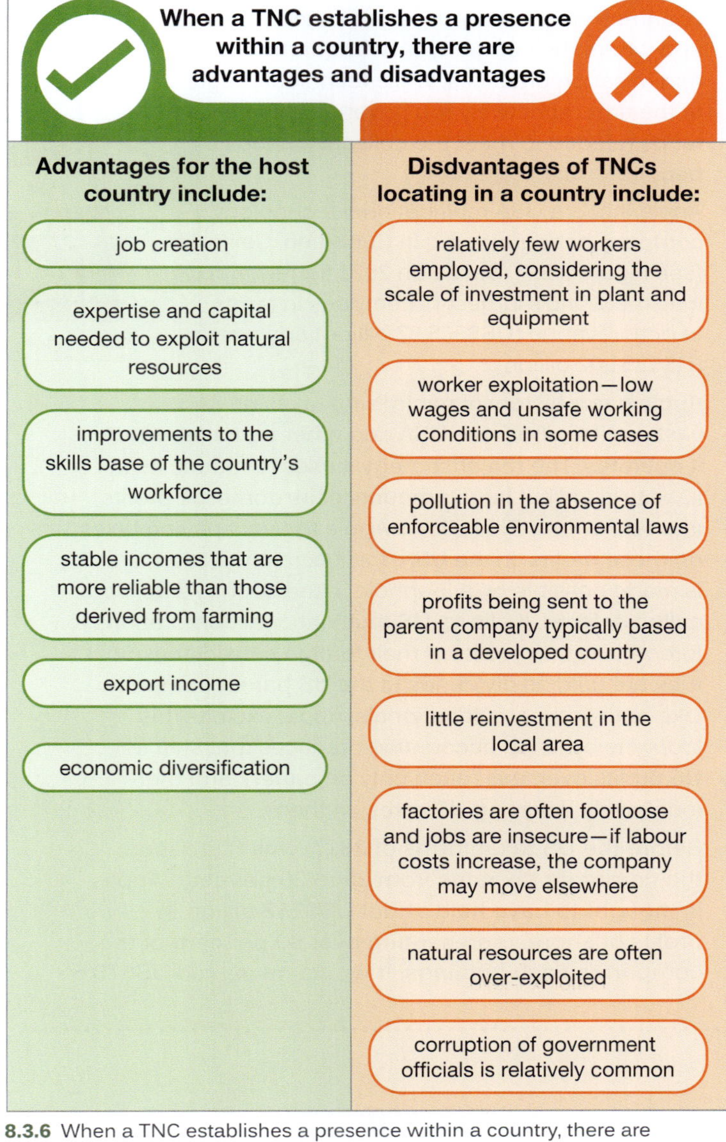

8.3.6 When a TNC establishes a presence within a country, there are advantages and disadvantages.

Small firms going international

Many experts, speculating about the future of large TNCs within the global economy, believe these corporations' influence will diminish. More small- to medium-sized firms might take advantage of the growth and accessibility of the global economy.

Factors enabling them to go international are:

- trade barriers have been removed. This has enabled small firms to gain access to markets that were once only available to large corporations
- access to computer and telecommunications technology is no longer restricted to large firms based on cost

- financial markets have been deregulated. This has given them access to capital investment
- consumer tastes have converged across borders as global television and online marketing campaigns promote an ever-increasing range of lifestyle products and options
- there is access to, and replication of, quality around the world. Smaller firms are often better placed to respond to changes in the patterns of consumption because they are often able to adapt production to shifts in demand more easily.

● **SPOTLIGHT**

Apple's tax minimisation

TNCs have often employed aggressive tax minimisation practices and make use of **tax havens** such as the Cayman Islands and the British Virgin Islands. They typically have complex corporate structures that are used to greatly reduce the amount of tax they pay in the countries where revenue is generated via sales and where the company is headquartered. Tax minimisation practices, while technically legal at the time, effectively rob governments of the funds needed to meet the collective needs of people.

Apple Inc. is a USA multinational technology company headquartered in Cupertino, California (see Figure 8.3.7). Based on 2021 figures, Apple was the world's largest IT company in terms of both revenue (US$365.82 billion) and assets (US$351.02 billion).

8.3.7 Apple, like many TNCs, actively seeks to minimise the tax they pay.

Apple has a history of minimising taxes on its vast profits in both the USA and other countries where it operates. The major strategy it uses to reduce its tax liability is to shift large amounts of its domestic profits into tax havens. This allows Apple to avoid paying taxes on these profits. In the USA, this tax minimisation strategy is made possible by a loophole in the country's tax code called deferral. It allows US multinational corporations to forego taxes on profits of their foreign subsidiaries until they are paid as dividends to the US parent company. Like many other multinationals, Apple exploits this loophole by using accounting practices that shift its US profits overseas (often only on paper), and then indefinitely deferring US taxes on them.

Before the US tax reforms of 2017, which reduced the corporate tax rates from 35 to 20 per cent, Apple is thought to have held about US$252 billion in profits offshore, representing over 90 per cent of the company's cash holdings. It would have paid US$78.6 billion in taxes if the profit had been repatriated back to the USA.

One tax haven used was the Republic of Ireland. Before a crackdown by the Irish government and the EU in 2014, Apple channelled vast amounts of revenue there, taking advantage of a very favourable taxation regime. When these loopholes eventually closed, Apple moved two of its subsidiaries to Jersey, a crown dependency of the UK, which makes its own laws. Not subject to most European Union legislation, it is a popular tax haven.

Gradually governments have collaborated to close the taxation loopholes used by TNCs. In Australia, new multinational anti-tax-avoidance laws increased the tax paid by TNCs including Apple. In 2020–2021, Apple Australia posted AU$11.8 billion in total revenue, but reported taxable income was just AU$578 million, resulting in tax of AU$173 million. While the tax rate paid is 30 per cent, the amount paid represented just 1.5 per cent of revenue.

Activities

Acquiring and processing geographical information

1 Identify the developments that made the emergence of global supply chains possible. What were the principal drivers of these changes?

2 Outline the impact that the development of global supply chains had on the geography of manufacturing. What has been the impact on consumer prices, the availability of consumer goods and households?

3 Outline the political and social impacts of economic restructuring in the USA.

4 Outline the growth of TNCs. How have these corporations changed the way they operate over time?

5 State how the proliferation of TNCs has impacted on the sovereignty exercised by nation-states.

6 Outline how small firms responded to the opportunities offered by global economic integration.

Applying and communicating geographical understanding

7 Study Figures 8.3.1 and 8.3.2. Using data from the graphs, write a report outlining the growth in the volume and value of world trade, 1950–2021.

8 Study Figure 8.3.5. Using data from the graph, describe the global distribution of the world's 2000 largest TNCs in 2019.

9 Conduct a class debate on the topic, 'TNCs are good for countries'. Based on the arguments presented, write an exposition arguing the case in favour and against TNCs.

10 Study the box, Spotlight: Apple's tax minimisation.

- a Why do TNCs like Apple engage in tax minimisation?
- b What strategies has Apple used to minimise tax?
- c How have governments responded?

UNIT 8.4

Drivers of international integration: Cultural imperialism

Everyone is constantly influenced by what surrounds them. It is almost impossible to avoid being positively or negatively impacted. A person's life is directly influenced by what they listen to, watch or do.

Cultural imperialism

Cultural imperialism is widespread. It involves the promotion or imposition of a culture, usually a politically powerful nation, over a less powerful one. The cultural dominance of economically and politically powerful countries determines the nature and cultural values of civilisations worldwide.

The principal source of cultural imperialism varies over time. In ancient times it was that of the Greeks and Romans. During the colonial era, the cultural influence of European colonial powers was imposed on their occupied lands, with that of the English being the most pervasive. Today, it's US cultural influences that dominate, most notably in the Western world.

US cultural imperialism

US cultural influence is so pervasive now that its imperialism is simply called Americanisation. US fast food, fashion, music, media, movies and technologies are now universal in contemporary global culture.

Americanisation is most apparent in the influence US culture and business has on other countries, including their media, cuisine, business practices, popular culture, technology and political techniques. US influences can be traced back over a century, but are now far more pervasive due to high-speed internet and the impact of US technologies and business models from companies like Apple, Alphabet (Google's parent company), Meta, Twitter, Amazon and Uber.

US culture is most apparent in popular culture (also called mass culture or pop culture). It encompasses the practices, beliefs and objects that are widespread in a society at one point in time. Elements include entertainment (music, film and television), sport, fashion, technology and politics.

The impact of Americanisation

The potential to crowd out local cultures is a concern of cultural imperialism. Some say embracing local cultural characteristics occurs naturally. For example, McDonald's introduced the McCafé concept in Australia in 1993 to cater to local coffee tastes. Australian coffee tastes, in turn, can be traced to post-World War II Italian immigration.

Americanisation also bled into the political and ideological sphere. Early twenty-first-century foreign policy was dominated by a **neoconservative** administration. At its core was spreading freedom, democracy and **free-market** capitalism. Some neoconservatives maintained that spreading US values would lead hostile countries, including Iraq and Afghanistan, toward embracing democracy. They also hoped these nations would join the USA to fight global terrorism and tyranny. Their aims were not achieved.

The proliferation of US-based social media platforms has proved problematic for many governments around the world. They are increasingly concerned about the privacy issues raised by such platforms, the tax avoidance strategies of these companies, and the role of these platforms in the distribution of fake news and the live-streaming of terror-related violence.

Film dissemination

Film and other media have become principal agents for cultural integration. The role they play in society is often defined by media selection, media usage and the effects of media engagement. They can have impacts on:

- agenda-setting
- knowledge acquisition
- issue framing
- stereotyping
- attitude formation (prejudices) and behaviour (discrimination).

While US movies influence a global mass culture, they are also integral to US culture. They, therefore, reflect the prevailing concerns, attitudes and beliefs of Americans. For example, the Academy of Motion Picture Arts and Sciences insisted that from the 96th Academy Awards, in 2023, the best picture qualifying films must meet inclusion standards. Their decision was in response to the Black Lives Matter movement.

The nature of the film industry has changed. Embraced before World War II, the cinema industry initiated an era in which movies ultimately became the dominant cultural form. In its heyday, cinema highlighted and reinforced national identities. It provided stories and images of people and settings that represented the culture. Internationally, it found diverse audiences beyond political boundaries, broadly influencing other cultures.

Internationally, Hollywood ultimately emerged as the strongest player (see Figure 8.4.1). Some were concerned that Americanisation—exported US cultural values, tastes, attitudes and products—would harm other cultures and identities. A few even argued that the USA was pursuing cultural domination. Others see the film industry's US domination as simply one outcome of a competitively successful production model. There are still major movie industries operating in a diverse number of countries (see Figure 8.4.2).

Country	Revenue (US$ millions)
USA	20926
China	6984
Japan	3349
United Kingdom	2382
Germany	2306
France	1705
Canada	1210
Australia	1048
South Korea	1040
Brazil	944
Mexico	814
Spain	717

Source: Statista

8.4.1 Filmed entertainment revenue in selected countries worldwide, 2020

8.4.2 Film director Lancelot Oduwa Imasuen on location in Nigeria. Nollywood is the world's third-largest movie industry by number of films produced.

Mass media

The role of the mass media—film, television, newspapers, magazines and online content—is a hotly contested issue. Early thinkers thought it exerted a powerful effect on its audiences. Debate has since ebbed and flowed, especially relating to the level, scope and implications of its influence. The media does, however, shape culture by exposing people to different cultural traditions and social constructs.

Social media emerged as a major disruptor. It provided alternative sources of 'news', making it increasingly difficult for traditional media (newspapers and television) to compete. Advertisers moved much of their advertising revenue to social media platforms. Streaming services, such as Netflix, HBO, Amazon, Spotify and Apple Music, disrupted the industry. They killed off the DVD, reduced free-to-air television audiences and greatly reduced music CD sales.

SPOTLIGHT

Social media spreads pandemic misinformation

During the COVID-19 pandemic, governments and medical officials scrambled to provide the public with accurate and timely information about SARS-CoV-2. Those efforts were sometimes undermined by the spread of medical misinformation and fake cures on some of the world's most popular messaging platforms.

Both WhatsApp (owned by Meta) and Twitter (now X) came under scrutiny for their handling of misinformation about COVID-19. These platforms were used to spread messages often containing a mix of accurate and misleading claims. The latter had to be debunked by medical experts. The problem became so acute that governments urged people to stop sharing information via these platforms.

Misinformation was typically spread via messages forwarded by friends or relatives. It often came from supposed medical professionals or a friend of a friend who worked in government. Some denied expert guidance (e.g. a declaration that social distancing is ineffective) or promoted actively harmful treatments. Unverified claims spawned mass panic, such as falsely claiming supplies of toilet paper were running out.

Activities

Acquiring and processing geographical information

1. Define cultural imperialism and outline what it involves.
2. Define popular culture.
3. Outline the principal concern about cultural imperialism. Explain whether this concern can be sustained.
4. Explain why the proliferation of US-based social media platforms has proved especially problematic.
5. Explain why the relationship between movies and culture involves a complex dynamic. Give an example of a contemporary development.
6. Outline how the focus of the film industry changed over the course of the twentieth century.
7. Explain why the role of the mass media remains a contested issue.
8. Outline the impact that media disrupters, including streaming services, are having on traditional media.

Applying and communicating geographical understanding

9. Study Figure 8.4.1. Using data from the graph, describe the extent to which the USA dominates the global film industry.
10. Access the latest data identifying the top 10 grossing movies of the last year. Does the data reinforce the dominance of US distributors evident in 2021?
11. As a class, discuss the importance of inclusion requirements such as those adopted by the Academy of Motion Picture Arts and Sciences. Are they justified? To what extent do they reflect changing cultural and social attitudes? Investigate the nature of any resistance to such measures.
12. Study the box, Spotlight: Social media spreads pandemic misinformation. Outline the impact of social media on the information available to the public. Investigate other examples of misinformation spread by social media. Share these with your class.

UNIT 8.5

Drivers of international integration: World cities

The process of economic and cultural integration is often simply referred to as globalisation. It has created a new urban system in which the role of the world's great cities has less to do with imperial power and trade facilitation than transnational corporate authority, international banking and finance, and the work of international agencies. Geographers refer to such cities as **world cities**. These cities emerged as the control centres of information and finance that collectively sustain and reinforce the world's economic and cultural integration.

World cities defined

World cities are the command-and-control centres of the increasingly integrated global economy. They are not necessarily the world's largest cities, but they are its most important, especially in relation to their economic and cultural authority. These cities control the flow of information, cultural products and finance, that sustain the world's economic and cultural integration. They are also innovation hubs and are rich in human capital (the skills, knowledge and experience held by an individual or population). These cities dominate popular culture through their powerful media outlets and creative industries. Their high art, fine restaurants and vibrant nightlife help attract and retain talented workers. They are magnets for migrants and visitors, adding to their diversity, another key strength.

While world cities tend to be large, size alone is not the determining factor. Several relatively small cities have global city status because of their dominance in an important sector of economic activity. San Francisco, USA, for example, qualifies because of its high-tech dominance, and Zurich, Switzerland, because of its role in global finance. Brussels, Belgium qualifies because its hosts the institutions that govern the European Union (EU) and the headquarters of the North Atlantic Treaty Organisation (NATO). New York, London, Paris and Tokyo are at the top of the hierarchy of world cities.

The emergence of world cities

World cities have existed for thousands of years. Historically trade was the principal determinant of their status. Trade, in turn, fostered innovation in transport, finance, law, language and communication. Cities such as Istanbul (Constantinople), Türkiye; Alexandria, Egypt; and Venice and Genoa, Italy, are cities whose status was enhanced by being strategic points on important trade routes. It enabled them to expand their influence beyond their own region.

The age of colonialism and imperialism (from the 1760s through to the mid-twentieth century), combined with the Industrial Revolution, ushered in a new era of world city authority. The capital cities and new industrial cities of the great imperial powers—Britain, France, Germany, Spain, Portugal, the Netherlands and Russia—became centres of global importance. For example, at the height of the British Empire, in the early twentieth century, decisions made in London held sway over 412 million people: 23 per cent of the world's population at the time; its territory covering 24 per cent of global land. Raw materials and markets for manufactured goods, plus national prestige and imperial ambition, motivated empire acquisition.

Today, globalisation (of which trade remains an important component) is the principal process determining world city status. Many of the cities actively engaged with the globalisation process by specialising in finance, professional and business services, media and communication industries. In the future, many will successfully transition to centres dominated by emerging industries—digital technologies, life sciences, climate, energy and water, urban services, infrastructure and robotics.

Globalisation and the rise of world cities

The emergence of modern world cities is closely associated with globalisation. Since the late 1960s, this process has been driven by:

- technological developments in transport and communications
- moves away from protectionist economic policies (often referred to as **trade liberalisation**)
- deregulated financial markets
- emerging, new, information-based forms of economic activity (the so-called 'new economy')
- emerging new ways of organising economic enterprises (e.g. outsourcing of business-related services and key stages in the manufacturing process)
- emerging TNCs as key players in the global economy
- an emerging global market for lifestyle-related commodities, created, at least in part, by the processes of cultural integration and the development of global media networks.

These have contributed to a rapidly expanding world trade and the relocation of labour-intensive manufacturing processes (typical of the 'old economy') to low-cost developing countries. Consequently, the functions performed by world cities have moved beyond the national scale to become increasingly international.

The emergence of world cities is, therefore, integral to the internationalisation of economic activity. It is also linked to the development of a spatially dispersed but globally integrated organisation of production and exchange. Consequently, they are important as centres of international finance, transnational business and international business services. The result is a new world system of cities acting as 'organising nodes'. In other words, they link regional, national and international economies into the global economy.

The nature of world cities

World cities act as the command-and-control centres of the global economy. Decision-makers, based in world cities, run the global economy. TNCs can operate in 60 or more countries and will most often have their headquarters, or at least their functional hub, in the heart of a world city. This is often in the city's traditional business district—Lower Manhattan in New York City, 'The City' in London, Marunouchi in Tokyo, or La Défense in Paris—Europe's largest dedicated business district. Clustered around this corporate core are the business services—including lawyers, accountants and consultants—that a corporation needs to function globally. Historically, corporations kept this expertise in-house. But the global economy is now so complex that few corporations have the knowledge to deal with it. Specialists are needed, so businesses providing global expertise to TNCs have grown in the hearts of global cities.

The characteristics of world cities

While no two world cities are alike, they draw on a subset of characteristics. They typically host:

- most of the leading global markets for commodities, investment capital, foreign exchange, equities (shares) and bonds
- clusters of advanced business services, especially those dealing in finance, insurance, real estate, accountancy and marketing
- concentrations of corporate headquarters—not just transnational corporations, but also major national firms and large foreign firms
- concentrations of national and international headquarters of trade and professional associations
- most of the leading NGOs and intergovernmental organisations (IGOs) that are international in scope. For example, the United Nations (UN) is based in New York and the United Nations Education Science and Cultural Organisation (UNESCO) in Paris

- the most powerful and internationally influential media organisations, news and information services and cultural industries (e.g. art and design, fashion, film and television)
- world-leading educational institutions, research institutes and think tanks that act as centres of new ideas and innovation in business, economics, culture and politics.

They also dominate the economy and trade of a large surrounding area.

Spatial distribution of world cities

Several organisations have sought to identify and classify world cities. Some classifications are complex. All use a slightly different mix of criteria. The most frequently used are those proposed by the Globalisation and World Cities Research Network (GaWC) and those published by Kearney, a leading global management consulting firm.

The GaWC has developed a classification system that measures the interconnectedness of cities. They use the terms Alpha, Beta and Gamma to classify the extent to which cities are integrated into the global economy (see Figure 8.5.1). Within each category there are two, three or four subcategories.

- **Alpha++** cities are vastly more integrated into the global economy than all other cities. London and New York are Alpha++.
- **Alpha+** cities are the eight cities that complement London and New York by filling advanced service niches for the global economy. They are Beijing, Dubai, Hong Kong, Paris, Shanghai, Singapore, Sydney and Tokyo.
- **Alpha and Alpha–** are the 13 and 21 cities, respectively, linking major economic regions into the world economy. The Alpha cities are Amsterdam, Brussels, Chicago, Frankfurt, Kuala Lumpur, Los Angeles, Madrid, Mexico City, Milan, Mumbai, Sao Paulo and Toronto. The Alpha– cities are Atlanta, Bangkok, Barcelona, Boston, Buenos Aires, Dublin, Istanbul, Jakarta, Johannesburg, Melbourne, Miami, Munich, New Delhi, Prague, San Francisco, Seoul, Taipei, Vienna, Warsaw, Washington DC and Zurich.

8.5.1 Global distribution of Alpha-ranked world cities

- **Beta** the 78 cities that link moderate economic regions into the world economy. They include Athens, Berlin, Cairo, Dallas, Houston, Manila, Montreal, Auckland, Brisbane, Geneva, Vancouver, Kolkata, Detroit and Perth.
- **Gamma** the 59 cities that link smaller economic regions into the world economy. They include Adelaide, Cologne, Osaka, St Petersburg and Belfast.

The pattern of world cities is in Figure 8.5.1. At the top of the hierarchy are the command-and-control centres of the global economy: New York, London, Tokyo and probably Paris. Beyond that point, rankings are less clear-cut. Cities such as Madrid, São Paulo, Sydney and, most recently, Beijing, link large national economies into the global system. Other cities have important, multinational roles. They include Milan, Hong Kong, Los Angeles, Frankfurt and Singapore. Cities such as San Francisco, Toronto, Zurich and Chicago bring important regional economies into the system. It is worth noting that very few world cities are in developing countries; the key is global economic function rather than population size.

The 2021 'Global Elite' ranking of world cities by Kearney uses these criteria to rank the cities: business activity, human capital, information exchange, cultural experiences and political engagement (see Figure 8.5.2).

Rank	City	Country
1	New York	USA
2	London	United Kingdom
3	Paris	France
4	Tokyo	Japan
5	Los Angeles	USA
6	Beijing	China
7	Hong Kong	China
8	Chicago	USA
9	Singapore	Singapore
10	Shanghai	China
11	San Francisco	USA
12	Melbourne	Australia
13	Berlin	Germany
14	Washington DC	USA
15	Sydney	Australia
16	Brussels	Belgium
17	Seoul	South Korea
18	Moscow	Russia
19	Madrid	Spain
20	Toronto	Canada
21	Boston	USA
22	Amsterdam	Netherlands
23	Dubai	United Arab Emirates
24	Frankfurt	Germany
25	Vienna	Austria

Source: Kearney

8.5.2 Kearney's 'Global Elite' ranking of world cities, 2021

Ranking	City
1	London
2	New York
3	Hong Kong
4	Singapore
5	Shanghai
6	Beijing
7	Dubai
8	Paris
9	Tokyo
10	Sydney
11	Los Angeles
12	Madrid
13	Toronto
14	Mumbai
15	Amsterdam
16	Milan
17	Frankfurt
18	Mexico City
19	São Paulo
20	Chicago

Source: GaWC

8.5.3 World city ranking according to level of connectivity, 2020

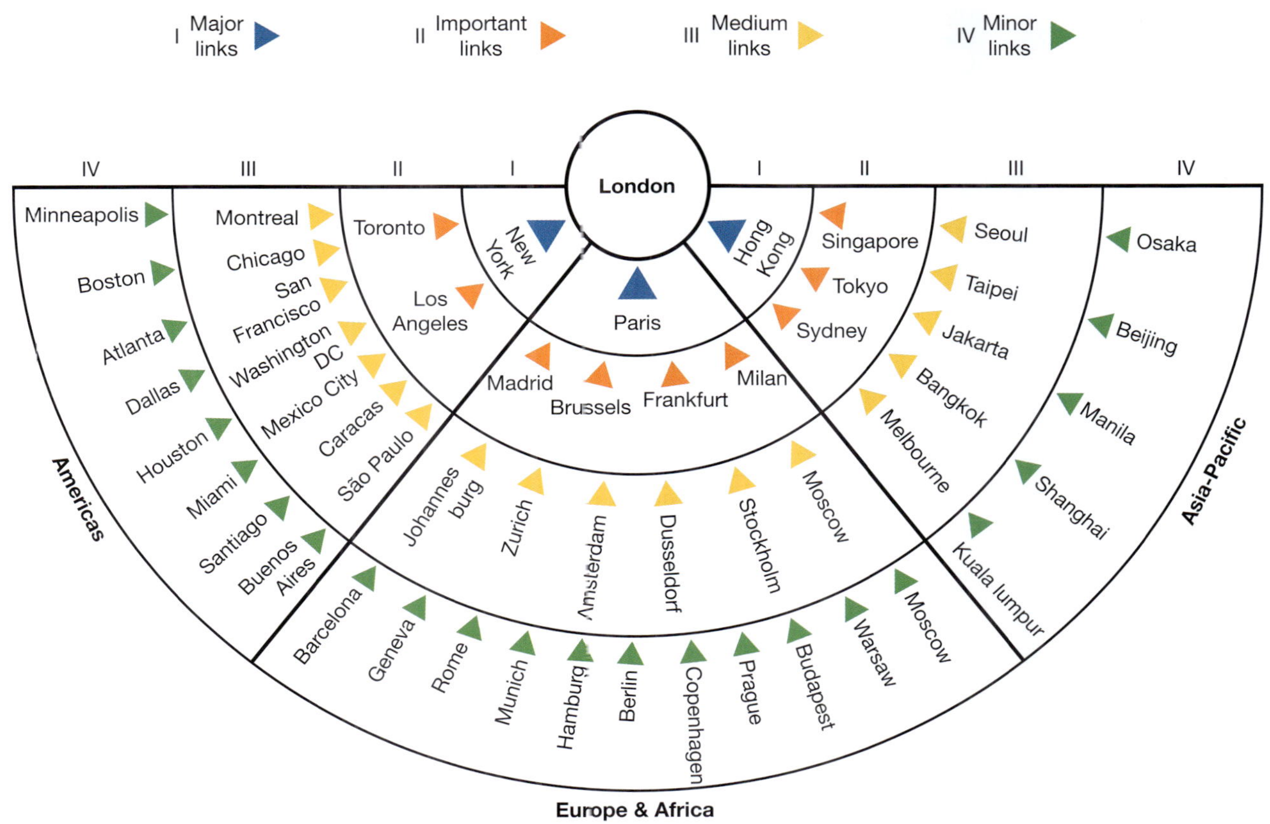

8.5.4 London's links with the rest of the world

Connectivity

Another way of looking at the status of cities is to examine their 'connectivity': a measure of the flows of information and knowledge between the international office network of global service firms. These firms are the basic agents of the world city's network formation. By quantifying such flows it is possible to rank cities based on connectivity and determine the strength of links between specific cities. Figure 8.5.3 shows their rank according to this connectivity measure. Sydney's rank of 10 puts it just ahead of Los Angeles and Madrid and just behind Paris and Tokyo. Sydney's strongest links are with major urban centres of North America and the financial centres of Asia: Hong Kong, Singapore and Tokyo. London's strongest links are with the North American cities of New York, Washington DC and Chicago, as well as Asian cities (including Hong Kong, Singapore, Tokyo and Beijing) and certain European cities, particularly those in Germany. London's links with other world cities are shown in Figure 8.5.4.

City	Number
Singapore	64
Amsterdam	67
Nagoya	70
Houston	74
Stockholm	74
Sydney	75
Zurich	79
Los Angeles	82
Taipei	90
Hong Kong	96
Chicago	105
Rhine–Ruhr	107
Seoul	114
Moscow	115
Beijing	116
Paris	168
Osaka	174
London	193
New York	217
Tokyo	613

Axis: 0 100 200 300 400 500 600 700

8.5.5 Distribution of global corporate headquarters, 2022 Source: McKinsey Global Institute

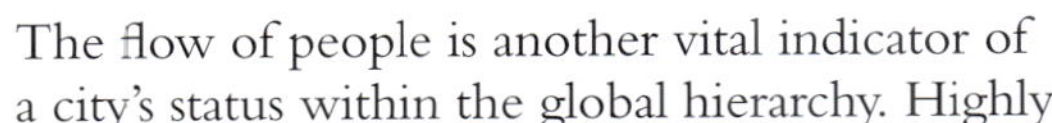

The flow of people is another vital indicator of a city's status within the global hierarchy. Highly skilled professionals are key operatives in the new economy. Their movement provides a useful insight into the relative importance of a city in the global economy. Such personnel working for global corporations will often move between the various branch offices of the company on a short-term or medium-term basis (see Figures 8.5.5 and 8.5.6).

Rank	Exchange	Market capitalisation (US$ trillion)
1	New York Stock Exchange, USA	25.15
2	NASDAQ, USA	18.99
3	Shanghai Stock Exchange, China	7.24
4	Euronext, Europe*	6.72
5	Japan Exchange Group, Japan	5.60
6	Shenzhen Stock Exchange, China	5.12
7	Hong Kong Exchanges, China	4.70
8	National Stock Exchange, India	3.28
9	LSE Group, UK	3.16
10	TMX Group, Canada	2.92
11	Saudi Stock Exchange, Saudi Arabia	2.67
12	Deutsche Boerse AG, Germany	2.18
13	NASDAQ Nordic and Baltics, Estonia, Latvia, Lithuania	1.96
14	SIX Swiss Exchange, Switzerland	1.94
15	Korea Exchange, South Korea	1.80

*Euronext: European stock exchange seated in Amsterdam, Brussels, London, Lisbon and Paris

8.5.6 Top 15 stock exchanges by market capitalisation, March 2023

The role of world cities

World cities developed beyond their former role as centres for international trade and banking to become:

- command-and-control centres of the global economy, in particular, the concentration and accumulation of international capital. The stock exchanges of New York, Tokyo and London influence the world economy
- preferred locations for finance and specialised service firms, including law, accounting, management consulting and advertising
- focused on advanced telecommunications technologies
- sites of production for new, innovative industries, especially the new information-based industries, which have replaced manufacturing as the dominant economic activity of such cities
- markets for the products and innovations produced. Those living in world cities are often the first to embrace the new technologies
- the base for media outlets with an international reach. For example, the New York-based News Corporation and Associated Press; the London-based BBC and Reuters; and the Paris-based Le Monde and Agence France-Presse
- powerful centres of cultural authority. The home of world-renowned cultural institutions, such as museums, art galleries, universities and research centres. They also tend to exert cultural leadership via major film festivals, theatres, orchestras, and opera and dance companies
- hosts of major sporting events.

There is close alignment between the characteristics of world cities outlined earlier and these roles.

● **SPOTLIGHT**

The 'big four'

London and New York, along with Paris and Tokyo, dominate the global hierarchy of world cities.

London (see Figure 8.5.7) remains a global financial capital, despite the United Kingdom no longer being the imperial power it was before World War II. This city hosts 193 corporate headquarters and is a popular location for the regional bases of TNCs. It is also the historical capital of the English language, which contributes to its status as a powerful media hub and major advertising centre. It is also the birthplace of the cultural, legal and business practices that define global capitalism.

Compared to New York, London has a time zone advantage for doing business with Asia and has the second-best global air connections of any city, with nonstop flights to 89 per cent of world cities outside of its home region of Europe. Only Dubai has better connections.

London is also a favoured place to live for the global rich who flock to the city's high-end retail precincts, fashionable restaurants, numerous theatres and royalty-enhanced social scene.

Along with these traditional strengths, London has emerged as Europe's top technology start-up centre. The city has more than 3000 tech start-ups, as well as Alphabet's largest office outside Silicon Valley.

New York (see Figure 8.5.8) hosts the world's top investment banks and hedge funds (firms engaging in speculation using credit or borrowed capital). The New York Stock Exchange's trading volume is nearly four times greater than Tokyo and 10 times that of London. There are 217 New York-based corporate headquarters.

Like London, New York is a global leader in media and advertising, the music industry (home to two of the big three labels), and one of the most important capitals of fashion and luxury retailing. With many iconic landmarks, international tourists spend more money in New York City than in any other city.

It has 120 higher education institutions and many research institutions, especially in medicine and biological and life sciences. There are over 2000 arts and cultural organisations, and more than 500 art galleries. Broadway theatre is one of the leading forms of English-language theatre in the world. The city is also one of the world's great sporting capitals.

Paris (see Figure 8.5.9) remains one of the world's great centres of creativity and innovation. For centuries, the city has attracted the world's brightest talent. It is one of the world's great cultural centres. Its immense contribution to the creative and performing arts has been complemented by its leadership in fashion and gastronomy (cooking).

Paris specialises in a diverse set of tradable industries, from financial and business services to advanced manufacturing and creative industries. Thirty-one of the world's largest 500 corporations are headquartered in Paris and the city hosts many world-class research universities. The focus is on high-tech employment and investment in research and development. Paris is the world's top tourist destination.

Tokyo (see Figure 8.5.10) has the most corporate headquarters (613) and is a principal financial market. Fifty-one of the world's largest 500 companies are based here, almost twice that of Paris (in second place). The stock exchange is the world's fourth largest in turnover. It is also a choice destination for many of the world's most accomplished chefs and restaurateurs, and innovators in fashion and design. Many of Japan's most prestigious universities are in Tokyo.

8.5.7 London

8.5.8 New York

8.5.9 Paris

8.5.10 Tokyo

● **SPOTLIGHT**

World cities as centres of cultural authority

A key element of globalisation is the role of world cities as centres of cultural authority. They can generate and spread ideas and values, and influence specific cultural processes. This link is most apparent with international competitions—the great global sporting spectacles. Coverage of major events and their host cities is seen worldwide thanks to communications technology. International media coverage of football (soccer) or Rugby Union World Cup competitions and Formula One motor racing, are examples of the contemporary relationship between world cities and global sporting culture.

Apart from Tokyo (1964 and 2020), London (2012) and Paris (2024) (see Figure 8.5.11), the choice of Olympic host cities focuses on those of regional importance to the global system. Host cities for the Summer Olympic Games are predominantly lower-order global cities with aspirations of advancement in function and status (e.g. Athens 2004, Rio de Janeiro 2016 and Brisbane 2032). Globalisation has increased the competition to host such events. Cities at various levels in the global hierarchy compete in an increasingly crowded global events market.

8.5.11 Paris won the right to host the 2024 Summer Olympic Games.

Activities

Acquiring and processing geographical information

1 Define, in your own words, the term world city. Why are these cities important? What role do they play in the global economy?
2 Outline the factors that led to the emergence of world cities.
3 Outline the relationship between globalisation and the emergence of world cities.
4 Outline how the nature of world city authority has changed over time.
6 Outline the characteristics of world cities.
7 Explain why Sydney and Melbourne are classified as world cities.
8 Outline the role of world cities.
9 Explain why world cities are described as centres of cultural authority.

Applying and communicating geographical understanding

10 Study the text under the heading, Spatial distribution of world cities, and complete the following tasks.
   a Describe the spatial distribution of world cities. Explain what is meant by the term connectivity.
   b Identify examples of world cities that:
      i link large national economies to the global economy
      ii have an important multinational role
      iii link important regional economies into the global system.
   c Study the ranking of world cities shown in Figures 8.5.1, 8.5.3 and 8.5.4. What world cities appear in all of the rankings? Which cities stand out as an exception in one or more of the tables?
   d Study Figure 8.5.2. Using an atlas, describe the global distribution of Alpha world cities. Explain the difference in the Alpha city classifications in the GaWC hierarchy of world cities.
11 Study Figure 8.5.4. Interpret the diagram and write a paragraph describing London's links with the rest of the world.
12 Study Figure 8.5.5. Interpret the data from the graph, and identify the dominant world cities in terms of the distribution of global corporate headquarters.
13 Study Figure 8.5.6. Interpret the data in the table, and identify the dominant world cities in terms of the distribution of the world's largest stock exchanges.
14 Study the box, Spotlight: The 'big four', and complete the following tasks:
   a Identify the principal world cities that constitute the command-and-control centres of the global economy.
   b Identify the features of London, New York, Paris and Tokyo that distinguish them as world cities.

UNIT 8.6

Drivers of international integration: Migration

Along with the role it plays in facilitating economic activity, migration is a key driver of cultural integration. In this context, cultural integration occurs when people from one culture adopt elements of another, while maintaining their own. Rather than casting aside their own culture, or retaining it and rejecting the new culture, they blend the two. In Australia, we call this multiculturalism. In the USA, they refer to a **melting pot** of many cultures coming together to form a more diverse culture.

Migrants bring elements of their culture into their new communities. These include their foods, languages, religions, music, arts, attitudes and traditions. The most readily recognised are restaurants and foods that come from other cultures. Migrants also adopt parts of their new culture as they adapt to local customs and ways of interacting (see Figure 8.6.1).

8.6.1 Australian university campuses, like many local communities, reflect the diversity of multicultural Australia.

Migration-based cultural integration can promote intercultural understanding and foster a greater sense of unity within communities. The host community benefits from experiences that they may not otherwise access. People can engage with the foods, traditions and arts of other cultures. This fosters respect and creates better informed and empathetic citizens.

Unlike **assimilation**, multiculturalism encourages migrants to maintain important elements of their cultural identity. They neither reject the new culture nor become absorbed by it. Instead, they integrate into it without losing the elements of their own culture.

Migration-based cultural integration increases diversity and enhances cultural awareness. It promotes respect for other cultures. It also enriches a person's life by exposing them to other cultures. In the workplace, individuals from diverse backgrounds can bring their unique perspectives, experiences, insights and knowledge to the tasks they engage in. School students gain respect and empathy for those who differ. Communities benefit from discovering new cultures, foods, sports and arts.

Activities

Acquiring and processing geographical information

1. Define cultural integration.
2. Explain what is meant by the term melting pot.
3. Outline what migrants contribute to the culture of the place in which they settle.
4. Outline the benefits of migrant-based international integration.
5. Explain how multiculturalism is different from assimilation.

Applying and communicating geographical understanding

6. As a class, discuss the evidence of cultural integration in your local community. To what extent, has your community's way of life been affected by immigration?
7. Conduct a class debate on the topic 'Cultural integration and multiculturalism are the preferred outcomes of migration'.

   To prepare for the debate:

   - write down all the arguments for and against the topic
   - conduct research to find sources to support your argument.

UNIT 8.7

Drivers of international integration: Tourism

One of the most important drivers of economic and cultural integration is global tourism (see Figure 8.7.1). Until the 1970s, tourism was largely restricted to those who could afford both the time and money to travel. Today most people in the developed world, and increasing numbers of people living in developing countries, engage in tourism during their lifetime. Tourism is now an accepted, even expected, part of the lifestyle of many people. The diversity of tourist experiences is also increasing, as people seek out new destinations and immerse themselves in different cultures. Tourism, however, is not impact-free. The massive numbers of people now involved in tourism have an impact on the environments and the cultures they engage with.

The extent of tourism's contribution to international integration was no more evident than in the speed at which COVID-19 spread globally. Tourism and international labour movements were the principal means by which the virus spread worldwide.

8.7.1 Tourists at a café in Quimper, France

Defining tourism

Tourism, broadly defined, involves the temporary movement of people to destinations outside their normal place of work and residence. It encompasses the activities tourists undertake while staying at their destination and the facilities created to cater for visiting tourists. International tourism involves travel to a country other than the one in which the tourist normally lives.

Tourism encompasses all those activities contributing to a tourist's movements from origin to destination. It involves consuming a wide range of goods and services provided by transport, tour operators, accommodation establishments, theme parks and attractions, entertainment and arts venues, museums and historical sites, travel agents, souvenir retailers and restaurants. Worldwide, tourism generates economic activity worth an estimated US$1.86 trillion and employs 333 million people. It involves vital cultural engagement. Tourists experience the benefits derived from interacting with other cultures and the places that are visited are exposed to new cultural influences. Both have positive and negative outcomes.

Cultural tourism

Increasingly, tourists choose to immerse themselves in local customs and routines. Cultural tourism includes educational tours, performing arts events, festivals, pilgrimages, visits to monuments, and the study of nature, folklore and art. The cultural tourist is likely to be a person who enjoys meeting local people and learning about their lifestyle. They are seeking educational experiences and use travel for personal growth and sincere appreciation of others.

Developing cultural tourism programs encourages destinations to celebrate and promote those elements that distinguish their communities. They provide opportunities for authentic cultural exchange between locals and visitors. Each impacts on the other.

The growth of international tourism

Much of global tourism's growth is due to the decline in travel costs. This was driven largely by developments in aviation technologies, especially high-capacity aircraft, beginning with the Boeing 747 in 1969, along with the rise of a global middle class with the discretionary income to travel (see Figures 8.7.2, 8.7.3 and 8.7.4).

The world's middle class keeps growing rapidly. By 2020, it encompassed half of the world's population—as either middle class or living in wealthy households. Out of the world's 8 billion people, 3.6 billion are considered middle class. This is a significant figure, especially considering that only a decade ago it was half that number (around 1.8 billion). The world's middle class is forecast to exceed 5.2 billion by 2030 (around 1.6 billion more than today), which will represent some two-thirds of the world's population.

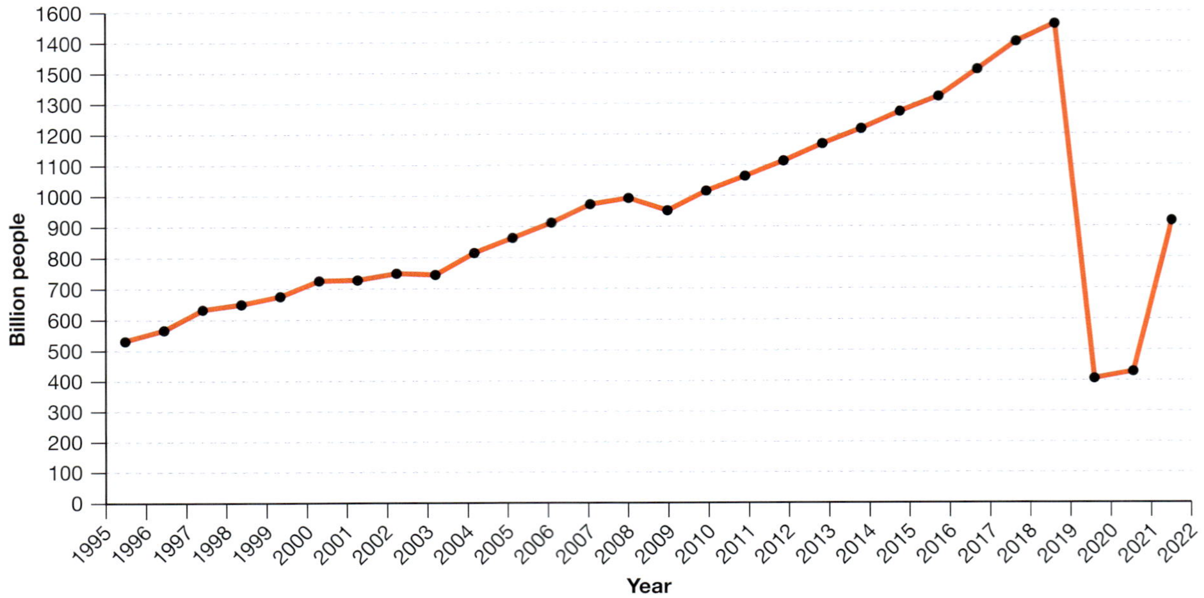

Source: World Tourism Organisation (UNWTO)

8.7.2 International tourist arrivals 1995–2022

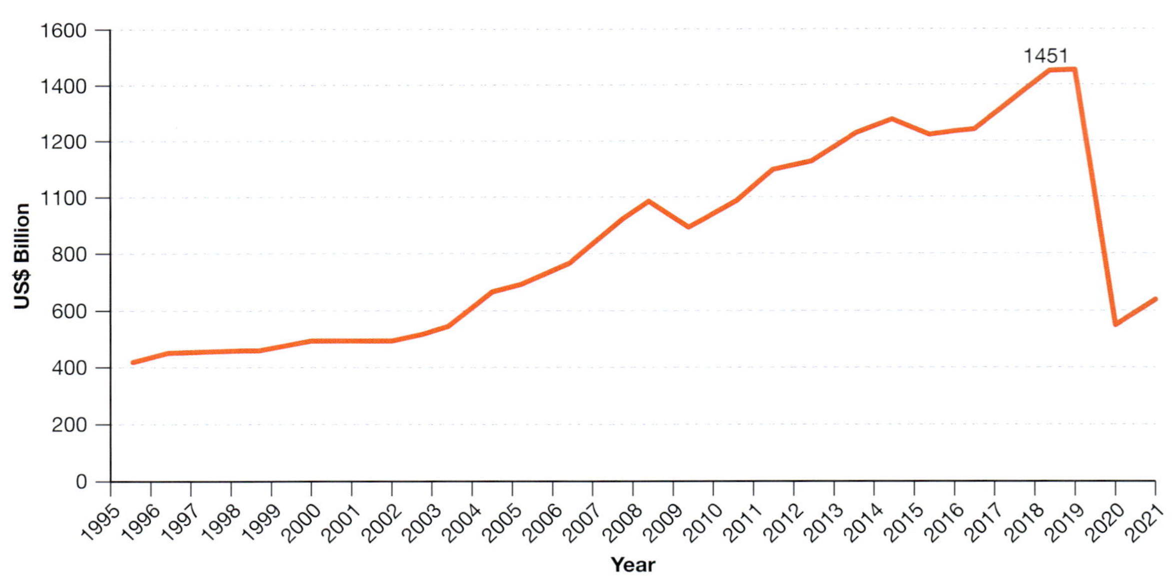

Source: UNWTO

8.7.3 The growth in international tourism receipts 1995–2021

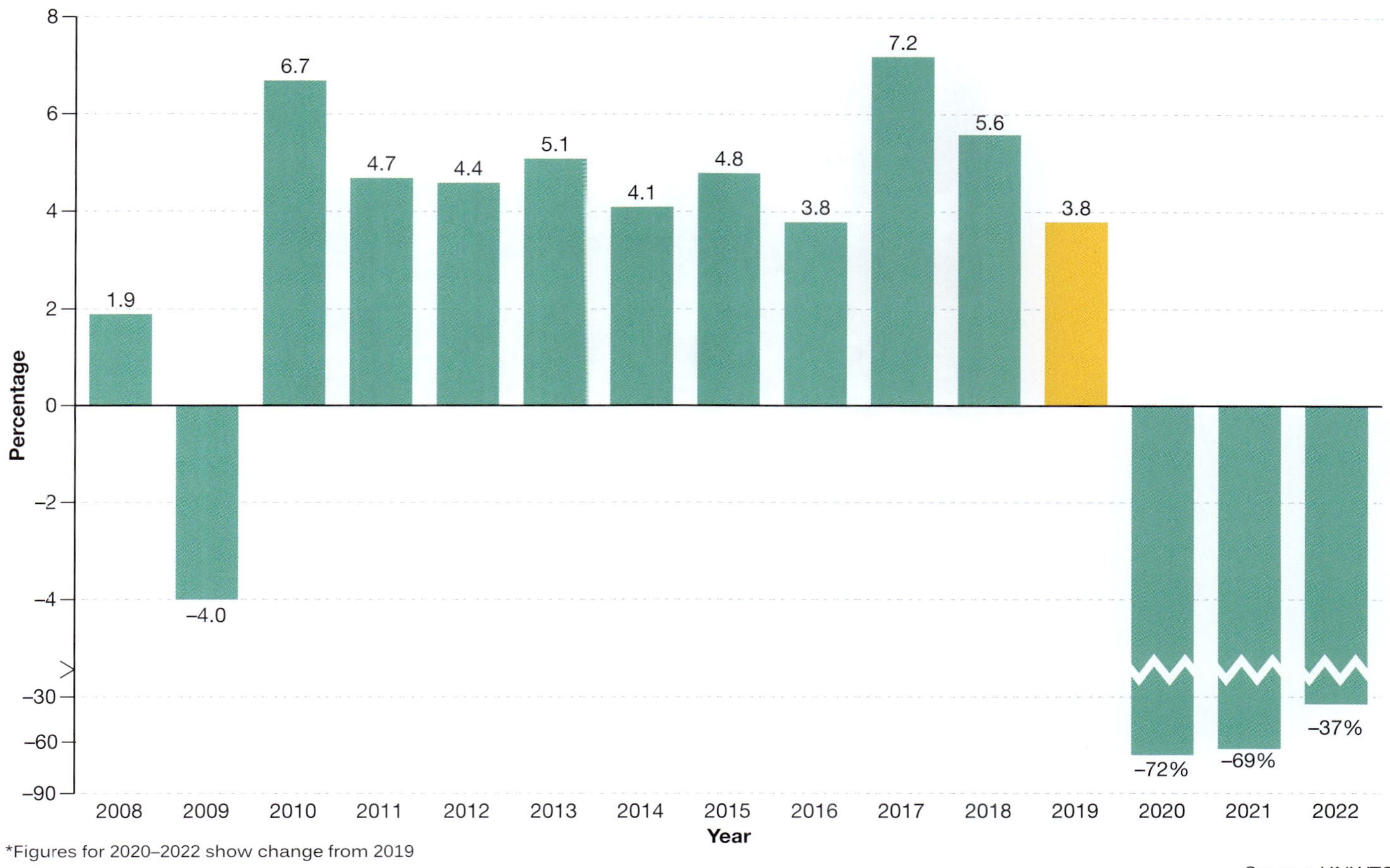

8.7.4 International tourist arrivals, annual percentage change, 2008–2022*

Tourism's benefits and costs

Global tourism growth has had pronounced social, environmental and economic effects. Until recently, attention was concentrated on tourism's economic impacts. This reflects the early optimism with which tourism was viewed. It was widely acclaimed as bringing a multitude of cultural and economic (generating wealth and employment) benefits.

The growth of tourism also prompted more perceptive observers. They raise questions concerning the social and environmental desirability of further expansion, especially on the massive scale of contemporary tourism development. Any growth in tourism will place increased demand on the biophysical and constructed environments, and require a large investment in tourism-related infrastructure. The impacts of such expansion must be considered or the attractiveness that initially drew tourists to a destination may be lost.

Social impacts of tourism

Social contact between tourists and residents can be mutually beneficial. The resident population can better understand visitors' customs, values and culture. They can also enjoy increased demand for traditional entertainment, crafts and music, helping to preserve parts of their national heritage. Tourists' patronage of museums, art galleries and theatres brings revenue that helps to maintain these facilities for local use. The tourists then go home with a better understanding of the host country, its culture and traditions.

Industry critics often suggest the transient social relationships inherent in tourism are incapable of producing any meaningful contact between visitors and residents. Increasing tourist numbers can shift attitudes in the resident population from initial euphoria to growing hostility (see Figure 8.7.5).

8.7.5 Protests against over-tourism and cruise-ship access in Venice, Italy, 2017

Tourism development may result in mutual misunderstanding, hostility and social tension as the tourists' demands affect residents' lifestyles. A significant and undesirable impact is the growth of prostitution, crime and gambling in the host population.

Activities

Acquiring and processing geographical information

1. Outline the relationship between tourism and international integration.
2. Define tourism. What does it encompass?
3. Explain tourism's economic importance.
4. Outline the nature of cultural tourism.
5. Explain why the growth of the world's middle class has been a key factor in the growth of international tourism.
6. Outline the benefits and costs of tourism.
7. Outline the social costs of tourism.

Applying and communicating geographical understanding

8. Study Figures 8.7.2 and 8.7.3. Using data from the graphs, write a paragraph outlining the trends apparent in international visitor arrivals and receipts since 1995.
9. Study Figure 8.7.4. Describe the trends in the annual percentage change in tourist arrivals since 2008. In which year did growth peak? Describe the impact of COVID-19 on international tourist movements.

APPLICATION AND CONSOLIDATION TASKS

Task 1: Infographic: What of the future?

Select one of the technologies featured in the Unit 8.2 box, Spotlight: What of the future? Conduct research and assess its potential in terms of economic and cultural integration. Present your findings as an infographic.

Task 2: Annotated visual display: TNCs

Select one of the TNCs listed in Figure 8.3.4, and investigate the nature and geographical distribution of its activities. Your display should include information that outlines:

- its history
- its key activities.

Include maps, graphs, diagrams and/or photos.

Task 3: Podcast: Cultural resistance

Prepare and record a podcast that discusses contemporary examples of how different cultural groups have sought to resist outside economic and cultural forces. It should be 4–5 minutes long.

Consider the history behind the issue, the different cultures and groups involved in the resistance, and the impacts of the issue on peoples and cultures.

Your podcast can include audio content such as music, snippets of interviews, etc.

CHAPTER

9 Population and resource consumption

The growth in the world's population is unprecedented. By the end of this century, the world's population is likely to exceed 11 billion, up from 7.96 billion in 2022.

Throughout history, increases in resource consumption and environmental degradation were largely a product of our efforts to achieve ever-higher standards of living for the rapidly growing human population. Consequently, the magnitude of the threat to the global environment is closely linked to human population size and resource use per person. Indicators of severe environmental stress include the loss of biodiversity, increasing greenhouse gas emissions, accelerated rates of deforestation and desertification, water shortages and growing food insecurity in many parts of the world.

This chapter investigates the growth, characteristics and distribution of the world's population, and the demographic processes shaping population change. It also examines the challenges arising from the changing size and distribution of population and the relationship between population and natural resources. The demographic characteristics of Nigeria and Italy are compared.

And it is a universal truism in development ... that if you want to change the society, educate a girl ... then she will have the empowerment and the economic freedoms, which come with that education ... and more choices about how and when to have her children, she will choose to have less children ...

Julia Gillard, former Prime Minister of Australia

9.0.1 The world's population continues to grow and is predicted to exceed 11 billion by the end of the century.

Chapter glossary

age structure percentage of the population (or number of people of each gender) at each age level in a population

child mortality rate the annual number of children under the age of five years who die per 1000 live births, also known as the under-5 mortality rate

continuous resources perpetual resources that will virtually always exist. They include solar energy, rainfall, wind energy, hydro-electricity, tidal power and geothermal power

demographer someone who studies the characteristics and changes in the size and structure of human populations

demographic changes changes in the size, composition, rates of growth, and density of population; changes to fertility and mortality rates; and changes to patterns of migration

demographic transition the theory that links the process of industrialisation to declining death rates followed by declines in birth rates

food security the state of having reliable access to a sufficient quantity of affordable, nutritious food

industrial agriculture the system of chemical-intensive food production, featuring large, single-crop farms and animal production facilities

infant mortality rate the annual number of deaths of infants under one year of age per 1000 live births

migration the act or process of people moving from one place to another with the intent of staying at the destination permanently or for a relatively long period of time

population dynamics the factors which contribute to populations over time

population structure the age and gender composition of a population, usually depicted as a population pyramid

rate of natural increase the percentage by which a population grows in a year; this is the difference between the birth rate and the natural death rate and excludes migration

renewable resources resources that will eventually be replenished if managed sustainably

replacement-level fertility the number of children a woman and her reproductive partner must have to replace themselves. The worldwide average is usually just above two children per couple because some children die before they reach their reproductive years

tipping point the point at which a series of small changes or incidents becomes significant enough to cause a larger, more important change

total fertility rate the average number of children a woman will have during her reproductive years

UNIT 9.1

The characteristics, growth and distribution of the world's human population

During the twentieth century, the world's population grew at a rate never before experienced. By 2000 there were 6 billion people, reaching almost 8 billion by mid-2022. The UN projects it to reach 9.75 billion by 2050 and 10 billion by 2055. This is a dramatic rise from 1900 when there were only 1.6 billion people, and 1950 when there were only 2.5 billion. It took all human history until 1804 to reach 1 billion.

Population growth

Every year total world population grows by about 83 million. Continued growth is inevitable because 25 per cent of the population is under the age of 15. This provides a built-in momentum for further growth, even as fertility rates (the average number of children a woman will have) continue to decline.

Calculating the number of years required to add each additional billion people helps to understand the rapid rate of increase. Figure 9.1.1 shows it took 2 million years to reach the first billion, 123 years for the second billion and only 12 years for the sixth billion.

Figures 9.1.2 and 9.1.3 show the growth of the world's population since 1000 CE. A semi-logarithmic graph helps determine the period of greatest growth.

There are signs the explosive growth of the twentieth century is beginning to slow, and the population could even stabilise before 2100. Even if the rate does stabilise or decline in the years to come, the population could still reach just over 11 billion humans by 2100.

Year (CE)	Population (millions)
1000	275
1100	306
1300	384
1400	373
1500	429
1600	486
1700	635
1800	919
1900	1571
2000	6073
2025	8899*
2050	9752*
2100	11 200*

*UN estimates

9.1.2 The growth in human numbers, actual and projected, 1000 CE–2050 CE

World population	Year reached	How long did it take?
1 billion	1804	2 million years
2 billion	1930	126 years
3 billion	1960	33 years
4 billion	1974	14 years
5 billion	1987	13 years
6 billion	1999	12 years
7 billion	2011	12 years
8 billion	2022	11 years

9.1.1 Adding billions to the human population

● **SPOTLIGHT**

How many people have ever lived on Earth?

Modern humans appeared around 50 000 BCE. Assuming counting starts then, and a constant growth rate applies up to modern times, approximately 106 billion people are estimated to have been born. This makes the current population roughly 6 per cent of all people who ever lived.

● SKILLS BUILDER

Graphing data with a huge range of values

Log-log and semi-logarithmic graphs are used to accommodate data with a huge range of values. They also allow us to judge the rate of change. The steeper the line, the greater the rate of change.

Semi-logarithmic graphs have a vertical scale that is graduated in a logarithmic progression. Equal intervals, or cycles, on the vertical scale increase geometrically (e.g. 1, 10, 100, 1000 and 10 000). The horizontal scale has a normal arithmetical progression.

Log-log graphs have both the vertical and horizontal axes graduated in a logarithmic progression.

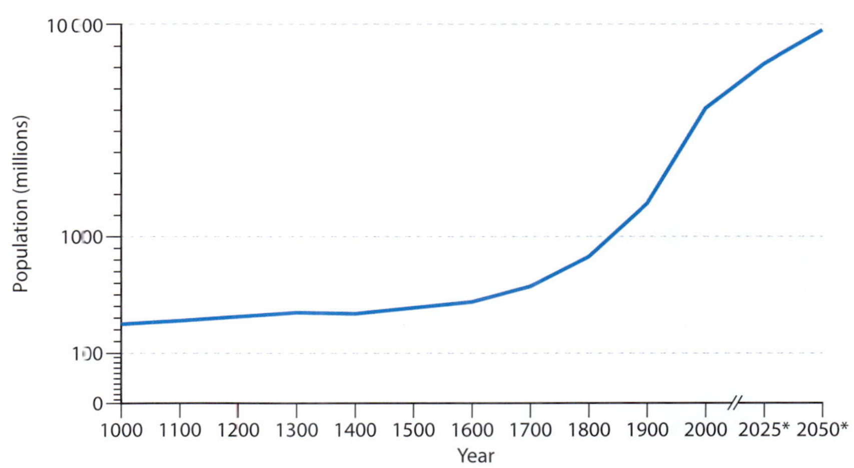

9.1.3 A semi-logarithmic graph of world population growth since 1000 CE

The rapid increase in the human population is an example of exponential growth. This occurs when some factor, such as population, grows by a constant percentage of the whole during each unit of time. It starts off slowly, but a few doublings quickly produce very high numbers (see Figure 9.1.5). After the second doubling, each additional doubling amounts to more than the total of all preceding growth (e.g. 1, 2, 4, 8, 16, 32, 64, 128, 256). Figure 9.1.4 shows the **rate of natural increase** in 2022.

Time unit	Births	Deaths	Natural increase
Year	133 990 000	67 100 000	66 890 000
Month	11 165 833	5 591 667	5 574 167
Day	367 096	183 836	192 260
Hour	15 296	7650	7636
Minute	255	128	127
Second	4.3	2.1	2.3

Figures may not add to totals due to rounding.

Source: UN Population Division

9.1.4 World vital events per time unit, 2022

9.1.5 Tandberg cartoon

The global pattern of population increase

The world map of population growth (Figure 9.1.6) shows the wide range of growth rates in different regions. Growth rates are highest in the places least able to cope. Developing countries have 83 per cent of the world's population, but account for 98.5 per cent of the annual increase—6.4 billion out of almost 8 billion in 2022. They also have the greatest proportional increase. The developed world's population of 1.3 billion will increase only marginally by 2050. Almost all will be in the USA, mainly due to immigration, from 333 million in 2022 to 375 million in 2050.

The rate of increase is linked to the fertility rate. **Replacement-level fertility** is around 2.1. If the fertility rate remains below this level for an extended time, a country's population will decline without a net increase from **migration**. Figure 9.1.7 shows the projected population change by county, 2022–2050.

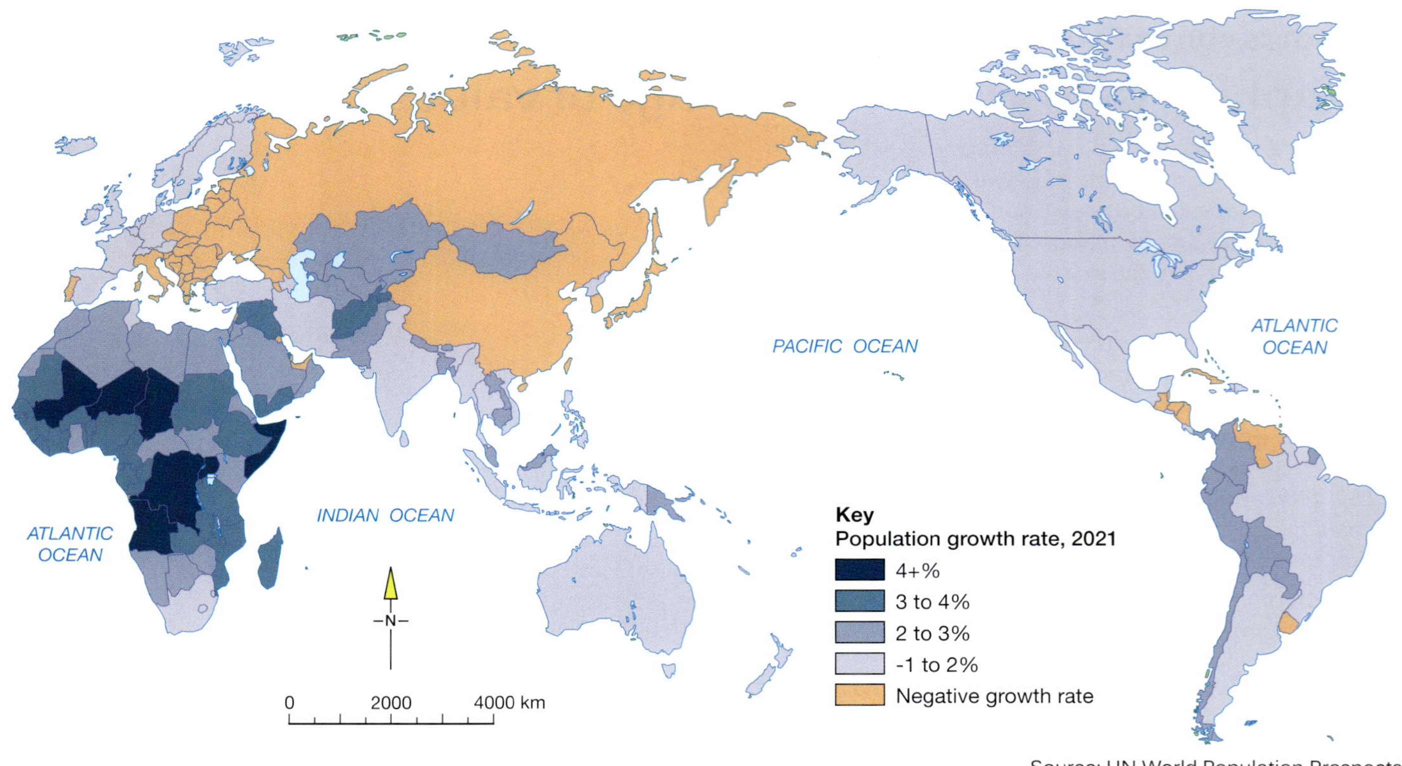

Source: UN World Population Prospects

9.1.6 Population growth rate, 2021

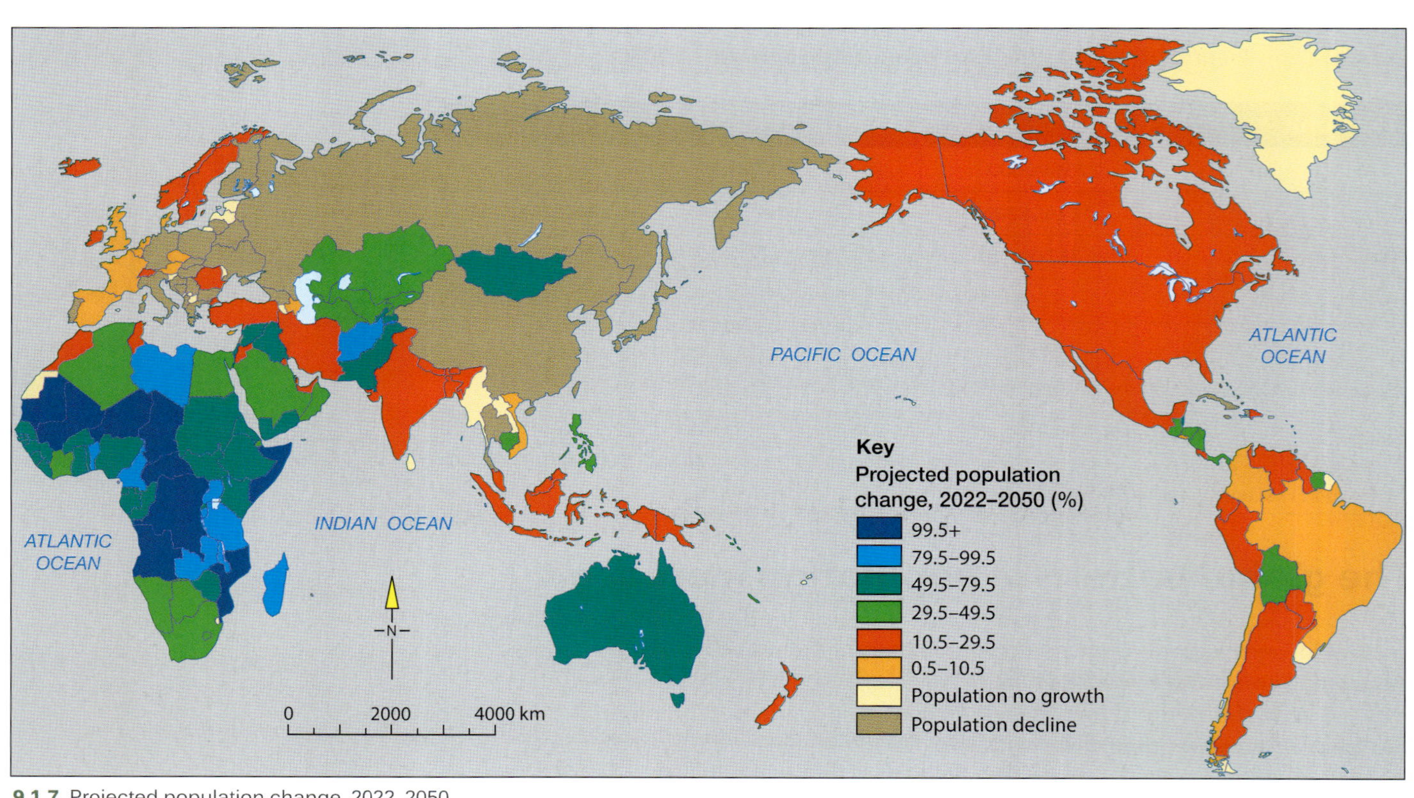

9.1.7 Projected population change, 2022–2050

Demographic trends

Demographic statistics, such as fertility rate, life expectancy and the percentage of the population under the age of 15, are used to predict future population growth.

In Africa, the population is expected to double between 2022 and 2050 (increasing from 1.4 billion to 2.5 billion in 2050). With fertility rates of up to 6.7 children per woman and 40 per cent of its population under the age of 15, the source for this increase is already in place.

In South America, population growth is relatively high, but uneven. Women in French Guiana, on average, have more than twice as many children (3.5) as those in Chile (1.4) and Brazil (1.6).

In Asia, growth rates vary significantly. China, now the world's second most populous country after India, experienced a dramatic decline in its fertility rate: from 6.5 in 1968 to just 1.2 in 2022 (see Spotlight: China's one-child policy on page 235 for more). Even so, its population of 1.4 billion is greater than the entire developed world. India's population of 1.43 billion continues to grow at an annual rate of 0.8 per cent. India's population will be almost 1.6 billion by 2035 and almost 1.7 billion is predicted by 2050, compared to China's 1.3 billion.

Over half of India's population is under 25 years of age and more than 65 per cent are under 35. The Philippines, which had 21 million people in 1950, is expected to grow to 158 million by 2050. Russia's population is expected to shrink from 144 million to 137 million. Japan's is expected to decline from 125 million in 2022 to 102 million in 2050.

In developed countries, the growth rate has slowed dramatically. Europe's fertility rate is now only 1.5, too few to replenish the population. It is expected to shrink from 742 million in 2022 to 724 million in 2050.

Figures 9.1.8 and 9.19 show selected demographic characteristics by region. Figure 9.1.10 illustrates the age–sex structure of the world population in 2022.

Region	Population mid-2022 (millions)	Projected population mid-2050 (millions)	Projected population change 2022–50 (%)	Total fertility rate	Percentage of the population <15 years	Percentage of the population >65 years	Life expectancy male/ female
World	7963	9752	22.5	2.3	25	10	70/75
Developed	1270	1297	2.1	1.5	16	20	75/82
Less developed	6694	8454	26.3	2.4	27	8	69/73
Africa	1419	2478	74.6	4.3	40	3	61/64
North America	372	424	14.0	1.6	18	17	74/80
Latin America and Caribbean	656	746	13.7	1.9	23	10	70/77
Asia	4730	5313	12.3	1.9	24	10	71/76
Europe	742	724	-2.4	1.5	16	19	75/81
Oceania	44	66	50.0	2.1	23	13	76/81

Source: Population Reference Bureau, World Population Data Sheet

9.1.8 World demographic trends by region, 2022

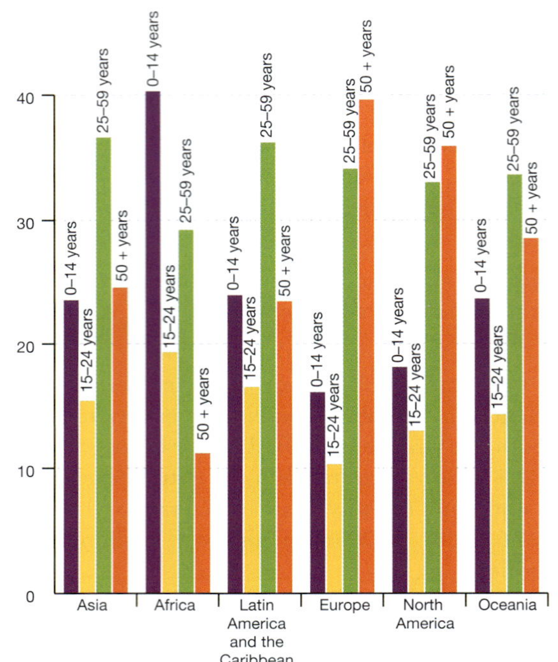

9.1.9 Percentage of population in broad age groups for the world by region, 2022

Source: Statista

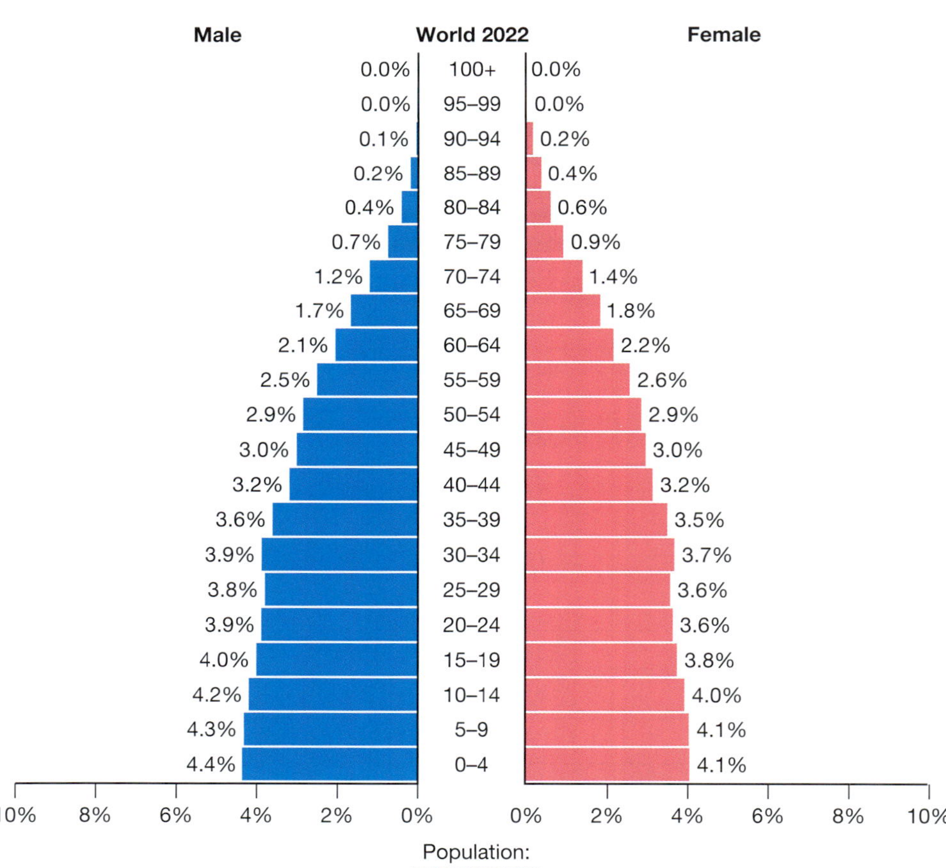

9.1.10 Age–sex structure of the world population, 2022

Source: UN Population Division

● SPOTLIGHT

The world's population is ageing

With sustained declines in fertility and mortality, the global population's transition to an older **age structure**, known as population ageing, will accelerate. Older people's (ages 65+) share of the global population increased from 5 per cent in 1960 to 10 per cent in 2022. It is projected to rise to 16 per cent by 2050, with ages 85-plus growing the fastest. Meanwhile, the proportion of children (ages 0 to 14) is falling, from 37 per cent in 1960 to 25 per cent in 2022, with a projected decrease to 21 per cent by 2050 (see Figure 9.1.11).

The timing and speed of age structure changes vary by country. Such changes have important social and economic implications. The age structure of a population affects national policy agendas and resource allocation. Countries with relatively high fertility and child dependency are challenged with investing the resources needed to develop young people's human capital (principally through education). Such investments provide an opportunity to reap the economic growth benefits of a larger, better-educated working-age population. Countries experiencing high old-age dependency must address the high costs of medical and long-term care needs of older people. They also need to invest in the wellbeing of, and future opportunities for, younger generations. Countries with high and moderate child dependency are generally found in the developing world, while those with high old-age dependency are found in the developed world.

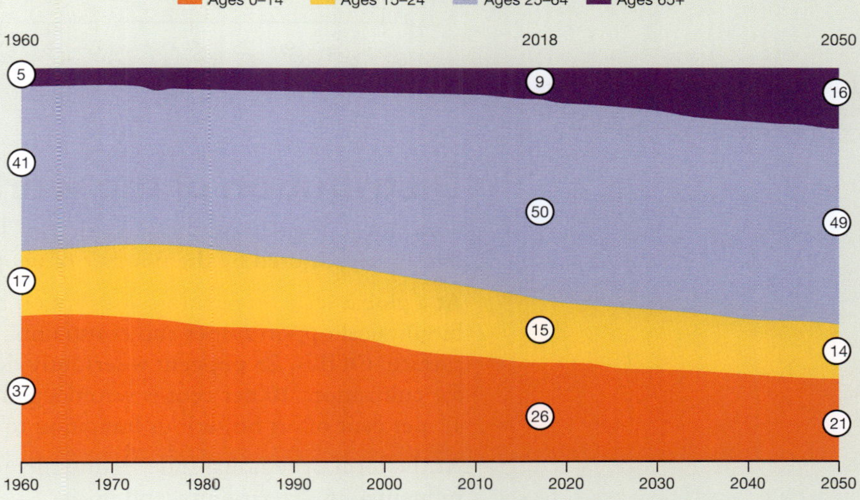

9.1.11 Cumulative line graph showing the percentage of population by age group, 1960–2050

Challenges posed by high rates of population growth

Many experts are concerned about the size of the world population increase. Another 3 billion people will degrade and even overwhelm the biophysical environment: the land's food-producing capacity and the other natural resources on which all life depends. Meeting their extra energy needs will accelerate the rate of climate change, especially if fossil fuel alternatives are not swiftly and fully embraced.

Others believe the population is a valuable resource. They claim that technology will enable humans to cope with the problems arising from any increase.

Declining rates of population growth

In 2022, the global population was growing at 0.9 per cent annually, down from 1.08 per cent in 2016 (see Figure 9.1.12.). Annual growth rates peaked in the late 1960s. The rate has more than halved since its peak of 2.19 per cent in 1963. It is currently declining and projected to continue to do so. Estimates put it at less than 0.5 per cent by 2050.

This means that the world population will continue to grow in the twenty-first century, but at a slower rate than in the recent past. Reasons include a declining fertility rate, decreased mortality (death) rate and increased life expectancy.

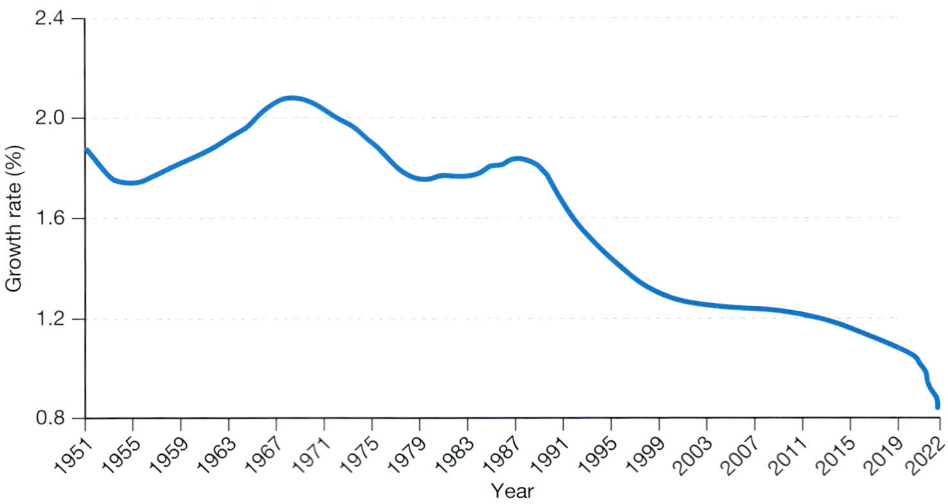

9.1.12 Yearly population growth rate in decline

Distribution of the world's population

The global population spread is uneven. There are substantial differences in population density—measured by the number of people per square kilometre.

At a global scale, population distribution and density are determined largely by biophysical opportunities and constraints. Land makes up just 30 per cent of Earth's surface. Of this, 28 per cent is seriously deficient in water, 22 per cent has soils too thin for cultivation, 10 per cent is waterlogged and 6 per cent is affected by permafrost. Only 11 per cent offers no serious limitation to settlement and agricultural land use.

At the local or regional scale, economic, political and social factors are likely to be influential in determining land use.

Most people live in the Northern Hemisphere and the developing world. Less than 12 per cent live in the Southern Hemisphere. The coastal margins of landmasses are more densely settled than inland areas, with 40 per cent living within 100 kilometres of a coastline.

In 2022, approximately 77 per cent of the world's population lived in Africa and Asia (excluding Russia), on only 40 per cent of the world's land area. Europe accounted for 9.3 per cent of global population, with a further 8.2 per cent in Latin America and the Caribbean, and 4.8 per cent in North America.

Figures 9.1.13 to 9.1.16 show the distribution of the world's population by continent, the world's 10 largest countries by population in 2022 and 2050, the density of the world's population and how the distribution of the world's population changed and is projected to change, between 1750 and 2150.

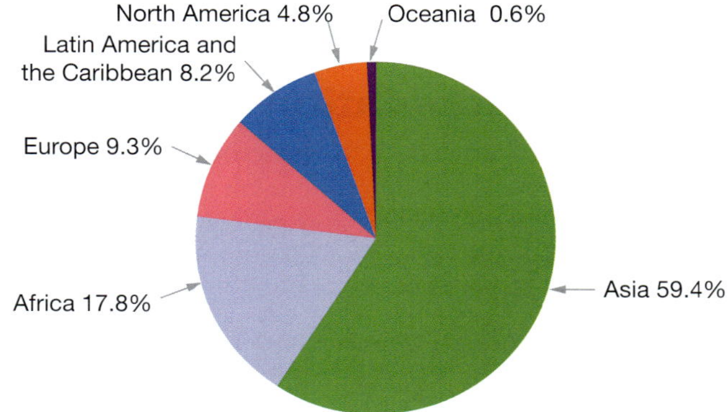

9.1.13 Distribution of the world population, 2022

2022		2050	
Country	**Population (millions)**	**Country**	**Population (millions)**
China	1437	India	1671
India	1417	China	1553
USA	333	Nigeria	378
Indonesia	276	USA	375
Pakistan	236	Pakistan	368
Nigeria	219	Indonesia	317
Brazil	215	Brazil	233
Bangladesh	171	Democratic Republic of Congo	218
Russia	144	Ethiopia	215
Mexico	128	Bangladesh	204

Source: Population Reference Bureau, World Population Data Sheet

9.1.14 World's 10 largest countries by population, 2022 and 2050

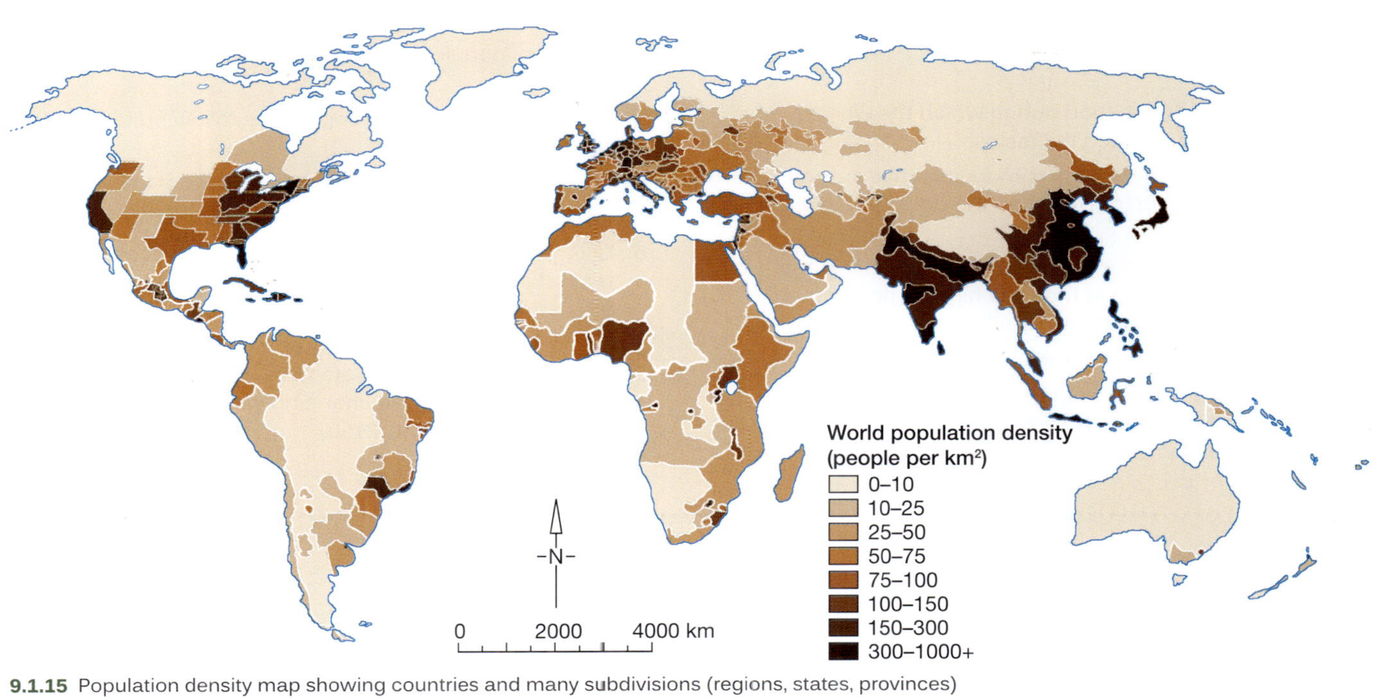

9.1.15 Population density map showing countries and many subdivisions (regions, states, provinces)

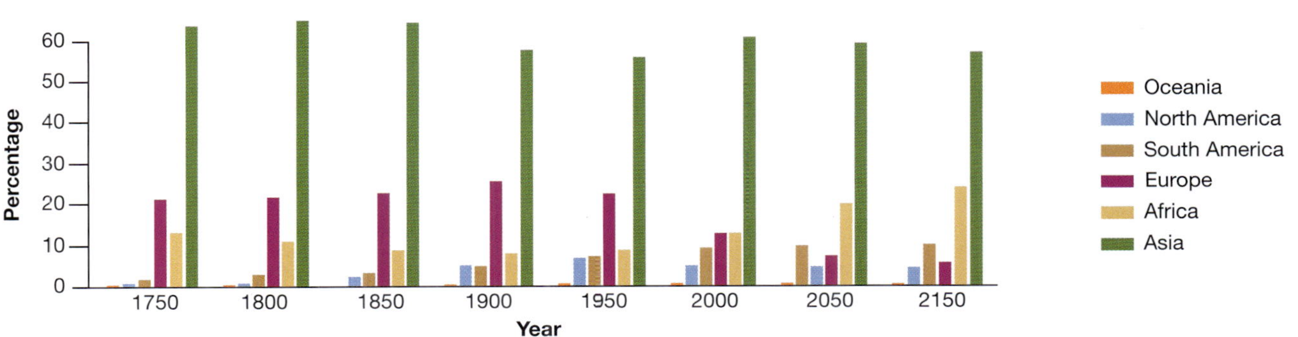

9.1.16 The changing distribution of the world's population, 1750–2150

Activities

Acquiring and processing geographical information

1 Explain what is meant by fertility rate and replacement rate. What occurs if the fertility rate remains below the replacement rate for a prolonged period?

2 Outline the key demographic trends and their implication for the populations of Africa, Latin America, Asia and countries of the developing world.

3 Outline the challenges posed by high rates of population growth.

4 Account for the slowing rate of world population growth.

5 Distinguish between the terms population distribution and population density. Outline the factors that determine the distribution of the world's population.

Applying and communicating geographical understanding

6 Using the information in Figure 9.1.2, draw a line graph showing the growth in human numbers. Use a scale of 10 millimetres to 1000 million people on the vertical axis and 10 millimetres to 100 years on the horizontal axis.

7 Study Figure 9.1.3.

a Identify the time period during which the world's population expanded the fastest.

b Explain why Figure 9.1.3 is more effective in showing world population growth than the graph you drew in activity 6.

8 Study Figure 9.1.7. Using an atlas, identify those parts of the world projected to have the highest percentage change in their population in 2021–50.

9 Study Figure 9.1.8 and complete these tasks.

a Identify the regions of the world in which the 0–14 age group is greater than the global average. Which region has the smallest share of its population under the age of 15?

b Identify the regions of the world in which the 65+ years age group is greater than the global average. Which region has the smallest share of its population over 65?

10 Study Figure 9.1.9. Outline the implications of this data for the demographic future of the selected world regions.

11 Study Figure 9.1.10 and complete the following tasks.

a Calculate the number of people under the age of 15.

b Calculate the number of people over the age of 60.

c Compare Figure 9.1.9 with 9.2.8. Identify the stage of the demographic transition the world has achieved given the structure of the population in 2022.

12 Study Figure 9.1.11. Outline the global pattern of age dependency. What generalisations can you make about the distribution of countries with high and moderate child dependency and those with high old-age dependency?

13 Study Figure 9.1.12. Using data from the graph, write a paragraph outlining the trend in yearly growth rates between 1951 and 2021.

14 Study Figure 9.1.14. Outline how the ranking of the world's most populous countries will change during 2022–2050.

15 Study Figure 9.1.15, then complete the following tasks using an atlas.

a Identify the most densely settled regions of the world.

b Which regions are the most sparsely settled?

c Using an atlas, compare Figure 9.1.15 with a physical map of the world. Describe the relationship between the nature of the biophysical environment and population densities.

16 Study Figure 9.1.16. Using data from the graph, write a paragraph outlining the changing distribution of the world's population, 1750–2150.

UNIT 9.2

Factors influencing population change

At a global scale, fertility and mortality are the determinants of population growth (or decline). At a national or regional scale, we need to add migration to our calculations.

Birth and death rates

The birth rate is the number of births during a specific period. The crude death rate is the number of deaths during a specific period. These two measures are usually expressed in terms of the number of births or deaths per 1000 people in a population per year. The rate of natural increase of a population is determined by calculating the difference between the crude birth rate and the crude death rate (see Figure 9.2.1).

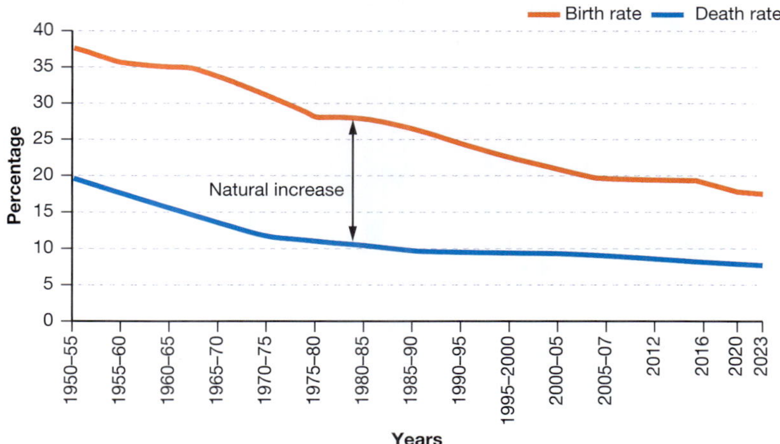

9.2.1 Trends in birth and death rates, worldwide, 1950–2023

Spatial pattern of births

Birth rates vary significantly worldwide. **Demographers** regard rates above 30 per 1000 people as high. The highest are in Africa and South-West Asia. Kenya, Tanzania and Uganda all had rates of 50 or more during the 1980s and 32 African countries still have birth rates over 30. These include Niger (44), Mali (43), Somalia (39), Burundi (41), Burkina Faso (37), Gambia (37), Guinea (36), Nigeria (35) and Mozambique (38). Several Central American and Asian countries are also in this category.

Fertility rates

The **total fertility rate** (TFR) provides insight into future demographic trends. For example, at the height of Kenya's population explosion in the 1980s, the average number of children born per woman was 8.1. In 2022, it was still high at 3.3.

Many countries experienced a steady decline in TFR in recent years. In 1970, China's TFR was 6.1, but by 2022. it was just 1.2. In one generation, India's TFR declined from 5.8 to 2.1, Egypt's from 7.2 to 2.5, Brazil's from 6.3 to 1.6 and Mexico's from 7.7 to 1.8. In comparison, Niger's TFR (at 6.7) still exceeds 4.3 (the rate for all of Africa).

The lowest fertility rates are in Western Europe. Many countries there have rates at or below 1.5 (e.g. Austria, Finland, Greece, Italy, Portugal, Luxembourg, Spain and Switzerland). Other countries with low birth rates are Australia (1.7), New Zealand (1.8), Japan (1.3), Canada (1.4) and the USA (1.7).

Moderately wealthy countries often have intermediate birth rates. This suggests a relationship between the birth rate and their level of economic development. But it is not the only factor. Cultural traditions and women's status play a role. In some male-dominated cultures, women have limited control over their fertility.

Generally, low birth rates are associated with higher standards of living. As material wellbeing increases, people tend to have fewer children. China is the exception, having achieved its low birth rate via strict population-control measures.

Overall, fertility has been declining over several decades. This explains the world population's growth rate decline from its peak of 2.1 per cent in 1962, to 1.23 per cent in 2022. The population replacement fertility rate is 2.1. About half of the world's population, in 83 countries, report below replacement-level rates. By 2050 over 130 countries, about two-thirds, are projected to have fertility rates below replacement level (see Figures 9.2.2, 9.2.3 and 9.2.4).

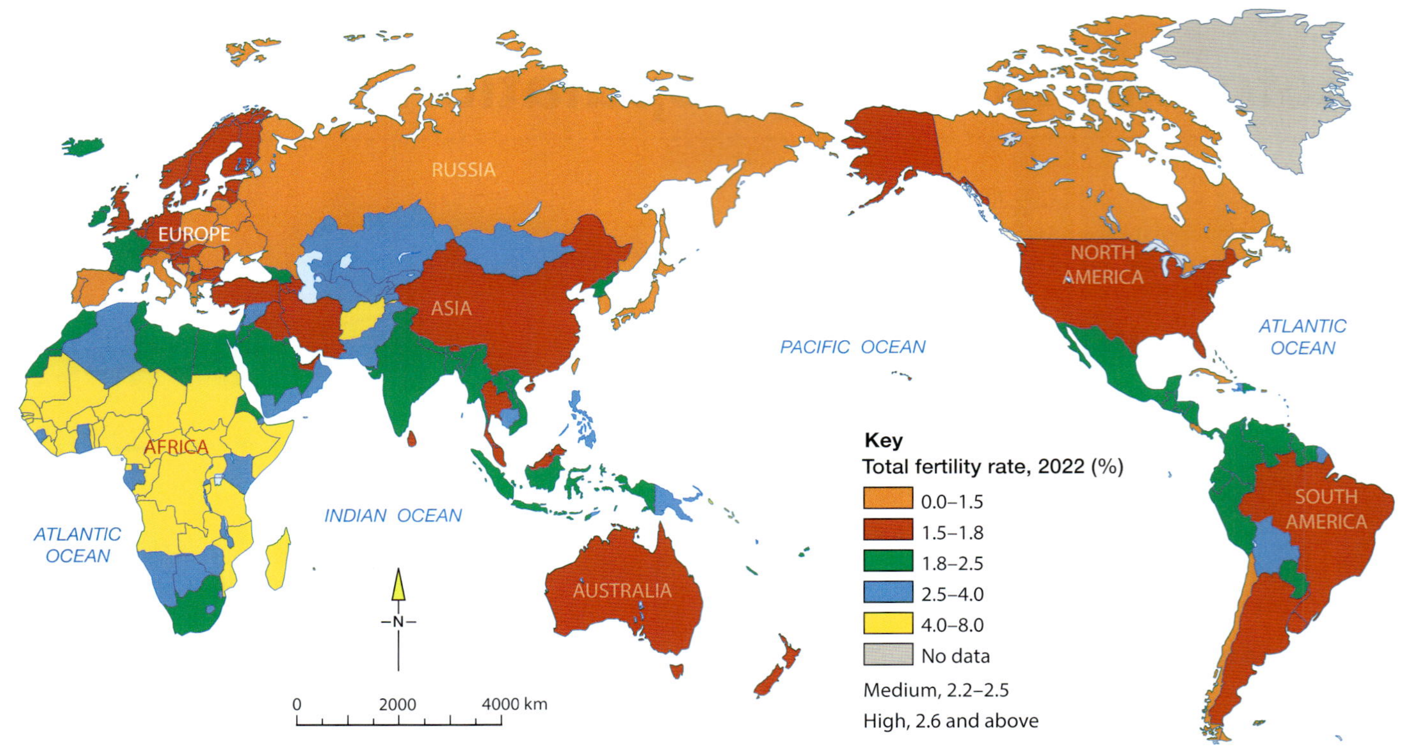

Source: Population Reference Bureau, World Population Data Sheet

9.2.2 Total fertility rate, 2022

Highest		Lowest	
Niger	6.7	South Korea	0.8
Somalia	6.3	Singapore	1.1
Democratic Republic of Congo	6.2	Taiwan	1.0
Chad	6.2	Andorra	1.0
Mali	6.0	Ukraine	1.0
Angola	5.3	China	1.2
Nigeria	5.1	Spain	1.2
Burkina Faso	4.7	Italy	1.3
Gambia	4.7	Japan	1.3
Burundi	4.3	Thailand	1.3

Source: Population Reference Bureau, World Population Data Sheet

9.2.3 Highest and lowest fertility rates, 2022

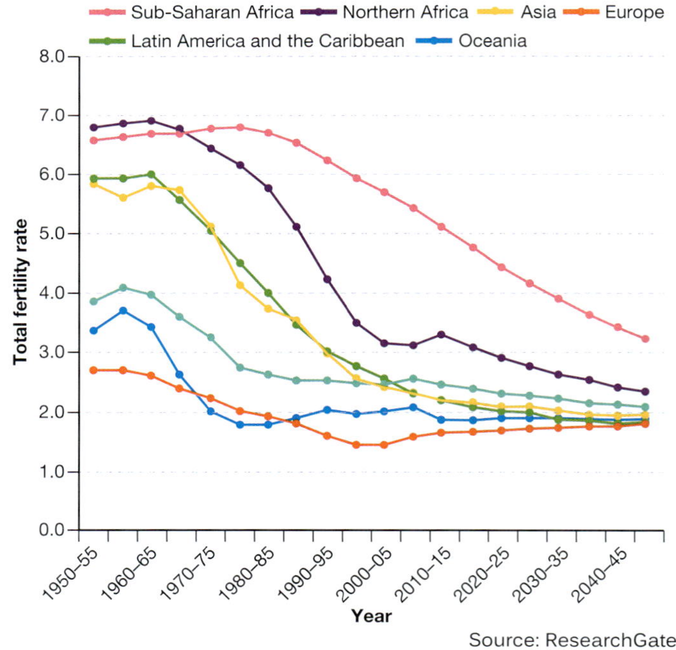

Source: ResearchGate

9.2.4 Trends in total fertility rate by region, 1950–2050 (actual and projected)

Factors affecting fertility rates

Many factors affect fertility rates.

- **Levels of economic and social wellbeing:** Fertility rates generally decline with increasing levels of development.
- **Infant mortality rates:** High infant mortality rates result in families having larger numbers of children in the expectation some will not survive infancy.

- **The importance of children as part of a family's labour force:** In many countries of the developing world, children are seen as an asset because of the labour they can contribute to subsistence farming practices. They also help collect water and fuel wood.
- **Levels of urbanisation:** Urban dwellers tend to have a lower fertility rate than people in rural areas.
- **Educational and employment opportunities for women:** Fertility rates tend to be high where women have little or no access to education and paid employment.
- **The average age of marriage:** As the average age of marriage increases, fertility rates tend to decrease. If marriage is delayed to the age of 25, a woman cuts her child-bearing years (typically ages 15–44) by 10 years and reduces her prime reproductive years (typically ages 20–29) by 50 per cent.
- **The cost of raising children:** The high cost of raising and educating children in developed countries has contributed to a reduction in fertility.
- **The availability of reliable methods of birth control:** Access to contraceptives tends to reduce fertility.
- **The availability of aged services and pensions:** These reduce parents' reliance on their children to support them in their old age.
- **Family size preferences:** In some societies, family size is influenced by a range of cultural factors, including religious beliefs and cultural traditions. These, in turn, are often linked to the status of women and the degree of control that women can exercise over their own fertility.

Mortality rates

The mortality (or death) rate (MR) is a measure of the number of deaths per 1000 people in a given year. Worldwide the MR is eight, but Africa has an MR of nine. In some African countries, it is over 10—Central African Republic (11), Chad (12), Lesotho (14), Nigeria (12), Somalia (12), South Sudan (11), Chad (12), Namibia (11) and South Africa (12).

Infant mortality rates worldwide declined from 198 per 1000 live births in 1960 to 29 in 2022. (The rate is four per 1000 in developed countries and 31 in developing countries.) They are highest in Sierra Leone, where 75 out of 1000 children born die before their first birthday. Rates are also very high in the Central African Republic (71), Somalia (71), Chad (65) and South Sudan (64). These are well above the whole of Africa rate of 47 and the worldwide rate of 29. Comparatively, Australia's infant mortality rate is only 3.2. Figure 9.2.5 shows the global pattern of infant mortality.

Several changes combined to slash overall **child mortality rates** (the number of children under the age of five years who die per 1000 live births) and increase life expectancy more generally without having the same dramatic impact on fertility rates. These include advances in medical science and public health, nutritional improvements and greater access to education. This means more young people survive to their reproductive years than in the past.

The overall decline in mortality rates is often hailed as a great accomplishment of human civilisation. Since the early 1950s, the global death rate has more than halved: from 19.7 to 7.7 deaths per 1000 population. The mortality rates for the developed and developing worlds are now 12 and 7, respectively.

In developed countries, improvements in medical science and public health came about slowly, so life expectancy increased gradually. In 1800, life expectancy at birth was about 35 years. By 1900 it had increased to 50 and by 1950 it had reached 66. In 2022 it stood at 70 for males and 75 for females (see Figures 9.2.6 and 9.2.7).

In developing countries, the rate of increase began much later, but was more rapid. By 1950 life expectancy had increased to about 50 years. It now stands at 69 years for males and 73 years for females, an increase of over half a year every 12 months. Most of the increase is due to the rapid adoption of medical technologies originating in the developed world.

These **demographic changes** mean fewer children die in infancy and people live longer.

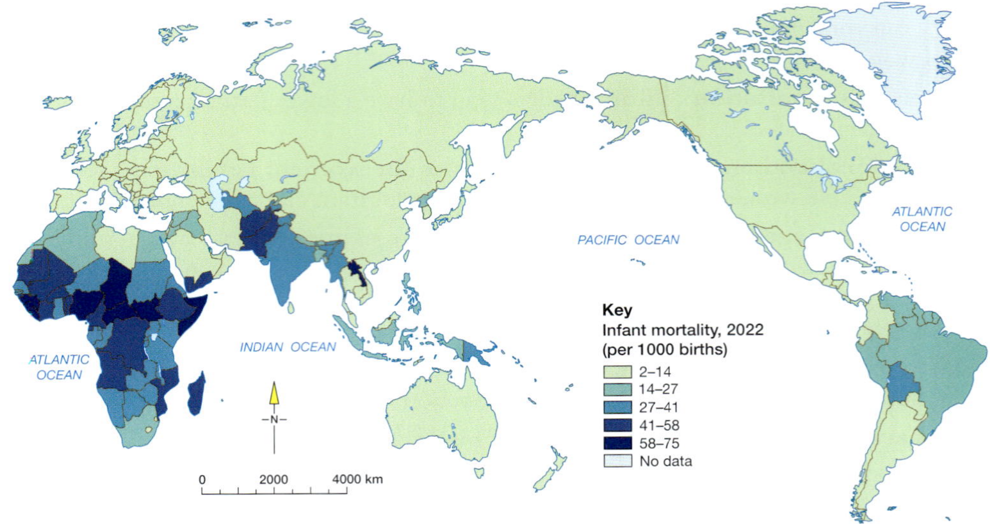

9.2.5 Infant mortality, 2022

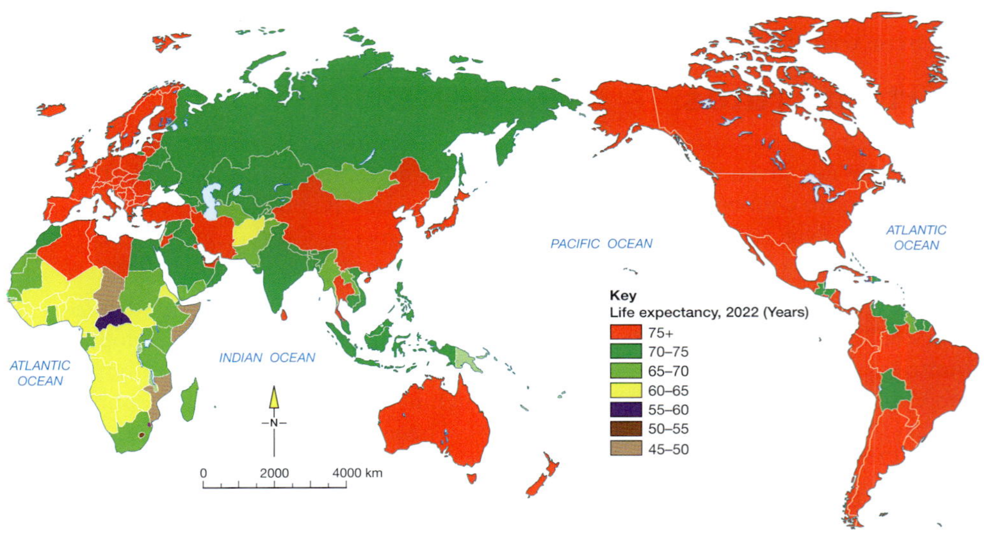

9.2.6 World map of life expectancy, 2022

Factors affecting mortality rates

Many factors affect mortality rates, including:

- nutrition standards
- standards of personal hygiene and effluent disposal (sanitation)
- access to safe drinking water and the incidence of infectious diseases
- access to medical and public health technology, including immunisation, antibiotics and insecticides.

High infant mortality rates usually indicate inadequate food (under-nutrition) and poor diet (malnutrition). Another indicator is a high incidence of infectious diseases, often contracted from contaminated water.

Population policies

The countries experiencing the most substantial fertility declines have implemented population-control policies. These are well-organised, government-initiated and emphasise birth control and family planning. Alone, such policies are unlikely to achieve sustained reductions in fertility. They must be associated with programs that promote economic and social development. Coercive policies are rarely successful. In India, in the 1970s enforced sterilisations conflicted with civil liberties and religious and cultural beliefs.

In many countries, family planning programs can be a highly emotive, cultural, religious and political issue. Culturally-sensitive programs are central to achieving sustained reductions in population growth. Such programs need to be combined with improved infant and maternal health care programs.

Highest		Lowest	
Country	**Life expectancy (years)**	**Country**	**Life expectancy (years)**
Japan	85	Lesotho	53
Switzerland	84	Chad	53
Singapore	84	Nigeria	53
South Korea	84	Central African Republic	54
Malta	84	Somalia	55
Iceland	83	South Sudan	55
Israel	83	Eswatini (Swaziland)	57
Sweden	83	Namibia	59
Norway	83	Côte d'Ivoire	59
Spain	83	Zimbabwe	59
Australia	83	Mozambique	59
France	82	Guinea	59
Canada	82	Mali	59
New Zealand	82	Democratic Republic of Congo	59
Italy	82	Benin	60
Taiwan	81	Burkina Faso	60
United Kingdom	81	Cameroon	60
Austria	81	Guinea-Bissau	60

Source: UN World Population Data Sheet

9.2.7 Countries with the highest and lowest life expectancy, 2020

SPOTLIGHT

China's one-child policy

China's Planned Birth Policy was introduced in 1979 and phased out, beginning in 2015. Its aim was to limit the country's population growth. Introduced as a temporary measure, it remained in place for over three decades. Officially, couples were limited to just one child. Second or subsequent pregnancies were met with fines, economic penalties and pressure to have the pregnancies terminated.

The policy was not an all-encompassing rule. It has, in practice, only applied to ethnic Han Chinese living in urban areas. Ethnic minorities and those living in rural areas were not subject to the policy. Nevertheless, it was very successful in reducing the population growth rate.

There were some unplanned demographic and social impacts. Because male children were often more highly valued than females, girls were sometimes neglected and abandoned. Female foetuses were often aborted and female infanticide occurred, creating a gender imbalance—118 males born for every 100 females. The worldwide average is 103 to 107.

The policy was effective at curbing China's population growth rate. At its peak, the fertility rate was over 7.5. By 2022, the TFR had dropped to 1.2, well below the population replacement rate of 2.1. To prevent too dramatic a population decrease, a special provision was introduced allowing millions of couples to have two children legally. To qualify, both partners had to be an only child.

● SPOTLIGHT

The demographic transition model

To explain the nature of population change experienced by countries, demographers developed the **demographic transition** model. It is based on the experiences of several European countries, Canada and the USA. It attempts to identify the different stages of demographic change and view their impact on total population (see Figure 9.2.8).

Future population projections assume we have entered Stage 4 of the demographic transition. Demographers assume that eventually the birth rate and death rate will reach equilibrium. This will occur several decades after reaching a point where couples average two children each. This two-child average is called replacement-level fertility because each couple replaces themselves in the number of people in a population. When the fertility rate is at replacement level, the two children born essentially replace the parents when they die. The replacement level is 2.1 children per family, accounting for child mortality.

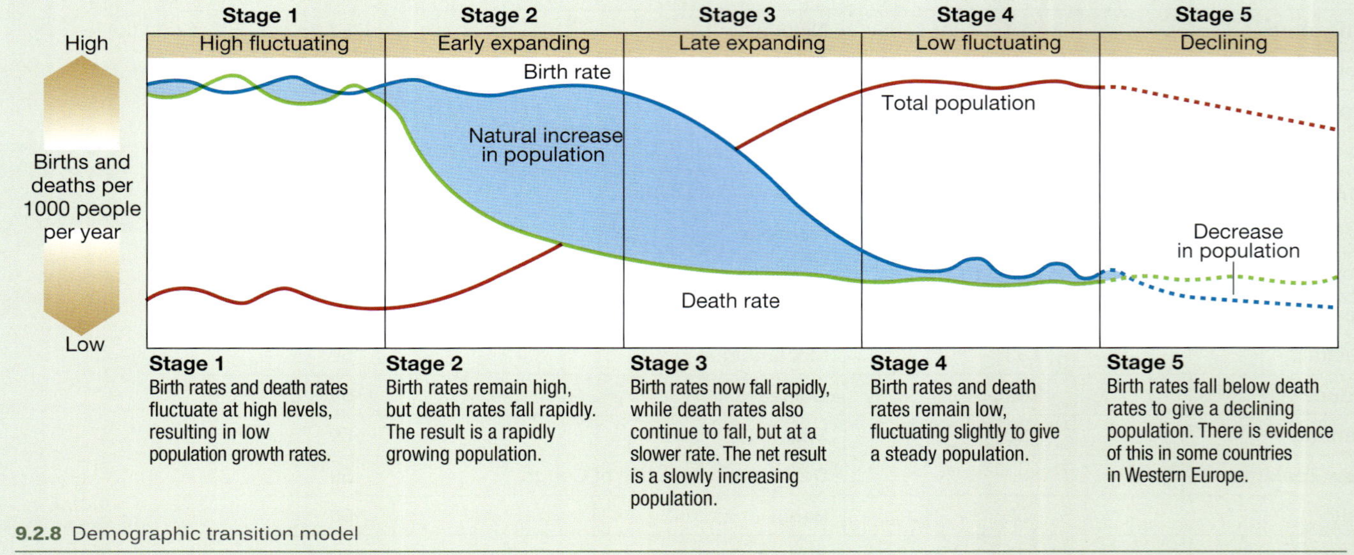

9.2.8 Demographic transition model

Population structure

By studying the age and gender structure of a population, some useful insights can be gained into the prospects for future population growth and a range of issues related to a population's economic and social needs. The demographic transition model shows only the rate of natural increase (or decrease) resulting from the difference in birth and death rates. In contrast, population pyramids show the impact of migrations, wars, major epidemics and gender imbalances.

● SKILLS BUILDER

Interpreting population pyramids

Population pyramids are bar graphs used to illustrate the age and sex structure of a population. The graph's vertical axis represents the various age groups of a population. The horizontal axis shows either the actual number or the proportion of the population for both males and females. Because each population pyramid represents 100 per cent of a particular population, comparisons can be made with the population pyramids of other populations.

Sometimes the horizontal scale shows the actual number of people in each age group. Before interpreting a graph, check the units of measurement used on the horizontal axis.

Populations are often divided into broader age groups based on dependency. The dependent parts usually include the 0–14 age group and the 65 and over age group. The changing proportion in each of the age groups provides valuable information about future population trends. For example, if the proportion of the population 65 years and over is growing, the population is ageing. If the proportion of the population 14 years and under is decreasing, the birth rate is declining, as is the rate of population increase.

Figure 9.2.9 shows some of the common population pyramid shapes and the conditions under which they develop.

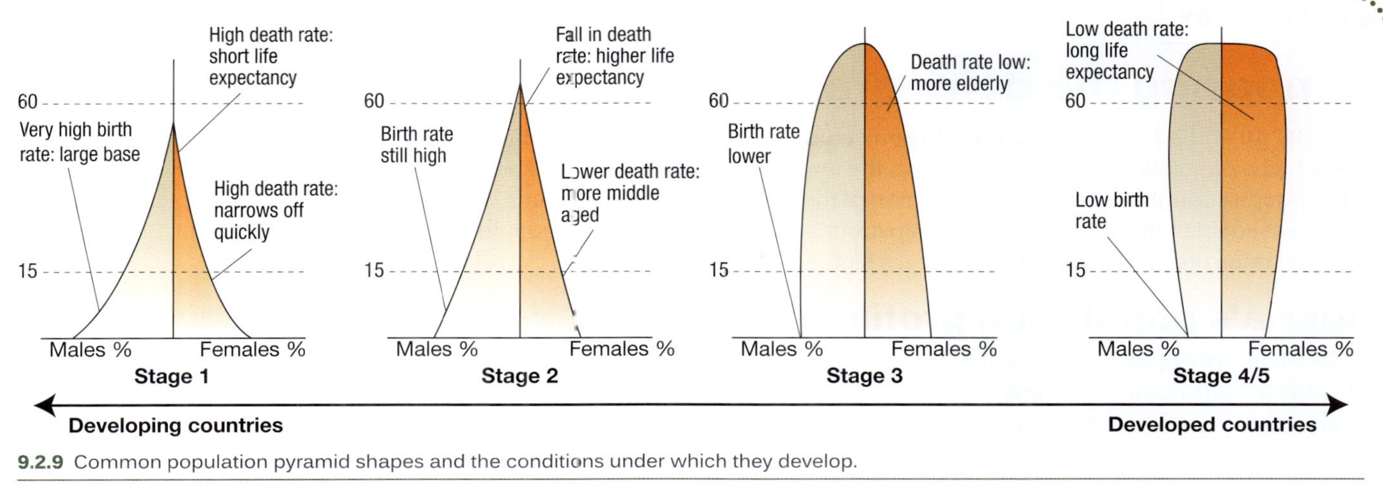

9.2.9 Common population pyramid shapes and the conditions under which they develop.

Activities

Acquiring and processing geographical information

1 Explain how the rate of natural increase is determined.

2 State where birth rates are highest.

3 Outline the global pattern of fertility. Where are fertility rates highest and lowest? How many countries have levels below the replacement rate?

4 Outline the factors that affect fertility rates.

5 Summarise where mortality rates are the highest. What factors affect mortality rates?

6 Account for the decline in mortality rates and increased life expectancy.

7 Identify the characteristics of a successful population policy.

8 Outline how the structure of a population provides insight into the prospects for future population growth.

Applying and communicating geographical understanding

9 Study Figure 9.2.1. Write a paragraph describing the trends in birth and death rates in the period 1950 to 2022. What has been the trend in the rate of natural increase?

10 Study Figure 9.2.2. Using an atlas, identify parts of the world with the:

- a highest fertility rate
- b lowest fertility rate.

11 Study Figure 9.2.4. Write a paragraph describing the trends in the total fertility rates (actual and projected) by region between 1950 and 2050.

12 Study Figure 9.2.5. Using an atlas, identify those regions of the world where infant mortality rates are the:

- a highest
- b lowest.

13 Study Figure 9.2.6. On which continent are the countries with the lowest life expectancy concentrated?

14 Briefly outline the specifics of China's one-child policy and assess its effectiveness. How has the policy been modified in recent years?

15 Write an explanation of the demographic transition model. Include references to the processes responsible for the trends apparent at each stage of the transition. What is meant by replacement-level fertility?

16 a What do population pyramids illustrate?

- b Using the populationpyramid.net website, find examples of population pyramids of countries typical of each stage of the demographic transition. That is, Stage 1 (expanding rapidly); Stage 2 (expanding slowly); Stage 3 (stable); and Stage 4/5 (declining).

■ CASE STUDY

Comparing the demography of Nigeria and Italy

Figures 9.2.10 and 9.2.11 show the **population structure** of two countries at different stages of the demographic transition. Nigeria has a population pyramid that shows the impact of high fertility and mortality rates. This is typical of many of the world's poorest countries. The large proportion of the population under 15 years will ensure high levels of population growth for the foreseeable future. Italy's population pyramid is typical of an ageing population in the developed world. Both birth rates and death rates are low and life expectancy is high.

Nigeria's population profile

In 2022, Nigeria's population was estimated at 218.5 million. This made it the most populous country in Africa and seventh in the world. One in 36 people call Nigeria home.

Nigeria has over 250 different ethnic groups and over 500 different languages. The Hausa-Fulani make up two-thirds of the population and outnumber every other ethnic group. Of these groups, most are Islamic. Other ethnic groups include the Nupe, Tiv and Kanuri.

The birth rate of 37 per 1000 population is well above the death rate of 12 per 1000, with 2.4 per cent rate natural increase. Total fertility rates remain high at 5.1. UN projections indicate the population will continue growing rapidly into the future. It projects a population of 377.5 million in 2050 and 410.6 million in 2100. By 2050, Nigeria will be the third most populous country.

Life expectancy for males is just 53 and females 54. Significantly, 43 per cent of the population is under the age of 15. Over-65s form just 3 per cent, but this group is projected to grow marginally to 4 per cent in 2050. Nigeria has a very young population; the size of the under-15 age cohort ensures numbers will keep expanding rapidly over this century as these young people begin having families (see Figure 9.2.10).

The population is unevenly distributed. The south and southwestern parts of the country are the most densely settled. Urban areas comprise 48 per cent of the population. Lagos is the country's largest city with 17.5 million people.

Nigeria struggles to meet the needs of this rapid growth. Only 68.5 per cent of the population have access to clean drinking water, with 31.5 per cent struggling for it. Similarly, only 29 per cent have access to improved sanitation. Children spend an average of nine years at school, with a national literacy rate of just 59.6 per cent.

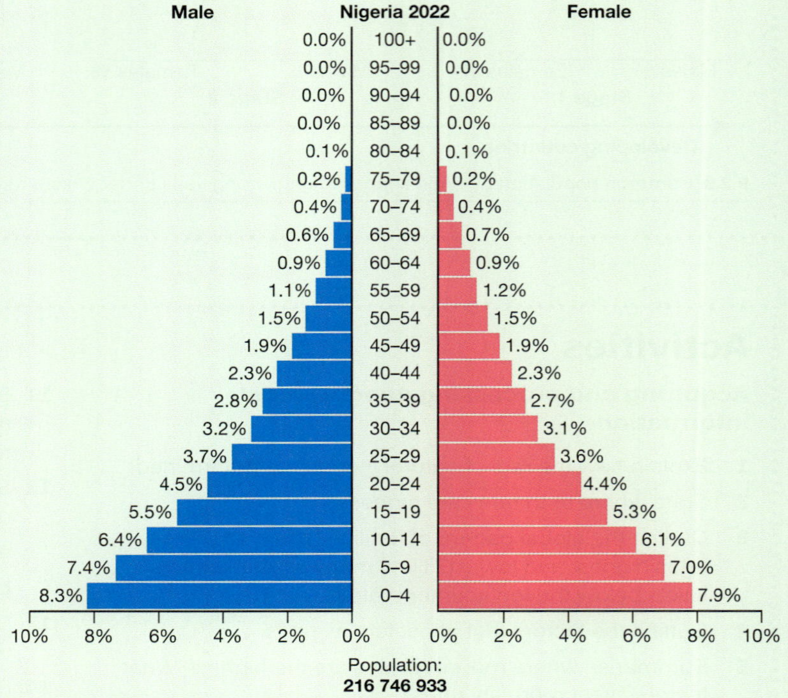

9.2.10 Population pyramid for Nigeria, 2022

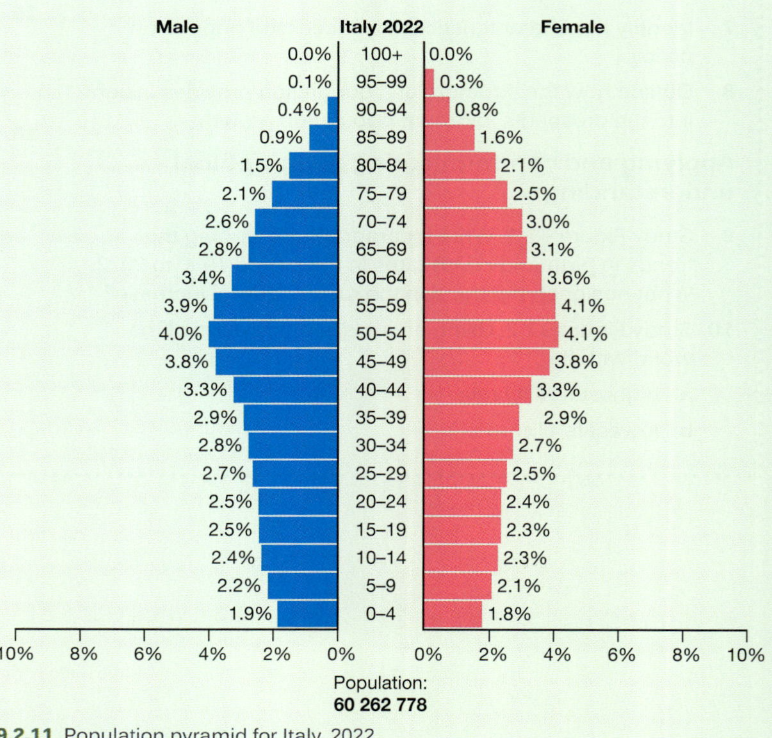

9.2.11 Population pyramid for Italy, 2022

Italy's population profile

Italy is a developed country and most Italians enjoy a high standard of living. In 2022, it's population was estimated at 58.9 million. This makes it the fourth most populous country in Europe (after France, the UK and Germany) and 23rd in the world. Just over 92 per cent are Italian-born. Of the rest, 1.1 million are officially registered Romanian citizens. There are also 500 000 Albanians and a similar number of Moroccans. An estimated 5 million-plus foreign residents include about 500 000 children born to foreign nationals.

Official figures exclude illegal immigrants, making the true numbers difficult to determine. It is estimated there are 670 000 illegal immigrants in Italy, many from Eastern Europe and North Africa. In recent years illegal immigrant numbers have increased sharply. A centre-right alliance government elected in 2018 cracked down on illegal immigration. The election of a conservative, centre-right government in 2022 resulted in an intensification of the campaign against illegal immigration.

Italy's birth rate of 7 per 1000 is below its death rate of 12 per 1000, giving it a rate of natural increase of –0.5 per cent. UN population projections indicate Italy's population entered its steady downward trend in 2016 with the annual population growth rate dropping to –0.1 per cent. It is projected to have a population of 53.7 million in 2050.

Life expectancy for males is 80 and for females 85. Only 13 per cent of the population is under the age of 15, while the over-65 population is 24 per cent. It is projected to grow to 35 per cent in 2050. Consequently, Italy can be described as having a rapidly ageing population (see Figure 9.2.11).

Italy's population is unevenly distributed. The Po Valley, in the country's north, is the most densely settled area with almost half the population. Seventy-one per cent live in urban areas. With 2.8 million people, Rome is Italy's largest city.

Activities

1 Study Figures 9.2.10 and 9.2.11. Complete the following activities.

- a Estimate the proportion of the population under the age of 15 for each of the populations in 2022.
- b Estimate the proportion of the population over the age of 60 for each of the populations in 2022.
- c Describe how the population structure of Nigeria is typical of many other developing countries.
- d Describe the future trends in Italy's population
- e Outline the potential for future growth in each population.
- f Explain how the future social needs of Nigeria and Italy will differ.
- g Explain why Italy might be concerned about its future labour requirements.

2 Draw up a table comparing the demographic characteristics of Nigeria and Italy.

Demographic characteristic	Nigeria	Italy

UNIT 9.3

Impacts of population change

The implications of rapid population growth need answers: how will the environment and humankind respond to this population explosion? Expert opinion is divided.

Environmentalists and ecologists regard the situation as potentially catastrophic. To feed the growing population, new agricultural practices that limit serious environmental damage are required. They also contend that natural resources and the biophysical environment, already burdened by resource exploitation, will simply collapse under future demand. Most concerning is global warming and the ensuing geopolitical consequences of large climate change population movements.

Economists sometimes argue that Earth can produce more than enough food to meet a rapidly-expanding population's needs. They assert that renewable energy investments will, by mid-century, slow down the rate of warming and technological innovations will help meet the challenges of rapid population growth, delivering higher living standards for most.

Rising populations present challenges. Producing adequate food and energy to support an increasingly sophisticated lifestyle is one; others relate to global inequalities. Challenges include:

- **environmental impacts:** interrelated issues of deforestation, desertification, land degradation, air and water pollution and climate change
- **economic impacts:** increased global inequalities and **food security**
- **social impacts:** population movements.

Environmental impacts

Population growth has a significant impact on the biophysical environment. In societies relying on subsistence, it leads to more intensive land use. The growing demand for food extends cultivation and grazing practices into increasingly marginal lands. Shortened fallow periods result, followed by accelerated erosion and declining soil fertility. A range of ecological and socioeconomic problems ensue. Seasonal flooding intensifies, droughts increase, rivers and coastal waters become silted and countless plant and animal species are lost.

Industrial agriculture emerged following World War II. It featured large, single-crop farms and animal production facilities. As a system of chemical-intensive food production, it met the needs of a growing population and was initially hailed as a technological triumph. There has been a growing realisation that the system is probably unsustainable. It has led to large-scale land clearing, which adversely impacts the environment, public health and rural communities.

9.3.1 Deforestation in the tropical forests of Malaysia

Deforestation

Deforestation is the temporary or permanent removal of large forest areas for agriculture, settlement or other uses. Almost half of Earth's old growth forest has been lost in the last 8000 years, most since 1950 when the population began rapidly accelerating.

Forest loss is concentrated in developing countries, particularly tropical Latin America, Indonesia, Malaysia and Africa (see Figure 9.3.1). Scientists are also concerned about the coniferous (boreal) forests being cleared in Alaska, Canada, Scandinavia and Russia. These make up about a quarter of the world's forested area.

At the current rate, up to 40 per cent of the world's remaining forest cover will be lost within 20 years. Alongside the short-term economic benefits, this causes gully and sheet erosion, topsoil loss and increased levels of dryland salinity.

The most acute deforestation is occurring in tropical forests, which absorb and store a third of the world's terrestrial carbon emissions via the carbon cycle. They now cover just six per cent of Earth's land surface, down from about 12 per cent. Most tropical deforestation has occurred in the past 70 years. Clearing these forests reduces their carbon absorption. Burning and clearing tropical forests accounts for 10–15 per cent of global greenhouse gas emissions. See Chapter 5 for more on deforestation.

Desertification

Desertification occurs when land productivity falls due to climate change, prolonged drought or human activities that expose topsoil to erosion. Only in extreme cases does it result in desert-like conditions. Deserts have expanded and contracted in response to climate variations over millennia. In some areas, land use practices have increased desertification's rate and extent. These include tree clearing (for fuel wood or charcoal), excessive ploughing and overgrazing (see Figure 9.3.2). Driven by a growing population, they leave the topsoil exposed. This leads to desertification, considered a form of land degradation. See Chapter G1 for more on desertification.

Land degradation

Land degradation is any adverse trend in the condition of the land, caused, either directly or indirectly, by human activities. These include anthropogenic (human-induced) climate change. It can result in a reduction or loss of biological productivity, ecological integrity or value to humans.

Forest degradation occurs in a forested area. Soil degradation directly affects the soil. Changes from natural processes such as volcanic eruptions and tsunamis are not considered land degradation.

Land degradation adversely affects over a quarter of Earth's ice-free surface. It impacts people's livelihoods, especially those who live in poverty in developing countries.

The impacts of land degradation extend beyond the land surface itself, affecting marine and freshwater systems, as well as people and ecosystems in places far away from the sites of degradation.

Land use changes and unsustainable land management practices, primarily agriculture, drive degradation. This reduces the economic productivity and integrity of croplands, pastures, forests and woodlands.

9.3.2 Land degradation caused by overgrazing

Poor land-management practices cause the loss of natural vegetation and soil deterioration. Erosion from agricultural fields is far higher than the soil formation rate, around 10–20 times (no tillage) to more than 100 times (conventional tillage).

For thousands of years, land degradation and land clearing were driven by agricultural land-use practices. Recent rapid population growth increased its rate and extent. However, long-term sustainable management is possible. Examples include terraced agricultural systems and sustainably managed forests.

Air and water pollution

The processes of urbanisation and industrialisation contributed to environmental degradation, overwhelming the ability of biophysical processes to cope, particularly with large numbers of people in concentrated areas. Severe atmospheric, water and noise pollution are common in most large cities (see Figures 9.3.3 and 9.3.4).

Air pollution is the presence of any chemical in the atmosphere in concentrations large enough to inflict harm on organisms, ecosystems or the climate. Almost any chemical in the atmosphere can become a pollutant if it occurs in high enough concentrations.

9.3.3 Urban air pollution, Jiujiang, China

9.3.4 Polluted urban waterway in a slum district of Jakarta, Indonesia

Air pollutants arise from natural and human sources. Natural pollutants are typically spread over the globe. They are diluted or removed by chemical cycles, precipitation and gravity. Most human-generated pollutants relate to industrialised and urban areas where people, cars, and industrial complexes are concentrated.

Air pollutants can be classified as either primary or secondary pollutants. Primary pollutants are the substances emitted directly into the air in high enough concentrations to cause harm, while secondary pollutants include ozone, nitrogen dioxide and 'acid' rain.

Water pollution is any change in water quality that can harm living organisms or make the water unfit for human use. It can come from point sources that discharge pollutants into water bodies at specific locations. These are relatively easy to locate, monitor and regulate. Nonpoint sources include chemicals such as those washed into drains from streets and carparks, lawns and construction sites, as well as fertilisers and pesticides used in food production. Nonpoint sources are difficult and expensive to control because they arise from diffuse sources.

Did you know?

Sources of up-to-date information about population include:

- Population Reference Bureau
- United States Census Bureau
- Australian Bureau of Statistics
- United Nations Population Fund.

Climate change

Earth scientists argue that climate change will accelerate land degradation processes through increased flooding, drought frequency and severity, intensified cyclones and sea-level rise. Outcomes will be affected by land management practices, permafrost thawing, coastal erosion and changing storm paths.

Economic impacts: Increased global inequalities

The developed countries make up one-fifth of the population yet consume 70 per cent of the world's energy, 75 per cent of its metals and 85 per cent of its wood. They also produce 80 per cent of the world's goods and services. The poorest one-fifth struggle on 1.4 per cent of goods and services.

Quality of life refers to an individual's overall wellbeing and that of a population. It is measured by income, housing and food quality, health care, education, transportation and communications. The standard of living is often based on per capita gross domestic product (GDP), with developed countries averaging US$40 078 and least-developed countries averaging US$2436. These figures do not reflect the extremes of poverty and wealth, such as Qatar, which has an average per capita GDP of US$130 475 and the Central African Republic, which has an average of US$700 per capita GDP. While criticised for not accounting for qualitative wellbeing, GDP does provide insight into relative affluence.

Environmental degradation related to population growth results from interactions between the environment and two groups. Affluent people in developed countries degrade the environment through high resource consumption and waste generation. The world's poorest degrade their resource base through necessity and lack of options. Their lack of urban infrastructure results in high levels of environmental degradation, especially water pollution. Rapid urbanisation and poverty see vast squatter settlements arising in the cities of the developing world (see Figure 9.3.5).

9.3.5 A squatter settlement in Mumbai, India

Food security

Assessing Earth's food-producing capacity is central to future population growth. Historically, agricultural production kept pace with population growth. Increases in the agricultural production required might extract an unacceptably high environmental cost. Food distribution also remains an issue if everyone is to have an adequate supply. Ensuring food security into the future is one of the world's greatest challenges.

The existing food system—the production, transport, processing, packaging, storage, sale, consumption, loss and wastage of food—feeds the world's population and supports the livelihoods of hundreds of million people. It is under growing pressure from many factors, including population growth, rising material standards of living, increased demand for animal-based products and climate change. These climate and non-climate sources of stress are impacting the four pillars of food security—availability, access, utilisation and stability.

A key question is, can the extra billions of people by the end of this century be fed? Since 1961, food supply per capita has increased over 30 per cent. This is principally from using more nitrogen-based fertilisers (about 800%) and irrigation (over 100%). Despite this, around 821 million people are undernourished and 151 million children under five years are stunted. Two billion adults are overweight or obese.

The UN's Food and Agricultural Organisation estimates about 50 per cent more food needs to be produced by 2050. This will create significant increases in greenhouse gas emissions and other environmental impacts, including deforestation and biodiversity loss. By 2050, the world's croplands must increase up to 20 per cent.

The rapid growth in agricultural productivity since the 1960s drives climate change, but it is also increasingly vulnerable to it because it directly impacts food systems and food security. Mitigating its impacts also potentially increases competition for agricultural resources.

Social impacts: population movements

Population movements have important social and economic implications for the places people depart from and go to; for example, the emergence of multicultural societies and the decline of rural communities as people move to cities. There are also the economic and social benefits derived from the international movements of guest workers. The plight of refugees (displaced persons) is another example of the social impacts of forced migrations.

Migration

Migration is moving from one place to another with the intention to stay there permanently or for a long period. The movement of people out of an area is called emigration, while movement into an area is called immigration.

It is broadly classified as internal migration (within a country) or international migration (across national boundaries). Voluntary migrations occur when people move to improve their economic and social wellbeing or for personal freedom. Forced migrations are usually due to circumstances beyond individual control, such as natural disasters, wars and civil unrest. These often initiate large-scale population movements.

Features of contemporary population movements include:

- the globalisation of migration
- an increase in the volume of migration in all regions
- a growing diversity in the type of migration—most countries with a migrant intake have a mix of immigration categories: labour, refugee, permanent settlement and family reunion
- an increasing proportion of women in all types of migration across all regions—this is particularly the case with labour migration and some refugee movements
- the increasing international mobility of highly qualified personnel
- the movements associated with economic and social change in newly industrialised countries.

International movements

Human migration has long been a part of history and is now a global process reshaping societies and cultures. Many contemporary migrations are related to economic, political and cultural links formed by globalisation, which will probably increase movements. Over 265 million people (3.4% of the world's population) live outside their birth country.

9.3.6 A contract worker on a cruise ship, an industry that relies on contract labour

Immigration affects regions and countries differently. Countries such as the USA, Canada, Australia and New Zealand are largely the product of European immigrants and their descendants. Many indigenous populations were dispossessed. The source of immigrants in these countries has changed over the last 40 years, with many coming from Asia. In the USA, they also come from Mexico, Central America, South America and the Caribbean.

International labour movements often involve workers sending money home to support their families. These income remittances represent one of the largest international capital flows at an estimated AU$860 billion per year. They play a vital role in the economies of the major labour-exporting countries such as India, the Philippines, Mexico, Nigeria, Pakistan, Bangladesh and Nepal.

Contract migrations involve contracted employment for a specified period and are believed to remit over AU$630 billion annually to the workers' homelands. It is estimated there are 50 to 60 million contracted foreign workers (see Figure 9.3.6).

Contract migrations fall into several categories.

- **Guest workers:** in some parts of the world (e.g. the Middle East's oil-rich states) there are more jobs than workers. Labour shortages drive governments to permit guest workers entry for a specified period, mostly within a rigid contractual framework. They are not permitted to settle or be accompanied by dependants. In some cases, guest workers are denied a range of civil and political rights, which can make them vulnerable to exploitation and mistreatment.
- **Business migrations:** globalisation spurred the international movement of skilled executives and professionals. Firms send personnel to overseas offices or joint ventures. Japanese companies, for example, have more than 110 500 staff in their foreign branches and 1 million Japanese travel abroad annually on short business trips.
- **Student migrations:** these also contribute to contract migration by studying at secondary schools and higher education institutions overseas. Both developed and developing countries see large numbers of their students migrating to study. Another is the student-exchange program. Over 400 000 international students were enrolled in Australia in 2020.
- **Seasonal movements:** many workers relocate to take up opportunities that combine work and leisure. They include fruit pickers, and staff in seasonal tourism areas like ski resorts during winter months or coastal tourist destinations in summer.

Internal migrations

One dominant population trend is the drift to cities. This leads to rapid growth in city sizes and the proportion of the population living there. There are two reasons for this growth—natural increase and rural–urban migration or urbanisation. The latter refers to the increasing proportion of people residing in towns and cities.

Changes in economic activity patterns can prompt shifts in population within countries. If sustained, they may eventually bring about significant changes in population distribution.

The shift in employment from manufacturing to the service sector within developed countries has altered the workforce distribution. Many ageing industrial areas struggle with outmoded methods of production, older plant and equipment, unionisation and congested urban infrastructure. It is challenging for them to compete with newer industrial centres in the rapidly growing economies of East Asia.

The industrial regions of the north-east USA and the Midlands of the UK experienced sharp declines in employment due to factory closures or relocations to countries of the developing world. Many displaced workers, typically those who were younger and more mobile, relocated to areas where work was readily available. Usually, this was by choice. Others remained in areas characterised by growing disadvantage—high unemployment, increases in crime and family breakdown, and urban decay. See Chapter 7 for more on economic restructuring.

Australia has experienced a shift of people from New South Wales, Tasmania and South Australia to Victoria, the Australian Capital Territory and Queensland. These regions had stronger growth in service sector employment over the last decade.

Climate change is likely to become a key future driver of internal and international population movements.

Activities

Acquiring and processing geographical information

1 Outline the various perspectives on the extent to which world population growth represents a major challenge.

2 Outline the impacts that population growth in societies pursuing a subsistence lifestyle has on the environment.

3 Explain what is meant by the term industrial agriculture. What are its characteristic features and what has been its environmental impact?

4 Define air pollution and water pollution.

5 Distinguish between natural and human-generated sources of air pollution.

6 Distinguish between point and nonpoint sources of water pollution.

7 Define the term quality of life.

8 Outline the extent of the inequality experienced by Earth's population.

Applying and communicating geographical understanding

9 Study the text under the heading Deforestation and complete the following tasks.
   a Explain what deforestation is. How extensive is it? Where has it been largely concentrated?
   b Identify the environmental impacts of deforestation. What is its relationship to climate change?

10 Study the text under the heading Land degradation and complete the following tasks.
   a Define land degradation. How extensive is it? What forms does it take?
   b Outline the principal causes of land degradation.
   c List the indicators of land degradation.
   d Outline the impact climate change has on land degradation.

11 Study Figures 9.3.3 to 9.3.5. Working in groups, brainstorm the problems faced by cities in the developing world that are experiencing rapid population growth. Share your group's points with the rest of the class.

UNIT 9.4

Population and natural resources

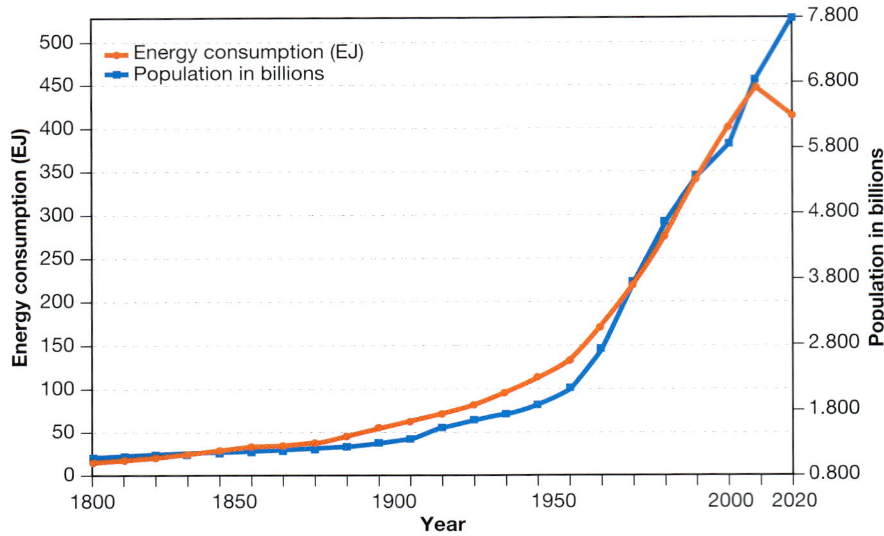

9.4.1 Population and energy consumption, 1800–2020. Increases in efficiency mean the two variables remain closely aligned despite per capita energy use growing significantly.

Economic growth and improved lifestyles are exhausting Earth's ecological wealth faster than it can be replenished. An area of productive agricultural land about the size of Tasmania is devastated each year. Three times that amount is eroded to the point that it is no longer commercially useful. Deserts increase annually by about 6 million hectares.

Water contamination and shortages are globally pervasive issues. The WHO found that at least 2.2 billion people do not have access to safe drinking water services. Nitrates and pesticides from agribusiness and wastewater are contaminating surface and groundwater. The disposal of toxic industrial wastes add to the difficulties.

Deforestation continues worldwide. Europe's forests have become the victim of environmental poisoning ('acid' rain). In the tropics, forests are harvested for commercial activities, such as palm oil and cattle grazing. Clear-felling is causing the largest loss of species in history. Rising standards of living and advances in technology also play roles in resource exploitation. As the economic circumstances of people improve, they aspire to higher material standards of living. Figure 9.4.1 highlights the close relationship between population growth and energy consumption.

Did you know?

Iron ore and coal are Australia's biggest exports, worth around $120 billion a year. The largest export markets are China (31%), Japan (13%) and South Korea (6%). This highlights the fact that the consumption of natural resources often takes place in places well beyond where they are sourced.

● **SPOTLIGHT**

Defining and classifying natural resources

A resource is something from the natural environment (water, air, trees, mineral ores and fuels) that is used to meet one's needs and wants.

The nature of resources varies greatly. Some appear to be continuously available, while others have limited supply and could be exhausted. Resources are usually grouped into four categories.

- **Renewable resources** will eventually be replenished if used. It theoretically includes forests, groundwater, wildlife and fish.
- **Non-renewable, or exhaustible, resources** have a finite quantity. Once used there is no possibility of reuse. Many of the world's energy sources are examples.
- **Recyclable resources** can be recycled after use. While mineral resources are limited, some can be recycled repeatedly into a wide range of products, such as scrap iron and steel. Materials often recycled include paper, aluminium cans, rubber, glass or plastic bottles, copper wire, silicon, gold, and lead from car batteries.
- **Continuous resources** are those that will always exist. They include solar energy and rainfall, wind energy, hydro-electricity, tidal power and geothermal power.

Global patterns in the distribution and consumption of natural resources

9.4.2 Paraburdoo mine, Western Australia's Pilbara region

Earth's physical processes create an uneven distribution of global resources. For example, rich basalt soils result from localised volcanic activity. Rainfall patterns depend on topography, prevailing winds and the location of the ocean and local currents. Coal deposits occur in Carboniferous-era swamplands dating back 345 million years. Other minerals are concentrated in specific regions through deposition or other geological processes.

Mineral ores have no consistent global natural distribution. Some countries are well endowed while others are not.

The world's iron ore production is dominated by Australia, Brazil, China and India. Together, they account for around 75 per cent of global supply. China's need for iron ore is so significant that in 2018–2019, the 375 million tonnes it produced met just one-third of its needs. It is a net importer, despite being one of the world's largest producers. Nickel production is dominated by five countries: the Philippines, Russia, Canada, Australia and New Caledonia, which together mined 63 per cent of the world's total production in 2016–2017. Other minerals are more widely available, including copper, lead and zinc (see Figure 9.4.2).

Figure 9.4.3 highlights the uneven distribution of selected mined commodities. In each case, their production is dominated by a handful of countries.

World's top 10 coal producers, 2020	
Country	**Production (million tonnes)**
China	3902
India	757
Indonesia	563
USA	485
Australia	477
Russia	400
South Africa	248
Kazakhstan	113.0
Germany	107.4
Poland	100.7
Other	590.0
Total	7741.6

World's top 10 iron ore producers, 2021	
Country	**Production (million tonnes)**
Australia	900
Brazil	380
China	360
India	240
Russia	100
Ukraine	81
Canada	68
Kazakhstan	64
South Africa	61
Iran	50
Other	228
Total	2532

World's top ten gold producers, 2020	
Country	**Production (million tonnes)**
China	368
Russia	331
Australia	328
USA	190
Canada	171
Ghana	139
Brazil	107
Uzbekistan	102
Mexico	102
Indonesia	101
Other	1539
Total	3478

Source: BP Statistical Review of World Energy

9.4.3 The uneven global distribution of selected commodities

● **SPOTLIGHT**

The global oil industry

The global oil industry includes the exploration, extraction, refining, transportation and marketing of oil-based products, especially petroleum. The industry's principal products are fuel, oil and petrol. Oil is also the raw material for many chemical products, including pharmaceuticals, solvents, fertilisers, pesticides and plastics.

The global pattern of production and consumption features significant differences between areas of production and consumption. To manage this, a vast global distribution system of tankers and pipelines has been developed.

The Middle East is the world's most significant oil-producing region, followed by North America (see Figure 9.4.4). The USA, then China are the world's two largest oil consumers. Figure 9.4.5 shows global consumption of oil by region. The Middle East, North America and the Commonwealth of Independent States (CIS) have increased their share of global production over 20 years. Figure 9.4.6 shows consumption trends. The Asia-Pacific region significantly increased its share of oil consumption from 1994 to 2019.

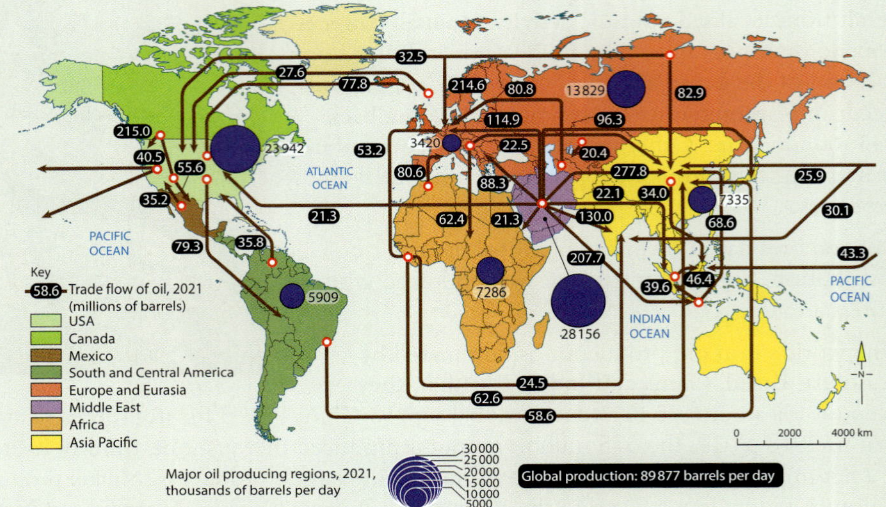

9.4.4 The global pattern of the production and distribution of oil

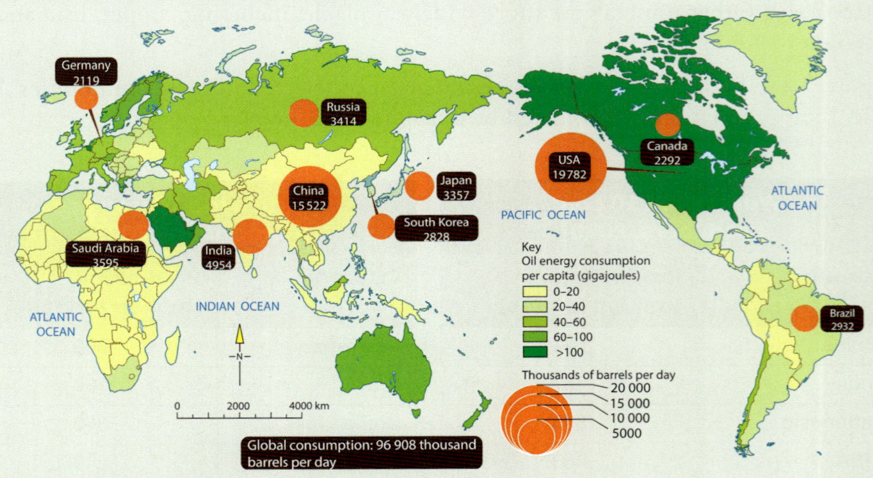

9.4.5 The global pattern of the consumption of oil

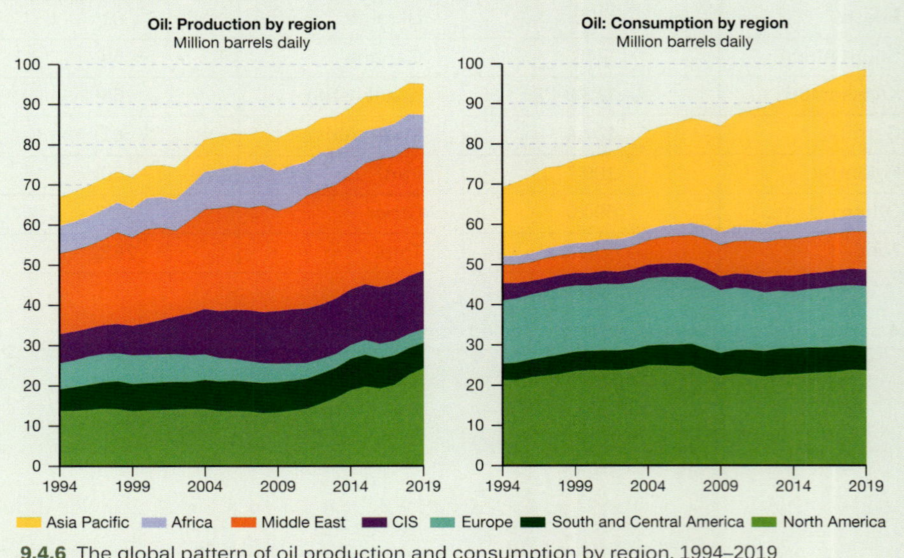

9.4.6 The global pattern of oil production and consumption by region, 1994–2019

Natural resources and the distribution of the world's population

9.4.7 The Nile River flows over 6600 km from its source to the Mediterranean Sea. For thousands of years, it has been a vital source of irrigation, transforming the dry area around it into agricultural land.

The uneven distribution of resources impacts world population distribution. Fertile soil distribution and rainfall greatly influence population concentrations. The most populated parts of a continent are around freshwater, with a few exceptions. River valleys, especially in mid-latitudes, are typically associated with rich, high-yielding, alluvial soils capable of sustaining large populations. The best examples are the Ganges and Yellow (Huang He) rivers. Other densely settled river valleys includes the Paraná/Paraguay River Basin (South America), the Nile (North Africa, see Figure 9.4.7), and the Kanto Plain around Tokyo, Japan, the world's largest city.

The uneven distribution of mineral ores disperses populations, often into remote areas. Australia's largest iron ore and gas deposits are concentrated in Western Australia's north-west. Such isolated deposits once forged new remote communities. The iconic Australian mining towns Broken Hill, Kalgoorlie and Mount Isa arose from an era when people settled permanently near their work. Newer mining communities, such as the Pilbara towns of Port Hedland, Newman and Karratha, still provide goods and services, but typically, they do not provide workers with a permanent home. Many are fly-in fly-out (FIFO) workers who live in company camps at work and regularly return home.

Population dynamics and levels of resource consumption

Population dynamics refers to how a population changes over time. It is closely linked to the level of resource consumption, and the wastes generated. The issue is more complex than just how many people live on Earth. What is happening within a population requires investigation. Therefore, a range of factors must be considered. These include population distribution (density, migration patterns and urbanisation), its composition (age, sex and income) and, importantly, consumption patterns.

A population's age structure can impact the environment. The current global population has both the largest proportion of young people (26% under 15 years) and the largest percentage of elderly people (9%) in history. The likely influx of young people into cities leads to intensified urban environmental issues.

In developing countries, rapid urbanisation often outstrips the pace of development from people seeking new opportunities. It leads to slums and squatter settlements, poor (if any) environmental regulation and higher levels of pollution. In developed nations, the increased concentration of people creates localised environmental problems. These include stormwater and vehicle exhaust pollution, and heat islands arising from dark, solid surfaces.

Life expectancy has an impact on resource utilisation. Since 1960 it has increased by about 20 years, a remarkable achievement. But this has a massive environmental toll—longer lives mean people now consume resources and produce waste for around 40 per cent longer than they once did.

The uneven distribution of income and wealth creates additional pressures on the environment. The very poor engage in unsustainable resource use to survive, such as burning rubbish or plastics for fuel or depleting scarce natural resources, such as forests and animal populations, to feed their families. Meanwhile, more affluent people consume disproportionately large amounts of resources through their cars, homes and lifestyle choices.

Challenges created by increasing resource consumption

As the world population increases, environmental degradation increases. So does further irreversible change. This is largely an outcome of efforts to raise living standards. Environmental stress indicators include increasing greenhouse gas emissions, ozone depletion and 'acid' rain, deforestation, desertification, loss of topsoil, loss of biodiversity, and water and food shortages in many regions.

Natural resources, including mineral ores, water, energy and fertile soils, form the basis of human life. However, their rapid depletion due to rapid population growth and rising standards of living, poses broad-scale challenges. These include climate change, shrinking freshwater reserves, fish stocks and forests, fertile land erosion and habitat loss leading to species extinction.

Humans now extract and use around 50 per cent more natural resources than 30 years ago, about 60 billion tonnes a year. People in affluent countries consume up to 10 times more than those in the poorest countries. Each day, one North American consumes around 90 kilograms of resources. In Australia it is about 48 kilograms; in Europe, 45 kilograms, and in Africa, only 10 kilograms.

To ensure Earth's future prosperity, people must adopt lifestyles that are more sustainable. This is essential to protect both the natural resource base and the planet's fragile ecosystems.

● **SPOTLIGHT**

Indigenous resource management

Indigenous peoples have a long history as skilful managers of natural resources, demonstrated by the fact that many have existed in balance with their environment, for tens of thousands of years. Some of these groups, including Aboriginal and Torres Strait Islander peoples, work actively to develop specific resources and focus on the conservation of species.

Many forms of resource conservation are part of Aboriginal and Torres Strait Islander cultures. The following are some of the techniques used by First Nations Australians to preserve or manage fish stocks:

- direction by elders (those knowledgeable about the local ecology)
- designated fishing areas for individuals and/or groups (see Figure 9.4.8)
- restrictions on species caught
- closed seasons
- specified methods of fishing
- various taboos concerning fishing
- total avoidance of fishing
- sacred pools.

The last approach has been adopted by some Aboriginal peoples in parts of inland Australia. The sacred pools, where fishing is prohibited, ensure that stocks remain to build up supplies in a river system once a drought has ended.

9.4.8 Stones piled up to form a large rock fish trap known as a Mayoorr, Dampier Peninsula, Western Australia

Impacts of resource exploitation on indigenous peoples

Many indigenous communities live in biologically diverse and resource-rich regions. There may be wealth-generating potential to extract oil, gas or other resources. The rights and interests of indigenous peoples have often been ignored. Colonisation legacies and modern-day exploitation and expropriation have both posed distinctive threats. International and national recognition of indigenous rights is growing and can inspire future action.

Natural resource exploitation by local entrepreneurs or transnational corporations has caused suffering in indigenous societies. Responsibility normally lies with the country's government. Reasons for inaction may include:

- no laws to protect the environment
- lack of effective control at a government level

- indigenous people having no effective voice in government
- laws being bypassed due to a corrupt system
- inadequate regulation monitoring, making it difficult to pinpoint responsibility
- inadequate penalties, especially if the potential rewards are high by comparison.

To local indigenous peoples, especially those living at a subsistence level, a resource-based project that degrades the environment may be devastating. They may be unable to respond for the following reasons:

- Small, isolated groups are politically insignificant on their own.
- Disputes within their communities, cultural differences, geography and poor communications may impede the establishment of a united pressure group.
- Their needs and wants may be simply ignored.
- They may be swayed by articulate politicians offering empty promises.
- Key local leaders are 'bought' (given inducements in exchange for support).

SPOTLIGHT

Rio Tinto blasts 46 000-year-old sacred site

In May 2020, Rio Tinto, the Anglo-Australian mining giant, destroyed a sacred rock shelter in Juukan Gorge, Western Australia. Its historical and cultural significance was well understood. Archaeologists believe that the cave is the only inland site in Australia to show signs of continual human occupation through the last ice age (see Figure 9.4.9).

9.4.9 The Juukan Gorge rock shelter before its destruction by Rio Tinto

Rio Tinto later revealed it had three alternative options to preserve the site, but chose to destroy it without informing the traditional owners of the alternatives. The blasting of the cave site enabled the company to expand its Brockman 4 iron ore mine.

The local custodians, the Puutu Kunti Kurrama and the Pinikura peoples, fought the decision without success. The destruction of the rock shelter brought widespread criticism to Rio Tinto.

Following a public outcry, Rio Tinto issued a statement that it 'deeply regret[ed]' the events at Juukan Gorge and unreservedly apologised to the Puutu Kunti Kurrama people and the Pinikura people. The company conceded that the destruction of the rock shelters should not have happened.

Resource exploitation and inequalities in human wellbeing

Inequalities in human wellbeing exist between and within countries for many reasons. The environment and climate can affect things like access to fresh water and the ability to grow food. Natural resources, such as oil and mineral ores, are important sources of economic wealth. The political, economic and social organisation of a country can have a big impact on its wellbeing.

● SPOTLIGHT

Defining poverty

These extracts illustrate a range of perspectives:

> Poverty is a denial of choices and opportunities, it is a violation of human dignity. It means lack of basic capacity to participate effectively in society. It means not having enough to feed and clothe a family, not having a school or clinic to go to, not having the land on which to grow one's food or a job to earn one's living, nor having access to credit. It means insecurity, powerlessness and exclusion of individuals, households and communities. It means susceptibility to violence, and it often implies living in marginal or fragile environments, without access to clean water or sanitation.
>
> United Nations

> Poverty is hunger. Poverty is a lack of shelter. Poverty is being sick and not being able to see a doctor. Poverty is not having access to school and not knowing how to read. Poverty is not having a job, is fear for the future, living one day at a time. Poverty is losing a child to illness brought about by unclean water. Poverty is powerlessness, lack of representation and freedom.
>
> World Bank

> Poverty is about more than inadequate income. It is also about a lack of fundamental freedom of action, choice and opportunity. It is about vulnerability to abuse and corruption.
>
> World Bank

Did you know?

About 1.2 billion people—16 per cent of the world's population—lack access to electricity. In Africa, only 45 per cent can access electricity, while in sub-Saharan Africa, this falls to just 35 per cent. In India, 19 per cent lack access to electricity.

While there are many ways in which resource exploitation can impact on human wellbeing, access to clean water is probably the most fundamental. Another is access to energy. Herein lies one of the great dilemmas of the modern age. In the past, meeting energy needs largely meant using fossil fuels. If everyone in the developing world had the same demand for energy as the developed world, the impact on global warming would be potentially disastrous. Fortunately, the shift to renewable (non-polluting) sources of energy provides solutions. Renewable energy usage has recently increased significantly. Figure 9.4.10 shows the changing share of primary energy by sources. Fossil-based fuels, especially oil and coal, have a declining share, while renewables, and to a lesser extent natural gas, have increased significantly.

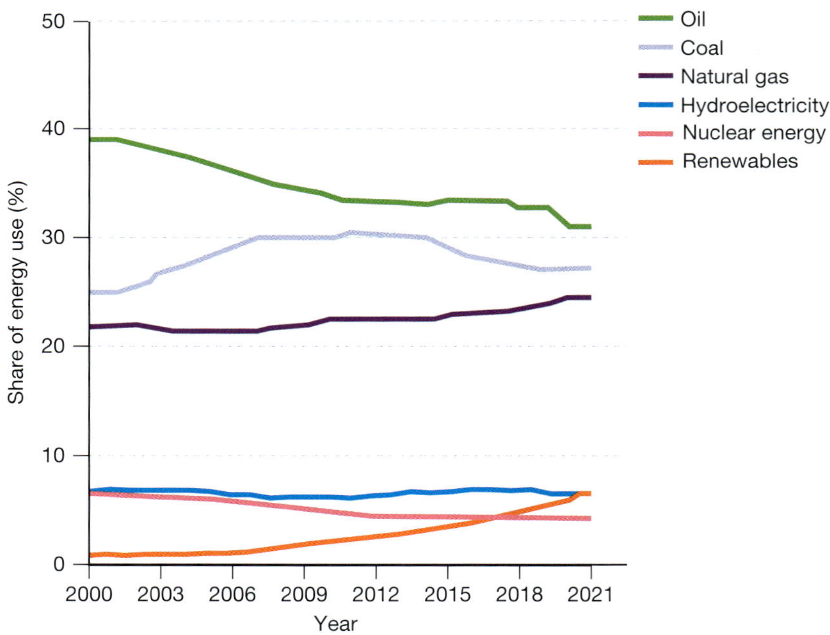

9.4.10 Percentage usage changes of global primary energy by source, 2000–2021

● SPOTLIGHT

Access to a safe source of water

Access to a clean source of water is essential for human wellbeing. A safe source is vital to functioning and maintaining health. It also plays a crucial role in food production. Unsafe water sources, poor access to basic handwashing facilities and unsafe sanitation are often linked. They are a leading risk for diarrhoeal diseases and can exacerbate malnutrition, especially in children. It is a leading risk factor for death, especially for those on low incomes.

Worldwide, an estimated 2.1 billion people (29% of global population) don't have access to safe drinking water and 666 million (9%) do not have access to an improved water source (see Figure 9.4.11). About 1.2 million people die annually as a consequence. Unsafe water sources cause six per cent of deaths in low-income countries (see Figure 9.4.12).

9.4.11 Children collect water from a well, Bafing, Côte d'Ivoire

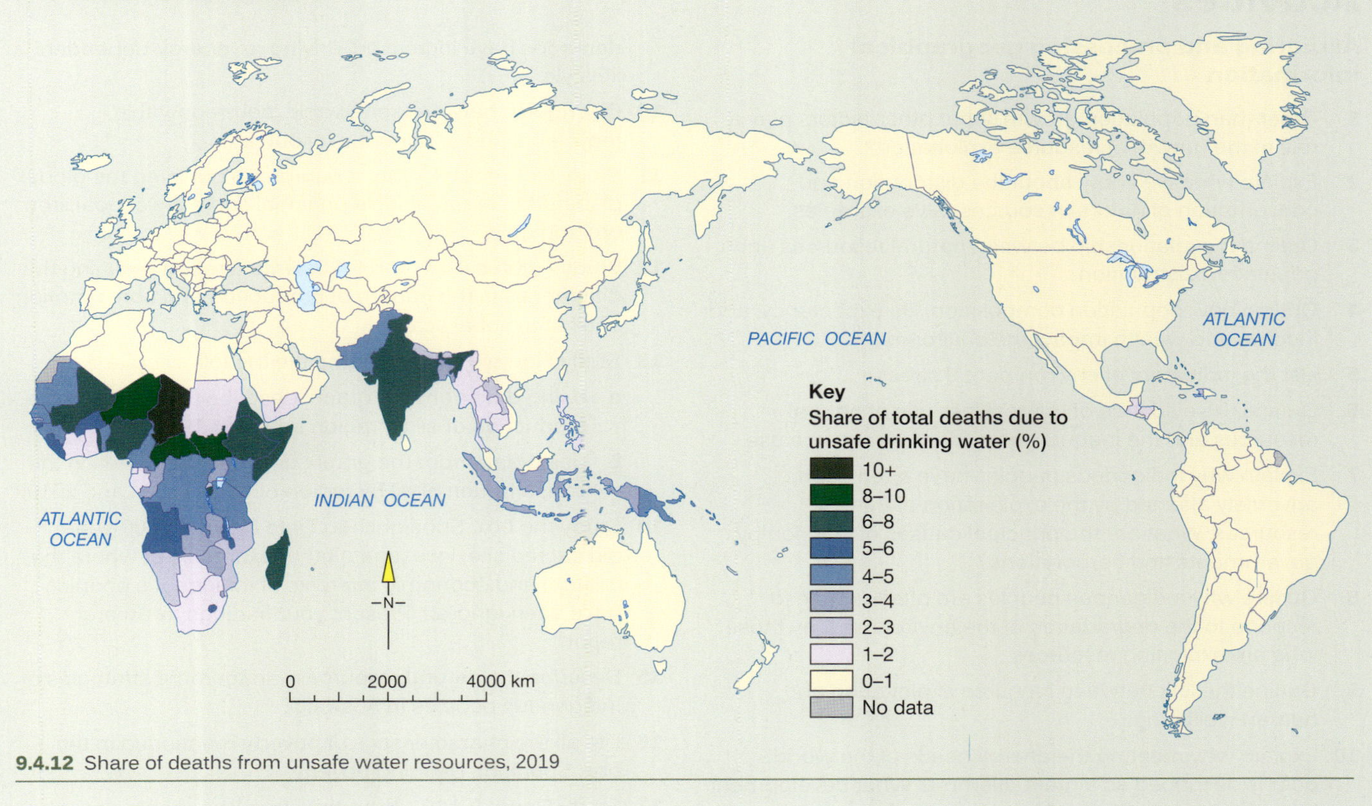

9.4.12 Share of deaths from unsafe water resources, 2019

Impact of resource exploitation on environmental degradation

A growing awareness developed in the 1960s of the impacts that rapid population growth and resource exploitation were having on the environment. Also concerning was the apparent indifference or inability to effectively control the resulting environmental degradation.

Not until the mid-1980s did the increased consumption of fossil fuels and build-up of atmospheric contaminants start growing as a global concern.

The 1987 Montreal Protocol was a landmark global commitment aimed at addressing the environmental impacts of human activity. It specifically targeted phasing out chlorofluorocarbon (CFC) production, the main class of chemicals primarily destroying the atmosphere's ultraviolet-screening ozone. The protocol still stands as a powerful example of what international cooperation can accomplish.

Similar international agreements are aimed at slowing, and ultimately reversing, the rate of global warming. Rising carbon dioxide levels are directly linked to increases in the consumption of coal, particularly for power generation and industry, and oil for transport. These are driven by population growth, rising material standards of living and technological advances. Although uncertainties remain around the extent of global warming, a growing consensus is that Earth's ecosystem is reaching a **tipping point**. Beyond it, rapid and uncontrollable warming will take place.

Activities

Acquiring and processing geographical information

1 Other than population growth, what other factors play a role in the growth of resource exploitation?

2 Explain what we know about the distribution and consumption of natural resources. Give examples.

3 Outline the relationship between natural resources and population distribution.

4 Outline how population composition, life expectancy, and income and wealth impact the environment.

5 List the indicators of environmental stress.

6 Outline the evidence of our reckless use of natural resources and the inequalities evident in resource use.

7 Explain why indigenous peoples have often been adversely affected by the exploitation of Earth's resources. What are the principal causes of inaction by governments and corporations?

8 Outline why indigenous peoples are often unable to respond to the degradation of the environment by those who are exploiting resources.

9 Outline the link between resource exploitation and human wellbeing.

10 Explain why meeting the energy needs of the world's poor is described as a great dilemma. What development offers hope for the future?

Applying and communicating geographical understanding

9 Study Figure 9.4.1. Describe the relationship between population and energy consumption. Speculate why this relationship is so closely aligned that our energy demands have increased, driving an energy-dependent lifestyle.

10 Distinguish between renewable, non-renewable, recyclable and continuous resources.

11 Study Figure 9.4.3. Write a paragraph outlining the global distribution of the world's principal suppliers of coal, iron ore and gold.

12 Study Figures 9.4.4 and 9.4.5. Write a report outlining the differences in the pattern of production and consumption of oil at a global scale.

13 Study Figure 9.4.6 and complete the following tasks.

- a Using data from the graph, describe the trends in the production of oil by region between 1994 and 2019.
- b Using data from the graph, describe the trends in the consumption of oil by region between 1994 and 2019.

14 Study the box, Spotlight: Rio Tinto blasts 46 000-year-old sacred site. Investigate other examples of where the cultural and economic interests of indigenous peoples have been ignored. Present your findings as an oral report.

15 Describe the natural resource management strategies of Indigenous peoples in Australia.

16 List all the characteristics of poverty mentioned in the box, Spotlight: Defining poverty.

17 Study Figure 9.4.10. Using data from the graph, describe the trends in the share of global primary energy by source, 1994–2019.

18 Study Figure 9.4.12. Using an atlas or online world map, identify those parts of the world where the share of deaths from unsafe water sources is greater than 4 per cent.

APPLICATION AND CONSOLIDATION TASKS

Task 1: Presentation: The population debate

Undertake research into the arguments of the two camps in a debate on the topic, 'How will the environment and humankind respond to this population explosion?'

Camp 1: Environmentalists and ecologists

Camp 2: Economists

Working in pairs or small groups, brainstorm other arguments and supporting examples that you could use in a debate about the issue.

When undertaking your research, ensure you cover:

- environmental challenges
- economic challenges
- social challenges

Report your group's findings to the class.

Optional: when presenting your report you can use multimedia, posters and/or other visual tools.

Task 2: Class debate: Technology is the solution

Conduct a class debate on the topic, 'Technological developments in the agricultural sector will enable humankind to cope with the rapid escalation in human numbers'.

Step 1: Identify the sources you want to use. Ensure you use a variety of reliable sources.

Step 2: Read and make notes, and write a summary of your findings.

Step 3: Organise the information in a logical order.

Remember: when undertaking research do not focus on your side of the argument, you need to know what the other side is arguing. This will help you rebut their arguments.

Task 3: Report writing: Population growth

Undertake research into the implications of rapid population growth. Select one of the following:

Environmental impacts:

- deforestation
- desertification
- land degradation
- air and water pollution
- climate change

Economic impacts:

- increased global inequalities

Social impacts:

- population movements

Your report should include visuals to support your findings (maps, graphs, illustrations, diagrams, etc.).

Task 3: Infographic: Implications of an ageing population

Create an infographic that identifies the factors contributing to the acceleration of population ageing.

- Show how much the 65+ years share of the population has increased since 1960. How much is it projected to increase by 2050?
- Show how much the 0–14 years age group has declined? How much is it projected to decrease by 2050?
- Outline the implications of the world's age structure for national policy agendas and resource allocation.

CHAPTER

10 Place and cultural change

Culture refers to a group's shared way of life. It includes things such as music, art, religion, language, folklore and traditions. It also involves forms of economic, political and social organisation, the ways people interact with the environment, and their attitudes towards other groups of people.

Culture occurs at a variety of scales, from local to global, and it is always changing due to migration, globalisation and modernisation—forces that can foster both cooperation and conflict among cultures.

The idea of a 'culture of place' recognises that places are more than just locations on a map, but are defined by their physical and human characteristics.

This chapter explores the elements that contribute to the development of the culture of place. Also scrutinised is mass consumer culture through a focus on media, fashion, brand images, sport and music.

Dubai is explored in a detailed study of cultural integration. Its spatial patterns and cultural characteristics are examined, as is the evidence of cultural integration at a global scale. The influence of economic, environmental, social/cultural and technological factors on cultural integration and the impacts of change are investigated. Opportunities to enhance environmental sustainability and human wellbeing are also examined.

Every choice someone makes to come to a festival like this, to learn about another culture, to experience the simple act of trying out a different food, seeing a different form of dance, hearing a different language, that is a choice, to open yourselves up to another culture, ethnicity and to diversity.

Jacinda Ardern, former Prime Minister of New Zealand

10.0.1 Local cultural traditions coexist with global technologies, Kyoto, Japan

Chapter glossary

Country often used by Aboriginal and Torres Strait Islander peoples to describe a specific place to which they are connected. The concept of Country is complex and incorporates cultural, physical, linguistic and spiritual features

cultural adaptation the modification of culture to incorporate aspects of another culture

cultural adoption the acceptance and integration of different cultural elements into one's own

cultural diffusion the dispersion, or spread, of different cultural elements between countries

culture the ideas, customs and social behaviour of a particular people or society

culture of place the particular characteristics of a place that makes it what it is

economic nationalists people who oppose globalisation or at least question the benefits of unrestricted free trade; they favour protectionism and advocate self-sufficiency

interculturality the interaction of people from different cultural backgrounds based on respect for existing ethnic and cultural differences; it also refers to a focus on the pluralist transformation of urban and public spaces

lifestyle the interests, opinions, behaviours and behavioural orientations of an individual, a group or culture

location is the position of a particular feature or place on the surface of Earth

locational disadvantage the inability to access services in an area and inadequate social and economic infrastructure, increasing inequality for individuals with low income and low socioeconomic status

morphology the functional form and character of an urban place

multiculturalism a system of beliefs and behaviours that recognises and respects the presence of people from all cultural backgrounds

place an area or location that has unique physical and human characteristics

place perception our awareness of places and the particular opinions we have about them

populist politician a politician claiming to champion the common person, usually by attacking a real or perceived elite or establishment

social isolation having minimal contact with others

sovereignty the supreme, unrestricted power to govern a state

space a specific geographical setting with distinctive physical, social and cultural attributes

UNIT 10.1
Culture of place

The concept of place

Places are parts of Earth's surface identified and given meaning by people. They vary in size and may be perceived, experienced, understood and valued differently. Descriptors include their **location**, shape, boundaries, features and environmental and human characteristics. Some characteristics are tangible, such as landforms and people. Others are intangible, such as scenic quality and **culture**. Each place has unique characteristics (see Figure 10.1.1).

10.1.1 The Snowy Mountains are an iconic Australian place.

The concept of **place** can be applied at any scale. A dynamic concept, it need not be fixed in time or space. Through globalisation, a place can change over time as its physical setting and cultures are influenced by new ideas or technologies. The degree to which places are interconnected is vital to understanding this concept.

The physical characteristics of a place make up its natural environment. They result from the interaction of geological, hydrological, atmospheric and biological processes. These include landforms, water bodies, climate, soils, natural vegetation and animal life.

Human characteristics come from ideas and actions. They include land use, population density, language patterns, religion, architecture, and economic and political systems.

Perception of places

Place perception refers to our awareness of places and the particular opinions we have about them. It is our feelings and interpretations about the characteristics of a place that help develop place perception. Place perception influences the decisions we make about a location.

However, our feelings are sometimes irrational and result in perceptions that are not always based on reality. One example is the decision we might make about living in an area affected by natural hazards, such as bushfires and tropical cyclones. Our perception of a place as attractive or desirable is very separate from our understanding of its hazard potential.

The perception of place is an individual thing. People can perceive the same place very differently. Therefore, an ordinary place can have a special meaning. The family holiday destination visited each year will be seen as special to that family, even years after they have stopped going there (see Figure 10.1.2). Outsiders may view your local neighbourhood very differently to you. Every neighbourhood is a place with signs and symbols that provide personal meaning and significance to those living there.

10.1.2 After spending past family vacations at Port Stephens, NSW, the author perceives it as a special place.

● SPOTLIGHT

Culture

Culture can be defined in general as the way of life of a group of people. Specifically, it can combine those elements of human existence such as traditions, customs, languages, belief systems, art, architecture, music, food and institutions that are passed down from one generation to the next. It includes the material goods the group creates and uses, and the skills it develops. It also includes territorial affiliation and shared history, political and education systems.

No cultures are static; they change over time. When cultures come into contact through migration, trade or the latest telecommunications technologies, they impact each other. Some believe Earth's cultural diversity is under threat. They fear cultures are being swamped by Western (mainly US) influences. Others see this as a culture-enriching process. Advanced communication technologies make it inevitable that cultures of all types will become more interrelated and interdependent.

While cultures have always evolved in response to contact with other people and their cultures, the pace of change has increased. In the past, cultural influences spread slowly due to the distances involved and the limitations of transport technologies. Today, because of the internet, the media, world trade and the ease (and increasing affordability) of long-distance travel, cultural influences can spread across the planet very quickly.

Human characteristics contributing to culture of place

Culture of place or sense of place relates to those characteristics of a place that make it what it is, or the meaning people ascribe to it. For example, it could focus on the relationship between people and the setting in which they live, such as Australia's Snowy Mountains (see Figure 10.1.1). Another definition is the emotional attachment someone has to an area based on their experiences. Related concepts include attachment to place, national identity and regional awareness.

Everyday routines and experiences within a particular setting result in people developing a set of shared meanings. These shared meanings often carry over into people's attitudes and feelings about themselves and the place in which they live. When this occurs, a culture of place develops. It also includes the symbolism they attach to the place and the character of a place as seen by outsiders. Sydney's Opera House and Bondi Beach are such symbols. They position Sydney in terms of a culture of place. Bondi is symbolic of the beach-loving lifestyle of Australians.

10.1.3 The architecture, noise, colour and vitality of the winter Vivid festival highlight Sydney's culture of place.

Culture of place develops through people's shared experiences. These include **lifestyle**, cultural diversity, daily routines, patterns of speech, dress codes, and the nature of the routine encounters and shared experiences in bars, restaurants, cafes and parks. It also includes a place's history, architecture, noise, colour, vitality and the presence of indigenous cultures (see Figure 10.1.3).

Culture of place is not static. It changes over time, as do the social values of the communities that interact within places. It's important to consider the past, present and possible future social values when investigating places. When designing places, and promoting a culture of place, it is crucial to understand the communities using the area and consider their ideas and needs.

Did you know?

Places can be identified by their absolute or relative locations. Absolute location is a specific point on Earth's surface expressed by coordinates, such as latitude and longitude. Relative location is expressed as a distance and direction from one place to another.

Interactions between natural and human characteristics of places

The **morphology** of places—their functional form and character—results from the interactions of natural and cultural processes. While each place is unique, it may have characteristics shared with other places. Thredbo illustrates how the interaction of natural and cultural factors results in a place that is unique, while it is also similar to other places (e.g. other Australian ski resorts).

The interplay between places' uniqueness and similarities is at the heart of the geographic inquiry into why things are found where they are.

● **SPOTLIGHT**

Thredbo Alpine Village

Thredbo ski resort is a place showcasing the interactions between biophysical and cultural factors. The biophysical factors shaping place include the topography, vegetation and climate. The cultural factors include the area's Indigenous past, the arrival of Europeans, Australia's cultural diversity, planning interventions and restraints, and recreational activities (see Figure 10.1.4).

The village's morphology (its functional form and character) is shaped by its physical setting. The village spans the Thredbo River, which passes through a steep-sided valley. To its west are the high plains of the Australian Alps, including Australia's highest peak, Mt Kosciuszko. The small area of relatively flat land on the valley floor proved vital to the site's use. As the village grew, the settlement spread up the steep valley sides. Recreational activities are concentrated on the south-eastern facing slope—the side receiving the least insolation.

The principal First Nations groups in the region are the Yaitmathang, the Wolgalu, the Waradgery and the Ngarigo. Aboriginal peoples typically moved into the High Country of what is now Kosciuszko National Park in the warmer months. In late spring and summer, the Alps became a place for ceremonial gatherings. The annual feasting on the Bogong moths that migrated to the region around spring was a major attraction. During winter, people would retreat to the lower-lying valleys and foothills.

European settlement dates from the 1820s. Graziers were attracted by the extensive grasslands and permanent water supply. Grazing leases were granted along the Thredbo Valley, but increasing concerns about the impact on the fragile alpine environment saw these revoked in the 1900s.

In the early 1950s, European migrants flooded in to work on the Snowy Mountains Hydro-Electric Scheme. They brought European culture here, including a love of snow-based recreational activities.

Czech immigrant Tony Sponar recognised the site's potential as a ski resort. A ski instructor in Austria, he envisaged Thredbo as an Australian equivalent.

The resort's establishment was challenging as the site lay within an environmentally protected area. In January 1957, the NSW State Park Trust granted Sponar's syndicate a lease option. In 1961, Lendlease acquired the lease and in 1987, Amalgamated Holdings Limited (now EVT Limited) purchased it. Unlike most Australian urban places, Thredbo is a company town with Kosciuszko Thredbo acting as both the resort operator and the local government authority.

10.1.4 Thredbo, NSW

Thredbo was developed with a distinctive European alpine village character, reflecting its immigration heritage. Strict planning controls now determine the look of all new developments. Being within a national park imposed an additional layer of environmental regulation.

Thredbo sees seasonal changes. From June to mid-September it is packed with winter sports enthusiasts; accommodation is scarce and expensive. In summer, it is Australia's premier mountain-biking destination and hosts themed events, including the Thredbo Blues Festival. It is also popular with Australian alpine environment enthusiasts, particularly those who walk up Mt Kosciuszko.

Activities

Acquiring and processing geographical information

1 Define place.
2 Outline the physical and cultural characteristics of place.
3 Define place perception.
4 Explain what we mean by the term culture of place. How does this develop?
5 Define morphology.

Applying and communicating geographical understanding

6 Study the box, Spotlight: Thredbo Alpine Village. Explain how the interaction between natural and cultural factors produces a unique culture of place.
7 Thinking about the place in which you live, discuss the natural and cultural factors that make it different from other places.

Factors contributing to culture of place

A range of factors contribute to culture of place. These include indigenous cultures, history, cultural diversity, architecture and lifestyle, noise, colour and vitality. These factors interact to influence how a place is perceived by people.

Indigenous cultures

Indigenous peoples have a view of the world that is distinct from the mainstream, non-indigenous worldview. Place (referred to as **Country** by Aboriginal and Torres Strait Islander peoples) is central to their being. Also of relevance is the observation that the world's indigenous peoples typically see themselves as a part of nature, not separate from it. They see themselves as custodians of the land. Their responsibility is to care for it for the benefit of future generations.

In terms of indigenous cultures, land, family, lore, ceremony and language are the five key interconnected elements. For example, families are connected to the land through the kinship system, and this connection to land comes with specific roles and responsibilities, which are enshrined in the lore and observed through ceremony. In this way, the five elements combine to create a way of seeing and being in the world that is distinctly indigenous.

Understanding how intricately interconnected these elements are, helps us to understand the damage done when colonisation occurred. When people are disconnected from their cultures, this has a deep impact on their sense of identity and belonging, which gives meaning and purpose to people's lives. Understanding this helps us to find appropriate ways to respond to the pain caused by colonisation.

Although there is diversity among indigenous and non-indigenous cultures, land, family, law/lore, ceremony and language all play a crucial role in shaping all our lives, regardless of our culture or heritage. Identifying what this looks like in our own life can help us develop empathy with others.

For example, consider how your life is shaped by the language you speak, your family traditions, the area where you live and how you approach significant life events such as weddings and funerals. As we grow in empathy and understanding, we begin to relate better to one another.

History

Each place (town or city) has its own internal geography, which is expressed in a mosaic of distinctive neighbourhoods and streetscapes. For example, it is possible to recognise rich and poor neighbourhoods, ethnic neighbourhoods, residential districts, commercial districts, industrial districts and those undergoing transition. Each place has its own character and story, the product of successive phases and development and demographic, social, cultural and political change. Each era in a place's history leaves its mark, in the layout of streets, the fabric of its buildings, the nature of its institutions and the cultural legacies of its residents.

The nature and character of places are often the unplanned outcome of processes occurring over long periods, through accruing successive generations of building activity.

This leaves traces that serve to structure subsequent building activity and provides both opportunities and constraints for building processes. These traces include street patterns, land subdivisions and titles, infrastructure development and past building construction. For example, factors shaping the morphology of cities such as Rome and Athens can be traced back to ancient times, with their historical ruins continuing to influence the functional form and character of both cities, and also their culture of place.

Sydney's central business district (CBD) street pattern is a legacy of the city's origins as a penal settlement. For the most part, the street pattern established by the First Fleet, in 1788, has survived to this day. Sydney's relatively narrow streets, and in part, its irregular street pattern, is a stark contrast to the broad avenues and connecting laneways of Melbourne's CBD.

Social factors also play a role in terms of the historical impulses that impact on culture of place. Social forms often find expression in the physical layout of our urban places, and conversely, new physical forms produce or reproduce various social forms. For example, well-landscaped public **spaces** encourage interactions. Conversely, low-density urban sprawl, poorly serviced by public transport and social infrastructure, can cause **social isolation** and **locational disadvantage**.

Cultural diversity

Culture shapes us and our identity and influences our behaviour. It plays an important role in defining culture of place. A vital aspect is population diversity. Australia's cities are culturally diverse (see Figure 10.1.5). Of all the world's cities, Sydney has the seventh-largest percentage of foreign-born individuals. Almost 32 per cent of residents were born overseas (most commonly the UK, New Zealand and China, followed by Vietnam, Lebanon, India, Italy and the Philippines). Just over 1 per cent are Indigenous and 16.9 per cent are Asian Australians. The most spoken languages are English, Arabic languages, Chinese languages and Greek.

Melbourne's culturally diverse population has almost 36 per cent born overseas (most commonly the UK, Vietnam, Italy, China and New Zealand, followed by India, Sri Lanka, Malaysia, South Africa and Sudan). The most spoken languages are English, Chinese languages, then Greek, Italian and Vietnamese.

This cultural diversity adds to the richness of the original residents' lived experience and shapes their perceptions of place. People from diverse cultural backgrounds bring with them their customs and traditions, their languages, religions, sports and pastimes. They introduce new foods and bring new skills into the economy.

10.1.5 Chinese New Year Celebrations, Sydney. Chinese people have been central to the culture of place in Australia since the gold rush days.

● SPOTLIGHT

European cities' historical legacy

European cities have a distinctive culture of place, with many distinctive features derived from lengthy histories. Some date from the Roman Empire (27 BCE–476 CE), including Florence, Italy (founded in 59 BCE), Paris, France (52 BCE) and London, UK (43 CE). Elements of Roman and medieval urban development are preserved in their street pattern and preserved ruins (see Figure 10.1.6).

In the historic cores of some older cities, the street layout reflects ancient patterns of rural settlements. The often-narrow streets derive from a time when hand- and horse-drawn carts were the principal means of transportation and urban development was often gradual and small-scale.

Plazas and squares are an important historical legacy that characterised Greek, Roman and medieval cities with their piazzas and marketplaces. They are still important meeting places, lined with restaurants, bars and cafes.

Cities such as Vienna, London and Paris bear enduring traces of an imperial past. Central Paris arose from Emperor Napoléon III's authority (1852–1870), the last of the French monarchs. He commissioned Georges-Eugène Haussmann to rebuild the French capital. The renovation involved vast public works. Medieval neighbourhoods that were deemed overcrowded and unhealthy were demolished. These were replaced by wide, tree-lined avenues, new parklands, squares and fountains. The street plan and distinctive appearance of Paris are central to its culture of place.

The absolute power of the Romanov monarchy is also evident in St Petersburg. Founded in 1703 by Peter the Great, St Petersburg became the Russian capital in 1713. Building the new capital was the most expensive of all of Peter's many imperial ambitions. Up to 40 000 people worked on building the city. Peter called the city paradise, but for many of the workers, it became a cemetery.

Conflict has shaped cities. The legacy of defended hilltops and city walls arose from the need for defence, seen in Salzburg, Austria. City walls limited—and shaped—the growth of modern European cities. World War II lay waste to many cities. The devastating bombing altered many cities, including Berlin, Dresden (see Figure 10.1.7), Warsaw and St Petersburg.

10.1.6 Rome's ancient ruins are evident throughout the city.

10.1.7 The bombing of Dresden in February 1945 left the city devastated.

Economic change transformed cities' functional form and character. The Industrial Revolution and the rise of empires gave rise to the industrial city and the great ports of Europe. From the economic restructuring process emerged post-industrial, service-based economies. Old industrial areas declined and advances in shipping technologies and cargo handling saw docklands become obsolete, with the land becoming available for urban renewal and regeneration (see Figures 10.1.8 and 10.1.9). See Unit 7.3 for more on changes to shipping technology.

10.1.8 and 10.1.9 St Katharine Docks, 1881 and 2018, from busy port to exclusive residential and commercial precinct and marina

● SPOTLIGHT

Interculturality and multiculturalism

Interculturality is an approach that focuses on the pluralist transformation of urban and public spaces. It goes beyond an emphasis on equal opportunities and respect for existing ethnic and cultural differences. It means recognising how important cultural diversity is, along with the rights of people from different cultural backgrounds to participate in building a common identity, one defined by diversity, pluralism and respect for human rights and fundamental freedoms. This is consistent with Australia's **multiculturalism** emphasis. Multiculturalism is a system of beliefs and behaviours recognising and respecting the presence of people from all cultural backgrounds. It acknowledges and values their sociocultural differences and encourages and enables their contribution within an inclusive cultural context.

Did you know?

Australia is home to people who identify with more than 270 ancestries.

Architecture and streetscapes

From the earliest civilisations, architecture has periodically transformed places, each era leaving its mark on the morphology of cities. This contributes to our perception of places.

Sydney has its distinctive architectural heritage—from the colonial buildings lining Macquarie Street to the grand sandstone structures on Bridge Street. More recent office and apartment towers were designed by some of the world's greatest architects. Added to this is one of the most famous buildings of the twentieth century, the Sydney Opera House, designed by the Danish architect Jørn Utzon.

The internationalisation of Sydney's architecture is increasingly apparent. A CBD building there could be in Dubai, New York, London or Shanghai. Modern architecture has a degree of ubiquitousness or universality. The world's great architects have designs featured in cities all over the world. An obvious example of cultural integration, the end product is a sameness or homogeneity.

Another important element of built environments is streetscapes—the visual elements of a street. These are its roads, bicycle lanes, buildings facades, signage, footpaths, bus shelters and street furniture, trees and plantings, and open spaces that combine to form a street's character. All add to our sense of place.

Lifestyle

Lifestyle embraces the interests, opinions, behaviours and behavioural orientations of an individual, a group or a culture. It is the typical way or style of living (e.g. rural or urban). The nature of the place a person lives in determines the available set of lifestyle choices.

10.1.10 Bondi Beach represents Sydney's lifestyle aspirations.

Many factors shape the nature of places. Degree of affluence plays a major role in the range of choices and lifestyles possible—greater affluence provides far more choices.

The natural environment determines the range of lifestyles available. In areas near the sea, a beach or surf-based culture or lifestyle can be adopted (see Figure 10.1.10). Those living in alpine regions may embrace a lifestyle based on snow sports.

The nature and proximity of the cultural environment also play a role. The richer the social and cultural environment, the more diversity of lifestyle choices is available. Generalising the concept of an Australian or European lifestyle is subjective. Though prone to stereotyping, focusing on such broad lifestyle categories can be useful in analysing how an individual's attitudes, way of life, values, or world views impact on their lifestyle.

● **SPOTLIGHT**

A new era for China's architecture

In 2020, the Chinese government directed architects, property developers and urban planners to end copycat building.

In their new era for China's cities, authorities have proposed measures to ensure all new buildings embody the spirit of their surroundings and highlight Chinese characteristics. A rejection of Western architectural trends, it seeks to impose a distinctively Chinese culture of place on rapidly growing cities.

A ban on buildings over 500 metres is a response to the current construction boom. Globally, of 10 completed buildings above 500 metres, half are in mainland China. They include the planet's second-tallest skyscraper, the twisting Shanghai Tower (632 metres) and Shenzhen's Ping An Finance Center (599 metres). Beijing's CITIC Tower and the Tianjin CTF Finance Center are the world's seventh- and ninth-tallest buildings.

This move is partly because tall buildings were used to brand developments and set cities apart. But the new restrictions are also about economics—China's skylines became littered with unfinished towers as economic growth slowed and developers ran out of money. Above a certain height, construction costs for very tall buildings increase exponentially with each additional floor.

The government's directive also imposes a ban on architectural plagiarism, imitation and copycat behaviour (see Figures 10.1.11 and 10.1.12). It provides incentives for architects to adhere to the new planning laws and regulations. There are warnings against demolishing historical buildings, traditional architecture and even old trees to make way for new developments, a move in keeping with the growing emphasis placed on heritage preservation in China.

10.1.11 and 10.1.12 The Beer Square, Boluo county, Huizhou city (bottom), China was built in 2012. It is a replica of the World Heritage-listed Austrian village of Hallstatt (top).

10.1.13 Brightly lit Times Square, New York City, USA

Noise, colour and vitality

The noise, colour and vitality of a place help to shape people's perceptions. Consider New York's Time Square in Manhattan (see Figure 10.1.13) or the Las Vegas Strip. Both places are full of colour and noise, attracting tourists. Gathering crowds provide a degree of vitality not always found in urban areas. The Opera Bar on the concourse to the Sydney Opera House is one favourite meeting point for young people. Sometimes an effect is periodic, associated with sporting events and festivals. Sydney Harbour on New Year's Eve and the winter Vivid festival both bring the Opera Bar to centre stage.

Street life

The street life of a place is culturally determined and influenced by the nature of its constructed environment. It can vary according to such variables as the time of day, day of the week, season and special events.

● SPOTLIGHT

Sydney's culture of place

Sydney's culture of place is defined by the nature of its biophysical environment, especially its climate, beaches and harbour setting. It also has its own distinctive multicultural character and a casual lifestyle. Residents enjoy sport and outdoor pursuits. It has its renowned architectural heritage, especially its iconic structures—the Sydney Opera House and Harbour Bridge—and unique streetscapes. The low-density urban sprawl yields a suburban lifestyle. It has an increasingly international outlook and status as a world city and popular tourist destination.

Within Sydney, there are several distinctive urban villages. Each has a unique culture of place. Examples include Double Bay and Paddington in Sydney's east; Mosman and Crows Nest on Sydney's North Shore; The Rocks (see Figure 10.1.14), Darlinghurst, Glebe and Millers Point on the borders of the CBD; Norton Street in Leichhardt; and Cabramatta in the city's west.

10.1.14 The Rocks exhibits a distinctive culture of place within Sydney.

On weekdays, Sydney CBD's streets and other public spaces are crowded with workers (many suburban commuters), shoppers and tourists. They gather in parks and squares, restaurants and food courts for lunch. After work (especially on Fridays), people visit bars to socialise. By night, theatregoers, tourists and inner-city dwellers patronise restaurants and bars. Luxury hotels are well patronised by visitors, and cruise ships and ferries ensure Circular Quay is in constant motion. The bars and restaurants lining Circular Quay are crowded, especially on weekends. On weekends, the CBD is generally quieter, but the city's department stores and other luxury retailers are full of shoppers. Sydneysiders join tourists to attend its galleries, museums and theatres. Places like The Rocks, Chinatown and Darling Harbour become crowded and alfresco dining is popular. The city's energy and colour are most vibrant during its great celebrations—New Year's Eve, Australia Day, Vivid Festival and The Festival of Sydney. At other times, the city is a place of commemoration, for example, Anzac Day, when crowds gather to remember the country's war dead. Sometimes people gather in demonstrations to exercise their democratic right to protest. The Domain attracts music lovers when hosting concerts and sports fans when it screens global sporting events. The atmosphere is relatively casual, a reflection of Australia's relaxed lifestyle compared to the world's other big cities.

Activities

Acquiring and processing geographical information

1 State the five interconnected elements of indigenous culture.

2 Describe the nature of the connections indigenous people have with the land.

3 Outline the impacts of colonisation on the sense of identity of indigenous people.

4 Outline the ways in which history leaves its imprint on the morphology of cities (their functional form and character).

5 Outline the ways in which social factors influence the nature and character of places. Give examples.

6 Define social isolation and locational disadvantage.

7 Outline the role of culture in determining culture of place.

8 Define interculturality and multiculturalism.

9 State what is meant by the term lifestyle and identify the factors that determine the lifestyle people are able to aspire to.

Applying and communicating geographical understanding

10 Construct a diagram to illustrate the range of factors that contribute to a sense of place.

11 As a class, reflect on the relationship non-indigenous people have with the environment. How does it differ from that of Indigenous Australians? Which of these relationships is the more sustainable?

12 Create a flow diagram or mind map outlining the morphology and changes over time in European cities. Include in your diagram:

- Roman and medieval urban development
- the narrow streets
- marketplaces and squares
- imperial power and absolute monarchy transforming cities
- the role that conflict has played on shaping the morphology and culture of place
- the impact of economic change.

13 Study the box, Spotlight: A new era for China's architecture. Outline the key elements of the Chinese government's directive outlining its architectural preferences. What did the directive seek to address?

14 Outline the factors that contribute to Sydney's culture of place.

■ CASE STUDY

A European lifestyle

While generalisations can be misleading, it's possible to make some broad observations about what constitutes a European lifestyle. The qualifier is that Europe is home to 747 million people across 44 countries, each with its own distinctive cultural heritage and traditions.

10.1.15 Piazza Navona, central Rome

10.1.16 The Carnival of Venice is world-famous for its elaborate masks. It ends with the Christian season of Lent, 40 days before Easter.

Urban places

Seventy-five per cent of Europeans live in cities and towns, but the reach of urban places extends well beyond their boundaries. Europeans have embraced an urban lifestyle and use city amenities, such as cultural, educational or health services, even if they live in regional areas. While cities are the drivers of Europe's economy and creators of European wealth, they depend heavily on regional resources to meet the demand for energy, water and food. They also provide a 'sink' for their waste.

While some larger European cities have high-rise apartment towers and commercial buildings, most have a relatively low skyline, largely due to their age. Their development usually pre-dates inventions like lifts and steel-reinforced concrete buildings. Some have building codes and heritage regulations to protect both significant buildings and places, as well as their traditional culture of place. More recent strong planning controls aimed at limiting urban sprawl help to protect the existing built environment.

Most people in European cities live in apartments at relatively high densities, especially compared with the low-density sprawl of car-dominated North American and Australian cities. European CBDs retain their focal position in residents' shopping and social lives. Their streets are full of noise, colour and vitality, especially in the early evening when residents leave their apartments to meet friends or take a stroll. Pre-dating the automobile age, streets tend to be narrow. They are often lined with shops, cafes, bars and restaurants, and double as social places for people to meet and interact. Italy's urban piazzas, for example, are important social spaces (see Figure 10.1.15).

There is strong neighbourhood stability in most European towns and cities. Europeans change residences half as often as Australians. Many occupy build-to-rent apartments that offer secure tenancy.

Consequently, friendships within neighbourhoods tend to be more enduring and relationships may extend over generations.

Festivals are important in defining the culture of place of European towns and cities. Long-standing events like the festivals of Bayreuth, founded in 1876, Salzburg in 1920 and Edinburgh in 1947, rub shoulders with new, cutting edge festivals. Their diversity is extraordinary—from the classical to the contemporary; from elitist to popular; from huge pop and rock concerts to micro-festivals hosted in remote villages (see Figure 10.1.16).

Public spaces also host a range of market events, such as weekly farmers' markets or annual Christmas markets. The latter are associated with celebrations during the four weeks of Advent. Originating in Germany, these markets are held throughout Europe and have become an important tourist attraction. Their history dates back to the late Middle Ages (see Figure 10.1.17).

10.1.17 Christmas markets, Nuremberg, Germany

Italian lifestyle

In Italy, although home-based entertainment (television and wireless technologies) has grown in popularity, public spaces remain central to the Italian lifestyle. Young and old meet friends on a near-daily basis, often in the cities' piazzas in the evenings. The young frequent the bars, cinemas and pizzerias. While social media and smartphones allow Italians to maintain ties with friends, face-to-face interaction is still highly valued. Often these interactions are facilitated via social media.

Football and food are the great constants of the Italian lifestyle. Italians are passionate about both. Work patterns revolve around the midday meal, even though the leisurely two-hour-long lunch break is disappearing. Bars and trattorias cater cheaply and quickly to casual diners. Evening meals are late in the day, with restaurants opening at around 7.30 pm. Despite a wide regional variety of foods, Italians share a love of noodles in the form of a wide variety of pasta (see Figure 10.1.19). In the south, pasta is often dressed with sauces made of olive oil, tomatoes, and spices. In the north, especially Piedmont, it is coated in cream, butter and cheese. The south is renowned for citrus fruits, olive groves and vineyards. Italy is also one of the world's largest wine producers. Every region is known for its wine, which is drunk with most meals.

10.1.18 Piazza del Duomo on Ortygia, Sicily, is a popular evening meeting place.

Football is a national obsession that inspires much public interest and media attention and is a means of expressing regional loyalties.

For most Italians, religion plays a much smaller role in daily life than it did last century. It is usually focused on Sundays or special celebrations such as Christmas and Easter. Older generations, especially in rural settlements, tend to be more involved and may attend mass daily.

The emphasis on conformity and a commitment to the institution of the family remains a key element. Grandparents, children and grandchildren often still live together in family units, although this is becoming less common, especially in cities.

10.1.19 Italians are passionate about food.

10.1.20 A Spanish tapas bar, Plaza Nueva, Bilbao

Spanish lifestyle

The Spaniards' lifestyle is not dissimilar to the Italians'. They too see the public realm as an important social space in which to meet and interact. Their daily structure differs from Australians. As in Italy, the evening meal is late in the evening. Many Spaniards have a light early breakfast and a snack (tapa) later in the morning. Lunch is the main meal of the day, and eaten anywhere from 1.00 pm to 4.00 pm at home or in a café. Because dinner is eaten late, another snack—*la merienda*—is eaten late afternoon. Eating out is common, the evening meal could be sandwiches, early on in a *salones de té* or *pastelerías*, or tapas (a variety of small dishes) and a beer in a bar (see Figure 10.1.20). They don't sit down to their evening meal until as late as midnight.

There are significant regional variations in food, including famous dishes, such as paella, tortilla, Spanish omelette, *gambas al ajillo* (prawns in garlic), *pisto* (vegetable stew), and cured meats, including jamon carved thinly off cured legs of pork, and spicy chorizo. They usually cook with fresh ingredients. Supermarkets are becoming common, but open markets are still a vital part of life. Typical of Europe, sport in Spain is dominated by football.

The extended family is also important. Large family groups are often seen in restaurants. While it is less common for three generations to live in the same household, people still turn to family in difficult times. While religion holds less significance now, church feast days are still marked by fiestas (festivals).

Activities

Acquiring and processing geographical information

1. To what extent is Europe an urban-based society?
2. Why do European city centres have relatively low skylines?
3. What type of housing dominates European cities? What impact does this have on peoples' lifestyles and the culture of place of European towns and cities?

Applying and communicating geographical understanding

4. What role do festivals and market-like events play in the culture of place of European towns and cities?
5. Draw a Venn diagram. Identify the similarities and differences between Italian and Spanish lifestyles.
6. What is special about the Italian and Spanish lifestyles?

UNIT 10.2

Influences on the culture of place: The media

The culture of a place is changeable and can be influenced by many different factors. **Cultural diffusion** describes the dispersion, or spread, of different cultural elements between countries. People confronted with new cultural influences can reject them outright, adopt the change, or find a way to adapt. **Cultural adoption** involves the acceptance and integration of different cultural elements into one's own. **Cultural adaptation** involves cultural modification to incorporate aspects of another culture. These processes occur across scales from local to global.

Media

Key elements of culture are mainly dispersed via the media. Technological advances in telecommunications have greatly expanded the reach, immediacy and influence of the global media giants, which are mostly US-based. It is often claimed that cultural integration is actually the Americanisation of global culture.

The media strongly influences the perception of places, on scales from local to global. Anti-social behaviour reported in a particular neighbourhood, for example, could adversely impact the perception of that place. At a global scale, a place's representation in the media can have an impact on the perception of a tourist destination or entire country. Multiple channels construct and communicate such perceptions. These channels include direct experience, word-of-mouth, audiovisual media, and, increasingly, social media. However, mass media remains the main way in which information about places is spread.

The changing media landscape

When television was introduced to Australia in 1956, it was conceived as a community and nation-building undertaking. In NSW, non-metropolitan centres such as Newcastle, Orange and Tamworth had their own stations, each dedicated to reporting local news and events. Sydney eventually boasted three commercial stations—Channels 7, 9 and 10. Nationally, Australia was served by the public broadcaster, the ABC (and later SBS). The ABC, like its counterparts in the UK (BBC) and Canada (CBC), was seen as a vehicle to promote national culture, identity and awareness. Over time, the regional broadcasters were absorbed into the three commercial free-to-air providers. The emphasis had shifted from local to national. The same trend occurred in the print media. Small, regional newspapers were consolidated into larger enterprises, the biggest being News Corporation. In 2020, regional newspapers across Australia either ceased publication or moved to online-only models. This was partly due to the collapse in advertising revenue accompanying the COVID-19 pandemic.

Perception of place is influenced by the media people engage with. Australia's concentration of media ownership is an ongoing concern to many Australians. It limits the perspectives people are exposed to. No other democratic country has its media dominated by only two key players: News Corporation and Nine Entertainment Co. Nine merged with Fairfax Media, the publisher of the *Sydney Morning Herald* and *The Age* newspapers, in late 2018. Many believe the concentration of media ownership narrows information sources. Those who disagree point to the diverse range of information sources available via new media. Changes to the federal government's foreign ownership laws, along with the US's purchase of Channel 10 by CBS, could significantly alter the Australian media landscape over the next decade.

Digital media

Australian media providers are increasingly being eclipsed by global providers of news and entertainment (see Figure 10.2.1). The old media (free-to-air television and

10.2.1 National media providers are being eclipsed by global providers.

10.2.2 New media technologies made remote schooling possible.

newspapers) is having to compete against pay television and internet-based streaming services that generate and broadcast their own content. Also emerging is the new media—computer and internet providers that transmit news and entertainment and enable people to interact online. It includes websites and blogs, streaming services (audio and video), chat rooms, online communities, social networking and sharing platforms. New media is dominated by largely US-based global players such as Apple, Alphabet (Google's parent company), Meta (Facebook, Facebook Messenger, Instagram and WhatsApp), X (previously Twitter) and YouTube (owned by Alphabet). Streaming service providers include Netflix, Disney, Amazon, Apple, YouTube, Warner Media (HBO), and Nine Entertainment (Stan). Except for the latter, all are US-based. Such social media platforms have attracted an ever-increasing share of advertising revenue at the expense of old media outlets (television, radio and newspapers).

Some new media and streaming service providers are subscription-based. Consequently, they are less dependent on advertising revenue. Zoom, which rose to a new level of prominence during the COVID–19 pandemic, is US-based video conferencing and online chat service. Zoom's cloud-based peer-to-peer software platform is used for teleconferencing, distance education, medical consultations and social interactions.

The rise of new media has increased the interaction between people at a variety of levels. It has allowed people to express themselves through blogs, websites, videos, photography, social media and other user-generated media. As new technologies emerge, the world becomes more globalised. Globalisation allows the world to be connected no matter what the distance is between users.

New media technology developments have made it possible to make friendships and sustain a sense of community through digital social places. During the COVID-19 pandemic, digital social spaces often became more prominent than physical places. Many people kept contact with family and friends via media such as Zoom and FaceTime. New media technologies have made it possible to break the relationship between physical place and social place. These technologies also played an important role in sustaining online learning during the pandemic (see Figure 10.2.2). They initiated and expanded remote working arrangements, making them more acceptable.

Global media networks

Disseminating film and media is far more complex than it once was. It occurs on a global scale and involves some of the world's best-known and most valuable TNCs. The transition from local to national to global was driven by technological innovation, deregulatory economic policies, global economic integration and an emerging transnational corporate culture. The process of change has resulted in an industry dominated by US-based corporations, including Meta, Disney and Alphabet. The world's largest media and entertainment providers are listed in Figure 10.2.3.

The corporations in Figure 10.2.3 set about constructing a global media market. Their objective is to ensure their products (and those of their advertisers) reach the largest possible number of consumers. They regard international borders as arbitrary obstacles. This has led to standardisation and homogenisation of world markets, which are characteristic of cultural integration.

The major media and entertainment providers have sought to expand vertically. They extend their control over all aspects of programming (production and film archives), distribution and transmission. Netflix and HBO are excellent examples of internet-based streaming services that spend vast sums to produce the content they broadcast.

Company	Country	Headquarters	Revenue (€ billion)
Alphabet (Google)	USA	Mountain View, California	217.84
Meta (Facebook) Inc.	USA	Palo Alto, California	99.7
Comcast	USA	Philadelphia, Pennsylvania	98.4
Tencent Holdings Inc.	China	Shenzen	73.43
Apple Inc.	USA	Cupertino, California	57.86
The Walt Disney Company	USA	Burbank, California	57.00
Amazon	USA	Seattle, Washington	53.21
Bytedance	China	Beijing	49.00
Charter Communications Inc.	USA	Stanford, California	43.70
Warner Media (AT&T) Inc.*	USA	New York City, New York	30.13
Microsoft Corp.	USA	Redmond, Washington	28.91

*Now Warner Bros Discovery

Source: Statista

10.2.3 The world's largest media and entertainment providers by revenue, 2021

Did you know?

The success of direct-to-consumer streaming services such as Netflix (222 million subscribers in late 2022) and Disney+ (152 million subscribers) highlights how important streaming has become to the media industry. Traditional TV networks such as Disney's ABC and Disney Channel are seeing their viewers and cable-TV subscribers shift to on-demand services. Such developments provide consumers with greater choice and expand the business opportunities for content providers. Free-to-air broadcasters, such as Australia's ABC, have responded by providing content that is free of charge through streaming platforms such as iView.

Global media companies took advantage of converging entertainment, information and telecommunications technology to move beyond mass media and into personalised media and greater individual choice. They now provide new media services—streaming and other telecommunication. These services utilise satellite and cable infrastructure and new forms of income, such as subscription.

Emerging global news providers increased the quantity of information now available, as has the proliferation of smaller web-based providers. The latter increased the range of perspectives available to all. In the USA this led towards an ideological polarisation, apparent even in mainstream media. This is exemplified by the conservative Fox News and the progressive CNN and MSNBC. In Australia, Sky News has become the voice of conservative views, while ABC and free-to-air television stations are more moderate. The more progressive newspapers include the *Sydney Morning Herald* and *The Guardian*.

CNN

CNN (Cable News Network) is a US news-based cable television broadcaster (see Figure 10.2.4). A subsidiary of Warner Bros. Discovery (WBD), CNN was founded in 1980 by US entrepreneur Ted Turner as a 24-hour cable news channel. Globally, CNN programming airs through CNN International. It is available in over 212 countries.

CNN is a good example of how the media has become detached from the restraints and influences of place and culture. Until the 1997 launch of CNN's Spanish-language news network, CNN en Español, the CNN News Group had largely ignored regional and local differences by providing a global English news service that aired the same content, regardless of the country its viewers lived in.

This concentration of views has become known in media circles as the 'CNN effect'. Audiences worldwide are presented with sober assessments of 'the facts' from a CNN perspective. Despite claiming otherwise, rarely does the CNN news examine an issue from a range of perspectives. US conservatives rail against the organisation, with former US President Donald Trump regularly dismissing the broadcaster as

10.2.4 The CNN Center in Atlanta, Georgia, USA, the broadcaster's world headquarters

'fake news' while Fox News was pro-Trump. Progressive Americans decry Fox News as a threat to national unity. They accuse it of contributing to the polarisation of US society for commercial gain.

The Walt Disney Company

A US$93 billion media conglomerate, one of the world's largest, this company's stated mission is to:

> ... entertain, inform and inspire people around the globe through the power of unparalleled storytelling, reflecting the iconic brands, creative minds and innovative technologies that make ours the world's premier entertainment company.

The company's media holdings include film studios, television networks, cable channels, associated production and distribution companies, and company-owned and operated television stations across two divisions—Walt Disney Television and ESPN (a sports-focused provider). These include the Disney Channel, the American Broadcasting Company (ABC), Freeform, National Geographic and FX.

Disney's other corporate activities include book, magazine and newspaper publishing; music recording; live stage shows; real estate development; major league baseball and ice hockey; retail stores; product licensing; and computer software and online services (see Figure 10.2.5 for revenues).

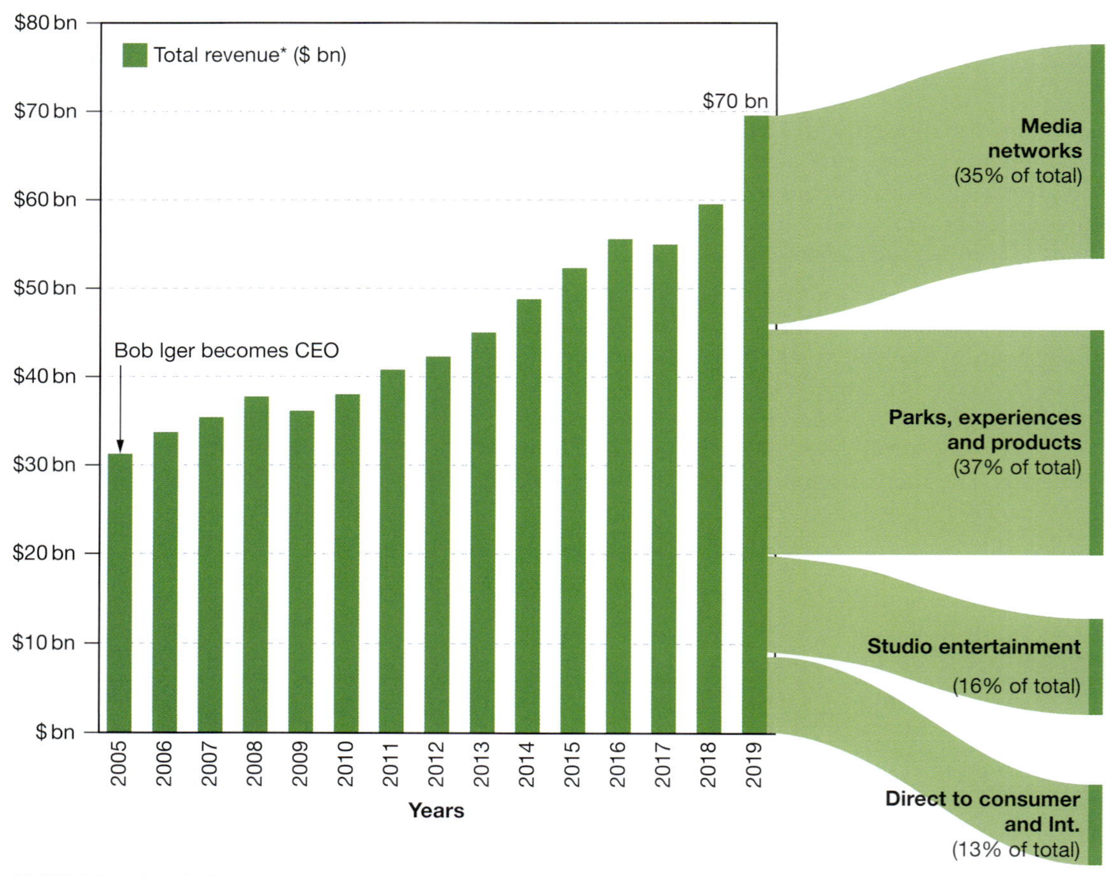

10.2.5 Disney's principal revenue streams

Disney's Parks, Experiences and Products is the division that brings Disney's stories, characters, and franchises to life through theme parks and resorts, cruise and vacation experiences, and consumer products. It has theme parks in the USA, Japan, China and France (see Figure 10.2.6). In 2018 (just prior to the COVID-19 pandemic), 153 million people visited its amusement parks. Disney's Vacation Club resorts (a vacation timeshare program) has more than 220 000 members.

Theme park	Attendance
Magic Kingdom (Walt Disney World), Florida, USA	20 860 000
Disneyland, Anaheim, California, USA	18 660 000
Tokyo Disneyland, Japan	17 910 000
Tokyo DisneySea, Japan	14 650 000
Disney's Animal Kingdom, USA	13 844 000
Epcot at Walt Disney World, Florida, USA	12 750 000
Shanghai Disneyland, China	11 800 000
Disney's Hollywood Studios, USA	11 260 000
Disney's California Adventure, USA	9 860 000
Disneyland Park, France	9 840 000
Hong Kong Disneyland, China	6 700 000
Walt Disney Studios Park, France	5 300 000

10.2.6 Disney theme park attendances, 2018

Did you know?

Disney has added a disclaimer to some of its old movies regarding racist stereotypes:

'This program includes negative depictions and/or mistreatment of people or cultures. These stereotypes were wrong then and are wrong now. Rather than remove this content, we want to acknowledge its harmful impact, learn from it and spark conversation to create a more inclusive future together.'

In March 2019, Disney acquired many of the assets of Rupert Murdoch's 21st Century Fox for US$71.3 billion. They included Fox's film and television studios, its US-based cable and satellite channels, and its entertainment and direct broadcast satellite divisions in the UK, Europe and Asia. Murdoch retained control of the Fox Broadcasting Company, Fox Television Stations, the Fox News Channel, the Fox Business Network and the Fox Sports channels. These were integrated into a new company, Fox Corporation, a transaction confirming Disney's status as one of the world's largest and most powerful media and entertainment corporations.

It has been argued that Disney represents the single most powerful and influential force in the globalisation of Western culture. In this century, the real power to promote and consolidate consumer capitalism will be held by the 'infotainment' industry: film, TV, music, ideas and information.

Fox Corporation

Fox Corporation is a US media company controlled via a family trust. Australian-born Rupert Murdoch was the chair until his retirement in September 2023, his son, Lachlan, has taken over as chair and CEO.

The company's assets focus on television, news and sports broadcasting. News Corporation has media assets in Australia and the UK. Principal assets include Dow Jones and Company (publisher of *The Wall Street Journal*), News UK (publisher of the *Daily Mail*, *The Sun* and *The Times*) and book publisher HarperCollins. News Corporation Australia's interests span newspaper and magazine publishing, subscription television (Foxtel and Binge), market research, DVD and film distribution, and film and television production trading assets. It has nearly 70 per cent of the country's daily metropolitan newspaper circulation and its only commercial 24-hour news channel.

The Murdochs' media assets typically project a conservative point of view. In the USA, Fox News championed Donald Trump during his presidency, while in Australia, Rupert Murdoch campaigned for retaining tax concessions that primarily benefited the wealthy, and the Murdoch-owned press adopted a negative position around proposed climate change action. He is seen as a political player not just a publisher.

Activities

Acquiring and processing geographical information

1 Distinguish between cultural diffusion, cultural adoption and cultural adaptation.
2 Define culture. Explain how it evolves over time.
3 State the role of the media in facilitating cultural diffusion. What has made it such an effective agent of change? How does it shape our perception of places?
4 Outline how the Australian media landscape changed over time. What has been the economic impact of integration on it?
5 Differentiate between old media and new media.
6 Describe the impact of the new media on interactions between people. How has it impacted the sense of community experienced by people?
7 State the factors that facilitate the emergence of global media networks. What are the objectives and motivations of such enterprises?
8 Outline the link between the emergence of global news providers and the ideological polarisation that is increasingly evident in countries such as the USA and Australia.

Applying and communicating geographical understanding

9 What is meant by the statement, 'CNN is a good example of how the media has become detached from the restraints and influences of place and culture'?
10 Assess the claim that Disney represents the 'single most powerful and influential force in the globalisation of Western culture'.
11 Study Figure 10.2.5. Using data from the graph, write a paragraph outlining Disney's sources of revenue.
12 Explain why Rupert Murdoch was sometimes accused of being a political player, not just a publisher. What views did he use his media organisation to promote?
13 Have a class discussion about the proposition that news organisations have a responsibility to present the news objectively.

UNIT 10.3

Influences on the culture of place: Fashion

Fashions serve as a reflection of time and place, and are determined by society, culture, history, economy, lifestyle and the marketing system. The market for fashion ranges from the world of couture to mass-produced, ready-to-wear clothing.

Fashion as a form of cultural expression

From a cultural perspective, fashion represents an important form of identity and self-expression. It plays a vital part in defining a person's identity. How someone dresses helps to communicate who they are or want to be.

Clothing has long enabled people to identify themselves publicly. Over the last 50 years or so, a wider array of subcultural groupings emerged. They visually mark their differences from the dominant culture and their peers by adopting elements of commercial culture, especially clothing. Fashion subcultures are groups organised around or based on certain clothing choices and/or personal appearance that render them distinctive enough to be recognised or defined as a subset of the wider culture. Many of these looks have been adopted, or adapted, from other cultural contexts.

The global fashion industry

The fashion industry is one of the world's most globally integrated. It is largely a product of the modern age. Until the mid-1880s, most clothing was custom-made by local dressmakers and tailors, or at home. The mass production of clothing in a range of standard sizes arose from the Industrial Revolution, the rise of capitalism, the development of the factory production system and the advent of the department store and other forms of retailing. The scale of production meant that the price of clothing and footwear fell, making it available to a wider range of people.

The fashion industry is a global undertaking. Clothing may be designed in one country, manufactured in another and sold worldwide. For example, an Australian-owned surfwear company might buy its cotton fabric from Chinese mills, sew and screen-print garments in Vietnam, and ship the product to distribution centres in Australia, the USA and Europe. A high-end fashion house might weave Australia's finest merino wool into cloth in Italian fabric mills before shipping it to tailors in China. They then transform it into expensive business suits using designs from Milan's fashion houses in Italy. Figure 10.3.1 shows the corporate headquarters of some of the world's well-known clothing brands. Figure 10.3.2 highlights the dominance of China and the European Union in clothing production.

Brand	Location of headquarters
Billabong	Australia (Burleigh Heads, Queensland)
Calvin Klein	USA (New York)
Chanel	France (Paris)
Country Road	Australia (Melbourne)
Diesel	Italy (Molvena)
Duchamp	UK (London)
French Connection	UK (London)
Gant	Sweden (Nacka Strand)
Guess	USA (Los Angeles)
Hugo Boss	Germany (Metzingen)
Juicy Couture	USA (Los Angeles)

Brand	Location of headquarters
H&M	Sweden (Stockholm)
Levi Strauss	USA (San Francisco)
Louis Vuitton	France (Paris)
Mango	Spain (Palau-solità i Plegamans)
Prada	Italy (Milan)
Quiksilver	USA (Huntington Beach, California)
Ralph Lauren	USA (New York)
Rip Curl	Australia (Torquay)
SABA	Australia (Sydney)
Tommy Hilfiger	Netherlands (Amsterdam)
Zara	Spain (Arteixo)

10.3.1 Familiar clothing brands and their countries of origin

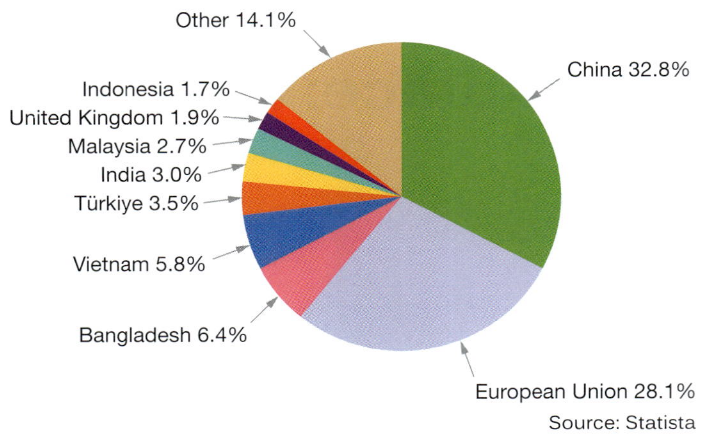

10.3.2 Leading clothing exporters, 2021

There are four key components of the fashion industry:

- production of raw materials: fibres, textiles and leather
- production process: designers and manufacturers
- marketing: advertising and promotion
- retailing.

Figure 10.3.3 illustrates the clothing manufacturing process.

Production: Raw materials

The cloth used to make clothes is either woven or knitted using a variety of fibres. They can be natural plant and animal fibres or synthetics. Animal-based fibres include silk, fur, wool, angora, mohair and alpaca.

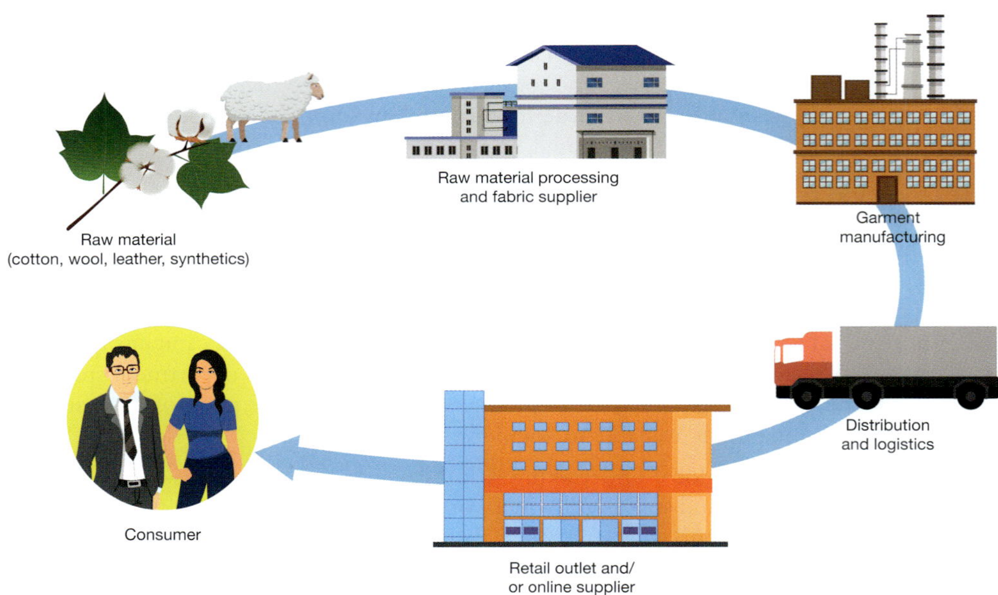

10.3.3 The clothing manufacturing process

Cotton is the most important plant-based fibre. Biophysical factors, especially climate, and production costs, determine where these fibres are produced. They also influence the distribution of non-fibre clothing materials, such as leather (an animal-based product) and rubber (a plant-based product).

Synthetic cloth is made from human-made fibres. Examples include polyester, nylon, rayon, spandex, latex and kevlar. These have many uses, some not possible with natural fibres. They can be used to create waterproof fabrics and stretch fabrics, often used in swimwear and underwear. The chemicals used to make them are derived from crude oil (the same oil that is refined into petrol).

Production process

Increasingly, clothing manufacturing is undertaken in the low-cost countries of East and South Asia, and Central and South America. Key elements of the creative process remain concentrated in the developed world, with clothing design still centred in the world's fashion capitals—Milan, Paris, London and New York. Global advertising campaigns also originate from agencies based in world cities.

● SPOTLIGHT

Can China still compete as the world's fashion factory?

Just over 30 per cent of the world's clothing is manufactured in China (see Figure 10.3.4). Yet the global clothing manufacturing map is changing fast, raising questions about its future status as 'the world's factory'. China's manufacturing sector faces higher costs at home, cheaper competition abroad and a global economic slowdown. The latter was driven, at least partially, by the COVID–19 pandemic. It compelled some Chinese firms to outsource lower-skilled production to cheaper labour markets offshore in Asia and Africa. Offsetting such developments is the increasing tendency for Chinese firms to invest in technological innovation, specialisation and quality upgrades. This contributes to the advanced manufacturing sector's development. So, while some garment and textile industry producers move up the value chain, others struggle with issues such as long-term sustainability and worker welfare.

10.3.4 Factories in Shengzhou, China, make 40 per cent of the world's neckties.

Transport logistics is a critical link between the centres of production and consumption. Fashions can change quickly, so designers and manufacturers must be able to respond to shifts in consumer demand. Once produced, garments and footwear need to be transported to distribution centres around the world. From these centres, the goods are distributed to retailers. Given the industry's highly competitive nature, it is vital to minimise the time from design to manufacture to retailer delivery.

Marketing

The global fashion industry is increasingly dominated by global brands. Worldwide marketing campaigns generate awareness. These seek to link the brand to the lifestyle aspirations of potential consumers. Glossy magazines targeting specific demographics, socioeconomic and social groups, are often used to promote particular brands. Television and social media-based campaigns and event sponsorship are used to raise brand awareness.

The move towards a homogeneous global marketplace and the shift to brand dominance offers several advantages to large corporations. It offers economies of scale in production and distribution. Marketing costs are lowered while the brand image remains consistent. They are able to respond quickly to changing fashions (consumer demand) and offer a competitive edge to online retailers. Lost is the rich local diversity in the way people dress.

Retailing

Clothing retail is also increasingly global. Australia's largest cities now host a range of international fashion retailers. Manufacturers and retailers adopt strategies that help them to grow their businesses, generate profits and respond to changes in consumer tastes.

The internet transformed how people purchase goods and services. It also challenged traditional retailers (bricks-and-mortar stores). Consumers can source their favourite brands from online retailers based anywhere in the world and have them home-delivered within days. This has led to a massive expansion of the logistics industry, built around the delivery of online purchases. The lockdowns associated with COVID–19 pandemic led to a massive expansion of online retailing. Eight out of 10 Australians shop online and one out of every 10 items is bought online. Clothing is the most common purchase.

● **SPOTLIGHT**

Zara: Responding to changes in consumer tastes

Zara is a Spanish clothing company with annual sales revenue of more than €18 billion. It built a retail strategy based on responding to its customer's fast-changing tastes. It developed a supply chain enabling new fashion deliveries as soon as a trend emerges. Zara's parent company Inditex is the world's largest clothing retailer.

Zara can deliver new products twice weekly to its global network of 1751 retail outlets in 44 countries (see Figure 10.3.5). This is over 10 000 new designs (or 20 new clothing collections) each year. It takes the company just 10–15 days to go from the design stage to having stock in stores.

Zara's whole approach is based on being responsive to customers' needs. Unlike many other clothing retailers, who subcontract manufacturing to Asian-based companies, Zara has 14 automated Spanish factories. Here, robots work around the clock cutting and dyeing fabrics and creating the basis of their final products. These go to a network of 300 small workshops in Portugal and Galacia in northern Spain, where the clothes are sewn.

10.3.5 Zara store in Guangzhou, China

Activities

Acquiring and processing geographical information

1. State how fashion retailing has been transformed by globalisation.
2. Outline the importance of fashion and how it represents a form of cultural expression.
3. Describe the nature of the global fashion industry as it exists today.
4. Identify and describe each stage of the components of the fashion industry.

Applying and communicating geographical understanding

5. Describe new recent fashion trends, who wears them, where they come from, and what the trends/public image says about the followers.
6. As a class, reflect on your fashion choices. To what extent is uniformity evident, or are there those who seek to express their individuality through the clothes they wear?
7. Study Figure 10.3.1. Complete the following tasks.
   - **a** Outline what the information provided tells us about the contemporary nature of the global fashion industry.
   - **b** Investigate one of the brands listed in the table. Briefly describe the nature of the brand, the market segment targeted, where it sources its fashions and the extent of its global reach.
8. Using Figure 10.3.2, identify the principal clothing exporters. What does this suggest about the nature and extent of economic and cultural integration in the fashion industry?
9. Outline the factors threatening China's status as the world's leading clothing exporter. What offsetting developments are helping to sustain China's competitiveness?
10. Identify at least 10 clothing retailers. Identify their target markets. What is unique about the retailing strategy adopted by each retailer? Which retailer appeals to you? Why?

UNIT 10.4

Influences on the culture of place: Brand image

Brand image is people's perception of a brand of a good or service. It is the unique bundle of associations that develop in customers' minds. It represents what the brand stands for and the set of beliefs held about it, along with consumers' perception of the product and its positioning in the market. Brand image conveys emotional value, not just a mental image. Consumers purchasing a product or service are also purchasing an image. They develop associations with the brand from which they form the brand's image. The world's most recognised brands include Apple, Google, Amazon, IBM, Visa, Coca-Cola, Nike, Disney, McDonald's, Pepsi, Microsoft, Samsung, Mercedes-Benz and Toyota.

Brand positioning

A couple of examples illustrate the importance of brands and their positioning.

Tesla manufactures electric vehicles. It has developed a cult-like following, despite its cars being more expensive than its competitors. In terms of branding, Tesla promotes its vehicles' exclusivity, design and quality. The luxury vehicles' long-range and eco-friendliness are emphasised. Its branding is so successful it has become the world's most valuable carmaker by stock market value. It is ahead of Toyota, even though Tesla delivered just 499 535 vehicles in 2020 compared to Toyota's 8.8 million.

SPOTLIGHT

McDonald's branding

McDonald's is a triumph of consumer branding. Founded in 1948 in California by George and Maurice McDonald, its exponential growth and success is largely attributed to Ray Kroc, who came across McDonald's restaurant and proposed to help the two McDonald brothers franchise their concept (see Figure 10.4.1). In just six years, Kroc built 100 McDonald's restaurants. In 1961, George and Maurice McDonald agreed to sell their franchise to Kroc for US$2.7 million. Kroc implemented numerous reforms, including the golden arches and the chain's mascot, Ronald McDonald. Kroc successfully positioned the McDonald's brand as a family-friendly, fast-food retailer.

10.4.1 The first McDonald's in Downey, California, USA

The McDonald's logo is readily recognised by people from a variety of cultures and places. A recent international survey found that 88 per cent of those interviewed recognised and could name its logo. Ronald McDonald (and Happy Meals) was introduced to tap into children's emotions and sensitivity. Ronald proved a powerful marketing tool, increasing McDonald's sales by 30 per cent relatively quickly. Nearly 40 years after his introduction, Ronald McDonald is still widely recognised by children.

McDonald's brand continues evolving, with new healthy options giving a nod to the obesity crisis and the McCafé sub-brand appealing to an increasingly sophisticated adult market.

By 2021, McDonald's was serving 70 million customers a day in more than 40 031 outlets in over 119 countries. It is the world's second-largest private employer with 1.9 million employees.

McDonald's success lies in the brand's ability to respond to consumer criticisms and adapt to the needs of new generations. It has successfully expanded its restaurants worldwide without losing control over quality, achieved via its strict franchise management model.

10.4.2 Louis Vuitton store, Hong Kong, China

Apple is one of the world's most readily identifiable brands. Its products are highly sought after by millions of loyal customers. In its branding, Apple showcases the same qualities in their products that their consumers like to think they have—an Apple buyer sees themself as innovative, imaginative and creative. Apple leaves price out of its branding, focusing instead on product value and its consumer relationships.

Louis Vuitton

Founded in 1854, Louis Vuitton is a leading international fashion house famous for its distinctive high-end products. These range from luggage and leather goods to ready-to-wear clothing and accessories. With over 460 stores in 50 countries, the Paris-based retailer is the world's most valuable luxury brand with revenue exceeding €15 billion in 2020 (see Figure 10.4.2).

The company is a subsidiary of LVMH Moët Hennessy—Louis Vuitton SE, a French-based multinational corporation formed in 1987 under the merger of Louis Vuitton and Moët Hennessy (which itself formed when the champagne producer Moët & Chandon and the cognac producer Hennessy merged).

LVMH controls around 60 subsidiaries. Each manages a small number of prestigious brands, 75 in total. The revenue of the entire group exceeded €64 billion in 2021. Dior, the French fashion label, effectively controls LVMH, with just over 40 per cent of the company's shares, giving it 60 per cent of the company's voting rights.

Louis Vuitton's leather goods are produced in their own workshops in France, Spain and the USA, while footwear and ready-to-wear clothing are produced in France and Italy. This partly explains the high cost of their goods.

The luxury brands controlled by LVMH include:

- fashion and leather goods: Fendi, Berluti, DKNY, Givenchy, Loewe, Kenzo, Fenty, Céline, Thomas Pink, Christian Dior, Marc Jacobs, Rimowa and Louis Vuitton
- perfumes and cosmetics: Dior, Sephora, Acqua di Parma, Guerlain
- wines and spirits: Veuve Clicquot, Moët & Chandon, Dom Pérignon, Krug, Hennessy, Cloudy Bay
- watches and jewellery: TAG Heuer, Hublot, Bulgari, Chaumet, FRED, Zenith.

Activities

Acquiring and processing geographical information

1. Explain the concept of brand image.

Applying and communicating geographical understanding

2. Identify three brands not mentioned in the text that you are familiar with. Investigate the brand images that the products' suppliers seek to promote.
3. Identify the brand image promoted by McDonald's. Outline the success of this branding. How has McDonald's branding changed over time?
4. As a class, discuss McDonald's branding. What makes it appealing to you as a consumer? How might McDonald's branding be improved? Who are McDonald's principal competitors and how does their branding differ from that of McDonald's?
5. Outline why Louis Vuitton is considered to be the world's most valuable luxury brand.
6. Describe how Louis Vuitton differs from other luxury brands.
7. State how many Louis Vuitton-controlled brands you recognise. Investigate how many of these have a retail presence in Sydney's CBD.

UNIT 10.5

Influences on the culture of place: Sport and music

Sport

Sporting competitions take place at a range of scales from local to global, from suburban sporting competitions to global events such as the Olympic Games. The telecommunications revolution enabled access to live sporting events from anywhere in the world.

World sport is increasingly dominated by TNCs, managers and marketers. Tradition, loyalty and community heritage count for less than in the past. Sport has become a product to be sold.

Teams and sportspeople are brands to be promoted. They are bought and sold on the transfer market. They are transformed into human billboards, carrying advertising on their clothing and sporting equipment. Sporting personalities often have several corporate sponsorships and product-endorsement deals.

Media networks keenly seek broadcasting rights for sports. They often attract bids of tens of millions of dollars. The naming rights of stadiums are for sale. Their playing surfaces get decorated with corporate logos for TV coverage. Virtual advertising occurs when the advertiser's message is temporarily projected onto any surface, including spectators. In the Sydney–Hobart yacht race, the yachts become floating billboards, ensuring maximum exposure of sponsors' names and products.

Two important factors are at play:

- **demographic change:** larger disposable incomes, shorter working weeks, earlier retirement, longer life expectancy and healthier lifestyles have boosted the global demand for sport
- **the communications revolution:** ever-more sophisticated TV coverage has made elite sport (and the advertising, sponsorship and marketing that accompany it) available to a much wider audience. For example, an estimated 5 billion people watched some of the 2022 FIFA World Cup, over half the world's population (see Figure 10.5.1).

10.5.1 Argentina's Lionel Messi lifts the trophy after winning the FIFA World Cup 2022 final against France.

The future of sport is closely linked to global media networks. Sports rate highly with the public, so they attract advertising, the principal source of income for media organisations.

Sport merchandising is also another readily identifiable indicator of cultural integration. Like their peers worldwide, Australian teenagers wear clothing and caps with the names, logos and colours of US basketball and football teams.

Sport in Australia

Sport is an important element of Australian culture. For many, it is tied to the image of ourselves and what we want the rest of the world to see in us. Some sports stars achieve celebrity status and earn large salaries and attract lucrative sponsorships. We watch and judge their performance, sharing their triumphs and failures (see Figure 10.5.2). Australia's participation in international sporting events gives many Australians an opportunity to express their patriotism. It is also a global passion, as evidenced by international football, cricket or rugby matches.

In Australia, sport was once considered to be the great social leveller. Achievement resulted from personal effort and ability, not inherited wealth or class. So sport symbolised equality. It evened out differences and raised what was seen as working-class to high culture. This is still evident, particularly with the infiltration of sports from other countries and events such as the Olympic Games, which attracts all classes in society.

10.5.2 Adelaide players celebrate winning the 2022 NAB AFLW Grand Final.

Most big sporting events in Australia now include the name of their main sponsor. The Melbourne Cup is the Lexus Melbourne Cup. Cricket's historic Sheffield Shield has been replaced by Marsh Sheffield Shield, referring to the competition's sponsor, Marsh McLennan, an international insurance broker. There is the KFC Big Bash League Cricket, the Kia Australian Open tennis and Rolex Sydney–Hobart Yacht Race. Nike sponsors The Matildas, as well as Australia's women's, basketball team and cricket teams, among others.

Not only can sporting personalities be hired, bought and sold, clubs are also a product. They can be created to fit a corporate vision or destroyed when they no longer fit a marketing strategy (e.g. rugby league clubs Perth Reds, Adelaide Rams and Hunter Mariners). In rugby league, many older community-based clubs were compelled to merge and, when they failed to comply with the new corporate vision, they were forced from the competition.

Due to cultural integration, there has been a subtle change in the variety of sports played in Australia. Within the past 30 to 40 years certain sports, including baseball and basketball, have taken off. Today, players from the US National Basketball Association (NBA), including Kawhi Leonard of the LA Clippers and LeBron James of the LA Lakers, are recognised worldwide due to slick marketing campaigns.

Drawing on this heightened popularity, Australia created its own Basketball Australia (BA). It has had some success in promoting players to the world stage, such as Ben Simmons (Philadelphia 76ers), Liz Cambage (Las Vegas Aces) and Lauren Jackson (Seattle Storm).

Music

As the world has become more integrated and interdependent, two somewhat contradictory trends in music have become apparent. First, is a sharp increase in the diversity of musical styles listened to globally. Second, is the increasing dominance of Western cultures (especially the USA and UK). This has led to the decline and, sometimes the loss, of many unique musical styles and practices. Other relevant observations are:

- music continues to play an important role in articulating identity and place
- music remains a crucial factor in cultural survival throughout our globalised world
- music, has, over time, developed into a commodity that is bought and sold.

Music is one industry that has been particularly at the mercy of marketing and globalisation. If artists are to succeed in this highly competitive industry, they usually need to be contracted to one of the large music corporations: Warner Music Group, Universal Music Group (which also owns EMI) and Sony Music Entertainment. These three companies have taken over vertical and horizontal control over most aspects of this industry. Warner is the world's biggest recording company, with a 25.1 per cent share of the world market, followed by Universal (24.3%) and Sony (22.1%).

The industry's globalisation means that Australians are mainly exposed to music on streaming platforms and radio stations, which mainly showcase artists from the USA or Britain or Ireland. Australian artists account for less than 40 per cent of the top 100 Australian album sales by volume, an achievement given that recorded music is the most concentrated global media market and Australia's share of global music sales is less than 2 per cent. This success can partially be attributed to radio stations such as Triple J and The Edge 96.ONE, which offer alternative music for the youth market.

Small bands find it difficult to compete and many of the local venues where bands once played can no longer afford to sponsor them. The only alternatives for truly competitive artists are to fund their own music production or compete for gigs as an opening act for internationally recognised performers.

Did you know?

Total revenue for the global recorded music market for 2021 grew by 18.5 per cent to US$25.9 billion.

Streaming revenue grew by 32 per cent to US$25.1 billion in 2021, for the first time accounting for more than half (56.1 per cent) of global recorded music revenue. This growth was driven by a 24.1 per cent increase in paid subscription streaming.

There were 524 million users of paid streaming services at the end of 2021, up from just under 487 million at the beginning of the year. Australia, a top 10 music market, recorded growth of 6 per cent.

● **SPOTLIGHT**

Streaming services transform the global music industry

In 2016, revenue from streaming exceeded that of physical music sales for the first time. By 2020 streaming revenue reached more than 70 per cent of total recorded music revenue (see Figure 10.5.4).

Early in the digital revolution, the music industry was one of the first major victims of digitisation and was decimated by pirating. In the USA, physical music sales (notably records and CDs) collapsed from US$14.6 billion in 1999 to just US$1 billion in 2020. The industry was further disrupted in 2010 when streaming arrived. Streaming companies such as Spotify and Apple Music have given consumers access to a vast collection of current music and an even bigger library of past hits for as little as US$10 per month, or a free version with advertising.

The big music companies act as gatekeepers between artists and streaming services. Six players, the largest being Spotify, dominate services (see Figure 10.5.3). A major advantage for artists is that sources from streaming are readily tolled, audited and monetised, providing them with income. Being digitised, streaming services provide higher profit margins than physical music sales.

The number of music streaming subscribers worldwide is projected to grow to over 1.2 billion over the next decade.

Provider	Users	Subscribers
Spotify	350 million	188 million
Apple Music	–	88 million
Amazon Music	–	74 million
Tencent Music	657 million	43 million
YouTube	>1000 million	80 million
Pandora	61 million	6 million

Source: *Sydney Morning Herald*

10.5.3 Music streaming platforms and the number of users and subscribers 2021

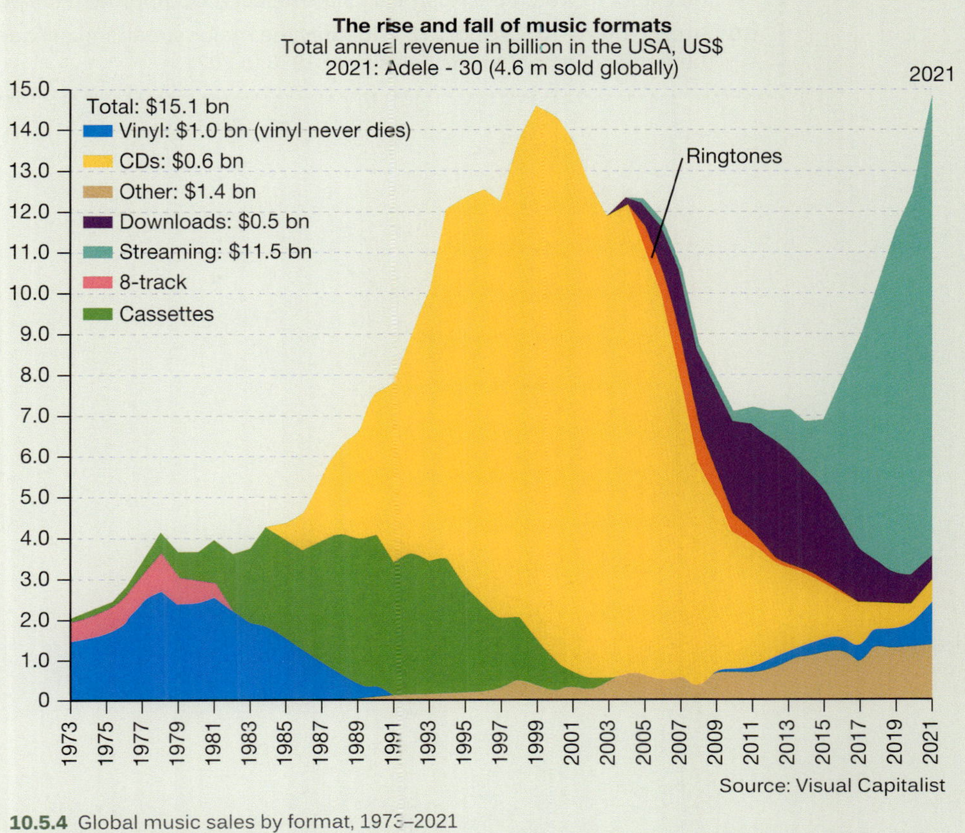

10.5.4 Global music sales by format, 1973–2021

Activities

Acquiring and processing geographical information

1. Explain how the revolution in telecommunications transformed our engagement with sport.
2. Outline how sport has been transformed by the process of cultural integration.
3. Identify and describe the factors that played a role in the globalisation of sport.
4. Identify the somewhat contradictory trends in music. What other trends are relevant?
5. Outline the impact of globalisation on the music that Australians are exposed to.
6. Outline the impact streaming has had on the music industry and musicians.

Applying and communicating geographical understanding

7. Construct a mind map explaining the impact of sport in Australia. Outline the role sport plays in Australia in the following areas:
    - defining culture of place
    - the impact on Australian society over time
    - the extent to which commercial interests have transformed sport
    - the cultural integration of sport
    - why US sports have made little headway in securing commercial TV coverage in Australia.
8. Working in groups, brainstorm and then share with the class the ways in which international sports:
    - have influenced Australia's culture
    - marketing has impacted Australian teenagers.
9. Access the websites of the National Rugby League, Australian Football League and Rugby Australia. Compare these with the internet site of the NBA in the USA. Assess the extent to which the Australian sites reflect a US approach to sports marketing.
10. Study Figures 10.5.3 and 10.5.4. Identify the major streaming services and describe the trends in global music sales from 1973 to 2021.

UNIT 10.6

The impact of change and responses to it

Cultural and economic change has a range of impacts. The focus here is on the loss or endangerment of indigenous and other languages, the homogenisation of landscapes and threats to diversity and **sovereignty**. Significantly, the process of cultural change has not been universally embraced. Over time, resistance to the excesses of integration has increased.

Loss or endangerment of indigenous and other languages

One of the best measures of cultural diversity, among others, is the state of the world's languages. Some are growing and becoming more widespread; others are fading. Many people blame the growing uniformity on technological advances in telecommunications and the global economy's emergence. When languages are lost there is an associated loss of culture, identity and even memory.

Today, just five languages—Mandarin Chinese, English, Russian, Spanish and Hindi—are spoken by 50 per cent of the world's population. This is despite there being 6511 languages. In 2021, the scientific journal *Nature* estimated that 1500 of these could be lost by the end of the century if there is no immediate intervention.

Unfortunately, Australia has one of the world's worst records for language loss. Of over 250 Indigenous languages in Australia, only 123 are still in use. Only about 12 remain relatively strong and being taught to children.

About a third of the world's languages are spoken by less than 1000 people. Linguists fear that many speakers may succumb to pressure to speak a mainstream language. Some people believe the loss of languages is inevitable and not necessarily a matter for regret. Media magnate Rupert Murdoch argues, 'The homogenisation of language is a force for global harmony and economic efficiency'. Reinforcing this view are the Murdoch-owned Indian satellite networks, which only produce programs in English and Hindi, ignoring the Indian subcontinent's language diversity. In China, Mandarin is spread across the country by the electronic media it dominates. It is the only official language, ignoring the linguistic diversity of the country. What this approach fails to recognise is that when we lose a language, we lose a whole way of perceiving the world. See Units 7.6 and 7.7 for more on languages.

Homogenised landscapes

Homogenisation refers to the process by which different places increasingly resemble those found in other areas. This process of homogenisation is apparent in both rural and urban settings.

Homogenisation of rural landscapes

Some rural landscapes retain their unique character, such as the rice terraces of Bali, Indonesia and the Longji and Honghe Hani rice paddy steps or terraces of China—some are over 500 years old (see Figure 10.6.1). Also notable are the vineyard, olive grove and cork tree landscapes of Europe. Many, however, succumb to the influence of the rapidly expanding worldwide markets for commercial agricultural goods and advances in agricultural technologies. Examples include the large-scale cropping and grazing practices used across North America, Australia, Russia, Brazil and Argentina (see Figure 10.6.2). Capital-intensive agriculture has made substantial inroads in more and more countries.

10.6.1 Longji hillside rice terraces, Guangxi, China

10.6.2 Combine harvesters work in a canola field, Saskatchewan, Canada

10.6.3 London's 'Shard' and Tower Bridge

10.6.4 Dubai's helical-shaped Cayan Tower

Some governments have acted to protect existing rural landscapes. Switzerland implemented measures to protect its characteristic rural landscape and traditional settlements. Some, including the Lavaux vineyard terraces, are protected by World Heritage listing. The Philippines has protected its Rice Terraces of the Cordilleras. UNESCO assigns a high value to rural landscapes that showcase historical features, traditional crops and local products, and the presence of architecture related to agricultural activity.

Homogenisation of urban landscapes

There is an increasing similarity between urban places regardless of their geographical location. For example, the CBDs of the world's cities are increasingly dominated by commercial and residential towers that have been largely influenced by the contemporary school of architectural design. These towers are typically clad in a glass curtain with polished stone structural elements. Only the occasional 'look-at-me' building breaks the conformity, such as London's Shard and Walkie-Talkie buildings; Dubai's Burj Khalifa and Cayan Tower; and New York's Jenga Tower (see Figures 10.6.3 and 10.6.4). Retail, commercial and industrial elements of urban landscapes are increasingly homogenised. Retail areas are dominated by global retail brands and one industrial park looks like any other.

Threats to cultural diversity and sovereignty

The traditional Australian hamburger features beetroot. The Big Mac, embraced by Australians, and available worldwide, features pickles, a nod to its American heritage. The Australian pizza, whether it be the meat-lovers or Hawaiian, would be a culinary outrage to Italians. The Big Mac is a cultural adoption; the pizzas, a cultural adaptation. Neither is a threat to cultural diversity and sovereignty. Yet they reveal how much everyday things are influenced by cultural integration.

Advanced technologies provide unprecedented access to other cultures. People can see, hear and experience once inaccessible elements of other cultures. However, misrepresentations, stereotyping and a loss of cultural identity are all possible consequences.

The negative impacts of cultural integration include the threat it poses to cultural diversity, such as:

- the promotion of mass consumer culture by TNCs—think global brands and fast food
- the increased prevalence of global media networks that can overwhelm local cultural influences
- a loss of individualism and group identity when cultural integration promotes the Western ideal of individualism and a homogeneous set of values and beliefs
- the promotion of Western culture and ideologies via computer-based technologies, especially social media
- in the field of education, curriculum and resources geared to the interests of dominant cultures, placing non-dominant cultures at a disadvantage.

Despite this:

- global media networks allow diverse cultures to promote awareness and provide public knowledge and understanding of their stories and identities
- global media networks can, if appropriately regulated, play an important role in preserving culture thus retaining diversity (e.g. local content rules)

- control of their own media outlets (where possible) allows specific cultural groups to exercise control over their own cultural property
- mass media can help to revitalise and restore cultural diversity (e.g. technology can be used to preserve language, customs and culture).

Sovereignty refers to the ability of a government to exert authority over a nation and its people. In democracies, this power is derived from the will of the people via the electoral process. Cultural and economic change can reduce a government's ability to exercise this authority.

Resistance to change

The process of cultural change has not been universally embraced. Over time, resistance to the excesses of integration has increased. For example, divisions have emerged in the developed world's democracies. These are especially apparent between those who have been disadvantaged or alienated by cultural and economic change and those who have benefited from the process. The disadvantaged include people living in manufacturing-dependent communities where unemployment increased sharply as factories closed and labour-intensive industries relocated to the low-cost countries of Asia and Central and South America. Such divisions have fuelled the polarisation and increased partisanship within the political sphere.

Economic nationalists and **populist politicians** have exploited people's fears and insecurities about the nature of economic and cultural change. People opposed to change often resent the growing gap between the rich and poor, and social and cultural changes that clash with their values and or religious beliefs. Some dislike it because it exposes them to various global forces and influences that alienate their children from their cultural heritage.

Another dimension is the rise of religious fundamentalism and extremism. This includes the formation of terrorist organisations. Religious fundamentalists seek to defend key theological and ideological beliefs, and traditional social and cultural practices, against the challenge posed by Western cultural and social trends. Conflict arising from religious fundamentalism is detailed in Chapter 11.

● **SPOTLIGHT**

Right-wing political activism

The developed world has witnessed a rise in the number of right-wing political movements. Examples in Australia include Pauline Hanson's One Nation. Its policies include the rejection of Islam and multiculturalism, and slashing Australia's immigration intake. They claim that climate change is a hoax and they aim to protect Australian manufacturing, Australia's cultural identity and 'traditional' family values.

Support for such views—despite coming from a minority—has encouraged several ultra-nationalist right-wing political groups to emerge, for example, Reclaim Australia and the United Patriots Front. Both are anti-immigration, anti-multiculturalism and anti-Islam (see Figure 10.6.5). Similar right-wing movements have emerged in Germany, France, Austria and the USA. Sometimes such groups have a significant political impact.

In Australia, conservative or right-wing worldviews are actively promoted by groups such as the Conservative Political Action Conference (CPAC) and the Institute of Public Affairs. Alongside them is a small group of right-wing radio shock jocks and conservative television presenters, especially those featured on Sky News's Sky 'after dark' programs. In Sydney, the shock jocks of Radio 2GB are among the most ardent promoters of conservative worldviews.

10.6.5 Reclaim Australia activists stage a protest in Melbourne's Federation Square

The rise of right-wing political groups is often countered by progressive (left-wing) activism. Sometimes competing protests result in violent clashes, one such conflict being the imposition of restrictions on civil liberties during the COVID-19 pandemic.

Activities

Acquiring and processing geographical information

1 Outline the trends in the diversity of the world's languages. What has been the impact on indigenous languages by these trends?

2 Outline the link between cultural and economic integration and the rise of economic nationalists.

3 Explain the link between cultural integration and the rise of religious fundamentalism.

4 Outline the issues that concern those opposed to cultural integration.

Applying and communicating geographical understanding

5 As a class, debate Rupert Murdoch's claim that the loss of languages is inevitable and not necessarily a matter for regret.

6 Write an exposition on the topic, 'The homogenisation of language is a force for global harmony and economic efficiency'.

7 Study the box, Spotlight: Right-wing political activism. Account for the rise of right-wing activism in Australia and elsewhere.

8 Undertake research by investigating contemporary examples of how different cultural groups have sought to resist outside cultural forces. Present your findings as a written report.

UNIT 10.7

Culture of place: Dubai

10.7.1 Dubai's striking urban landscape

Dubai is one of the seven emirates that make up the United Arab Emirates (UAE). The UAE has a population of 9.9 million, of which 2.96 million live in Dubai.

Over the past 50 years, Dubai has emerged as a unique example of cultural integration. It has developed a distinctive culture of place and is a melting pot of diverse cultures. It managed to accommodate often conflicted cultural tendencies in pursuit of its economic and geopolitical ambitions.

Vast amounts of money transformed this small desert emirate into one of the world's largest construction sites, a major centre of commerce and finance, an important transport hub and popular tourist destination. Oil revenues initially accelerated the city's development, but today account for less than 5 per cent of revenue and represent less than 1 per cent of GDP. The economy now relies on trade, tourism, aviation, real estate and financial services.

Founded in the eighteenth century as a fishing and pearl-diving village, Dubai has emerged as a thriving commercial centre. It is famous for its vast shopping malls, lavish resorts and stunning, ultra-modern architecture. It has the world's tallest building, the Burj Khalifa, and the world's largest human-made harbour (see Figure 10.7.1).

Central to Dubai's emergence as a cosmopolitan global city has been its successful integration of immigrant groups. The UAE's indigenous population is relatively small but wealthy. Immigrant expatriates account for about 85 per cent of the population and are central to this transformation. They are typically employed in industries of little interest to Emiratis. There is a relaxed accommodation of cultural differences.

UAE's geography and governance

The UAE is on the Persian Gulf on the north-east end of the Arabian Peninsula (see Figure 10.7.2). It consists of seven emirates (Abu Dhabi, Ajman, Dubai, Fujairah, Ras Al Khaimah, Sharjah and Umm Al Quwain), each governed by a ruler. Together, the rulers form a Federal Supreme Council. One serves as the President of the UAE.

The climate of the region helps to shape the way of life experienced here. Dubai has a hot desert climate with two distinct seasons—summer and winter. Summer is from late April to early October. It is hot—regularly above 38°C—and being coastal the humidity can be uncomfortable outdoors. Rainfall is scarce but the windy conditions create frequent dust storms. Winter is from late October to early April, offering the most pleasant weather, ideal for outdoor activities. Any rainfall occurs in winter. Most of Dubai's water comes from desalination plants.

Dubai's topography is dominated by fine, white sandy deserts and a flat coastline. East of the city are darker reddish sand dunes. Farther east are the rugged and largely uninhabited Hajar Mountains. The desert is the emirate's most popular tourist attraction. Four-wheel drive dune bashing, quad bike adventures and camel safaris are popular.

Demography

A key feature of the city of Dubai's demography is its cultural diversity. Only about 10 per cent of the city's population are UAE nationals. The rest are expatriates. Of these, about 85 per cent are Asian, mainly Indian (51%), Pakistani (16%), Bangladeshi (9%) and Filipino (3%). There are also over 450 000 Westerners (about 5% of the population), with the British forming the largest group (about 100 000 people).

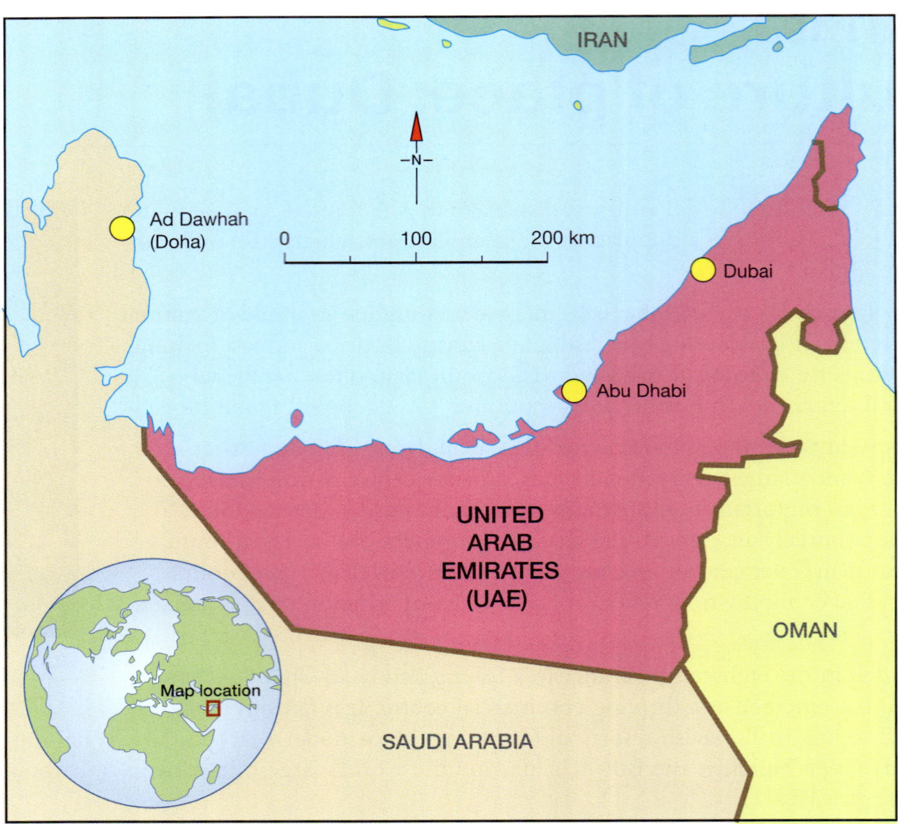

Figure 10.7.2 The United Arab Emirates

As a whole, about 10 per cent of the country's 9.9 million people are UAE nationals (1.98 million). The rest of the population (7.9 million) is made up of expatriates, as outlined in Figures 10.7.3 and 10.7.4.

Males make up 72 per cent of the population and women just 28 per cent. The gender imbalance is due to the large male-dominated construction workforce. Most are expatriate workers from India, Pakistan and Bangladesh (see Figure 10.7.5).

With such cultural diversity, multiculturalism is readily acknowledged and, to a degree, celebrated. The UAE is one of the more liberal Arab nations. Nationals recognise the rights of all ethnicities within an very limited human rights framework. Additionally, UAE nationals and the government readily accept this cultural diversity, enjoying the economic benefits arising from it.

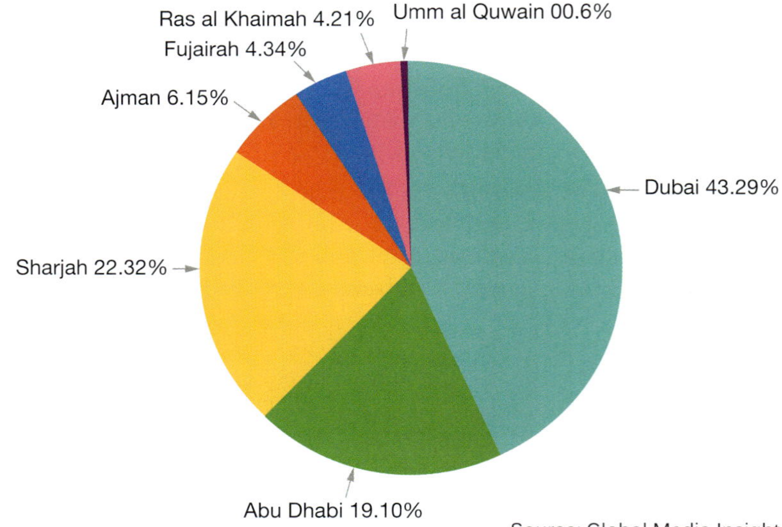

10.7.3 Population distribution of the UAE, 2023

Country of origin	Millions	Percentage
India	2.80	27.49%
Pakistan	1.29	12.69%
Bangladesh	0.75	7.40%
Philippines	0.57	5.56%
Iran	0.48	4.76%
Egypt	0.43	4.23%
Nepal	0.32	3.17%
Sri Lanka	0.32	3.17%
China	0.22	2.11%
All other countries	1.82	17.94%
Total expat population	9.00	88.52%

Source: Global Media Insight

10.7.4 UAE expatriate population, by country of origin, 2023

10.7.5 Immigrant contract labourers from Bangladesh, India and Pakistan work on a Dubai construction site

Cultural diversity

Islam and the Arab way of life are central to Dubai's cultural character and culture of place. The influence of Islamic and Arab culture is evident in the city's architecture and lifestyle—its religious observances, customs, cuisine, clothing and music. While Islam plays a significant role in society's fabric, and there are constraints on freedom of speech and gender-based roles, efforts have been made to build a more socially inclusive and tolerant city for non-Muslims.

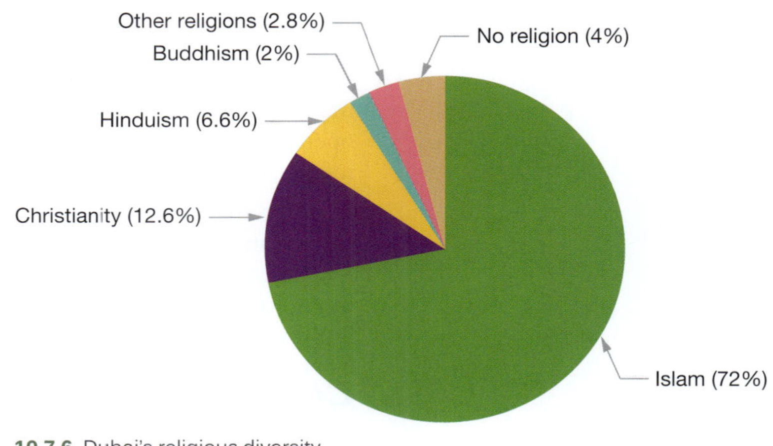

10.7.6 Dubai's religious diversity

There are limits to this tolerance. During the month of Ramadan, it is illegal to publicly eat, drink or smoke between sunrise and sunset. The law applies to Muslims and non-Muslims. The sale and consumption of alcohol is tightly controlled. Adult non-Muslims are permitted to consume alcohol in licensed venues, typically within hotels, or at home if they have an alcohol licence. Places other than hotels, clubs and specially designated areas are typically not permitted to sell alcohol.

Emiratis typically wear traditional clothing. Men wear an ankle-length, loose-fitting garment made of white cotton, known as a *kandora* or *dishdasha*. A *ghutra* covers the head and is held in place by the *agal*, a black cord. Traditionally, women wear an abaya—a long, black, flowing light coat. It goes over their clothing, often Western clothes, or a *jalabeya*, a traditional long-sleeved full-length dress. A *shayla*, a black scarf, covers the head.

For non-Emiratis and visitors to Dubai, all clothing types and cultural preferences are generally acceptable. Swimwear is permissible on beaches, pools and spa areas, and for water-based entertainment. Beyond these, female visitors are expected to avoid revealing too much skin. In places of religious worship, conservative dress is expected. In mosques, women's clothing must cover shoulders, arms and legs plus headscarves. Premarital sex and homosexual acts are punishable under Sharia law.

Religions include significant numbers of Hindus and Buddhists, while Christians form a minority (see Figure 10.7.6). Religious minorities are free to practise their faith. Christian festivals are celebrated—shopping malls and hotels are lavishly decorated at Christmas (see Figure 10.7.7). Residents of all religious faiths eagerly participate in the

10.7.7 Christmas-themed decorations in the Dubai Mall

celebrations. The focus is generally on the commercial aspects of Christmas rather than the religious.

While Arabic cuisine is widely available, Western-style fast food, South Asian and Chinese cuisines are popular.

Dubai's global city status

Dubai is a middle-ranking global city and the product of one of the most remarkable economic transformations in modern history. The WSP Global Inc., one of the world's largest professional services firms, gives Dubai a global city index ranking of 16, just behind Sydney. The index's place-based criteria include measures such as housing, public realm, urban green space, social infrastructure and climate change. Mobility-based criteria include infrastructure and public transport, logistics and freight productivity, and pedestrian and cycling infrastructure. Also considered are urban-based systems, such as power generation systems, water treatment and distribution, waste management and technology-based measures (connectivity and infrastructure, fixed internet speeds and feeds, mobile internet, open data, information and data security, planning and policy). See Unit 8.5 for more information on world cities.

The Economist's Global Liveability Index ranks cities based on stability, health care, culture and entertainment, education and infrastructure. It ranks Dubai 74th out of 140 cities. Such a score means that 'day-to-day living is fine, in general, but some aspects of life may entail problems'. Dubai also featured in the list of *The Economist*'s most-improved cities.

Connectivity

One of the world's most connected cities, Dubai is a major hub for air and sea travel. The International Airport (see Figure 10.7.8) consistently tops the list of the world's best airports for flight connections, facilities and passenger satisfaction. It serves 140 airlines and 270 destinations. It is the world's busiest airport, as measured by international passenger traffic. In 2019, just prior to the COVID-19 pandemic, the airport handled 86.4 million passenger movements and 373 261 flights.

10.7.8 Dubai International Airport is the world's busiest international airport.

The state-owned airline, Emirates (founded in 1985), is one of the world's largest international airlines and a key player in the emergence of Dubai as a global city. Each week Emirates operates more than 2400 flights from its Dubai airport hub to over 152 destinations in 80 countries across six continents. The airline has more than 262 aircraft including Boeing 777s and the world's largest fleet of Airbus A380s. Emirates has 49 000 employees.

Dubai has plans for a second airport at a cost of US$32 billion. It aims to be the world's biggest airport by 2050, capable of handling up to 255 million passengers annually. The initial phase of the project will increase the airport's capacity to 130 million passengers per year, and the total development will cover an area of 56 square kilometres.

State-of-the-art port facilities transformed Dubai into the region's logistics hub. Named Jebel Ali, the port is the world's ninth busiest, and contains its largest human-made harbour. It is the biggest, and by far the busiest, port in the Middle East.

Tourism

Seventeen million tourists visited Dubai in 2019. The principal sources, in order of relative size, are India, Saudi Arabia, the UK, China, Oman, Russia, USA, Germany, Pakistan and the Philippines. Over 700 hotels and resorts accommodate them, with some of the world's most lavish resorts found here. Many are destinations in themselves.

Dubai's tourist attractions include the Burj Khalifa, the Dubai Mall and Fountain (see Figure 10.7.13), Palm Jumeirah and the famous Atlantis Resort (see Figure 10.7.9) and the emirate's desert dunes (see Figure 10.7.10). Theme parks include IMG Worlds of Adventure, Legoland Dubai, Motiongate, Bollywood Parks, indoor ski slopes (see Figure 10.7.11) and waterparks (see Figure 10.7.12).

Dubai's climate is a major attraction for European tourists keen to escape cold winters.

10.7.9 Dubai's famous Atlantis, The Palm resort is built on the reclaimed land of Palm Jumeirah.

10.7.10 The UAE's dunes are a popular tourist attraction.

10.7.11 Ski Dubai's indoor ski area

10.7.12 The Aquaventure waterpark at Atlantis, The Palm

10.7.13 Dubai's famous fountains and the Dubai Mall (left) seen from the top of the Burj Khalifa

Activities

Acquiring and processing geographical information

1. State what is unique about Dubai as an example of cultural integration. What has been used to drive Dubai's transformation into a major, well-connected, destination with a unique culture of place? How has this changed over time?
2. Outline the distinctive features of Dubai's population. How do these features contribute to Dubai's culture of place?
3. Describe Dubai's location. Speculate to whom its location has proved beneficial, in terms of its ambition to be a global hub in the aviation industry.
4. Outline how the climate and topography of Dubai have helped to shape the emirate's culture of place.
5. Outline Dubai's demographic characteristics. How does the demographic diversity of Dubai contribute to its unique culture of place?
6. Examine the extent to which the Arab culture impacts what people wear in Dubai.
7. Outline the extent to which non-Islamic religions are accepted in Dubai.
8. Describe Dubai's status (ranking) as a world city.
9. Outline the connectivity of Dubai with the rest of the world. How does this connectivity reflect the economic ambitions of Dubai's rulers?
10. Outline the role tourism plays in Dubai's emerging culture of place. What are the emirate's major tourist attractions?

Applying and communicating geographical understanding

11. Using Google Earth™, explore Dubai. Locate and view the following landmarks:
    - Dubai International Airport
    - Jebel Ali (port facilities)
    - Burj Khalifa and the Dubai Mall
    - Palm Jumeirah
    - Atlantis, The Palm.
12. Study Figure 10.7.4. Using data from the graph, write a report outlining the diversity of Dubai's expatriate population.
13. As a class, discuss the observation that Dubai is an open and tolerant society.
14. Access the Department of Foreign Affairs and Trade's (DFAT's) Travel Advisory website. What advice does the site give to Australians travelling to Dubai?
15. Determine if any of your classmates have been to Dubai. Ask them to outline their impressions. How does it differ from their way of life in Australia?
16. Write a report outlining Dubai's unique culture of place.

APPLICATION AND CONSOLIDATION TASKS

Task 1: Advertising campaign: Your local area

Reflect on the cultural diversity of the place you live in and its culture of place, then prepare an advertising campaign highlighting its uniqueness.

Consider the signs of diversity evident in the area and to what extent you believe this diversity determines your local area's culture of place.

Brainstorm examples of noise, colour and vitality that contribute to the culture of place there.

Your advertising campaign could be presented as one of the following:

- an annotated visual display
- a short ad for traditional media (TV or radio).

Task 2: Report writing: Sport or media

Select one of the following options to investigate, then prepare a written report.

Option 1: Research which companies sponsor major events in your city or town. Consider events such as sporting events, the arts, cultural events and educational activities. Research how much money each of these companies spends on sponsorship and advertising and how many of them are foreign-owned.

Option 2: Study Figure 10.2.3 and select one of the media and entertainment providers listed. Investigate the nature and extent of their activities.

Your report can include visuals (maps, graphs, diagrams, photos) to use during your presentation.

Task 3: Podcast: Musician or music genre

Select one option below to investigate, prepare and record a podcast. Your podcast should be 4–5 minutes long.

Option A: a song or artist that has been marketed without backing from music labels, or has used the internet to market their music in an innovative way.

Option B: the origins and spread of a music genre that appeals to you.

Your podcast can include audio content, such as music, snippets of interviews etc.

Task 4: Prepare an oral report: Islam in Europe

Investigate the social and political tensions associated with the spread of Islam in Europe.

Include in your report:

- the history of Islam in Europe
- an overview of the social and political tensions

Select one country and investigate specific events.

Your report can include visuals (maps, graphs, diagrams, photos) that you would use during your presentation.

CHAPTER 11 Political power and contested spaces

Political geography, also called geopolitics, examines the relationships between politics and its geographical context. It explores the forces shaping the political processes and outcomes—social, economic and cultural. These forces take place within a particular context, spatial or geographical, and include resources, environments, ideology, ethnicity, class, culture and religion.

The collapse of Communism in Europe and the rise of globalisation at the end of the twentieth century heralded a promising new era, one where political ideas would converge, promising lasting peace. Instead, the twenty-first century brought new conflicts and a growing cultural divergence, including the rise of isolationist politicians like Donald Trump in the USA and movements such as Brexit in the UK. These highlight the growing discontent of some countries with the forces of globalisation.

This chapter examines the nature and character of nation-states, different political systems and ideologies, as well as power blocs. Political conflicts and their impacts and how nation-states respond to conflicts is covered, with particular reference to the dispute over the South China Sea.

Geopolitics is a form of mental mapping that divides the world into friends, strategic allies and mortal enemies.

Professor Klaus Dodds, University of London

11.0.1 Leave campaign supporters celebrate the day Britain officially left the European Union, 31 January 2020.

Chapter glossary

autonomy the right or power to govern oneself

colonialism a government policy of imperial conquest or the founding of colonies to gain new territories for occupation to exploit the resources and people of that territory

community a group of people with a strong shared interest about which they communicate regularly; communities vary in size; examples include a club, a neighbourhood, a town, a political movement or a globally scattered group of scholars

dictatorship a form of government in which democratic politics and decision-making are suppressed and replaced with a self-appointed, usually military, autocratic ruler who governs by personal decree rather than by constitutional means

ethnicity the distinctive combination of racial and cultural attributes of a group that distinguishes it as a separate people

federation a political unification of formerly separate states or colonies, in which the former states retain some sovereignty and surrender some sovereignty to the new federal government; it encompasses a constitutional division of powers between the two levels of government, state and federal

fundamentalism the reduction of a religion or theory to its most simple, basic form and the elevation of that simplest form to be the absolute truth about the world and human behaviour

geopolitics the processes of interaction of politics and geography and the study of politics in its geographical context; geopolitics attempts to explain the impact political tension and conflict have on geography, economy, society, culture and population, globally and on other levels

ideology a set of ideas about how the economy, society, culture and institutions are and/or should be organised and function

nation a group of people with a common and distinctive racial, national, religious, linguistic or cultural heritage

political power the power to govern states and their peoples and thus control the destinies of states and peoples

state a territory that has a well-defined boundary and is autonomously and effectively governed by a sovereign government; also known as a country or nation-state

superpower a country that has the power and capacity to influence events on a global scale

totalitarian the political system in which absolute power is vested in a single party or dictator

UNIT 11.1
Geopolitical characteristics

Political activity is primarily aimed at gaining and using power. The geographical location and resources of a country or region significantly influence its political activity and power dynamics. In turn, **political power** results from the control of land, resources and people, on both local and global scales. The exercise of political power often has a profound influence on the geographical environment.

Nation-states and territories

A state is an autonomously (independently) governed territory also known as a nation-state or country. It has a boundary defined by legal regulations policed by established powers (see Figure 11.1.1). Each nation-state has a government authorised to create and enforce laws. The government also has the power to regulate the movement of people within and across its borders, as well as to control transactions involving money and goods.

States recognise each other's existence and engage in official interactions via international relations. There are nearly 200 states or countries with equal **autonomy** and legal power to govern their territories. Most are members of the United Nations (UN), with South Sudan admitted in 2011 as the 193rd and newest member state. The relative power of states in the geopolitical system remains uneven, which is a major cause of tension.

11.1.1 The world is divided into nation-states separated by borders.

● SPOTLIGHT

Nations and nation-states

The terms **nation** and nation-state are often used interchangeably, but they represent different concepts. Nations, or nationalities, are defined more by shared ethnic features, such as language, race, religion and culture. Some states contain multiple nations. For example, Canada has three distinct nations within one **state**. The First Nations peoples are its indigenous population. A significant French-speaking minority is centred in Quebec province. The third nation of English-speaking, mostly British-heritage people, makes up the bulk of the population.

For centuries, people of various nations have desired to break away from large nation-states to form independent nation-states. This has often escalated into conflict, sometimes violent clashes There are many examples. The Kurdish people are a nation with no nation-state whose traditional lands extend across parts of modern-day Türkiye, Syria, Iraq and Iran. Following regional unrest, particularly in Syria, Kurdish forces gained a strong foothold in the region, but their progress was halted by a Turkish invasion. In Spain, Catalonia and its capital Barcelona have long sought independence from Spain (see Figure 11.1.2). The people speak a unique language and practise customs and traditions quite different from the Spanish. Independence was overwhelmingly supported by votes in 2014 and 2017, but the Spanish central government deemed them invalid. Many feared Catalonian independence would incite other Spanish regions to seek independence.

11.1.2 The Spanish region of Catalonia has long sought independence.

Political systems and ideologies

A political system is the set of institutions that constitute the government. As a constitutional monarchy, Australia's set of political institutions differs from the USA, which is a republic. Australia's head of state is the monarch, a hereditary position represented by the Governor-General. In the USA, the President is the head of state, an elected position.

Nation-states generally fit into one of two broad political system categories: unitary states and federated states. Unitary states have a central government that controls most government functions. Most nation-states use this type of system. Federated states share power between different levels of government. Australia is a federation, with power divided between the federal government and state and territory governments. Other examples of federations include the USA, Canada, Switzerland and Malaysia.

Each nation-state's structure may differ, but there are five main types: democracy, republic, monarchy, communist and **dictatorship** (see Figure 11.1.3). Many countries have features of more than one type of system. For example, many democratic nations are republics, while others are constitutional monarchies.

Ideology is a system of ideas and beliefs about how the economy, society, culture and institutions should function and/or be organised. Ideology is crucial in determining a nation-state's political system. Ideologues, strong advocates for an ideology, often have a utopian or idealistic worldview, believing their ideology would make the world a better place. Political systems tend to be pragmatic and societies combine different ideologies and organisational systems. Nevertheless, some governments rigidly maintain an ideology and pressure their citizens to conform to this view.

Political system	Main features
Democracy	Allows individual citizens to participate in the political process. Democratic countries are representative democracies—their people elect representatives, who make laws on their behalf.
Republic	Government is subject to the will of the people and can be changed by them. Usually, the head of state is not a monarch.
Monarchy	Hereditary leaders, born into their roles. Absolute monarchs have unrestrained power (e.g. Saudi Arabia, Oman, Brunei, Qatar and Swaziland). Most are constitutional monarchies, with powers limited by the constitution and parliament as the main source of power (e.g. the European monarchies, Japan, Australia, Nepal and Jordan).
Communism	These constitute only a few countries, notably China, also Vietnam, Cuba, Laos and North Korea. One party, the Communist Party, is allowed to exist. Strong central power controls most aspects of society. Some communist countries allow private businesses, but they are subject to close control, especially in China.
Dictatorship	Power held by an individual or a small group with loyalty from the military providing considerable power. Leaders are not accountable through fair elections, constitutions or parliaments. They control all aspects of life. Private businesses do exist, but are tightly controlled to benefit the leaders.

11.1.3 Main political systems

During the late nineteenth and into the twentieth centuries, capitalism, communism and fascism emerged as the world's most influential ideologies (see Figure 11.1.4). The conflict between them drove much of the tension and warfare of the last century. These ideologies and the geopolitical landscape they created continue to shape tensions around the world.

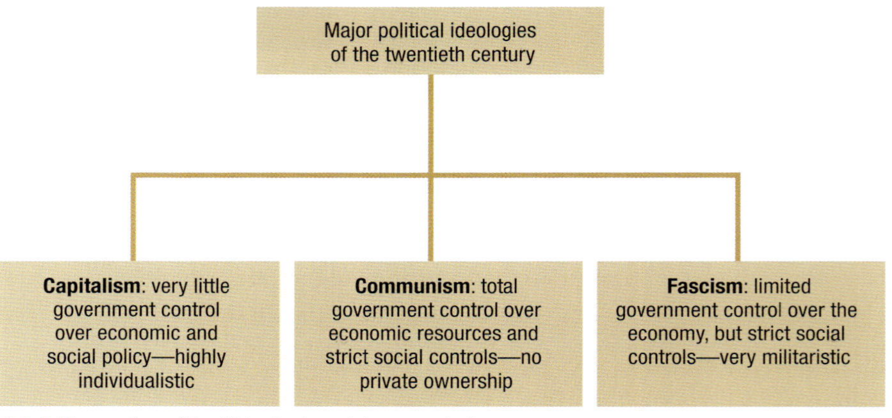

11.1.4 The main political ideologies of the twentieth century

Capitalism

Capitalist ideology advocates private ownership of all resources and productive activity, with the government playing a minimal role in the economy. The allocation of wealth, income and products within the economy should occur in a free market where purchasing power determines an individual's entitlements. Emphasis is on the individual, with a call for 'small government', one with limited impact on people's lives.

Communism

Communism requires all resources and productive activity to be collectively owned and controlled by the state to benefit the people. Private ownership of businesses is abolished and wealth and income are distributed according to need rather than purchasing power.

Communism emerged in Russia in the early twentieth century when the Bolshevik Party seized control and established the Soviet Union (USSR). After World War II it spread to dominate parts of Asia (including China), Africa and Eastern Europe (see Figure 11.1.5). Communism became focused on individual leaders, resulting in Stalinism, named after Joseph Stalin, who ruled the Soviet Union as a dictator. In North Korea, Kim Jong-un rules the purportedly communist nation as a dictator.

Fascism

Fascism combines an extreme form of nationalism with tight government control on social issues to create social harmony. It originated in Italy in the 1920s under Benito Mussolini. Fascism advocates a powerful militaristic state where individual rights are subordinate to the good of society. The media and social organisations are tightly controlled and any form of dissent is outlawed.

Germany's Nazi Party adopted the most extreme version. They advocated extreme racist social policies that led to the Holocaust. They exterminated what they called undesirables, such as Jews, homosexuals, communists, Romani and people with mental illnesses or disabilities. Over six million people of Jewish descent and a further one million other supposedly less-superior people were killed.

After World War II, fascism as a political system generally faded away, although it persisted in some South American nations and Spain until the 1980s. In the 2000s right-wing extremist parties with elements of fascism have risen, particularly white supremacist groups in Western-style democracies (see Figure 11.1.6). They are often fuelled by anti-immigration views that distort traditional values. These groups use the internet and social media to spread their views. This can lead to violence, as committed in the 2019 mosque attack in Christchurch, New Zealand, where 51 worshippers were killed and 49 more injured by a lone Australian gunman. He had been influenced by anti-migration propaganda on white supremacist websites. In 2020, the Director-General of Security at the Australian Security Intelligence Organisation (ASIO), Mike Burgess, warned that these groups pose a threat to Australian security and may engage in terrorist activities to achieve their aims.

Extreme nationalism

Nationalism holds the collective will of the people and state as outweighing that of the individual. Extreme nationalism, seeing one's nation as far superior to others, breeds extreme racism, conflicts, intense disunity and even civil wars. Yugoslavia was a country made up of several distinct ethnic groups with diverse languages, cultures and religions. Nationalism led to the conflict between the groups resulting in a bitter civil war from 1991 to 2001 (see Figure 11.1.7). Over 100 000 people died.

Extreme religious ideology

Extreme and radical religious ideology has become a major issue in the twenty-first century, surpassing political ideology as a cause of tension. However, it extends religious beliefs to radical extremes, encompassing ideas on how societies should be structured politically, socially and economically.

Many believe in error that such radical religious ideology is only an issue of the Islamic faith. In fact, it is present in virtually all religions. In Northern Ireland, differences between Catholic and Protestant Christians fuelled conflict. **Fundamentalist** Hindus committed terrorist acts against Christian and Muslim minorities in India. In the USA, fundamentalist Christian groups have attacked gay people, abortion clinics and people of other faiths. Radical Jewish groups have initiated violence against Muslims in Israel and the Middle East.

11.1.5 Mao Zedong was Communist China's founder and remains a revered figure throughout the country.

11.1.6 Anti-government protest by Greece's Golden Dawn, a far-right, neo-fascist party

11.1.7 Extreme nationalism led to a bitter and destructive civil war in the former Yugoslavia.

Power blocs

Power blocs are groups of nations with a common cause, often formed for security reasons. Alliances between countries are common for wartime protection. The North Atlantic Treaty Organization (NATO) is a powerful military alliance between 32 countries including the USA, Canada, Türkiye, the UK and many other European nations. Membership increases each member state's collective military power.

Power-based alliances are increasingly being transformed into trading blocs, where countries agree to reduce trade barriers between member states. Some countries belong to multiple blocs. Bilateral trade agreements between specific countries are common. Australia is a member of several trading blocs and economic alliances.

● **SPOTLIGHT**

The European Union

The European Economic Community (EEC) was formed in 1958 to promote trade. In 1993, it was renamed the European Union (EU) and its aims broadened to include promoting peace and greater cultural understanding across member states. With 27 members, the EU is the world's largest and most integrated trading bloc. Member states share a common customs and immigration zone, and many EU laws apply across the Union.

A European Parliament and court system coexist with local political and legal institutions. The EU negotiates trade agreements with non-member states, benefiting smaller EU members with greater negotiating power. Nineteen member states use the Euro as their currency.

The EU has faced issues, particularly when Britain, an EU member since 1961, left in January 2020 (see Figure 11.1.8). Its departure, known as Brexit, was hotly debated in Britain causing great political upheaval. The pro-Brexit voters won in a narrow majority, despite many attempts to prevent it. They argued EU membership meant Britain had lost too much control to the EU Parliament and its bureaucracy.

11.1.8 Pro- and anti-Brexit demonstrators in London in 2019

Activities

Acquiring and processing geographical information

1 Outline how political power is achieved.
2 Define the term nation-state.
3 Differentiate between a unitary and federated state.
4 Explain the term ideology.
5 Describe the key features of a power bloc.
6 Explain the advantages countries acquire by being part of a power bloc.

Applying and communicating geographical understanding

7 Write a paragraph on the distinction between a nation and a nation-state.
8 Refer to Figure 11.1.3. Using the internet and other research, identify an example of a country for each type of political system. Give a short explanation of why the country fits the type of political system.
9 Complete this table summarising the key features of different political ideologies.

Ideology	Main features
Capitalism	
Communism	
Fascism	
Extreme nationalism	
Extreme religious ideology	

Western influence

The West's economic, cultural and political influence remains a major source of tension. Many fundamentalist Islamists view Western culture as corrupt and contrary to their religious teachings, believing it must be destroyed. A long-standing religion-based conflict over the status of Israel compounds this. Many in the Middle East oppose the predominantly Jewish State, contesting its ownership of land inhabited by Islamic Palestinians. Support for Israel in the West is strong, particularly in the USA, Israel's key ally.

Fundamentalist religion is blamed for the extreme forms of terrorism that emerged in the early twenty-first century. The attacks on New York's World Trade Center (see Figure 11.1.10) and Washington's Pentagon building in 2001 marked a new wave of terrorism that persists to this day. The USA and its allies, including Australia, responded by invading Afghanistan and later Iraq, leading to further conflict. Terrorist attacks on Western targets include Bali in 2002 and 2004, four Madrid trains in 2004, and the 2005 London bombings. More recently, there have been attacks across Europe, especially France, Spain, the UK and Belgium, prompting massive security operations (see Figure 11.1.9).

11.1.9 French soldiers patrol Paris landmarks in 2016 following terrorist attacks.

■ CASE STUDY

Religious fundamentalism

At the start of the twenty-first century, al-Qaeda and its leader Osama bin Laden were the focus of attention regarding radical Islamist terrorism. This group perpetrated the World Trade Center attacks (see Figure 11.1.10). Another group, originally associated with al-Qaeda was to emerge by 2014 as a much greater force—ISIS.

11.1.10 World Trade Center attack, New York City, September 2001

ISIS

ISIS, also known as the Islamic State (IS) or Daesh, sought to create its own country based on strict Islamic laws taken from the eighth century which included brutal punishments, public executions and severe repression. They took advantage of the chaos of Iraq after the US invasion in 2003 and the civil war in Syria between a disparate mix of anti-regime forces and the ruling dictator Bashar al-Assad. By 2016 their forces were beaten back and their influence waned. They still operate through affiliated groups engaging in terrorist activities, particularly in less stable regions in eastern Africa and the Middle East.

11.1.11 Trump supporters at an anti-abortion rally in San Francisco, 2020

Christian nationalism

Christian nationalism emerged in the USA as a movement that believes the nation is defined by Christian values. They believe its Christian identity is under threat and must be preserved at all costs. Since European settlement, Christianity has played a key role in US society.

As politics became more fractured, radicalisation at both ends of the political spectrum led to this movement growing to prominence after the 2020 US presidential elections. These were marred in controversy after the incumbent Donald Trump claimed the election was rigged. His main supporters were right-leaning Christian groups and the controversies around the election further radicalised some elements within this group.

There is no single Christian nationalist group, but rather shared views and objectives. They restrict what they believe to be unchristian activities, such as same-sex relationships, transgender rights and even contraception. In mid-2022 the US Supreme Court overturned a long-standing decision on access to abortion, seen by many as an example of the Christian nationalist ideology impacting American life (see Figure 11.1.11). Some US states banned abortion, even in cases of rape, incest or where the mother's health is at risk.

Did you know?

The Cost of War project hosted by Brown University in the USA estimates that US spending on the War on Terror (2001–2023) was US$8 trillion and 929 000 people died due to violence associated with the wars in Afghanistan, Pakistan, Yemen, Iraq and others. More than 20 million people were displaced.

Activities

Acquiring and processing geographical information

1. Explain the role of the West in the rise of religious fundamentalism.
2. Outline the aims and objectives of ISIS.
3. Describe how the conflict in Iraq and Syria assisted in the rise of ISIS.

Applying and communicating geographical understanding

4. Briefly outline the nature and objectives of Christian nationalism.

UNIT 11.2

Influences on political tensions and conflict

Conflicting ideologies

Political systems are based on different ideologies. Conflict sometimes arises when these ideologies clash, often leading to internal conflicts, such as civil wars. One of the most famous is the US Civil War in the mid-1800s. Fundamentally it was a clash of ideologies, including opposing views on slavery. More recent conflicts include the Korean War in the early 1950s and the Vietnam War in the 1960s and 70s, which involved clashes between communist and pro-Western democratic views.

State sovereignty

Sovereignty means to have autonomy or independent power to achieve desired outcomes, such as the power to make and enforce laws. In a geopolitical context, the term 'state' can be used interchangeably with the term 'country'. A state has a defined territory, a government and authority to enter relations with other states. Most importantly, other countries must recognise the existence of a country.

A significant consequence of statehood is sovereignty, the right of the country to be autonomous and make its own decisions. Sovereign states have the power to control their people and territory through domestic laws and policies. They also have the right to engage with other nations in international laws and organisations.

SPOTLIGHT

Somaliland

Some regions are not nation-states because they do not have sovereignty. Somaliland is one, occupying a slice of Somali territory on the southern shores of the Gulf of Aden. A peaceful region, Somaliland is not a country, though it calls itself one. To the rest of the world, Somaliland is a self-governing autonomous region within Somalia.

Despite having a functioning government, police force and military, it does not have sovereignty and is not recognised as a sovereign state by other countries (see Figure 11.2.1). Its independence requires recognition from its neighbours and most are members of the African Union (AU). A large regional inter-governmental organisation, the AU has frequently stated that Somaliland's recognition can only occur with the consent of Somalia. A bitter civil war has raged across Somalia for more than 25 years. Its extremely fragile central government fears recognition of Somaliland would create further division in the country with other regions pushing for independence, including Puntland, Jubaland and Hiranland. As a result, Somalia is exerting its right to sovereignty by not allowing Somaliland to break away to form its own nation-state.

11.2.1 Somaliland has its own currency, but its sovereignty is not recognised.

● **SPOTLIGHT**

Russia's invasion of Ukraine

Ukraine is a large country in Eastern Europe that shares a long border with Russia. Following the break-up of the Soviet Union, Ukraine regained its independence, but maintained close economic, cultural and religious ties with Russia. In the post-Soviet era and into the early twenty-first century, many Eastern European countries once closely aligned to Russia drifted towards the West. This shift saw most Eastern European powers join the NATO military alliance, which Russia views as its main rival and threat. Ukraine remained neutral throughout this time and maintained a more pro-Russian view. However, this began to change in 2008 with proposed moves for Ukraine to join NATO. This alarmed Russia as Ukraine was a buffer between major NATO nations and Russia itself (see Figure 11.2.2). Internal politics in Ukraine saw this move put on hold. Tension grew between those who wanted to see Ukraine move into the EU and NATO, those who wanted Ukraine to remain unaligned, and those who wanted Ukraine to be more aligned with Russia.

11.2.2 NATO members as of mid-2023

The eastern regions of Ukraine, those closest to the Russian border, have traditionally had a large population of Russian speakers and tend to be more pro-Russian. In 2014, a separatist movement erupted in the Donbas and Luhansk regions, with armed conflict breaking out and some territory falling to separatists. Although never officially acknowledged, these separatist forces were supported by Russia. At the same time, Russia supported separatist movements in the Crimean Peninsula in the south, which was subsequently annexed.

11.2.3 Russia bombed Kyiv, Ukraine, in early 2022.

In February 2022, Russia invaded Ukraine, calling it a 'special military operation'. Russian President Vladimir Putin claimed the attack was to protect Ukraine's Russian speakers from the neo-Nazi fascist government that was engaging in genocide against them. In reality, the attack was about regime change to ensure that a pro-Russian and anti-West government was installed.

Russian troops crossed into Ukraine from the Russian ally nation of Belarus, attacking the Ukrainian capital Kyiv from the air (see Figure 11.2.3). Tanks and troops raced towards the city in a bid to capture the government and end the war within days. The invasion was very poorly planned and Western nations responded with extensive sanctions and military aid, including sophisticated

11.2.4 A destroyed Russian T-72 tank in eastern Ukraine

anti-tank missiles (see Figure 11.2.4). These had a devastating impact on the poorly trained Russian troops.

As of mid-2023, the war continued with both sides fighting brutal battles often ending in stalemate. Despite Russia's supposedly superior military resources, Ukraine held firm, launching a significant counter attack in June 2023, with material support from the West, and repelled attack after attack. Total casualties are unknown, but according to some sources exceed 200 000 deaths, including thousands of civilians.

Russia's goal of minimising NATO expansion failed. Finland, which shares a 1340-kilometre-long border with Russia, joined NATO in April 2023 after decades of neutrality, citing concerns of Russian aggression. Sweden, another formerly neutral country near Russia, was accepted into NATO in July 2023.

In September 2020, Ukraine formally applied to join NATO. All member countries agreed Ukraine should join eventually, but NATO has not given Ukraine a timeline or a formal path to membership.

Sovereignty and climate change

State sovereignty can hinder international relations and allow countries to act in their own interests. The USA is the second biggest producer of greenhouse gases, and the 2016 Paris Climate Agreement was signed by then-President, Barack Obama, but it was not ratified by the US Congress. Unapproved by Congress, it was not made law within the US. In June 2017, newly elected President, Donald Trump, announced the US would withdraw from the agreement, citing sovereignty as a key reason. He argued withdrawing would reassert the USA's sovereignty and prevent it from weakening. All international agreements like this one rely on nation-states to act together towards a common global goal.

Regime change

Regime change has been a common method of exerting influence over other countries throughout history. It often involves countries seeking to oust the government of another country and install a new one favourable to their views. It is a common source of conflict. Newly installed governments are often referred to as puppet governments or regimes, as they are effectively controlled by another power. In the colonial era, it was common for European nations to replace local leaders with those more favourable to the colonists.

Changes in international power structures

After World War II, the USA and Soviet Union were the world's dominant powers. The collapse of the Soviet Union in the late twentieth century left the USA as the only **superpower**. The twenty-first century has seen new powers emerge seeking to exert influence and control, particularly China.

With the world's second largest population and second-largest economy, China gradually extended its influence throughout the world through soft diplomacy. It uses widespread aid and development programs to exert influence, a tactic used by powerful nations for centuries. China has also massively increased the size of its military power in recent decades and has been more willing to use it to exert influence. This is seen in the spread of Chinese militarism in the South China Sea.

Global media

The media plays a crucial role in holding governments and leaders accountable, informing the public about their actions, both positive and negative, through media reports. This influences public perception, making the media and its controllers highly influential.

When the COVID-19 pandemic began in 2020, conservative Fox News was criticised by many for downplaying the disease's severity, aligning with the policies and views of then-President Donald Trump. It pressured governments to take no social distancing actions. Some presenters claimed the pandemic was a hoax. This was in direct contrast to health authorities, which were reporting very serious concerns about the spread of the disease.

Social media's growth saw the emergence of citizen journalists, amateurs who report through blogs, social media, video channels and websites not bound by the same regulations as mainstream media. Their reporting can misrepresent the truth and they are more easily influenced by political agendas and sponsorship. Some argue the growth of citizen journalism helps break the control of media moguls. Others express concern that the lack of control and accountability makes news reporting less reliable. It is susceptible to 'fake news', where political opinions and outright lies are dressed up as real news. Technology is also playing a role in the creation and dissemination of fake news, including what is termed 'deepfake' content, making it harder to detect fake or incorrect content.

Technological change

The rapid growth of internet-based businesses in the twenty-first century transformed society. It has changed the way individuals interact and lets them easily purchase goods and services from other countries. This has disrupted usual government regulations and controls.

Social media emerged as a powerful tool to overcome these controls, as seen in the Arab Spring. A series of protest movements across North Africa began in the early 2010s, using social media to organise protests and communicate with news agencies. They spread their messages globally to disrupt government controls, leading to the overthrow of **totalitarian** regimes, notably the fall and assassination of Libya's Muammar al-Gaddafi. Similarly, in 2019, pro-democracy protesters in Hong Kong used social media to spread their messages, garnering worldwide support, and to anonymously organise protests, making it hard for authorities to identify the leaders (see Figure 11.2.5). A semi-autonomous region of China, Hong Kong has far fewer internet restrictions because until 1997 it was a British colony. As a major centre for global finance and trade it attracted a large foreign population.

11.2.5 Pro-democracy protesters in Hong Kong, June 2019

11.2.6 In 2020, Dr Bruce Aylward led the joint mission between WHO and China in response to the COVID-19 pandemic.

Intergovernmental organisations

Large-scale global organisations, such as the League of Nations (established after World War II, the forerunner to the UN) have fuelled the notion of a world government. With increasing globalisation, international organisations have played more important roles in governance. In the last half of the twentieth century, many international organisations were created, notably the United Nations, but also the United Nations Educational, Scientific and Cultural Organization (UNESCO), the World Health Organization (WHO), the World Bank and many others. Global legal systems were created to deal with disputes between nation-states, such as the International Court of Justice, and the International Criminal Court was established to deal with international crimes, such as genocide.

These organisations and their international laws and agreements have become powerful forces in controlling government actions. Their focus is on international relations and peaceful dispute resolution. For example, the International Atomic Energy Agency, regulates nuclear technology worldwide and plays an integral role in limiting countries' opportunities to build nuclear weapons.

These international bodies also help coordinate global responses to major events and issues. In 2020, WHO was a major player in coordinating the response to the COVID-19 pandemic on a global scale, providing information and guidance to countries on the best ways to respond (see Figure 11.2.6).

Australian membership

Australia is a party to around 1000 international treaties and agreements. Some of these are between Australia and just one other country, called bilateral treaties. Multilateral treaties are between Australia and multiple other countries. By joining these voluntary agreements, Australia gains the benefits that flow from their membership. It also has to abide by their stated obligations. In this way, these organisations can limit the power of the nation-state, but they also provide many benefits, the reason any country agrees to participate.

The Australian Department of Foreign Affairs and Trade (DFAT) explains on its website the benefits flowing to Australia by participating in international agreements and treaties:

> Australia participates in treaty-making because it is in the national interest to do so. If the projection of military and economic power were the main means by which national objectives were pursued, Australia would be vulnerable. Our geographic isolation and small population would be seen as a weakness. Nations, particularly states with a relatively small population such as Australia, benefit from a world where interaction between countries takes place within a framework based on fair, agreed and transparent rules as agreed in treaties. Australia is not a member of any single rigid regional grouping; rather, we build global or regional alliances and through them, seek to influence the standards by which international relations are conducted.

Impact on sovereignty

The expanding size and authority of international institutions, such as the EU and UN, raises significant questions for the governments of their member states:

- Should states participate in the building of regional and international alliances?
- Does integration into regional and international organisations mean countries will increasingly lose control of their affairs?
- Can countries afford not to participate in international organisations in an era when the power of the state is giving way to the power of multinational arrangements and structures?

Conflicting points of view exist. One is that institutions such as the International Monetary Fund (IMF), World Bank and World Trade Organization have too much influence over the decision-making processes of governments worldwide and these organisations reinforce the highly uneven global distribution of political and economic power. The alternative view suggests there is ample proof that membership provides a range of benefits for member states and being left out can have serious adverse consequences.

Britain's exit from the EU on 31 January 2020 is a case in point. It created many issues for Britain. As a non-EU member, it is no longer able to use EU-negotiated agreements, so it needs to make trade arrangements directly with other countries. One viewpoint sees long-term negative and significant economic consequences for Britain because it must negotiate on its own and not as part of the more powerful EU group. Another view points out the decision will enable Britain to have a more independent voice in matters such as migration and foreign policy.

Non-governmental organisations and pressure groups

Non-governmental organisations (NGOs) are typically non-profit and aim to exert influence on nation-states around specific issues. They rely on individuals for support and membership. Governments cannot become members, although they often support their work. Key areas in which they operate include:

- environmental protection and ecological sustainability (Greenpeace, the World Wide Fund for Nature (WWF) (see Figure 11.2.7), and the Australian Conservation Foundation
- social justice and human rights (Human Rights Watch, the International Committee of the Red Cross (ICRC) and Red Crescent, and Amnesty International)
- development and disaster relief (World Vision, Oxfam and Save the Children).

Strategies used by these organisations range from letter-writing campaigns to using their membership base to engage in protests, boycotts and other activities. NGOs use social media to spread information and gather support.

11.2.7 WWF anti-poaching campaign

Due to their diverse views, NGOs face limited ability to lobby governments. Some are highly critical of foreign aid agencies, while others are highly dependent on them for their financial survival. Nevertheless, NGOs continually find new ways to achieve their aims and convey their messages. Some have recognised the potential of trade measures to influence environmental practices.

Some NGOs derive strength from having a formal role in multilateral bodies, such as the UN, the Organization for Security and Co-operation in Europe, and the EU. Article 71 of the UN Charter says that the United Nations Economic and Social Council 'may make suitable arrangements for consultation with [NGOs]'. NGOs also make it possible for contact and information exchanges between conflicting parties. These may take place across borders without involving governments. As an accepted part of international relations, by influencing national and multilateral policy-making, NGOs are becoming increasingly important.

Dispossession of land

Land is fundamental to human welfare. Throughout history, political tension and conflict have arisen from its control and use. Access to land has always been essential to human survival and over thousands of years agricultural societies rose and fell based on their ability to control large territories. Territory ownership enabled agricultural production and any surpluses not required for the family would feed urban populations and large armies. Territory also provided the means for large numbers of people to work the land as slaves or peasants, pay taxes and be conscripted into armies. These traditional reasons for territory control continue to cause political tension and conflict in agricultural societies.

With the rise of industrial societies, land took on other attributes besides agriculture. It became the source of vital raw materials and energy, such as wood for fuel and construction, minerals for industry and fossil fuels such as coal and oil for energy.

Resource ownership and control

Secure access to raw materials for industry and energy has become even more critical. Those who control strategic raw materials like gas and oil hold significant geopolitical power. Modern industries require expanded markets for their cheap products, so controlling the territories where consumers live can help exclude their competitors. Early industrialisation by Western European countries in the nineteenth century led their governments to intensify efforts to control the world's territories. Tension and conflict arose from European societies' need for food, industrial raw materials and markets. This pushed them to expand their control across the planet.

In modern **geopolitics**, access to secure food supplies remains a vital consideration. Despite their small territories, Hong Kong and Singapore have prospered through industrialisation. Their educated populations are rich in human resources and the communications revolution contributed to their success. However, if their external transport links were cut, they would be vulnerable as most of their food, water and electricity is imported. Similarly, access to land and water is fundamental for Israel, which has fought a series of wars to secure territory to build a state, grow food and house its rapidly growing population.

● SPOTLIGHT

Conflict over water

Earth's natural resources include minerals, soil, water, plants and animals. Water is increasingly a source of tension and potential conflict among countries that share water resources. There can be competing demands on the water from rivers that flow through multiple countries.

Tensions are growing between China and India over the 3000-kilometre-long Brahmaputra River. The river originates high in the Himalayan Mountains in Tibet, under China's control. The Brahmaputra River drains a catchment of almost 600 000 square kilometres across China, Bhutan, India and Bangladesh (see Figure 11.2.8). Both India and China have engaged in dam building to provide hydro-electric power for their massive populations and growing industries. Both countries also plan to divert water from the river for irrigation and drinking water. India has over 120 proposals for dams along the river and its tributaries. Control of the water in the river remains a point of dispute between China and India, with the less-powerful nations Bhutan and Bangladesh sidelined from any discussions.

11.2.8 Water resources of the Brahmaputra River catchment area

Migration

Migration remains a cause of tension when there are differences in **ethnicity** or culture. It can be a major cause for discrimination, distrust and conflict. Immigration-induced tension can arise from the non-traditional intermingling of people.

This can occur when new immigrants move into a territory traditionally occupied by one race, such as West Indian immigrants arriving in Britain, and Algerian immigrants in France. It can also exist within settler societies, such as Australia, the USA and Canada. In Australia, many ethnic groups have faced discrimination and harassment on arrival.

Since the second half of the twentieth century, mass migrations occurred with people moving from war-torn nations to settle in new countries. From 2014 onwards, conflict in North Africa and the Middle East, especially Syria, led to large numbers of refugees fleeing towards Europe.

11.2.9 An anti-migration rally in Finland, with pro-migration protesters in the background

Figure 11.2.10 A picnic to protest anti-migrant policies at the Mexico-US border wall

Arriving in the south, they headed northwards to countries such as Germany. Their influx led to increasing anti-migrant sentiments within many European nations (see Figure 11.2.9). From this arose anti-migrant movements and in some extreme cases the re-emergence of right-wing, neo-Nazi parties in countries including Austria.

In other parts of the world, conservative politicians have used anti-migration sentiments to bolster their own political support. Former US President, Donald Trump, used anti-migrant rhetoric to win support from voters. Most notably, Trump spoke out against migrants from Mexico and Central and South American nations. He famously said he would build an impenetrable border wall between Mexico and the USA and make the Mexican government pay for it. Neither occurred, but it did garner him both considerable support and great criticism (see Figure 11.2.10).

Activities

Acquiring and processing geographical information

1. Explain the term state sovereignty.
2. Explain how state sovereignty can hinder global action on important issues.
3. Describe the role of the media in geopolitics.
4. What is citizen journalism? What has its impact been on politics?
5. What is an inter-governmental organisation?
6. Describe how inter-governmental organisations can help to limit geopolitical tensions.
7. Explain why Australia is a strong advocate for inter-governmental organisations.
8. Describe the features of an NGO.
9. Explain how NGOs gain power and influence.
10. Describe the influence of migration on geopolitics.

Applying and communicating geographical understanding

11. What features does Somaliland have that make it appear as a nation-state?
12. Outline why Somaliland is not considered a nation-state.
13. Explain why it is unlikely that Somaliland will become a sovereign state.
14. Write a paragraph on how global power structures have changed in the twenty-first century.
15. Using examples, prepare a report on the influence of technology on geopolitics.
16. As a class, discuss the advantages and disadvantages of membership of intergovernmental organisations.
17. Write a short report on the influence that the control of land and resources has on geopolitics.
18. Read the box, Spotlight: Russia's invasion of Ukraine. Complete the following activities.
    a. Outline the historical relationship between Ukraine and Russia.
    b. Explain the reasons for Russia's invasion.
    c. At the time this book was published, the war was ongoing. Conduct your own research into the current status of the conflict and write a short summary of the current situation.

UNIT 11.3

Impacts and responses to political tensions and conflicts

Migration and mobility of people

Wars and famines are two of the main causes of large-scale migration, but poverty, persecution and economic opportunity are other major causes of dislocation.

Forced migrations

Forced migrations of people often take place before, during and after a war. Finding safe havens for refugees has been one of the major global geopolitical issues since World War II. In the twentieth century, many wars were started with the express purpose of expelling and/or capturing large numbers of particular people with the goal to either 'ethnically cleanse' a territory or resettle a population. In more recent decades, tens of millions of refugees have fled from conflicts. Many are unable or unwilling to return to their homelands. These displacements often cause major humanitarian relief challenges for aid agencies, host countries and the UN. In recent years, conflict in Syria and parts of North Africa led to mass migrations of people towards Europe, with many making perilous journeys in small boats across the Mediterranean Sea (see Figure 11.3.1).

11.3.1 Refugees arrive in Lesvos, Greece, from Turkey, escaping the conflict in Syria.

Voluntary migrations

Once conflicts are settled, many people remain dissatisfied with the outcome and look for ways to migrate elsewhere. Despite not being displaced by the conflict, their movement is a longer-term consequence of it. In the 1970s and 1980s, millions of Vietnamese and Cambodians fled to South-East and East Asian countries. Vietnamese immigration to Australia has been an enduring result of the communist victory over the USA and its allies, including Australia, in the Vietnam War.

Political outcomes of conflict

Almost all geopolitical tensions and conflicts result in political changes. Some may simply involve a change of government by a peaceful election, others can involve the violent overthrow of a government by revolutionary upheaval or the establishment of a new constitutional basis for government.

Due to the Arab Spring uprisings, in 2011 several governments across North Africa fell. In Libya, a popular uprising began in early 2011 following the arrest of human rights activists. Libya had been under the control of Muammar al-Gaddafi for decades. Like other dictators across the region, he used violence and repression to maintain control. In response to the protests, there was violence and further repression by government forces. Protests quickly developed into a full-scale armed rebellion. By the end of 2011, al-Gaddafi had been removed from power by anti-government forces with the help of air power from Western nations, including the USA (see Figure 11.3.2).

11.3.2 A man stomps on a portrait of Libyan dictator Muammar al-Gaddafi after his overthrow.

Territorial outcomes of conflict

Many events resulting in political change also brought about territorial changes through forcibly redrawing boundaries. Since the end of the Napoleonic Wars in 1815, the big powers have been willing to intervene in European borders, just as they have in the rest of the (colonised) world. They have occasionally redrawn boundaries in attempts to accommodate ethnic distributions, national rivalries and aggressive national attitudes. These forced boundary changes rarely succeed in permanently solving such problems.

Social and cultural outcomes of conflict

A geopolitical crisis can result in social structures being reorganised. In some conflicts, particularly those resulting from social class inequality, there can be quite profound alterations in a society's class and family composition. This is particularly true following revolutions. Marxist-inspired revolutionary movements and upheavals (such as in Russia, China, Vietnam and Cuba) enormously affected long-term social structures. Entire social classes disappeared, especially landlords and capitalists. New classes of bureaucrats and party officials rose to prominence.

With the demise of the communist regimes of Central and Eastern Europe, the capitalist class has returned, but not the landlord class. The social structure of the former Central Europe communist countries today differs vastly from the more established capitalist societies of Western Europe. Across the communist nations of Asia, particularly China, tension developed between the political ideologies of communism and a desire for economic development. This tension led to a reinvention of communism towards greater acceptance of capitalist economic principles while also maintaining communist political principles. This shift has seen the emergence of a wealthy elite class of business entrepreneurs and a successful middle class in China.

Cultural reorganisation can also be the outcome of a conflict. Sometimes cultures, or their key aspects (such as languages), can be suppressed and even disappear due to the outcomes of geopolitical conflicts. The suppression of minorities and language bans and other forms of cultural expression can have lasting significance. In many parts of the world, ethnic cultures disappeared as a result of **colonialism** and/or assimilation policies. Against this, indigenous peoples are currently fighting back vigorously and multiculturalism policies are common in many Western countries. Yet there are still strong movements to suppress minority cultures in many other countries.

Responses to political conflict

The resolution of political tension and conflicts takes place at many levels. It involves many forces and can have many outcomes and can only be resolved in a lasting manner by addressing and overcoming the deep-seated causes.

Many attempts to resolve conflict fail because the causes are not understood. Adding to the complexity, many wars are deliberately started and pursued by leaders who have definite objectives in mind. Outside forces, such as the major powers (or the major powers acting on behalf of the UN), have often failed in their attempts to intervene in or mediate conflicts. This failure arises from their inability or unwillingness to grasp the complex social causes involved and the motives of some combatants. The causes of conflicts are complex and so too are lasting resolutions.

A global government

Examining geopolitical tensions and conflicts suggests a need to argue for some form of limited global governance, one largely concerned with trying to prevent and resolve violent disputes. This is exemplified by the League of Nations, founded in 1919, and the UN in 1945. Whether or not such global governance would have an effect depends on:

- the willingness of states to surrender some sovereignty
- the organisation having sufficient force and resources to command compliance with decisions and to finance whatever changes may be necessary, such as the resettlement of large populations
- the organisation having sufficient wisdom to understand and act on the causes of conflicts.

These three provisos are significant enough to start a global **federation**, but member states have not yet shown any will to contemplate this idea.

There have been some limited steps towards global governance in the economic and environmental spheres. In geopolitics, there are already many limitations on the powers of states to act independently of the international political **community**. As globalisation increases its pace, the need for certain states to take joint environmental and defence measures grows, particularly against pollution, nuclear weapons and military aggression. As a result, further steps towards global governance will occur. However, the re-emergence of nationalism in the twenty-first century indicates that forces are seeking to diminish global cooperation.

International treaties

Often conflicts are resolved or even prevented with treaties, which are special agreements between nation-states. To be valid, they must be voluntary and ratified by the parliament of the countries involved. This means each parliament must pass a law accepting the conditions of the treaty.

Treaties can prevent tensions from becoming conflicts. Australia has several treaties with its neighbours. Some are designed to prevent conflicts, as with the 1995 Australia–Indonesia Treaty. Treaties have also been created to limit the use of weapons, such as the 1997 Ottawa Treaty, seeking to ban anti-personnel mine use (see Figure 11.3.3), and the 1972 Biological Weapons Convention, banning biological weapon use. Such treaties are often vital for smaller countries, which are otherwise more vulnerable.

International agencies

There are two main types of international agencies tasked with helping to prevent or resolve geopolitical conflicts:

- intergovernmental agencies, many of which are part of the UN. These agencies bring together member nation-states with the aim of resolving conflicts and disputes (e.g. UN Peacekeeping missions).
- non-governmental organisations (NGOs), which are public, non-profit, cooperative agencies mainly relying on financial donations. They are typically concerned with alleviating the consequences of conflict, such as supporting refugees who have fled conflict zones.

11.3.3 An anti-personnel mine in Colombia, banned by the Ottawa Treaty

The UN also engages in civilian aid work, notably by assisting refugees. The UN High Commissioner for Refugees has had a monumental task in recent years with the massive influx of refugees caused by natural disasters in Türkyie and Syria, civil war in Sudan, and drought and famine across the Horn of Africa.

Nongovernmental organisations (NGOs)

There are many kinds of NGOs. Those providing relief (such as CARE, Oxfam and Médecins Sans Frontières) are not concerned with resolving disputes so much as alleviating their consequences. This may, in turn, effectively prevent a further outbreak of conflict. NGOs can also play a key role in bringing geopolitical matters to widespread public attention, as in the case of the Rohingya crisis in Myanmar.

Partitions and separations

When tensions and conflicts are caused by inter-ethnic hostility or ideological divisions, the partition (division) of a territory and the forceful separation of potential combatants can prevent further violence and even produce permanent peace. Throughout history, there have been numerous examples of partitions. Some have been successful, while others yield prolonged conflict. For example, the partition of Korea into North and South Korea at the end of the Korean War in 1953 left a legacy of great tension.

Activities

Acquiring and processing geographical information

1. Differentiate between forced and voluntary migration.
2. How does conflict contribute to migration?
3. Explain the term genocide.
4. Explain the potential social and cultural outcomes of conflict.
5. Explain why many attempts at peace fail.
6. Explain the concept of a global government.
7. Outline what an international treaty is.
8. Differentiate between intergovernmental and non-governmental agencies.
9. What is a partition?

Applying and communicating geographical understanding

10. Using examples, write a short report about how international treaties can help promote peaceful resolutions to conflict.
11. In a small group, conduct research into one intergovernmental and one non-governmental agency. Outline the goals and objectives of the organisation and describe one project it is currently undertaking.
12. As a class, discuss the advantages and disadvantages of partitions as a solution for resolving conflict. Summarise your discussions using graphics, such as a mind map or Venn diagram. You may find it useful to use the internet to research an example of a partition to help with this.

UNIT 11.4

Contested space: The South China Sea

Spatial patterns and characteristics of the South China Sea

Covering an area of 3 685 000 square kilometres, the South China Sea is a key strategic, and highly contested, region. It is bordered by China, Indonesia, Cambodia, Vietnam, Malaysia, Singapore, Brunei, the Philippines and Taiwan. All these nations have competing claims over different parts of the sea, its islands and reefs (see Figure 11.4.1). The sea lies within a region that has emerged in recent decades as one of the world's greatest economic hubs, giving it high strategic importance.

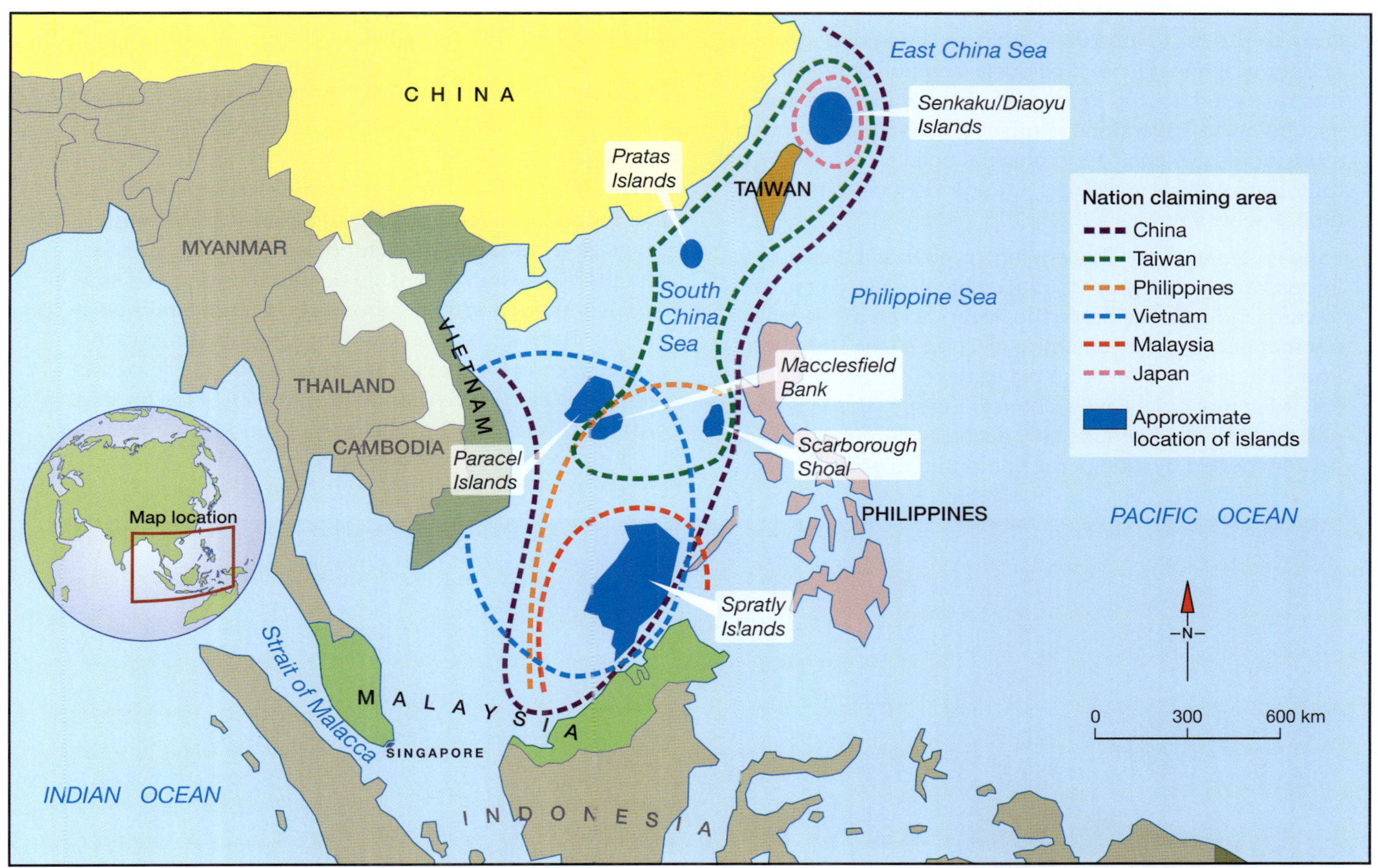

11.4.1 Countries with competing claims over the South China Sea

The South China Sea is a thoroughfare for the factories of China and South-East Asia to supply the rest of the world. The region is also rich in resources, including oil. Although Japan and South Korea do not border this sea, they rely on the region's fisheries for seafood and their fishing fleets' economic viability. Consequently, the sea is vital and those that control it gain much power.

Did you know?

More than $5 trillion worth of trade passes through the South China Sea annually.

Influences on the South China Sea dispute

While its role as a trade route between China and the rest of the world is paramount, there are many other factors influencing the territorial dispute in the South China Sea.

All this geography creates a strategic threat to China. Most countries bordering the South China Sea and its vital trade routes are allied to the USA, China's great geopolitical rival (see Figure 11.4.3). The US has significant military bases dotted across the region. It maintains a huge military force, including long-range bombers, aircraft carriers and nuclear submarines in Guam, a small US-controlled island in the Pacific about 5000 kilometres east of China.

● **SPOTLIGHT**

China's geographical dilemma

Although vast, China's physical geography limits its access to the rest of the world. This makes the country's global trade position quite vulnerable. Its economy and social structure rely on outside trade.

From the 1980s to the 2020s, China re-imagined its entire economy. It rejected its previously isolationist policies that restricted access to the rest of the world. It relaxed its ideological objections to private ownership, although strict repressive social and political policies remain. Trade routes are central to China's engagement with the global economy.

Land routes are very challenging, given that the vast, impassable Himalayan Mountains block western routes towards India. To the north, the frozen waste of Siberia is another barrier. A key objective of China's One Belt, One Road initiative is to recreate the trade routes of the ancient Silk Road that once linked China to Europe. This is complex and problematic because only a small volume of trade can be transported by road or rail compared to going by sea.

Sea routes are vital, making the South China Sea critical. Again, geography plays a role. To China's east lies the Pacific Ocean and its trade routes to North America. Japan controls the line of islands stretching between southern Japan towards the Philippines. Japan's powerful navy could block China from accessing this easterly route. Furthermore, the USA maintains a large military force on the Japanese island of Okinawa. This is directly in the middle of the gap between Japan and the Philippines.

All this makes the routes southwards towards Malaysia and Indonesia even more vital. However, the South China Sea is shaped like a funnel, with the Philippines and the Malaysian Peninsula acting to narrow the sea. At its southern end, ships are forced to turn either west or east. Turning west, ships must navigate through the extremely narrow Strait of Malacca, under 3 kilometres wide near Singapore (See Figure 11.4.2). Ships turning east must navigate through myriad islands that make up the Indonesian archipelago.

11.4.2 Ships lining up to pass through the narrow Strait of Malacca

The United Nations Convention on the Law of the Sea (UNCLOS) is an international agreement negotiated in 1982, which came into force in 1994. Every country with claims in the South China Sea is a UNCLOS signatory. For centuries, countries have had control over territorial waters. These typically extend 12 nautical miles (22 kilometres) from the coastline and effectively extend a country's border that far out to sea. A feature of UNCLOS was the creation of an Exclusive Economic Zone (EEZ), which grants the right to exploit all the resources in the sea for an area 200 nautical miles off a nation's coast. In the event of two areas overlapping, the distance is split in the middle between the two competing nations.

11.4.3 US Marines train with Filipino troops

Strategic islands and reefs

With the advent of the EEZ, control of small islands and reefs suddenly became critical. The area around the islands, up to 200 nautical miles, now fell under the control of the country laying claim to the island. Islands and reefs that were originally considered to be of no value or no economic or strategic significance suddenly took on great importance as the countries governing them were able to claim all the resources in their surrounding sea.

Early in the twenty-first century, China began claiming many of these islands and reefs as part of their territory. In response, other nations throughout the region began occupying other islands and reefs, including the highly contested Spratly Islands, where numerous countries claim possession (see Figures 11.4.5 and 11.4.6). The Philippines dispatched a World War II-era ship, the *Sierra Madre*, to the low-lying Ayungin Shoal (also called the Second Thomas Shoal) (see Figure 11.4.4). A small naval contingent constantly crews the vessel to assert Filipino sovereignty over the shoal. The vessel is monitored by Chinese vessels and its resupply ships are often harassed in a bid to stop the occupation. For example, in August 2023 a major diplomatic incident took place when a resupply vessel from the Philippines was harassed by Chinese naval ships. The Chinese ships used water cannons and close maneuvering to stop the Filipino ship from reaching the outpost. After two weeks of attempts the resupply mission was finally successful.

11.4.4 The Filipino ship, the *Sierra Madre*, on Ayungin Shoal

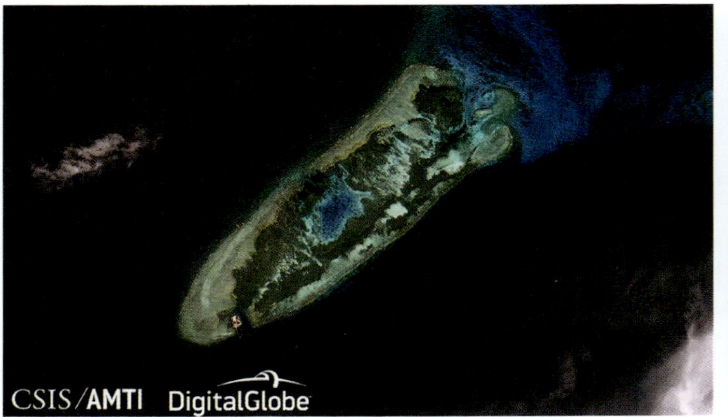

11.4.5 and 11.4.6 Fiery Cross Reef is part of the Spratly Islands in the South China Sea. Satellite images show the island before the massive development (left), including an airbase built by the Chinese (right).

11.4.7 The US warship USS *Kidd* exercising freedom of navigation in the South China Sea

A key feature of the South China Sea dispute has been the militarisation of the region, especially by the Chinese. This has seen some of the reefs and islands transformed into military facilities and huge land reclamation projects that extend their land area. This makes space for military bases, and, in some instances, air bases. China has argued that their works are designed to improve the living conditions of the people deployed to the islands and are not aimed at projecting military power.

Impacts of the dispute

The EEZ relates to economic control, but China has argued that by controlling the islands and reefs in the region they form part of their territorial waters. By this, they mean they lie within their borders, so ships and aircraft from other nations would need permission to enter the areas around the islands and reefs. This is highly controversial and has led to other nations, including Australia, launching what is called freedom of navigation naval deployments and military aircraft flights (see Figure 11.4.7).

Freedom of navigation deployments are a means of rejecting Chinese claims in the South China Sea. When they take place, they attract fierce criticism from China, which views them as hostile acts. Chinese authorities argue that they breach Chinese sovereignty and are dangerous and could lead to accidents as planes and ships play a game of cat and mouse across the South China Sea, as shown in Figure 11.4.8.

Others claim it is China that represents a danger within the region. For example, mass protests broke out in the Philippines in June 2019 (see Figure 11.4.9) after a Chinese coast guard ship collided with a Filipino fishing boat. In April 2020, Vietnam launched official complaints after a Chinese naval ship sank one of its fishing boats after deliberately ramming it in a disputed part of the sea.

11.4.8 Vietnamese and Chinese coastguard ships shadow each other in a disputed section of the South China Sea

11.4.9 Anti-Chinese protests in the Philippines, 2019, after a fishing boat was sunk

Environmental consequences

The dispute in the South China Sea also creates significant environmental impacts. There are more than 600 coral, 3000 fish and 1500 sea sponge species found within the sea. The islands are home to several turtle, seagrass and mangrove species and provide habitats for sea birds.

The occupation of once uninhabited islands has dramatically impacted these fragile ecosystems. China's program to expand reefs and islands through vast land reclamation projects has had a devastating impact on the environment. Several square kilometres of land were artificially created by smothering several reef systems. Air and water pollution have increased with the growing presence of permanent and semi-permanent settlements throughout the region.

Responses to the dispute

The South China Sea remains one of the most contested places worldwide. It has the potential to lead to all-out conflict, although many experts believe this is unlikely. A war within the region would greatly disrupt trade, which would significantly impact China's economy. This, in turn, would have an impact on the country's social cohesion and perhaps threaten the rule of the Chinese Communist Party.

International law provides an alternative pathway for resolving the dispute. All of the countries with claims in the South China Sea are signatories to the UNCLOS. The UNCLOS nominates the Permanent Court of Arbitration (PCA) as the place for disputes over compliance with the Convention. In 2013 the Philippines launched a case in the PCA against China.

The effectiveness of international law is limited by state sovereignty. Countries can refuse to participate in legal proceedings and there is a limited ability for international courts to enforce their rulings. The rulings do, however, provide legitimacy to the country's claims and add weight to their complaints.

● **SPOTLIGHT**

The South China Sea arbitration

In January 2013, the Philippines launched a case against China in the PCA. The basis of its case was that China was not abiding by its obligations under the UNCLOS. The Chinese government used the state sovereignty rule to refuse to recognise the Court's authority to hear the case. Nevertheless, the PCA continued to hear the Philippines' complaint.

As the case only dealt with the UNCLOS the court did not make any ruling relating to the claims of sovereignty over islands in the South China Sea. Among the issues the PCA dealt with were allegations by the Philippines that China's land reclamations and dredging projects were damaging to the environment and that China was disrupting lawful fishing and other economic activities by the Philippines. A key part of the case was also whether China's assertion to historic rights over the vast areas it claims in the South China Sea were valid.

In 2016, the PCA delivered its verdict. In a 501-page decision, the five judges found that China's claims to the territory violated the UNCLOS. Furthermore, it found that many of the rocks and reefs China claimed gave it an EEZ did not meet the definition under the UNCLOS to produce an EEZ. It also found that China was illegally interfering with the Philippines' EEZ.

While the case was, in a legal sense, a resounding victory for the Philippines, China has consistently refused to recognise the PCA's right to hear the case, let alone its actual ruling. Nevertheless, the ruling has placed further pressure on China to abide by its international law obligations.

Regional cooperation

While much of the attention in the South China Sea dispute centres around China's actions, other countries in the region are also in dispute with each other. Malaysia and Thailand resolved territorial disputes in the sea through negotiation and an agreement to jointly develop resources within the disputed area. Thailand has entered a similar arrangement with Vietnam. These negotiated agreements are seen as a model for the region.

Some commentators have also cited Australia's arrangement with Timor-Leste (East Timor). In 2018, Australia and Timor-Leste signed the Maritime Boundary Treaty. The Treaty marked the end of a long-running dispute between the two nations over their maritime boundary in the Timor Sea. The Treaty came about after conciliation (a type of mediation) by the PCA. DFAT noted on its website:

> As two democratic nations and close neighbours, Australia and Timor-Leste have highlighted the value of international law, and particularly UNCLOS, in the international rules-based system. Our joint success through the conciliation sets a positive example for the region and the international community. It highlights Australia's commitment to international law …

The positive example for the region and international community can be interpreted as a criticism of China's reluctance to engage with international laws, such as UNCLOS and the PCA.

The Association of South-East Asian Nations (ASEAN) also has an important role in resolving the dispute. All countries with disputed claims in the South China Sea, except China and Taiwan, are ASEAN members. Evidence is growing that the countries bordering the sea are forming a bloc, one with the potential to put united pressure on China. This was seen in 2020 when a Vietnamese fishing boat was sunk by Chinese vessels. The Philippines was quick to offer support to Vietnam after it suffered a similar incident in 2018. In 2020, at a conference of ASEAN leaders in Jakarta, Indonesia, plans were developed for a Code of Conduct relating to the use of the South China Sea.

Activities

Acquiring and processing geographical information

1 Explain the economic significance of the South China Sea.

2 Describe the purpose of the United Nations Convention on the Law of the Sea.

3 Differentiate between territorial waters and EEZ.

4 Describe the militarisation of the South China Sea.

5 What is the purpose of freedom of navigation cruises and flights?

6 Outline the risks associated with freedom of navigation activities.

7 Describe the environmental consequences of the South China Sea dispute.

8 Outline the impact of international law on the dispute.

9 Explain the role of regional cooperation in resolving the dispute.

Applying and communicating geographical understanding

10 Assess the importance of physical geography in understanding the South China Sea dispute.

11 Examine Figures 11.4.5 and 11.4.6. Describe the transformation of Fiery Cross Reef. As a class, discuss what impacts such a transformation could have had on the environment.

12 With a partner, discuss whether freedom of navigation flights and cruises serve any worthwhile purpose. Share your ideas with the class.

13 Briefly summarise the PCA case at the South China Sea Arbitration.

APPLICATION AND CONSOLIDATION TASKS

Task 1: Prepare a speech: A modern conflict

Take on the role of a political adviser to the Minister for Foreign Affairs and prepare a speech on a twentieth- or twenty-first-century conflict.

Include in your speech:

- the history of the conflict, including key events
- the main groups, including their political ideologies
- the consequences and outcomes of the conflict
- the political and territorial outcomes of conflict.

Your speech can include visuals (maps, graphs, diagrams, photos) that you would use during your presentation.

Task 2: Infographic: NGOs

Conduct research into a non-governmental organisation and create an infographic to inform the public about their work.

Your infographic should include:

- the history and establishment of the NGO
- an outline of the main characteristics and objectives of the NGO
- an explanation of some of the current work being undertaken by the NGO
- an analysis of the work of the NGO.

Task 3: Podcast: South China Sea dispute

Prepare and record a podcast on the current status of the South China Sea dispute. Your podcast should be 4–5 minutes long. Your podcast should include:

- a brief history of the dispute and its impacts
- a summary of the current situation
- predictions on the situation.

Your podcast can include audio content, such as music, snippets of interviews etc.

SECTION

3 Human–environment interactions

So profound and widespread has been the footprint of humanity that scientists now talk of a new epoch in the history of planet Earth—the Anthropocene. This is defined as the period dating from the beginning of significant human impact on Earth's ecosystems, including, but not limited to, anthropogenic climate change.

The start date of the Anthropocene is hotly debated. Some scientists date it from the beginning of the agricultural revolution, around 12 000–15 000 years ago. Others argue that it commenced in the mid-twentieth century, when rapid population growth, combined with the unprecedented expansion of humanity's technological capacity, greatly expanded the impact of human systems on the natural world.

Content focus

In this section of the text, students have the opportunity to apply their knowledge and understanding of Earth's natural and human systems via the study of either bushfires, as an example of a contemporary natural hazard, or COVID-19, as an example of a contemporary ecological hazard or climate change. In doing so, students study the relevant spatial dimensions, natural processes, the nature of human interactions and change, and the continuing or emerging challenges, opportunities and responses for the selected study. They also assess the effectiveness of people and organisations in managing one challenge or event resulting from the interaction between natural systems and human systems at a selected place.

In this section

CHAPTER

12 Changes to Earth's natural systems

At the beginning of the twentieth century, there were just 1.6 billion people on Earth. Pollution and environmental degradation were problems, but were mainly local. The world still seemed vast and large areas remained virtually untouched by people's activities. Just over 120 years later, the world's population has surpassed 8 billion. The environmental problems that have resulted from this rapid growth, and the rising material lifestyle expectations that have accompanied it, now impact the entire planet. How we manage these environmental challenges is critical to the future wellbeing of humanity and Earth's ability to sustain life as we know it.

This chapter focuses on changes to Earth's natural systems over time and how these impact land use and land cover. Of particular interest is the role of climate change. The issues covered draw on material explored in greater detail elsewhere in the book, especially in Chapters 4, 5, 6 and G1. These chapters delved into detail on land cover change, deforestation, glaciers and ice sheets, and desertification.

The truth is, the natural world is changing. And we are dependent on that world. It provides our food, water and air. It is the most precious thing we have and we need to defend it.

Sir David Attenborough, British naturalist and broadcaster

12.0.1 A polar bear standing on top of a melting glacial iceberg in Norway.

Chapter glossary

anthropogenic caused or influenced by humans

biodiversity all the different kinds of living organisms within a given area

deforestation the clearing of a forest or a stand of trees

fuel wood timber collected and burned for heating or cooking in the residential setting, especially in developing countries

global warming the unusually rapid increase in Earth's average surface temperature over the past century, primarily due to the greenhouse gases released by people using fossil fuels

UNIT 12.1
Natural and human-induced change

Changes to Earth's natural systems result from disturbances caused by natural ecological processes or human activity. Ecological factors encompass a variety of natural disruptions, such as fires, floods, droughts, storms, volcanic eruptions and earthquakes. Long-term natural changes can occur in climate or via disruptions to the water cycle, energy flows and nutrient cycles, plant succession, reproduction and regeneration mechanisms, and plant–animal interactions.

Human-induced, or **anthropogenic**, changes arise from overpopulation, pollution, fossil fuel use (see Figure 12.1.1) and **deforestation**. Such changes have negative consequences. They trigger climate change, soil erosion, poor air quality and scarcity of drinkable water. Of all these, climate change is the most pressing issue. Scientists believe we are reaching a tipping point. Once exceeded, Earth's systems will undergo significant and enduring changes. Crossing this critical threshold can lead to large, often irreversible, changes in natural systems.

12.1.1 Fossil fuels still supply most energy needs.

Evidence of climate change in the contemporary world

The evidence of anthropogenic climate change in the contemporary world is clear. Global temperatures are increasing, ice sheets and glaciers are melting, sea levels are rising, the ocean is acidifying and extreme weather events are increasing in frequency and severity. Earth has been warming at an alarming rate for several decades, and this warming is expected to continue for the rest of the century. Chapter 13 covers this in depth.

The 10 warmest years on record have occurred since 2010. The top two, three and four are 2016, 2019 and 2020; with 2023 the hottest yet. Globally averaged, the surface air temperature has warmed by over 1°C since reliable records began in 1850. Each decade since 1980 has been warmer than the last, with 2011–20 being around 0.2°C warmer than 2001–10.

Evidence for the causes of climate change over time

Climate change applies to any change in the planet's climate, either permanently or over a long period. Such changes result from two related processes: anthropogenic climate change and natural climate change. Anthropogenic climate change is caused by the activities of people, notably fossil fuel burning, while natural climate change results from the natural climate cycles that have been a feature of Earth's history for millions of years.

Natural climate change

The natural cycles driving climate are determined by incoming and outgoing energy. Also playing a part are the tilt of Earth's axis, natural oceanic warming and cooling cycles, and variations in the Sun's energy output and in volcanic activity levels. Also, throughout Earth's history, there have been at least five major ice ages, causing increasing glaciation. The earliest was about 2 billion years ago and the most recent about 2.6 million years ago, ending about 12 000 years ago. More recently the Little Ice Age occurred from the sixteenth century through to the nineteenth century, followed by a natural warming stage (see Unit G2.2 for more on the Little Ice Age).

SPOTLIGHT

Climate change evidence: An Australian perspective

The Australian Bureau of Meteorology's 2022 *State of the Climate report* highlighted the impacts of anthropogenic climate change from an Australian perspective. The Bureau's key observations were:

- Australia's climate has warmed by an average 1.47°C since 1910.
- Sea surface temperatures have increased by an average 1.05°C since 1900, making extreme heat events more frequent.
- Rainfall from April to October has decreased by around 15 per cent in south-western Australia since 1970 and around 10 per cent since the late 1990s (see Figure 12.1.2). Rainfall and streamflow have increased since the 1970s across parts of northern Australia (see Figure 12.1.3).
- Extreme fire weather has increased and the fire season has lengthened.
- The number of tropical cyclones observed in the Australian region has decreased.
- Snow depth, snow cover and the number of snow days have decreased in alpine regions since the late 1950s.
- Oceans around Australia are acidifying and have warmed by more than 1°C since 1900, contributing to longer and more frequent marine heatwaves.
- Rising sea levels around Australia are increasing the risk of inundation and damage to coastal infrastructure and communities.

12.1.2 Severe drought from 2017 to 2019 hit large areas of Australia

12.1.3 Flooding in Rochester, Central Victoria, 2022

● SPOTLIGHT

Spatial variations in the rate of warming

Climate change has a spatial dimension, as some places are warming more slowly than others. Scientists have found that the Arctic is warming over four times faster than elsewhere, and sea ice is rapidly disappearing. Europe is the fastest-warming continent, where recent heat waves have killed thousands. In other places, warming is much slower than the global average—about 1.1°C since the mid-twentieth century.

Rather than casting doubts on climate change science, these relatively cooler regions provide valuable environmental insights. Where global warming's influence is weakest is primarily due to air pollution, the hole in the ozone layer and melting ice. All are human-induced.

India, for example, is one of the slowest-warming places due to its air pollution. Sulphur dioxide, a pollutant released from burning sulphur-containing fuel (e.g. coal, oil or diesel) actually cools localised areas. It reduces and spreads incoming solar radiation and stimulates cloud formation. Carbon dioxide, in contrast, disperses quickly into the atmosphere. India has experienced less than 0.5°C of temperature increase compared with its average temperatures from 1951 to 1980 (see Figure 12.1.4). This is less than half the global average. Yet India does not escape periods of extreme heat. The subcontinent's southern and coastal regions experience tropical humidity that creates oppressive and dangerous conditions during summer.

Elsewhere, eastern Antarctica has warmed more slowly than western Antarctica. Scientists determined these differences are due to the atmosphere's ozone 'hole'. This thinning in the atmospheric layer acts as a shield from harmful ultraviolet sunlight. While agreements to eliminate the use of ozone-depleting chemicals shrank the hole, its impacts are likely to persist for decades. Another reason is that the cold water flowing from collapsing ice sheets has cooled sections of the Southern Ocean around Antarctica (see Figure 12.1.5).

The cold water flowing from the melting glaciers and ice sheets is also affecting the temperature of the North Atlantic Ocean, resulting in a lower rate of temperature increase (see Figure 12.1.6).

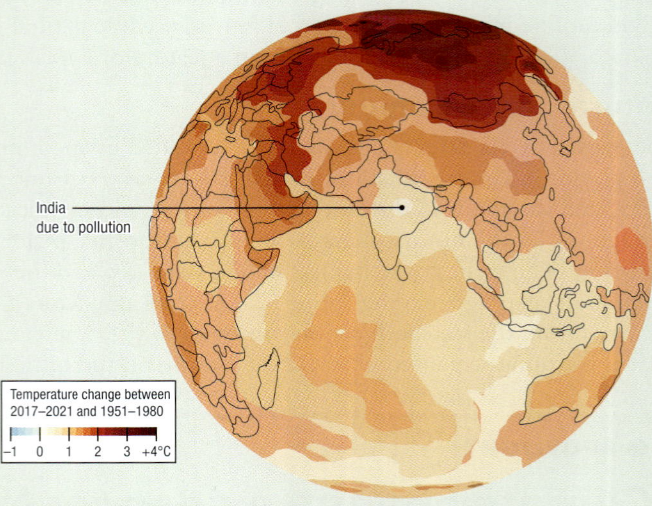

12.1.4 India is warming at a lower rate than the global average.

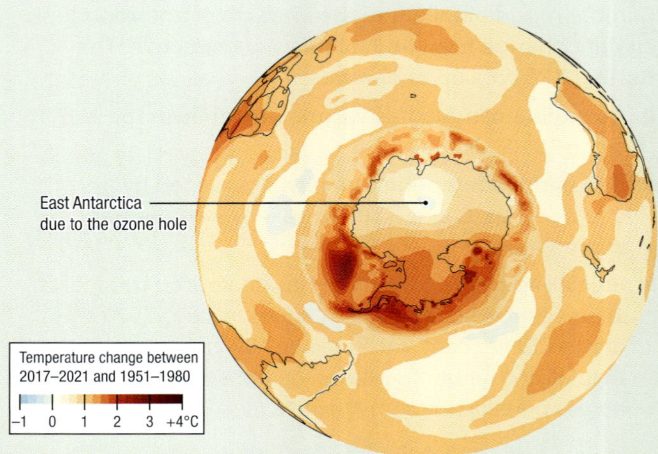

12.1.5 Meltwater has slowed the warming rate in parts of the Southern Ocean around Antarctica.

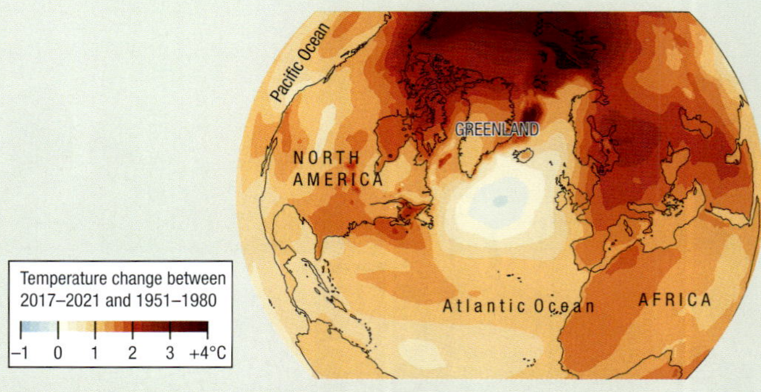

12.1.6 Cold water from Iceland's melting glaciers and ice sheets slowed the warming rate of the North Atlantic Ocean.

None of this accounts for the accelerated rate of warming that Earth is currently experiencing. The reasons lie with the impacts of a rapidly growing human population and their energy demands.

Anthropogenic climate change

Anthropogenic climate change is directly linked to the use of fossil fuels (oil, gas and coal), aerosol releases and the land cover changes associated with deforestation and agriculture. Since the Industrial Revolution began (around 1760), there has been a sharp rise in global temperatures. The scientific consensus is that this spike in temperature is caused by fossil fuels. Burning them releases heat-trapping carbon dioxide (CO_2) into the atmosphere. The carbon dioxide levels remained below 300 parts per million for at least the past 400 000 years, but since 1950, concentrations have increased rapidly to more than 422.97 parts per million, as of April 2023. Climate change is now associated with the process of **global warming**. This is defined as the unusually rapid increase in Earth's average surface temperature. Climate change is examined in greater detail in Chapter 13.

Activities

Acquiring and processing geographical information

1 List the ecological factors responsible for changes in Earth's natural systems.
2 Outline the human-induced changes responsible for changes in Earth's natural systems.
3 Explain what is meant by the term tipping point.
4 Outline the evidence of anthropogenic climate change.
5 Define the term climate change. What does it result from?
6 List the causes of natural climate change.
7 Outline the causes and consequences of anthropogenic climate change.
8 Define the term global warming.

Applying and communicating geographical understanding

9 Study the box, Spotlight: Climate change evidence: An Australian perspective. Write a report outlining the impacts of climate change in Australia.
10 Study the box, Spotlight: Spatial variations in the rate of warming. Explain why some parts of the planet are warming at a slower rate than others. What does this tell us about the nature of human–natural systems interactions?
11 Study Figures 12.1.4 to 12.1.6. Using an atlas, identify the parts of the world with above-average rates of warming and those with below-average rates of warming.

UNIT 12.2

Land cover change at a global scale

The impacts of human activity surround us. The most apparent of these are changes to Earth's land cover. Land cover change and the processes of deforestation, desertification, melting glaciers and retreating ice sheets are covered in depth in Chapters 4, 5, 6 and G1. This unit revisits these vital issues before investigating one example of human-environment interactions in detail. There is a choice of climate change in Chapter 13 or bushfires as an example of a contemporary natural hazard in Chapter 14, while COVID-19 is offered as an example of an ecological hazard in Chapter 15.

Deforestation

Deforestation is the process of clearing forested land (see Chapter 5). Throughout the history of human civilisation, forests have been cleared to make space for agriculture and animal grazing, to accommodate urban expansion and to provide wood for fuel, construction and industrial processes.

Deforestation has greatly altered global land cover. Australia, for example, has lost nearly half its forest cover in the last 200 years. In Europe, forest cover has been reduced from about 80 per cent 2000 years ago to just 34 per cent today. In North America, almost half of the forests in the eastern part of the continent were cut down between the 1600s and 1870s. Only 20 per cent of China remains covered in forest. Much of the world's farmland was once covered with forest.

The highest rate of deforestation is now occurring in tropical rainforests (see Chapter 5). Road construction in previously inaccessible regions has opened them up to exploitation by logging companies, cattle ranchers, and oil palm and rubber tree plantation operators.

Deforestation increases the carbon dioxide released into the atmosphere. Trees take in carbon dioxide from the air for photosynthesis and store the carbon in their wood. With fewer trees, carbon dioxide intake is reduced. More carbon dioxide accumulates in the atmosphere, which accelerates global warming. When forests are burned, their stored carbon is returned to the atmosphere as carbon dioxide.

Deforestation also impacts **biodiversity**. The planet's forests are home to numerous animal and plant species. When forests are logged or burned, many species are driven to the brink of extinction. Deforestation can leave soils more prone to erosion.

Deforestation need not be permanent. Forests in many areas are returning due to conservation efforts.

SPOTLIGHT

Agreement reached to protect Earth's lands and oceans

At the 2022 UN Biodiversity Conference (COP15) in Montreal, Canada, 190 countries approved a UN agreement to protect 30 per cent of the planet's land and oceans by 2030. The agreement also included 23 environmental targets to fight biodiversity loss, which if left unchecked, poses a severe threat to food and water supplies and species survival. Currently, just 17 per cent of terrestrial and 10 per cent of marine areas are protected. Biodiversity is declining worldwide at unprecedented rates. Scientists estimate that 1 million plant and animal species are at risk of extinction, many within decades. The last extinction event of such magnitude was 65 million years ago. It resulted in the extinction of the dinosaurs.

Desertification

As global temperatures rise and the human population increases, desertification increases. This is the permanent degradation of once-arable land. Human-caused land degradation in areas with low or variable rainfall is particularly concerning. These semi-arid lands account for more than 40 per cent of the world's land surface area. Around 2 billion people live on semi-arid land vulnerable to desertification. As many as 50 million people could be displaced if desertification remains unchecked.

While land degradation has always occurred, the pace has accelerated rapidly in recent decades. This has been driven by a range of factors, including fuel wood harvesting, farming and stock grazing. During these activities, trees and other vegetation are cleared, animal hooves compact the soil and crops deplete the nutrients stored in the soil. This contributes to soil erosion and the land has a reduced capacity to retain water. Both impact on the land's ability to support food production.

The lands subject to desertification are widespread and span more than 100 countries. The wellbeing of some of the world's poorest and most vulnerable people is at risk. The most impacted are those who depend on subsistence farming, which is common in affected regions (see Figure 12.2.1).

Did you know?

Tree rings, ocean sediments, coral reefs and layers of sedimentary rock contain evidence of Earth's past climate variations. Air bubbles trapped in glacial ice also provide a history of greenhouse gases stretching back over 800 000 years. The chemical make-up of this ice gives scientists insights into average global temperatures over time.

Scientists estimate that over 75 per cent of global land area is already degraded. Perhaps 90 per cent will have been degraded by 2050. Around 4.18 million square kilometres are degraded annually, with Africa and Asia being the most seriously affected. Anthropogenic climate change is probably accelerating the process of desertification.

12.2.1 Desertification in the Namib desert, a coastal desert in southern Africa

The consequences of desertification are far-reaching. When land is transformed into a desert, its ability to support surrounding populations of people and animals declines sharply. Crops don't grow, water becomes scarce and habitats are lost. Food scarcity becomes an issue, as do waterborne diseases. Chapter 5 examines desertification in detail.

Retreating glaciers and ice sheets

A most alarming impact of climate change is the changing distribution and thickness of the ice. The snows of Mt Kilimanjaro in Africa have melted by more than 80 per cent over the past 100 years. Arctic sea ice has thinned significantly over the past half-century. Its annual extent has declined by about 10 per cent over the past three decades and the edges of Greenland's ice sheets are thinning. The Himalayan glaciers are retreating so fast that scientists now fear most glaciers in the central and eastern Himalayas could disappear by 2035. In the Northern Hemisphere, the freshwater ice breakup occurs nine to ten days earlier than it did 150 years ago. The autumn freeze takes place ten days later. In parts of Alaska and Siberia, permafrost thawing has, in places, caused the ground to fall by as much as 5 metres. From the Arctic to the Antarctic, from Switzerland to the equatorial glacier of Indonesia, massive ice fields, glaciers and sea ice are disappearing at alarming rates.

When ice sheets and glaciers melt, the water released flows to the sea and oceans. There, it warms and expands, raising the average global sea level between 10 and 20 centimetres over the past 100 years. Glaciers and ice sheets are examined in detail in Chapter 5.

Did you know?

Earth's oceans absorb about 30 per cent of the CO_2 released into the atmosphere. As levels of atmospheric CO_2 increase through human activity, such as the use of fossil fuels and changing land use, the amount absorbed by the ocean increases. This decreases the ocean's pH and it becomes more acidic.

● **SPOTLIGHT**

Using satellite imagery to monitor glacier and ice sheet retreat

Using satellite imagery, scientists mapped the position of glacial fronts in 2000, 2010 and 2020. They found that 85 per cent of the Northern Hemisphere's glaciers had retreated in these two decades, accounting for a loss of about 7500 square kilometres of ice. Of the 1700 glaciers identified, almost all of them are retreating to some degree due to climate change.

Figure 12.2.2 shows the changes occurring in marine-terminating glaciers along the periphery of Greenland's vast ice sheet. They account for 62 per cent of the losses. In Northeast Greenland, ice loss was especially noticeable, particularly the Zachariae Isstrøm, which lost 1453 square kilometres from 2000 to 2020 (see Figure 12.2.3). Areas marked with large red and brown dots lost the most ice; those with large blue dots gained the most.

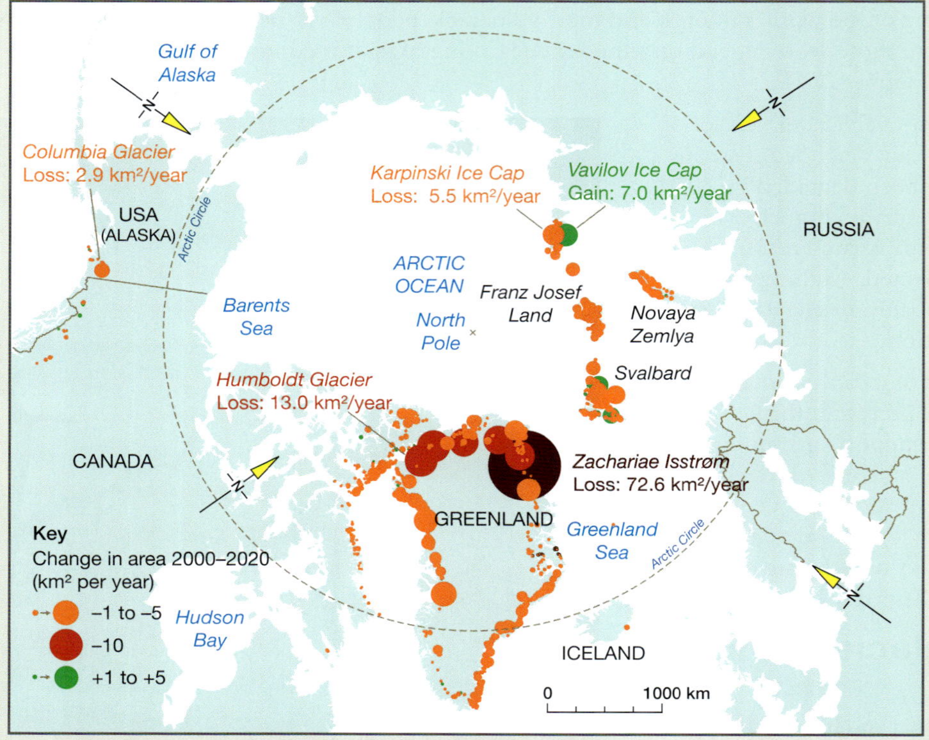

12.2.2 Glacial retreat in the Northern Hemisphere 2000–2020

12.2.3 Melting ice sheet in Greenland

Activities

Acquiring and processing geographical information

1. Explain what deforestation is and identify its causes.
2. Outline the extent of deforestation. Where is most evident today?
3. Outline the impacts of deforestation.
4. Outline the causes, extent and impacts of desertification.
5. Outline the causes, extent and consequences of the retreat of Earth's glaciers and ice sheets.
6. Explain how scientists can gain insight into past climates.

Applying and communicating geographical understanding

7. Study the box, Spotlight: Agreement reached to protect Earth's lands and oceans. Outline the significance of the 2022 UN Biodiversity Conference.

APPLICATION AND CONSOLIDATION TASK

Task: The Anthropocene epoch

The impacts of humans on Earth are so pronounced and widespread that scientists now talk of a new epoch in the history of planet Earth—the Anthropocene.

Either:

Conduct a class debate on the topic, 'The impacts of humanity are so profound it's now impossible to study natural systems in isolation'.

Or

Write a discussion outlining the arguments for and against the proposition that there is no longer such a thing as a 'natural system'.

CHAPTER

13 Climate change

Climate relates to the long-term patterns of weather. While weather varies from day to day, over the long term it is possible to determine the average or typical weather patterns. For example, it is known that the summer months in northern Australia are typically wet and humid, while in southern Australia, they are drier. Instead, southern Australia experiences wetter conditions during the winter months.

When generalised patterns like these have been observed over long periods, they become the climate of the region. However, the climate is not stable. Over very long periods, measured in thousands—and even millions—of years there have been substantial changes to Earth's climate.

Very recent increases in global temperatures, rising sea levels, melting glaciers and shrinking areas of sea ice provide mounting evidence the global climate is changing over an extremely short period. Most scientists have found convincing signs that humans caused these changes. Fossil fuel use and unsustainable farming are just two activities that release heat-trapping gases into the atmosphere.

This chapter investigates the spatial and temporal characteristics of climate change on a global scale. It also studies the evidence for climate change that results from natural variability versus human-induced changes. Finally, it focuses on the challenges, opportunities and responses of, and to, climate change in Costa Rica.

So we must do all that is humanly possible to combat this [climate change] challenge to humanity ... There is still time.

Angela Merkel, former Chancellor of Germany

13.0.1 A coal-fired power station releases pollution into the atmosphere

Chapter glossary

carbon cycle the sequence of processes by which carbon compounds are interconverted in the environment, involving the incorporation of carbon dioxide into living tissue by photosynthesis and its return to the atmosphere through respiration, the decay of dead organisms, and the burning of fossil fuels

carbon sinks natural storages of carbon, including oceans, forests and very long-term storages, such as coal, oil and gas deposits

climate long-term patterns in weather conditions

climate change sceptics people who do not believe that human activity is causing climate change

climate refugees people forced to flee their homes due to the effects of climate change

decarbonisation moving to substitute carbon-emitting technologies and activities with non-emitting approaches, such as renewable energy

deforestation the clearing and removal of forests to make way for human land uses, such as agriculture and urban development

fossil fuels energy sources formed in past geological times from organic materials; examples include coal, petroleum and natural gas

global temperature anomaly differences in temperatures from the expected typical climate patterns

greenhouse effect the increase in temperature on Earth created by the Sun's radiant heat passing through the atmosphere, in the same way the glass of a greenhouse creates a warmer temperature inside it

human-induced climate change changes in climate brought about by the action of humans; also called anthropogenic climate change

net-zero emissions a state when carbon emissions equal the capacity of carbon sinks to absorb the emissions

ocean acidification the change in the chemical make–up of the world's oceans, resulting in reduced pH levels, caused by the oceans absorbing more carbon dioxide

sea surface temperature the temperature of the top 1 millimetre of the ocean

sequester to isolate, bind or absorb

solar radiation energy received from the Sun

UNIT 13.1

Spatial and temporal characteristics of climate change

Climate scientists using many different sources of evidence have shown that Earth's climate is warming and that it is the result of human actions. While there remains political debate over climate change, there is virtually no debate within the scientific community about the extent and cause of this rapid warming. It is clear it will have consequences for all flora and fauna, not just humans.

Although climatic variations are expected, thorough research shows the rapidity of the changes between the mid-twentieth century and now are inconsistent with naturally adjusting climate patterns. This current timeframe coincides with significant global industrialisation and the widespread use of carbon-based **fossil fuels**, such as oil, gas and coal.

The global data

NASA data shows that Earth's average temperature has increased by 1.14°C since the late-nineteenth century. Most of this warming took place in the last three to four decades. For example, since 1880, 19 of the 20 warmest years on record took place between 2000 and 2020. More recently, the average temperatures globally in July 2023 were the hottest ever recorded.

Figure 13.1.1 graphs global temperature anomalies between 1850 and 2021. This data was collected from several different and scientifically reliable sources. They include the Hadley Centre and Climatic Research Unit (HadCRU), the US National Oceanographic and Atmospheric Agency (NOAA), the European Centre for Medium Weather Forecasts (ECMWF), and Berkeley Earth (a non-governmental scientific organisation). These anomalies relate to how much hotter or colder the annual average temperature is worldwide compared to the expected long-term climate pattern. Since the 1950s there has been a marked upward trend, with all years since the 1980s recording warmer-than-expected average annual temperatures worldwide.

By extracting the data supplied by Berkley Earth and the UK Hadley Centre, the **global temperature anomaly**, using 1951–1980 as a base, is even more apparent, as seen in Figure 13.1.2. Figures 13.1.1 and 13.1.2 both demonstrate the upward trend in global temperatures.

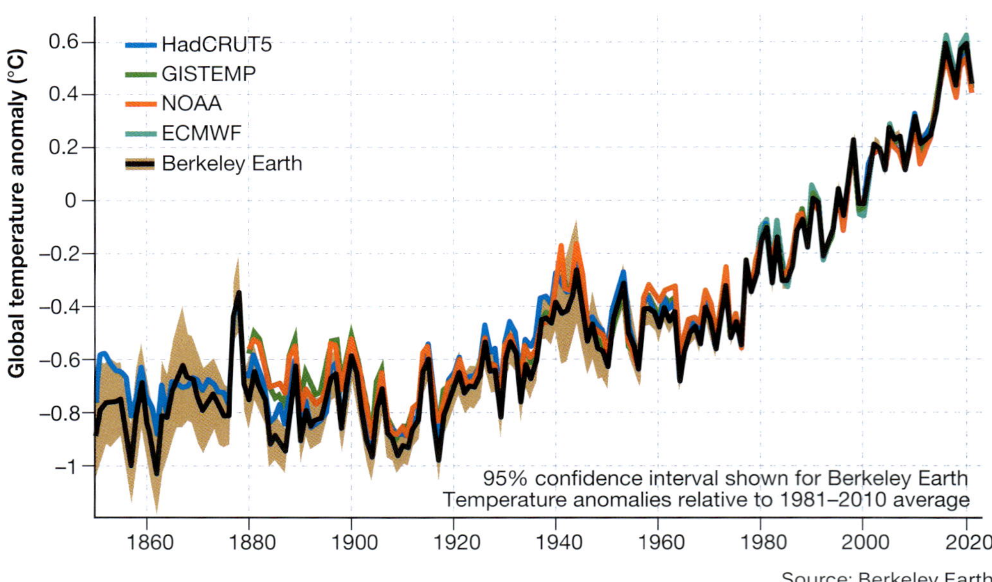

13.1.1 Global average temperature anomalies as recorded by several different scientifically reliable sources, 1850–2021

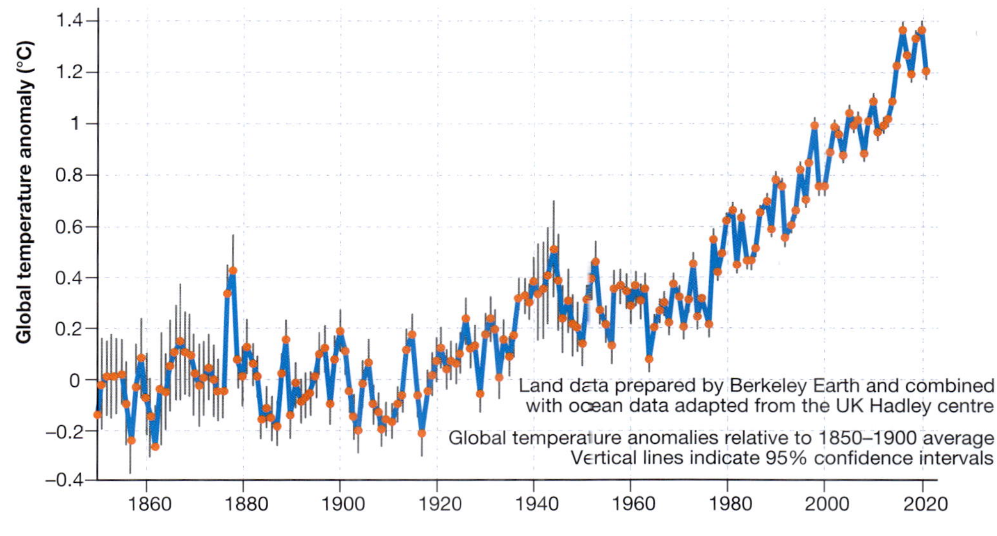

13.1.2 Global average temperature anomalies as recorded by several different scientifically reliable sources, 1850–2021

This warming is already having an impact on Earth's climate. Much of this additional heat is being absorbed by the world's oceans. Scientists measure **sea surface temperature** by taking the temperature of the top millimetre of the ocean's surface. There is a considerable amount of data going back to the mid-1880s. So, scientists can develop a long-term understanding of average temperatures. Sea surface temperatures vary due to many factors. For example, large-scale changes in temperatures in the Pacific Ocean are associated with a cyclical change to ocean currents, known as the El Niño and La Niña climate cycles.

However, over the long term, the surface of the oceans has been getting warmer (see Figure 13.1.3). Changes in ocean heat content from 1960 to 2019 confirm this warming trend (see Figure 13.1.4). A key consequence is the increasing frequency and intensity of storms. This was seen in 2017 with Hurricane Irma, the most powerful Atlantic hurricane in history to date, which was followed just a fortnight later by Hurricane Maria, another very powerful storm. Similarly, three large typhoons struck East Asia in less than a fortnight in late August and early September 2020. This highly unusual event saw mass flooding, landslides and widespread damage. These storms also sank the live-export ship, *Gulf Livestock 1*, killing 41 of the 43 crew members on board.

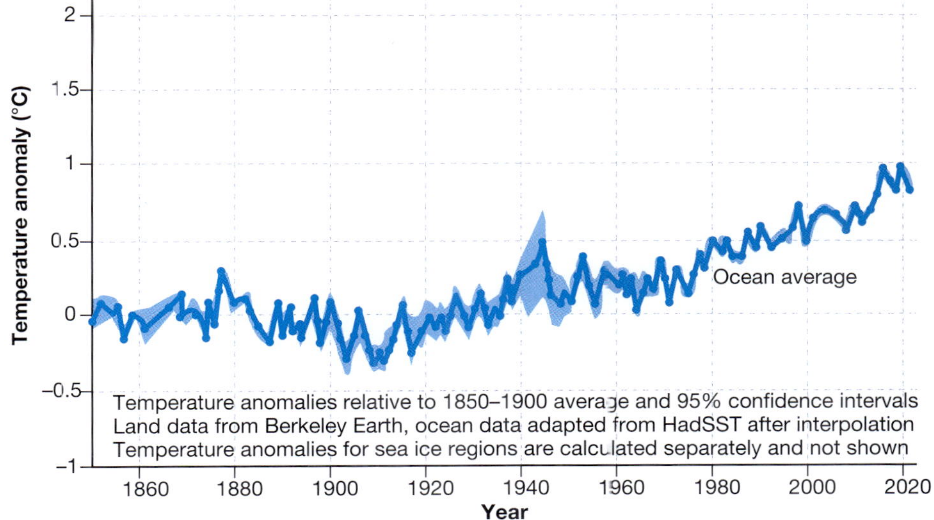

13.1.3 Changes in sea surface temperatures, 1880–2021

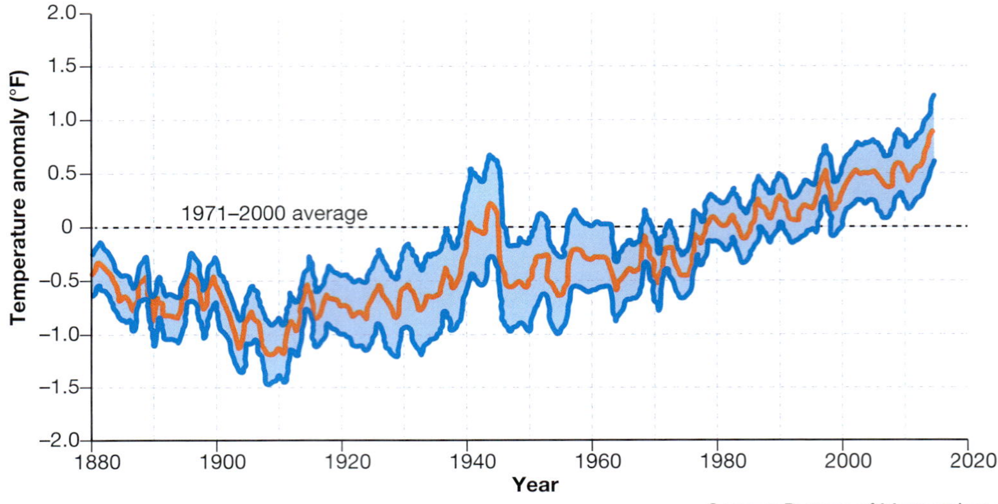

Source: Bureau of Meteorology

13.1.4 Average global sea surface temperature, 1880–2015. Shading indicates confidence in the range of estimates, which has increased significantly since 2005.

Climate change is a global phenomenon; nowhere on Earth escapes the impact of its effects. Figure 13.1.5 is a unique map prepared by NASA using data collected from many different sources and locations. It measures how far above or below temperatures were, compared to the average. It reveals that very few parts of the world show a cooler temperature. Nearly the entire globe has had temperature increases. The most considerable warming is in the north, especially in the Arctic region.

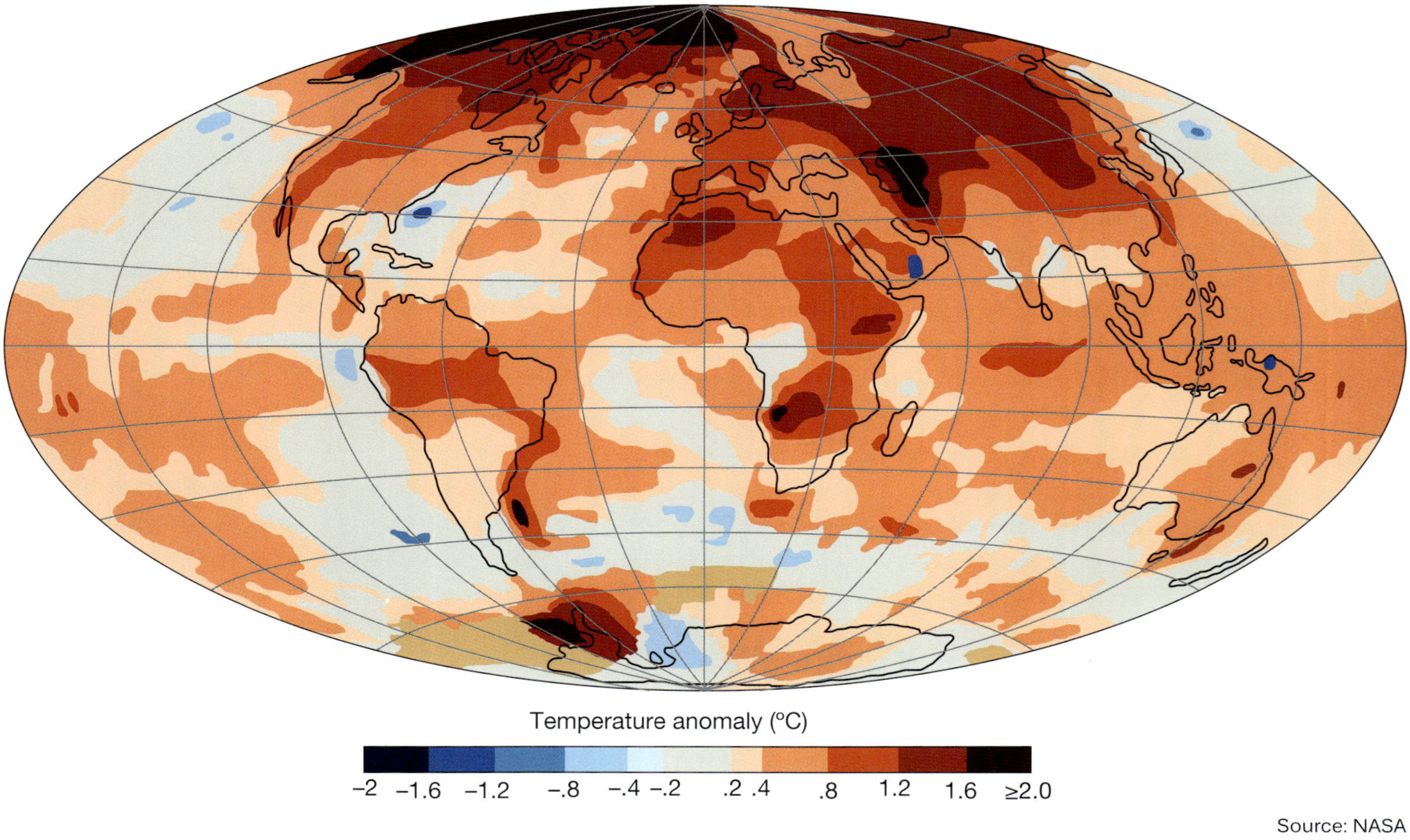

Temperature anomaly (°C)

−2 −1.6 −1.2 −.8 −.4 −.2 .2 .4 .8 1.2 1.6 ≥2.0

Source: NASA

13.1.5 Surface temperature anomalies on a global scale

Climate change in Australia

Regular and reliable climatic data have been collected in Australia since the early twentieth century. The Australian Bureau of Meteorology is responsible for collecting and analysing this data. Figure 13.1.6 shows the annual temperature anomaly for Australia between 1910 and 2019. The greatest ever anomaly was recorded in 2019. It is clear the Australian climate has warmed significantly since the 1980s.

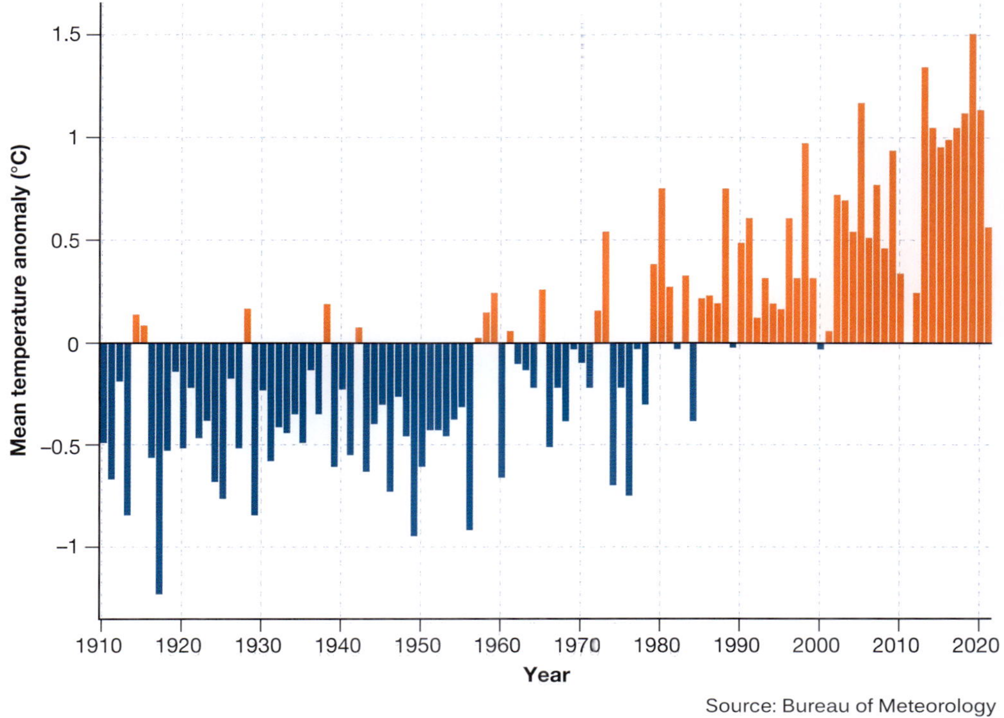

13.1.6 Annual mean temperature anomaly 1910–2021

Figure 13.1.7 shows the spatial extent of the temperature anomalies across Australia. This map was developed by the Bureau of Meteorology. It shows virtually the entire continent saw an increase in the average surface temperature by at least 0.15°C to 0.2°C each decade from 1970 to 2021.

The frequency and number of extreme heat events occurring in Australia are also sharply increasing (see Figures 13.1.8 and 13.1.9).

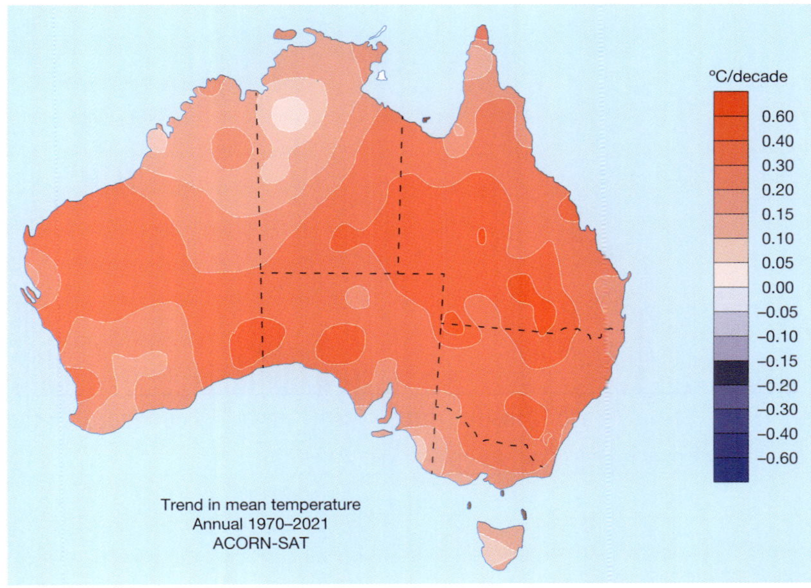

13.1.7 Temperature anomalies 1970–2021 in Australia

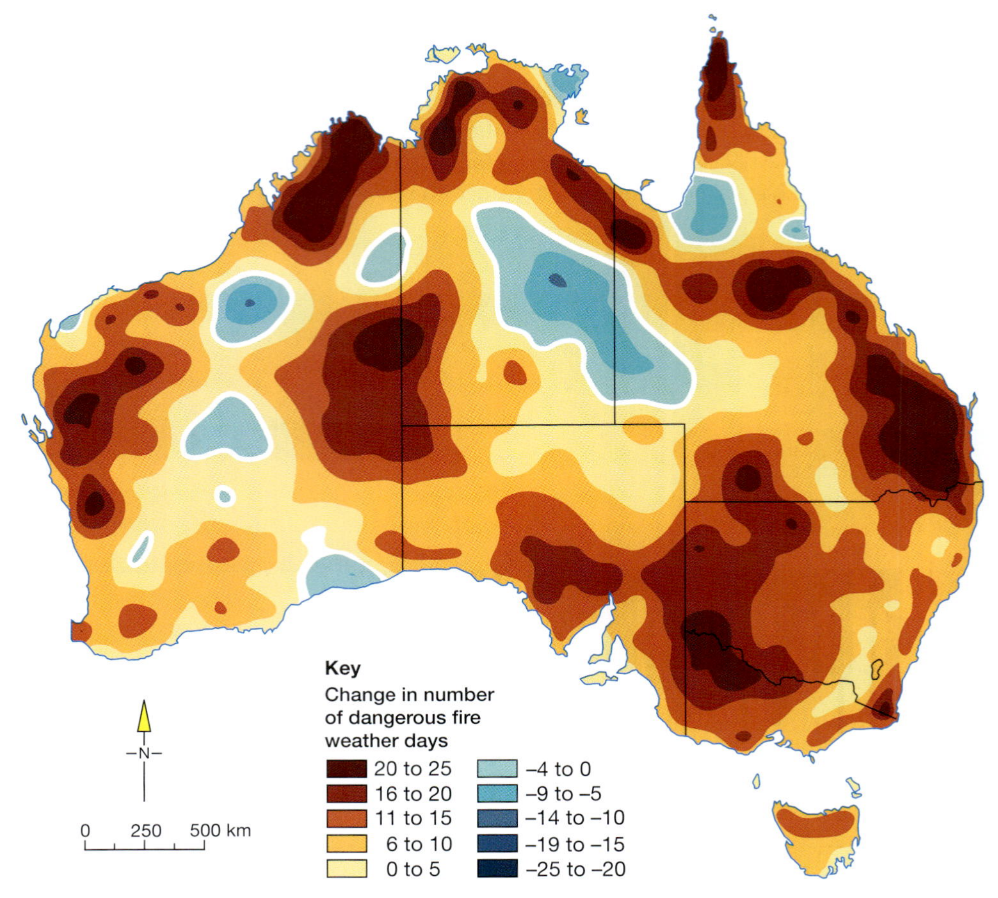

13.1.8 The number of days with dangerous conditions for bushfires has increased across much of Australia.

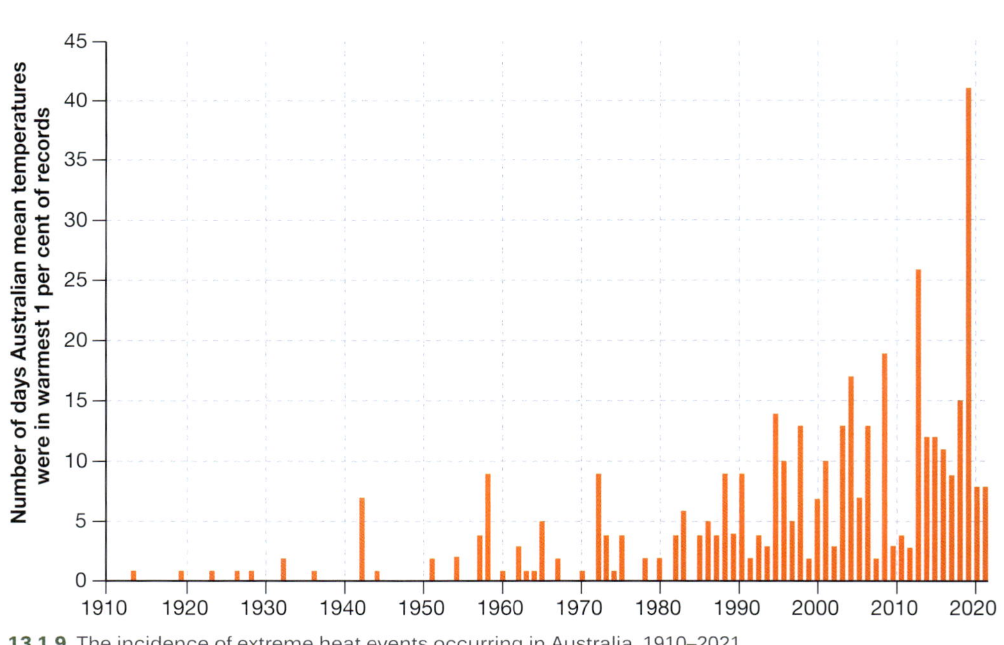

13.1.9 The incidence of extreme heat events occurring in Australia, 1910–2021

Activities

Acquiring and processing geographical information

1. Outline the nature and extent of climate change. What is its principal cause?
2. Prepare your own definition for the term climate change.
3. State what a temperature anomaly is.
4. Describe how sea surface temperatures are measured. How much has the temperature of the world's oceans changed?
5. Explain the link between sea surface temperatures and storm events.

Applying and communicating geographical understanding

6. Study Figures 13.1.1 and 13.1.2. Using data from the graphs, explain the trends in global temperatures from the mid-nineteenth century to the present day.
7. Figure 13.1.1 displays multiple sources of data. Explain how this adds to the reliability and usefulness of this graph.
8. Study Figure 13.1.4. Using data from the graph, describe the trend in ocean heat content 1960–2015. What is the trend over time in the level of confidence in the estimates made?
9. Study Figure 13.1.5. Write a short description of the global patterns of temperature anomalies.
10. Refer to Figures 13.1.6 to 13.1.9. Using this data, write a short report outlining climate change in Australia. Make specific reference to:
    - the annual mean temperature anomaly
    - the spatial pattern of temperature anomalies
    - the spatial patterns of increases and distribution of dangerous conditions for bushfires
    - the frequency of extreme heat events.

UNIT 13.2

Environmental and human causes of climate change

Earth's climate fluctuates over very long periods, measured over tens of thousands to millions of years. These fluctuations have, over time, dramatically changed its environment, ecosystems, and its flora and fauna. However, the change experienced in recent decades differs vastly from what occurred in the distant past. The rate and magnitude bear no resemblance to naturally occurring climate change; they are far more rapid and occurring on a global scale.

Natural climate change

Many factors drive Earth's climatic patterns. Ocean temperatures and currents are crucial, as is the distance from the Sun, **solar radiation** and the influence of solar storms and the chemical composition of the atmosphere. These factors evolve and change, making the climate change.

13.2.1 The Brunt Ice Shelf in Antarctica. During ice ages, massive ice sheets covered large areas of Earth's surface.

Ice ages

Periods of intense cold are regular cycles. Scientists have discovered evidence for at least five significant ice ages and many minor ones. In the mid-1850s, scientists thought the ice ages resulted from changes in the geometry of Earth's orbit around the Sun. It wasn't until the 1960s that they could take deep sediment cores from below the ocean and on land. They used these core samples to develop ultra-long-term climatic histories of the planet. They revealed periods of glaciation, times when ice sheets and glaciers formed due to very cold climatic conditions (see Figure 13.2.1) and periods of warming when the glaciers and ice sheets melt, sometimes known as terminations.

Collaborating with astronomers and geologists, Earth scientists were able to show a strong correlation between these periods of freezing and warming and the shape of Earth's orbit around the Sun along with its tilt. They aged an ice age's termination at about 900 000 years ago. This corresponded with a slight change in the tilt of Earth towards the Sun. The tilt increased the amount of insolation—the amount of solar radiation (energy in the form of heat) that reached the surface, warming its icy surface. This led to the ice melt of the termination.

Further studies reveal relatively regular cycles of glaciation due to very cold temperatures, followed by terminations during periods of warming. The length of terminations varies, with some lasting a few thousand years and others tens of thousands of years. The science behind the length of a termination is still unknown, but science indicates these ice age cycles—periods of glaciation and then termination—run between 40 000 and 100 000 years.

Some **climate change sceptics** argue the current climate change is part of this natural cycle of warming, as part of the last ice age termination. This is disputed by climate scientists. They argue that the speed of the current climate change is far too rapid to be associated with an ice age termination.

Volcanic activity

Very large volcanic eruptions have the power to influence climate, but this is typically over relatively short intervals. An eruption injects vast quantities of gas and ash into the atmosphere. The ash quickly falls back to the surface, but the volcanic gases remain in the atmosphere (see Figure 13.2.2).

These gases include sulphur dioxide, which cools the atmosphere, while gases such as carbon dioxide have a warming effect. Once in the atmosphere, sulphur dioxide turns to sulfuric acid, which quickly condenses into a fine aerosol. Its chemical structure reflects solar radiation from the Sun into space before it can reach Earth's surface. US Geological Survey research found volcanic eruptions during the twentieth century may have reduced temperatures by around 0.25°C for one to three years. In recent history, the eruption of Mt Pinatubo in the Philippines on 15 June 1991, released a staggering 20 million tonnes of sulphur dioxide into the atmosphere. The eruption led to a cooling of the world's climate of around 0.75°C for two years.

13.2.2 Karymsky, a volcano on the Kamchatka Peninsula, Russia, ejects large amounts of gas into the atmosphere.

Carbon dioxide is one of the main atmospheric gases contributing to global warming. Many geologists and volcanologists believe that many millions of years ago, in the deep geologic past, volcanoes may have released vast quantities of carbon dioxide. It significantly changed the global climate and potentially led to mass extinctions. This is still a topic of intense scientific debate.

However, it is agreed that there is no record of warming events associated with volcanic eruptions in recent history, quite unlike the many recorded cooling events associated with sulphur dioxide releases. The amount of carbon dioxide released by volcanoes annually is less than 1 per cent of the amount released by human activities. By way of comparison, the eruption of Mt St Helens in Washington State, USA, in 1980 released 10 million tonnes of carbon dioxide into the atmosphere in just nine hours. The US Geological Survey revealed that global human activity releases this amount every 2.5 hours. See Chapter 4 for more on the Mt St Helens eruption.

Human-induced climate change

Naturally occurring climate change is gradual over long time frames—tens or hundreds of millions of years. **Human-induced climate change** is climate change that is brought about by the actions of humans. This climate change has a far more rapid rate of change and operates on a global scale.

Disruption to the carbon cycle

Carbon is a key building block in the structure of all living things. Without carbon, life as we know it, could not exist on Earth. The **carbon cycle** is a naturally occurring process that cycles carbon through the biophysical environment. Carbon is released into the atmosphere in the form of carbon dioxide and is removed from the atmosphere and stored by natural processes. This system is highly effective at cycling carbon dioxide throughout the environment and storing excess carbon in storages, known as '**carbon sinks**'. Figure 13.2.3 shows how humans now play a crucial role within the carbon cycle. Human activities disrupt the natural carbon cycle. This disruption is the single biggest cause of the climate change crisis facing the world.

There are two key aspects relating to carbon transfer within the carbon cycle. The first involves the biosphere, the living aspects of the biophysical environment—flora and fauna. Animals respiring (breathing) release carbon dioxide as does decaying organic matter. This atmospheric carbon dioxide is removed via photosynthesis, the process that plants use to convert solar radiation into the energy they use to live (see Figure 13.2.4). Oxygen is released as a by-product. This natural cycle of release and absorption is essentially in balance, with the amount of carbon dioxide released being equivalent to what is absorbed via photosynthesis.

The second key aspect of the cycle is an exchange of carbon between the atmosphere and the ocean. Within the oceans, water is cycled vertically from the shallow surface waters to the deep waters towards the seafloor. Water is also cycled horizontally and

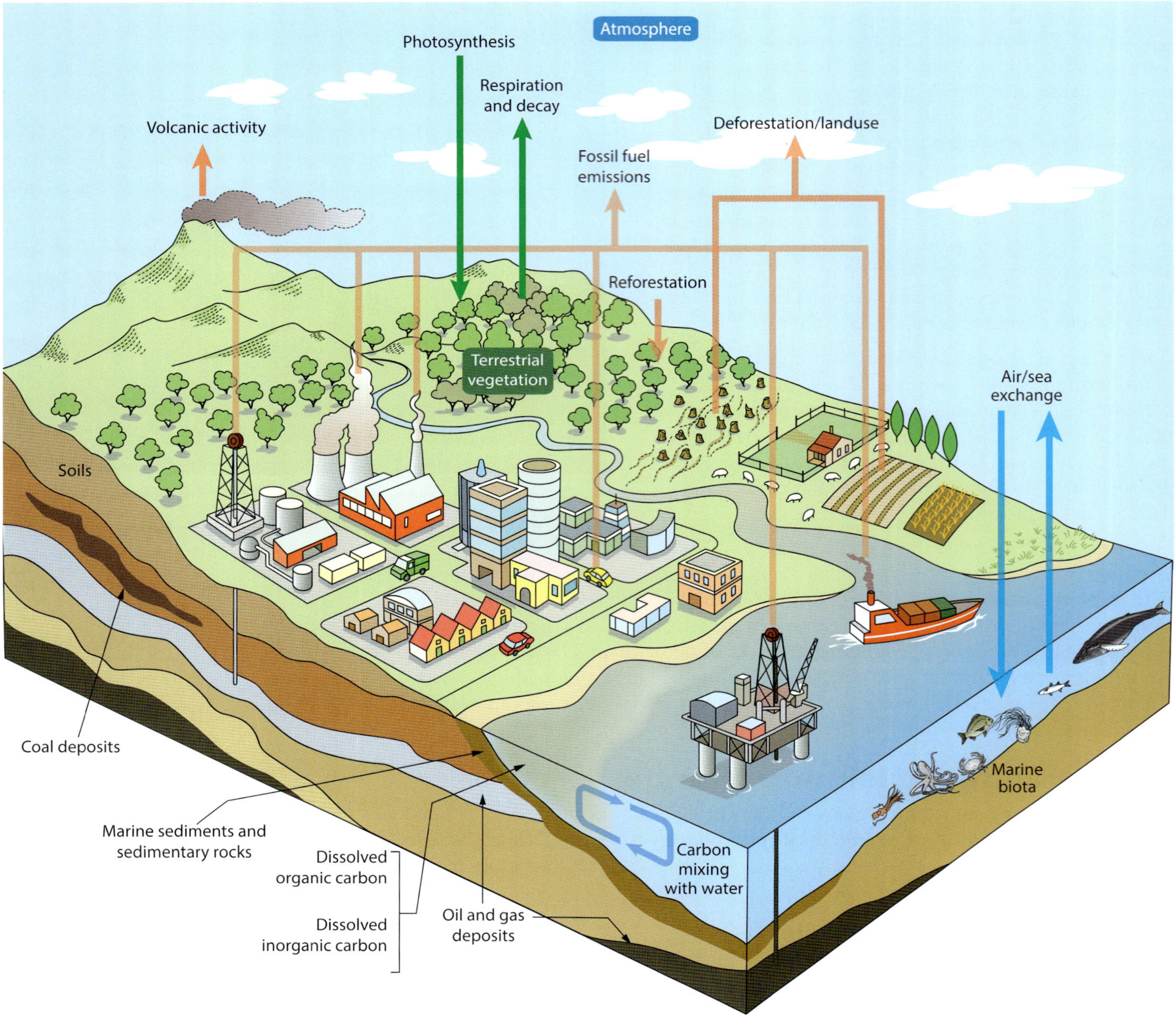

13.2.3 The carbon cycle including human interventions

13.2.4 Plants absorb CO_2 from the atmosphere through photosynthesis

moves vast distances via ocean currents. In shallow waters, carbon dioxide is exchanged with the atmosphere, while in deep waters it is stored. As the deep waters move vertically towards the surface, they bring carbon dioxide with them. This allows its exchange, and the reverse occurs as the shallow waters move deeper. This process is also naturally in balance with the movements counteracting each other.

Long-term carbon storages exist within the carbon cycle. These are found in ocean sediments and sedimentary rocks. Carbon is also found in deposits of coal (carbon in a solid state), oil (carbon in a liquid state) and natural gas (carbon in a gaseous state). These deposits were laid down millions of years ago when decaying organic materials were buried. The three states were created by varying degrees of pressure, with coal forming under the most pressure and gas the least.

These storages help the carbon cycle to remain in balance. This is because they take carbon out of the cycle, ensuring there is less carbon being exchanged with the atmosphere. They effectively 'lock' carbon out of the atmosphere. By **sequestering** carbon this way, the atmosphere evolved into its natural state, where the normal situation is a balanced carbon cycle.

Like all naturally occurring cycles, the carbon cycle is finely tuned. It is a balanced system where inputs and outputs equal each other. However, human activity is the third and non-naturally occurring part of the modern carbon cycle. Human ingenuity found ways to access the carbon locked away and use it as fuel sources (see Figure 13.2.5). These include oil, gas and coal, and are collectively known as fossil fuels. Carbon as coal, oil or gas, is highly flammable. As its stored energy is easily released in the form of heat it can be used to generate energy.

13.2.5 Coal is a major fuel source as it is highly combustible

When the carbon from these ancient deposits is burned, carbon dioxide is released into the atmosphere. Burning coal releases carbon that was once locked away into the carbon cycle. At least 2300 gigatonnes of carbon dioxide are known to be stored in all fossil fuels (a gigatonne is one billion tonnes). Some estimates put this figure as high as 5000 gigatonnes. Using fossil fuels activates the stored carbon and adds it back into the carbon cycle. As more carbon dioxide is added, the capacity for the cycle to absorb this carbon back into the system becomes increasingly reduced. What was once a balanced system becomes unbalanced as the carbon, in the form of carbon dioxide, accumulates in the atmosphere.

> **Did you know?**
> When carbon is placed under intense pressure, diamonds are created.

Figure 13.2.6 reveals that the amount of carbon dioxide in the atmosphere reached 410 parts per million in 2019. Scientific research on core samples shows that over the last 800 000 years, carbon dioxide had not risen above 300 parts per million until 1950. So, over the last 70 years it has increased nearly 30 per cent. Such a rapid increase can only be explained by the massive growth in fossil fuel use since global industrialisation began in the late-nineteenth century. The disruption of the carbon cycle has significant impacts on the chemical structure of the atmosphere which, in turn, affects climate.

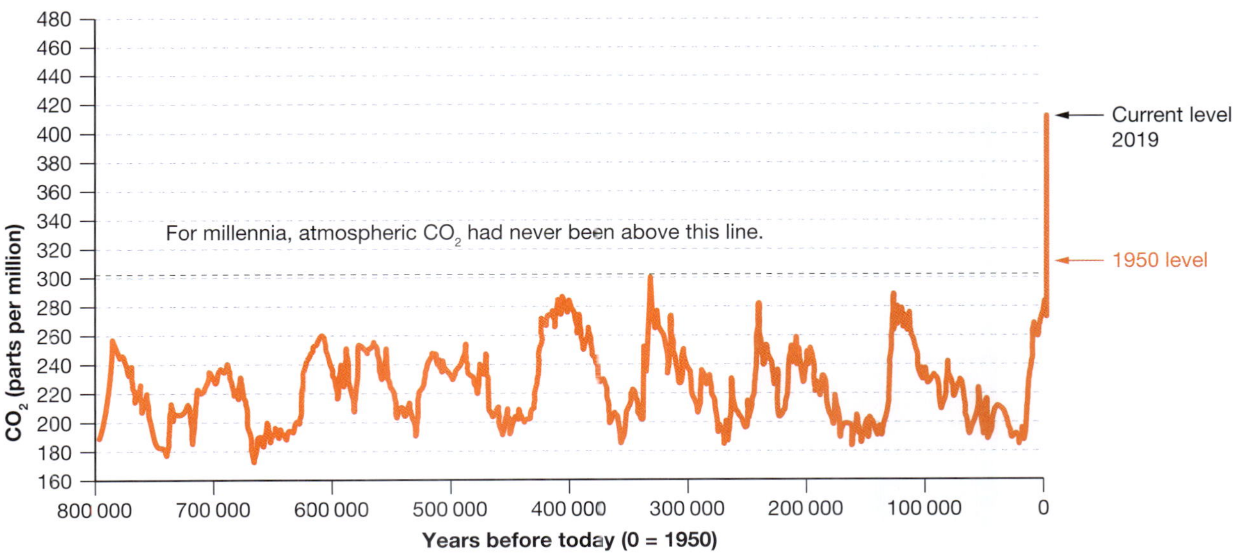

Source: NASA

13.2.6 Level of CO_2 in Earth's atmosphere over the last 800 000 years

13.2.7 Deforestation reduces the capacity of the biosphere to absorb carbon from the atmosphere.

Human activities have also decreased the capacity of carbon to be absorbed by the biosphere. Widespread land clearing and **deforestation** have significantly reduced the extent of carbon-absorbing forests. Forests still cover around 30 per cent of Earth's land surface, but they are being reduced at a rapid rate (see Figure 13.2.7). In the 25 years between 1990 and 2015, for example, 1.3 million square kilometres of forest was removed, mostly to make way for agriculture. This is an area bigger than South Africa and four times bigger than Italy. See Chapter 5 for more on deforestation.

The greenhouse effect

The Sun is Earth's main source of energy. This energy is received as incoming short-wave solar radiation (insolation), which controls the planet's climate, weather and water cycle. Because sunlight is converted into chemical energy via photosynthesis in green plants, the Sun supports all life on Earth.

Only 24 per cent of incoming solar radiation reaches Earth's surface directly. Another 21 per cent reaches the surface as diffuse radiation. Incoming radiation is transformed into heat energy at the surface. As it heats the ground, it radiates long-wave or infrared energy back into the atmosphere. Water vapour and carbon dioxide absorb 94 per cent of this energy, creating a natural **greenhouse effect**, shown in Figure 13.2.8. This traps some of the heat, and without it Earth would be much colder than it is now.

Human disruption to the carbon cycle is leading the way to a very significant increase in the amount of carbon dioxide in Earth's atmosphere. Several other gases also contribute to the greenhouse effect. These include nitrous oxide (N_2O), methane (CH_4), water vapour (H_2O) and ozone (O_3). Carbon dioxide is efficient at absorbing solar radiation, so its increased presence in the atmosphere is leading to an enhanced greenhouse effect. It means more of the solar radiation becomes trapped in the atmosphere rather than returning to space. As greenhouse gases increase so does global warming.

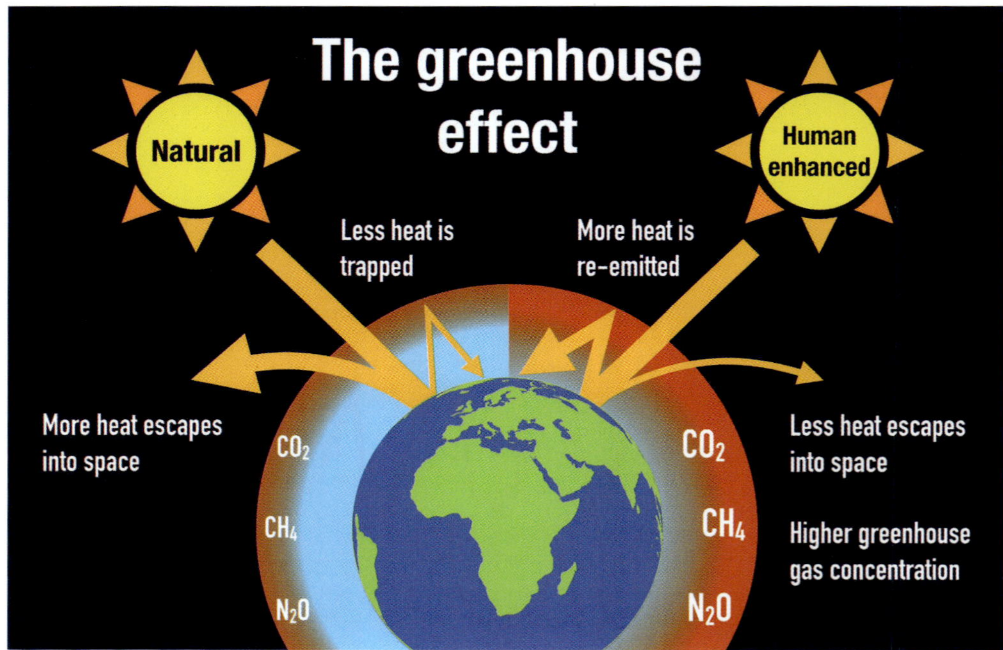

13.2.8 Natural versus human-induced greenhouse effect

Activities

Acquiring and processing geographical information

1 Define the concept of natural climate change.
2 Explain what an ice age is.
3 Outline the relationship between the Sun and ice ages.
4 Explain why the current period of climate change is not associated with an ice age termination.
5 Describe the impact of volcanic activity on climate change.
6 Compare the rate of carbon emissions from volcanoes to human-related activities.
7 Explain the concept of human-induced climate change.
8 State what a carbon sink is.
9 Explain the role of the biosphere in the carbon cycle.
10 Outline the role of the oceans in the carbon cycle.
11 Explain the main impacts of humans on the carbon cycle.
12 Explain the greenhouse effect.

Applying and communicating geographical understanding

13 Prepare a short discussion on whether the current period of climate change could be the result of naturally occurring climate variations.
14 Study Figure 13.2.3. Write a short description of the operation of the carbon cycle.
15 Study Figure 13.2.6. Explain the trends shown in the graph and use the data to describe the impact of human activity on carbon levels.
16 Study Figure 13.2.8 and the related text. Complete the following activities.
  a Outline the operation of the natural greenhouse effect.
  b Explain how the greenhouse effect is enhanced by emissions of greenhouse gases by humans.
  c What is the impact on the climate of an enhanced greenhouse effect?
20 In addition to carbon dioxide, methane and nitrous oxide are other significant greenhouse gases. Using the internet, conduct research on the types of human activities that lead to an increase in these gases in the atmosphere.
21 Create a flow chart to demonstrate how human activities disrupt the carbon cycle and enhance the greenhouse effect.

UNIT 13.3

Environmental and human impacts of climate change

Impacts on environments and natural systems

Climate change has significant consequences for Earth's natural systems. Interrupting the normal carbon cycle operation is the most significant cause of climate change. As the climate changes, ecosystems become stressed and altered. This impacts the flora and fauna that have evolved to inhabit an ecosystem and its unique conditions.

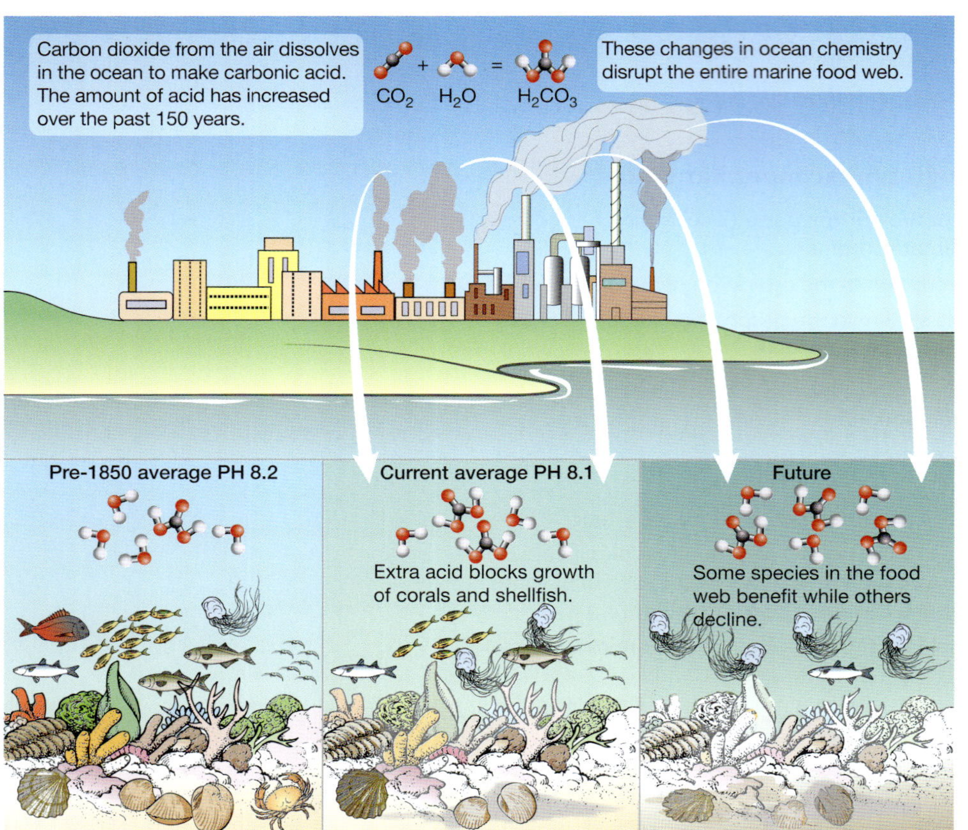

13.3.1 The oceans absorbed about half of all CO_2 pumped into the atmosphere in the last 250 years.

Ocean acidification

The world's oceans are the major carbon sinks. They absorb and store huge quantities of carbon from the atmosphere. The recent massive increase in carbon dioxide means more and more carbon is entering the world's oceans. This is beginning to change the chemical structure of the ocean's salt water.

Recent scientific research backs this up. Since the mid-eighteenth century, the pH of the ocean's saltwater has declined so much that it is now 30 per cent more acidic than it was in 1750. This rate of acidification is around 100 times more rapid than during the last 650 000 years.

Ocean acidification has significant consequences for marine life, as shown in Figure 13.3.1. As the chemical structure of the water changes, the calcium carbonate dissolved in it is reduced. This is a key ingredient used by many marine creatures to create their shells and skeletal structure (see Figure 13.3.2). The potential change is quite severe. Marine scientists estimate the amount of calcium carbonate across the world's oceans could be halved by 2100 if climate mitigation is not pursued. This will have devastating consequences for ocean ecology. As pH levels fall, many species will experience stunted growth, poor reproduction outcomes and increased vulnerability to disease. This affects animals across the ocean food chain, beginning with tiny pteropods that form the main food source for krill. These diminutive crustaceans are the main food source of baleen whales, such as humpbacks.

13.3.2 Rising ocean acidification impacts shelled animals, including marine molluscs like this nautilus.

Impacts of natural disasters

The focus of natural disasters tends to be on their human impacts. However, these disasters also have significant consequences for the natural environment. Climate change is increasing the prevalence and intensity of climate-associated natural disasters, particularly drought, floods, bushfires and cyclones.

The devastating bushfires that raged across southern Australia in 2019 and early 2020 were among the worst recorded. These fires followed years of prolonged drought, which was made more intense by climate change. In NSW alone, 5.4 million hectares were burnt, almost the size of Croatia, including 37 per cent of the total area of national parks and 42 per cent of state forest areas. The areas affected by the fires are home to 293 threatened animal species and 680 threatened plant species. Twenty-five per cent of the state's suitable koala habitat was burnt out (see Figure 13.3.3).

Across southern Australia, 60 000 koalas were estimated to have been killed in the fires. The WWF states up to 3 billion animals of all kinds were killed in the fires or died afterwards due to loss of habitat and food. The risk of starvation was so extreme that authorities used helicopters to airdrop food, including carrots and sweet potatoes (see Figure 13.3.4), across this vast area to provide animals with nutrition.

13.3.3 The 2019–2020 bushfires burnt out koala habitat on Kangaroo Island, South Australia.

13.3.4 Helicopters dropped emergency food to surviving animals following the 2019–2020 bushfires.

Did you know?

In 2023 temperatures rose to record-breaking levels worldwide as an El Niño event combined with human-induced climate change. It was the warmest year since records began in the mid-1800s, with July the world's hottest month on record and possibly the warmest in 120 000 years.

Extreme global temperatures contributed to heatwaves, wildfires and heavy rainfall worldwide. Wildfires raged across Canada, Greece, Italy and Spain. In early August 2023, wildfires broke out in the US state of Hawaii, mainly on the island of Maui. The wind-driven fires prompted evacuations, caused widespread damage, killing over 100 people and leaving others unaccounted in the devastated town of Lāhainā. Floods affected large parts of South and East Asia.

● SPOTLIGHT

Impacts of climate change on people and communities

Climate change has already had a direct impact on people and communities. It has prompted radical changes in agricultural production because many crops will not grow in a hotter climate. Politicians often cite the high cost of taking steps to reduce climate change as the key reason for delaying mitigation. Climate change sceptics support this stance. However, inaction is actually far more costly to the economy. For example, reduced agricultural production has massive economic consequences. It also has had significant impacts on health and wellbeing.

● SPOTLIGHT

Climate change threatens the ski industry

The global ski industry has felt the impact of climate change. Few other industries are more vulnerable. Snow skiing is entirely at the mercy of the weather, and warmer temperatures mean less snow. In the USA alone the industry is worth around US$20 billion per annum. In France, in 2019, more than 170 000 people relied on the ski industry. It accounts for six per cent of Austria's economic output. In 2017, a warm winter delayed the start of the ski season in the French Alps by several weeks; a similar delay in Austria cost the industry as much as €300 million (around AU$480 million) (see Figure 13.3.5).

13.3.5 The ski fields of Courchevel, France, early January 2017. A warm winter meant most snow on the ski runs had to be made using snow guns.

Research from Colorado State University and the University of New Hampshire in the USA found significant decreases in snowfalls. There has been a 41 per cent drop since the early 1980s in Colorado's famous snowfields. The warmer winters could cause property values in many ski fields there to fall by as much as 55 per cent by 2050. They will significantly shorten or even eliminate ski seasons in the USA. The average ski season in Colorado is already 34 days shorter than in the early 1980s.

The story is similar in Australia's ski fields. Unreliable snowfalls and a relatively short skiing season are just two in a range of challenges facing the industry. A steady decline in natural snowfalls in Australia's alpine region represents a long-term threat. The maximum annual snow depth has declined since at least 1954 (see Figure 13.3.6). Fortunately, advances in snowmaking technologies have kept pace. Resorts have been able to maintain a generally good snow cover throughout the shorter season.

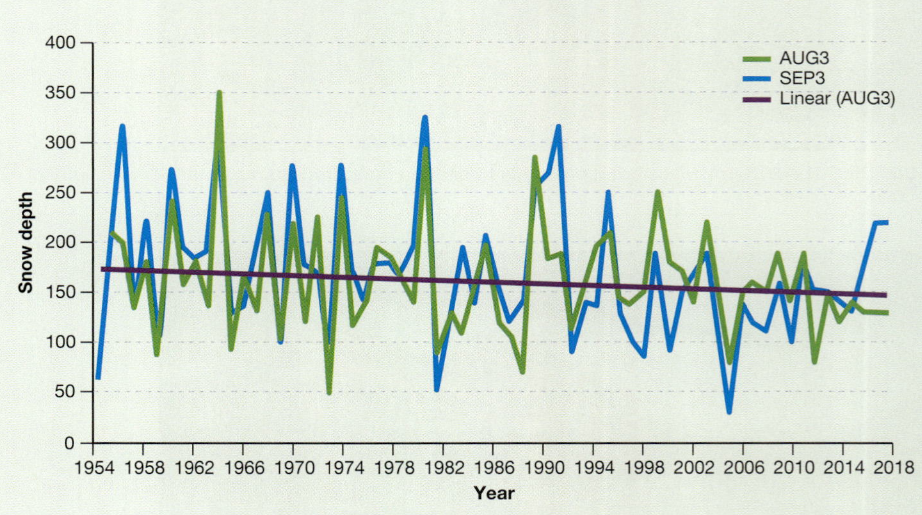

13.3.6 Record of snow depth taken at Spencers Creek in the Snowy Mountains, on 3 August and 3 September each year from 1954–2018

Health impacts

The WHO estimates that between 2030 and 2050 climate change will cause an extra 250 000 deaths from malnutrition, malaria and heat stress. A key consequence of climate change will be the widening spread of many tropical diseases. For example, a warmer world will see the habitat of mosquitoes expand considerably. Mosquitoes spread numerous diseases, including malaria, dengue fever, zika and West Nile viruses (see Figure 13.3.7). These are already very common in the tropics, but warmer temperatures will make mosquito control increasingly difficult in regions further away. Across the tropics, currently 400 000 people die annually from malaria.

Ironically, warmer temperatures may reduce the prevalence of some diseases in the tropics. For example, mosquito species carrying the malaria virus are more common when typical temperatures do not exceed 25°C too often. However, these mosquitoes will become more common in currently cooler areas.

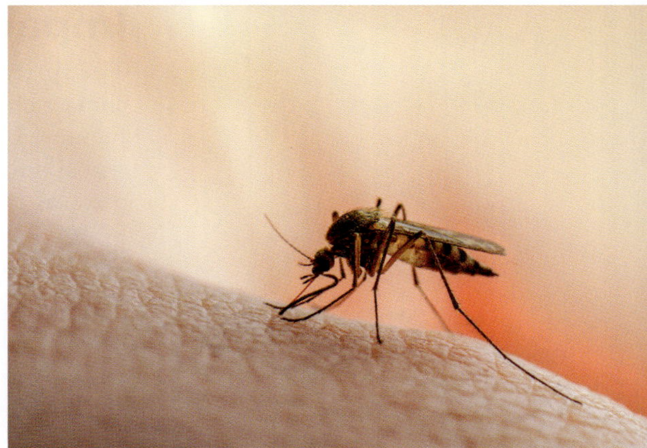
13.3.7 Climate change will expand the range of disease-carrying mosquitoes

Climate change will have an impact on the prevalence and range of animal diseases. For example, Nipah virus, which typically infects pigs and horses, was first detected in Malaysia in the late 1990s. When transmitted to humans it is fatal in 75 per cent of cases. Since 1998, more than 700 people have died from Nipah. Although not yet present in Australia, CSIRO's scientific modelling shows that climate change will create conditions in Northern Australia that will support its spread. So it is likely Nipah will be found in Australia by 2050.

Q Fever is another disease expected to expand with warming temperatures. Currently found only in North Queensland, it is typically carried by infected kangaroos, wallabies and bandicoots. Habitat loss and the impacts of climate change on the natural environment have pushed these species into urban areas. This is increasing the potential to transmit this highly infectious disease from animals to humans.

Heatwaves

Worldwide, climate change is increasing instances of extreme heat. Heatwaves increase deaths among the vulnerable, including the elderly and those with underlying health conditions. Heart attacks and respiratory illnesses spike during heat waves. Around 70 000 deaths were attributed to an intense heatwave across Europe in 2003. As the number of extreme heat waves increases, so will the related health impacts. Heatwaves also significantly impact food production. Research from the WHO estimates that rising temperatures will increase the prevalence of death due to malnutrition by 3.1 million per year over the next 30 years. This impact will be predominantly in developing countries.

Waterborne diseases

Climate change also increases the prevalence of waterborne diseases. For many people in the developing world, drought-reduced access to safe drinking water is forcing them to use contaminated water, exposing them to diseases. Equally, flood events contaminate drinking water, increasing health risks. The intensity and frequency of both droughts and floods will increase with climate change.

Natural hazards

There are many increases in the direct risks to human health and safety due to climate change. The prevalence and intensity of weather-related natural hazards has tripled since 1960. The threat of natural hazards is increasing, including wildfires, storms, flooding and droughts. WHO reports that these events account for around 60 000 deaths per year.

13.3.8 Thick smoke from bushfires blanketed Sydney for days in the 2019–2020 summer, increasing health risks.

Natural hazards present an enormous risk to human health and wellbeing. During the intense bushfires of 2019–2020, air quality in fire zones reached dangerous levels. Health researchers examining the impact found there were 417 deaths directly attributable to the bushfire smoke, along with thousands of hospital admissions (see Figures 13.3.8 and 13.3.9).

Health impact of bushfire smoke	Number of cases
Deaths	417
Cardiovascular hospital admissions	1124
Respiratory hospital admissions	2027
Asthma hospital admissions	1305

Source: Borchers, Arriagada et al., *Medical Journal of Australia*

13.3.9 Health impacts of the 2019–2020 bushfires (NSW, Vic, Qld and ACT) 1 October 2019–10 February 2020

Impacts on crop production

The agricultural sector is one of the most vulnerable parts of the global economy to climate change. This is because crop production is so dependent on climatic conditions. Figure 13.3.10 shows the predicted changes in crop yields for the four main staples of the global human diet, rice, wheat, potatoes and corn, by 2050. Due to a warming climate, patterns of production are expected to dramatically change over the coming decades.

Figure 13.3.11 reveals some areas will experience increased yields, especially Canada, northern Asia and Europe. These areas currently experience much colder average temperatures, so they have shorter growing periods, making their yields smaller. Climate change will bring warmer temperatures, moderating this average, thus increasing the growing season. However, other large parts of the world will experience significant declines in crop yields because the temperatures will increase too much. This escalates the prevalence of drought and heat stress on crops. It will affect all of India and much of China, two countries expected to have a combined population exceeding 4.1 billion by 2050. Large areas of Africa, South and Central America and much of South-East Asia will also suffer from reduced yields. Globally, the pattern will show a decrease in crop yields.

The developing world will bear the brunt of this impact, those parts of the world least equipped to deal with climate change, particularly given their high reliance on agricultural production. WHO and the United Nations Food and Agricultural Organization (FAO) predict that deaths associated with malnutrition brought on by increasing food scarcity directly as a result of climate change will increase by 2050. Subsistence farming communities are most vulnerable to changes in yields and increasingly unreliable weather patterns. These are farmers who grow only enough, or little more, to feed their own families.

In Australia, some areas will become more arable—able to support agriculture—as climate change increases rainfall, particularly in southern regions. However, it will also increase the frequency and intensity of droughts, making agricultural production less reliable and more variable than it already is.

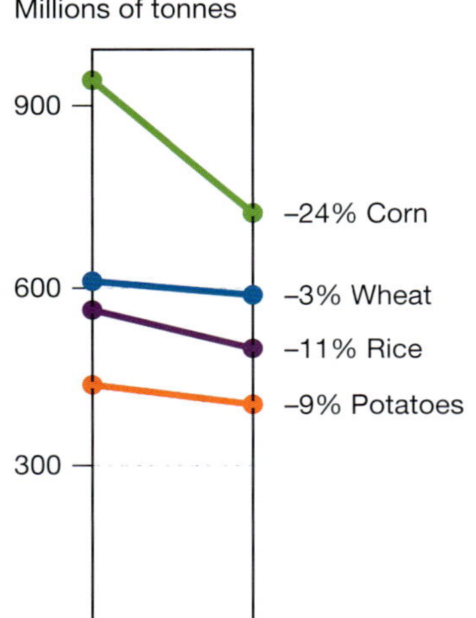

13.3.10 Predicted changes in global crop yields 2000–2050

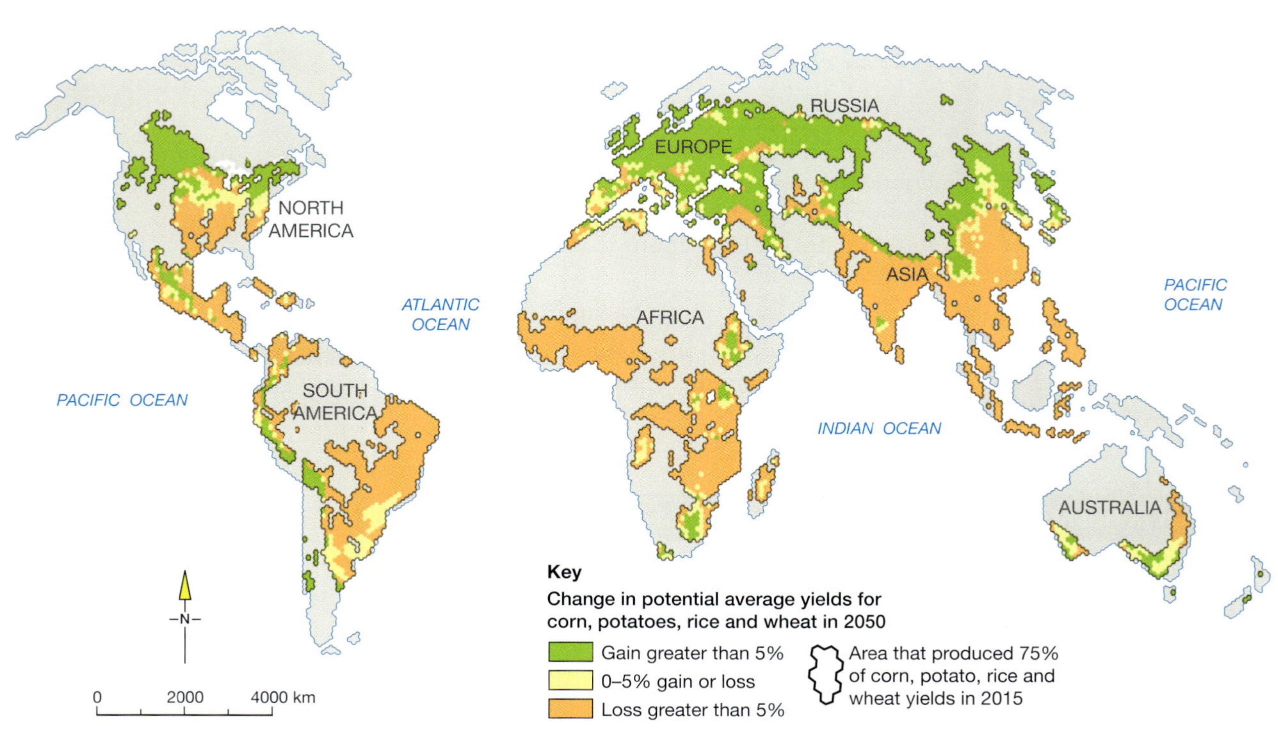

13.3.11 Impact of climate change on crop production

Activities

Acquiring and processing geographical information

1. Outline the process of ocean acidification.
2. Describe the consequences of ocean acidification.
3. Outline the impact of the 2019–2020 bushfires on the natural environment.
4. Outline the impacts of climate change on human health.
5. Explain how the incidence of malaria may be affected by climate change.
6. Explain the link between climate change and increased health impacts of natural hazards.
7. Outline the impact that climate change is projected to have on food production.

Applying and communicating geographical understanding

8. Using specific examples, prepare a short report on the impacts of climate change on the natural environment.
9. Study Figure 13.3.1. Write a report outlining the impacts of acidification on the world's oceans.
10. Using the internet, conduct research into the current status of the global ski industry. Describe the last ski seasons in Australia and Europe.
11. Study Figure 13.3.6. Using data from the graph, describe the long-term trend in snow depth at Spencers Creek.
12. Prepare a simple mind map showing the impacts of climate change on human health and wellbeing.
13. Study Figures 13.3.8 and 13.3.9. Write a paragraph outlining the impacts of the 2019–2020 bushfires on human health.
14. Refer to Figure 13.3.10. Describe the impact of climate change on crop production.
15. Using the text and Figures 13.3.10 and 13.3.11, write a brief report on climate change and food production.

■ CASE STUDY

The impact of climate change on the Great Barrier Reef

Coral reefs are one of the ecosystems most vulnerable to climate change. A key reason is that corals are highly specialised and require quite specific conditions to survive. Temperature is one of the most essential conditions. Coral will only grow in areas where the sea temperature does not fall below 17° to 18°C and rarely exceeds 33° to 34°C.

The Great Barrier Reef is the world's most extensive reef system. It is actually a collection of more than 2000 reefs stretching 2300 kilometres from Papua New Guinea's Fly River in the north down the Queensland coast to Fraser Island (K'gari) in the south. The reef system covers a massive 348 000 square kilometres. It has been identified as one of the environments most at risk of damage from climate change.

Coral reefs are complex structures. They mostly consist of rock, but the living reef is built by tiny animals called polyps. These primitive organisms consist of little more than a digestive sac or stomach and an outer skeleton of limestone (calcium carbonate). The polyp pulls food into its mouth through a ring of tentacles. Within the polyp live symbiotic algae called zooxanthellae. The polyp and the zooxanthellae mutually benefit each other, with the zooxanthellae producing sugars and oxygen through photosynthesis while the polyp provides nutrients to the zooxanthellae. The zooxanthellae give corals their colours.

The Great Barrier Reef lies in the Coral Sea. Figure 13.3.12 displays data from the Bureau of Meteorology on the summer sea surface temperatures there. The last two decades experienced higher temperatures, with 2016 having a record high. This increase resulted from the enhanced greenhouse effects. The Bureau estimates 93 per cent of this additional heat is absorbed by our oceans, making them warmer.

13.3.13 Healthy corals on the Great Barrier Reef

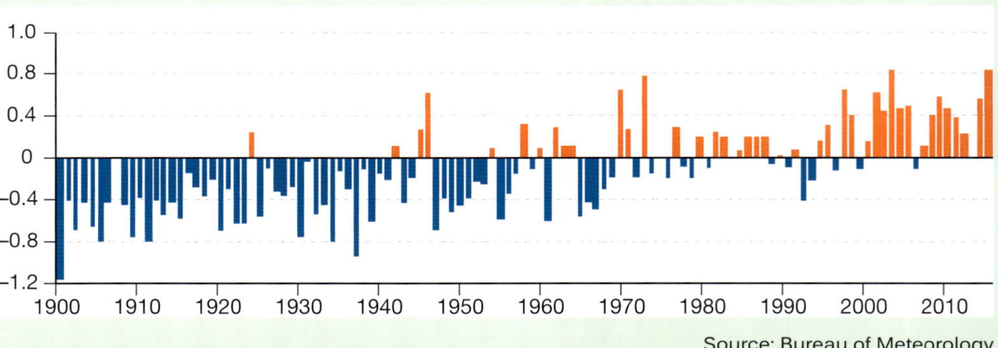

Source: Bureau of Meteorology

13.3.12 Coral Sea summer temperature anomalies

The higher sea temperatures directly impact the Great Barrier Reef. When the waters warm too much, the coral polyps become distressed. For reasons not yet fully understood, the distressed polyps expel the zooxanthellae. The most obvious sign of this is that the corals lose their colour, known as coral bleaching. This occurs over time, as polyps die without the zooxanthellae to provide them with energy (see Figures 13.3.13 and 13.3.14).

13.3.14 Bleached coral on the Great Barrier Reef

The worst coral bleaching event on the reef occurred in 2016. Record sea surface temperatures saw 29 per cent of shallow reefs across the Great Barrier Reef bleached, with the warmer waters in the north most affected. Researchers estimated 93 per cent of reefs across the system had some bleaching in 2016 (see Figure 13.3.15). In the south where there was little bleaching, many of the reefs were able to recover quickly, but coral mortality rates in the northern sections were very high.

Coral also has specific requirements for light and salinity. Like other plants, zooxanthellae require light to photosynthesise. Corals also require relatively stable salinity (saltiness) levels. When salinity levels fall, zooxanthellae die. If the level is too high, coral growth is stunted.

A key consequence of warmer sea temperatures in the Coral Sea is an increase in the frequency and intensity of tropical cyclones. Cyclones can directly damage the reef by causing storm waves, but they also bring huge volumes of rainfall. Rainfall from large storms and tropical cyclones impacts the reef in two ways:

- An increase in rainfall causes widespread erosion across inland areas. Large river systems, such as the Burdekin, Fitzroy and Gregory rivers drain huge areas of Central Queensland into the Coral Sea and therefore onto the Great Barrier Reef. In times of flood, they carry vast quantities of sediment (see Figure 13.2.16). This increases the turbidity (cloudiness) of the water, reducing the light that can penetrate the water, which affects the zooxanthellae.
- The vast quantities of fresh water that flow down the river systems during floods change the salinity levels of the reef waters and this also affects the zooxanthellae.

A feature of the Great Barrier is the existence of significant fringing reefs. These are reefs that grow close to and parallel to the shoreline. They are particularly at risk from the above-noted impacts. In 2017 a large Category 4 storm, Tropical Cyclone Debbie, brought huge winds (up to 260 km/h) and widescale flooding. Figure 13.3.9 shows the sediment plume of the Burdekin River after the storm, indicating the extent of the impact.

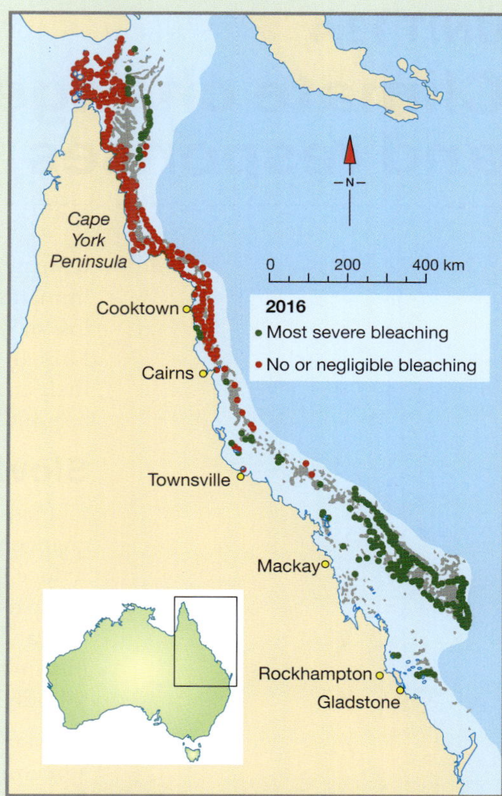

13.3.15 Extent of coral bleaching on the Great Barrier in 2016, the worst on record

13.3.16 Sediment plume of the Burdekin River in North Queensland before and after Tropical Cyclone Debbie in 2017

Activities

Acquiring and processing geographical understanding

1. Explain why coral reefs are so susceptible to climate change.
2. Describe the relationship between zooxanthellae and coral polyps.
3. Refer to Figure 13.3.12. Analyse the trend patterns shown in the graph.
4. Explain the cause of coral bleaching.
5. Outline the impact of rising sea temperatures on the Great Barrier Reef.
6. How do tropical cyclones affect the Great Barrier Reef?

UNIT 13.4

Climate change: Challenges, opportunities and responses

Climate change presents enormous challenges. It is already threatening natural and human systems. If left unchecked, over the twenty-first century it will result in further loss of ecosystems. The consequences for communities around the globe will be dire. Human ingenuity created the technology that could unleash the energy stored in fossil fuels and created climate change. Human ingenuity can also create opportunities to respond to the climate crisis and solve the problems it presents.

Slowing the rate and magnitude of change

Climate scientists agree that immediate action is vital to avoid the worst effects of climate change. In 2015, the world's leaders met in Paris, France, to begin setting targets to reduce greenhouse gas emissions. Known as the Paris Agreement, this international compact came into effect in 2016. Its overall goal is to keep global temperature increases below 2°C and preferably below 1.5°C compared to pre-industrial levels (usually measured in the mid-1700s). Under the agreement, the signatory countries committed to individual actions and targets.

Like any international agreement, the Paris Agreement is far from perfect. Such agreements are by their nature voluntary, so any country can withdraw its support. There are also limited mechanisms to take action against those countries that fail to meet their targets. In 2017, recently elected President Donald Trump announced the USA's withdrawal from the agreement. With the USA being the world's largest greenhouse gas emitter, this obstructed the agreement. The USA later returned to the agreement following Joe Biden's inauguration as president in 2021. It was a key election pledge, one supported by his voters.

Net-zero emissions

Several countries have gone much further than their Paris Agreement targets. Costa Rica was the first country to adopt a 2050 **net-zero emissions** objective. Since then, a number of countries, including major economies such as the UK, France, Spain and Australia, have also committed to net-zero emissions objectives. It means their emissions will be equal to or less than the capacity of carbon sinks to absorb them. The twenty-seventh Conference on Climate Change, held in Sharm el-Sheikh, Egypt, in late 2022 saw further commitments towards net-zero emissions. Several developed nations also committed to providing financial support for less-developed nations to move towards a zero emissions target.

Crucial to meeting these targets is the transition away from a reliance on fossil fuels towards renewable energy sources. This includes using energy sources, such as solar and wind power, which produce zero emissions during their operation. These technologies are becoming more advanced and efficient. A key concern with renewables has been their reliability. For example, solar energy is impacted by cloud cover and wind energy by calm days. However, the advent of very large batteries (including pumped hydro) enables energy storage to overcome these issues.

Transportation is another key contributor to emissions. Road transport accounts for around 12 per cent of all greenhouse gas emissions. The development of electric vehicles and those that use alternative fuels is well advanced (see Figure 13.4.2). In 2021, 6.4 million electric cars were sold globally. While a small proportion of the more than 80 million cars sold that year, electric car sales are increasing rapidly. As charging infrastructure improves, and as more governments create incentives to purchase such vehicles, the proportion is expected to grow.

● **SPOTLIGHT**

Solar energy in Australia

Australia is an ideal location for solar energy. Australian households adopting small rooftop solar systems have been key drivers in increasing the proportion of renewable energy in Australia's electricity sector. Even so, it is the large-scale projects that will have the greatest impact on emissions. In mid-2018, the Bungala solar project in South Australia began production (see Figure 13.4.1). Its location near Port Augusta is ironic because it was once famous for coal production and shipment. The semi-arid environment there is ideal for capturing solar energy.

The site was expanded further in 2020 and became fully operational. It has the capacity to provide 570-gigawatt hours, which is enough to provide power to around 113000 homes annually. More than half the state's power needs are met by renewable energy and the Bungala solar project is estimated to reduce carbon emissions by more than 500000 tonnes annually. The installation of battery storage is also helping to ensure the state overcomes any reliability concerns.

13.4.1 The Bungala solar energy project near Port August, South Australia

Carbon sinks and technological advances

Other strategies focus on increasing the capacity of carbon sinks. This is useful for industries that are presently unable to eliminate carbon emissions. The shipping industry contributes around 10 per cent of global greenhouse gas emissions. Although advances in engines have seen emissions cut, there are few options for reducing this to zero without dramatically reducing the capacity of ships to carry cargo. A similar issue arises in the airline industry. Adding more carbon sinks, like reforestation projects, to absorb the carbon emissions, will help to reduce impacts from these sources.

The agriculture sector is another significant source of greenhouse gas emissions. Several technologies are being developed to reduce this impact. For example, biochar is a soil supplement that significantly increases the capacity of the soil to absorb carbon from the atmosphere. It is produced when organic materials, such as crop waste, manure and grass, are heated at high temperatures in a low-oxygen process called pyrolysis. Australia is leading the world in biochar research. It has the potential to play a role in dealing with climate change.

13.4.2 Electric vehicles, known as EVs, are rapidly growing in popularity.

● SPOTLIGHT

Seaweed cattle supplement reduces methane

While carbon dioxide takes most of the attention relating to greenhouse gases, methane is 28 times more potent. Fortunately, overall methane emissions are much lower than carbon dioxide. Livestock raised for food is one of its principal sources. By 2022, livestock accounted for around 15 per cent of the total global greenhouse emissions. One cow's digestive system produces the same emissions per year as a typical car.

13.4.3 A food supplement made from *Asparagopsis* seaweed can almost eliminate livestock emissions of methane.

CSIRO researchers have been investigating strategies to reduce these emissions. *Asparagopsis*, a native genus of Australian seaweed, could radically reduce emissions. By adding a supplement made from the seaweed into normal cattle feed, it is possible to reduce their methane emissions by 99.9 per cent (see Figure 13.4.3). The supplement required is small, around 0.2 per cent of the cattle's diet, and can easily be used in dairy farms and feedlots. It would reduce emissions by 5 million tonnes annually and up to 500 million globally. By finding a way to use the supplement for cattle grazing on grass in paddocks, the emissions reduction could be many times greater.

The CSIRO has partnered with a commercial company to begin cultivating the seaweed on a large scale. It involves the Narungga Nation Aboriginal Corporation in South Australia.

Minimising and mitigating risks

To mitigate risk is to reduce its impact. Climate change mitigation programs acknowledge that climate change is already having an impact, also that some consequences cannot be eliminated and require action to deal with them. For example, changes in Australia's bushfire management approaches now consider the impacts of climate change in creating lengthier and more deadly bushfire seasons.

13.4.4 Floating homes in the Netherlands, a risk mitigation against rising sea levels

Similarly, many businesses now have climate change mitigation strategies to help them deal with climate change consequences and changes in government policies that result from it. For example, the Swedish auto giant Volvo decided that electric vehicles will make up half of its global car production. Following announcements by governments across Europe and some US states that all new cars are required to be electric by 2035, Volvo stated that it anticipated a rapid end to petrol cars.

Technology and human ingenuity are at the core of climate change mitigation. A key consequence is sea-level rise. This places coastal communities at extreme risk from inundation and in some extreme cases may result in permanent flooding. This is of particular concern to low-lying nations, such as the Netherlands. For centuries the Netherlands has had to deal with flooding from the sea. See Figure 13.4.4 and Unit G2.3 for more on the Netherlands' responses to flood risks.

● SPOTLIGHT

Tuvalu, the sinking nation

As a highly developed and wealthy nation, the Netherlands can use technology to mitigate the worst effects of sea-level rise. However, small island nations do not have this capacity. With a population of around 12 000, the tiny island nation of Tuvalu is one of the world's most vulnerable places to climate change (see Figure 13.4.5). Located 5100 kilometres north–east of Australia, Tuvalu is just 26 square kilometres. It is made up of nine low-lying islands, with the country's highest point being less than 5 metres above sea level.

As sea levels rise, Tuvalu is sinking beneath the waves. Two of the country's islands are almost entirely gone as waves continually erode the beaches. Fresh water is contaminated by salt water and the islands' food-producing gardens are shrinking and being lost to salt water. Fish, the main protein source, is becoming contaminated by algae that are expelled as coral becomes bleached. Infected fish are eaten by humans causing ciguatera poisoning, which leads to diarrhoea and fever. The local hospital deals with around 10 cases a week. Tuvalu's future is bleak. It has used its membership of the UN to pressure governments to act on climate change. Plans for extensive sea defences around the island have been developed. The Tuvaluan Government has even called on New Zealand and Australia to be prepared to accept them as **climate refugees** if their country becomes uninhabitable. This is the ultimate mitigation strategy for the nation and will become their last resort if sea levels are allowed to continue to rise.

13.4.5 Funafuti in Tuvalu, a country vulnerable to sea-level rise

Varying perspectives

Despite widespread scientific agreement about the causes of climate change, there still remains some disagreement. Most particularly, it relates to the disruption to the carbon cycle. Some people outrightly deny the science, believing climate change is not happening at all, or argue it is a sinister conspiracy theory. An internet search highlights the prevalence of such ideas. Websites dedicated to this viewpoint manipulate data or misreport it to create 'evidence' to support their views.

Climate change sceptics may accept that climate change is occurring, but argue it is associated with natural processes, not human causes. There is a far wider consensus that human activities cause climate change than previously. Very few politicians still deny it. The Yale Program on Climate Change Communication carries out an annual poll in the USA around people's views on climate change. In 2020, the number of Americans stating they were alarmed by climate change more than doubled from the 2015 figures, shifting from 11 per cent to 26 per cent. In 2020, the number of people who were 'concerned' by climate change sat at 28 per cent. The number of those 'dismissive' of climate change fell from 15 per cent in 2010 to 7 per cent in 2020. Research by the Australia Institute in mid-2020 randomly surveyed Australians, asking, 'Do you believe climate change is occurring?'. It found 79 per cent of respondents agreed, compared to a figure of 20 per cent in 2013. It was the highest result ever recorded; it came after the devastating 2019–2020 fires.

13.4.6 Protestors in New Zealand in late 2019 oppose the government's zero carbon emissions goal

While there is growing consensus that climate change is occurring, there remains disagreement on the best ways to tackle it. Although polls showed Australians were

13.4.7 Young people face the long-term consequences of climate change and are at the forefront of movements tackling the issue.

concerned about climate change, there was less agreement about whether climate change should be dealt with regardless of the economic costs. Some are concerned that taking action, such as moving towards zero carbon emissions or carbon taxes, will result in job losses and huge economic costs (see Figure 13.4.6). Many argue that society needs to transition away from fossil fuels gradually, for example, by moving from coal towards natural gas, which produces far fewer carbon emissions.

Others argue that inaction will have far greater economic impacts than action. They point to the enormous costs of more intense and frequent natural hazards, lost crop and food production, and threats to human health. All are associated with climate change. This is without even considering the impact on the natural environment and the associated potential costs. For example, the economic impact to the ski industry is well known. Also well-documented are the impacts of losing natural resources, such the Great Barrier Reef, which imperils the tourist industry as well as the environment. There is also a generational equity question. There is an increasing divide between younger and older people. Younger people argue that they will bear the consequences of unabated climate change (see Figure 13.4.7). It is they who will be facing the environmental, social, health and economic consequences of climate change long after those who are currently in power have died.

SPOTLIGHT

Media promotion of climate denialism

Despite growing public support for action on climate change, some media giants continue to provide a platform for climate change deniers. Many highly conservative news outlets downplayed the role of climate change in the devastating Australian bushfires of 2019–2020. Some even falsely claimed the fires resulted from arson. They also ignore newsworthy stories about the threats posed by a warming planet, while giving voice to those who claim that climate change is a hoax and/or a globalist conspiracy.

In an effort to replicate the commercial success of the sensationalist cultural conservatism promoted by USA's Fox News, News Corporation's Sky News Australia has deployed a raft of conservative, climate change-denying commentators.

Sky News's night-time line-up focuses on disseminating conservative, opinion-based content that resembles right-wing talk radio in terms of its rhetoric. While its television audience is relatively small, it reaches a much larger number of viewers via a content partnership with digital platforms such as YouTube, Microsoft News and Facebook.

Critics claim it was Rupert Murdoch's objective to energise a conservative political base by providing a steady feed of reactionary material focusing on often fabricated leftist threats. Sky News provides a platform for limelight-seeking politicians and fringe members of the scientific community. These include conservative politicians who support the coalmining industry.

The Murdoch-owned media outlets are, however, not alone in promoting climate change denialism. Sydney's most commercially successful AM radio station, 2GB, is well known for promoting far-right political causes in the search for advertising revenue. Its very conservative audience has its prejudices reinforced. While the station is now less strident in its climate change denialism, shock jocks continue to ridicule climate scientists and actively promote the interests of the coal industry.

Activities

Acquiring and processing geographical information

1. State the goal of the Paris Agreement.
2. Outline the limitations of the Paris Agreement.
3. Define the term net-zero emissions.
4. Outline the growth in electric vehicle sales.
5. Explain what biochar is.
6. Explain the concept of climate change mitigation.
7. Describe the strategies used by climate change deniers.
8. Assess the arguments in favour of limited action against climate change.

Applying and communicating geographical understanding

9. Using the internet, conduct research into the extent to which Australia is meeting its Paris Agreement targets.
10. Research biochar. Prepare a short report on its potential to deal with climate change.
11. Study the box, Spotlight: Seaweed cattle supplement reduces methane. Complete these activities:
    - a Outline the impact of livestock on climate change.
    - b Describe the role of *Asparagopsis* in reducing livestock emissions.
12. In small groups, brainstorm strategies that could be adopted in your local community to mitigate the risks of climate change. Share your ideas with the class.
13. Study the box, Spotlight: Tuvalu, the sinking nation. Complete the following activities:
    - a Explain why Tuvalu is so vulnerable to climate change.
    - b What are the current impacts of climate change on Tuvalu?
    - c Outline the mitigation strategies being developed for Tuvalu.
14. Using examples, write a report outlining strategies to reduce the impacts of climate change.
15. Collect a small media file of examples of commentary opposing action on climate change. Using your knowledge of the topic, assess the accuracy of the claims made.
16. As a class, discuss why you think there is a growing consensus around climate change.

■ CASE STUDY

Costa Rica: Managing the effects of climate change

Costa Rica is a small nation of just 51 000 square kilometres in Central America. It straddles the narrow land bridge connecting North and South America with coastlines on both the Pacific Ocean and the Caribbean Sea (see Figure 13.4.8). With a population of just over 5 million, Costa Rica is considered a small power, yet it has become one of the world's most influential nations in terms of climate change mitigation.

In late 2019, Costa Rica made global headlines when it received the United Nations Champion of the Earth Award. This is the UN's highest environmental honour. It recognised the progress Costa Rica has made towards protecting its environment and tackling climate change.

13.4.8 Costa Rica

Costa Rica's vulnerability to climate change

As a small nation located between two oceans, Costa Rica is uniquely vulnerable to the effects of climate change. It is highly susceptible to tropical storms and hurricanes. Significant portions of its landmass are flat coastal plains, making it vulnerable to sea-level rise.

The Intergovernmental Panel on Climate Change (IPCC) listed Costa Rica as a climate change hotspot. This means it is highly vulnerable to significant impacts over the next 50 years. If global climate change is left unchecked, it is expected to result in temperatures in the tropical nations increasing by between 3°C and 6°C compared to the recorded averages in 1960–1990. Furthermore, increasing sea surface temperatures dramatically increases the risk of severe weather events. The frequency and intensity of both floods and droughts are expected to increase as weather patterns become less predictable and more extreme.

This was seen in a severe drought that affected much of the country from late 2018 into 2019. The north-western province of Guanacaste received just 25 per cent of its annual rainfall. NASA research using special satellites found that vegetation across the province was under severe distress due to high temperatures and lack of rainfall.

Costa Rica's actions

Costa Rica is a small country and contributes very little to global carbon emissions. Yet the country has taken a leading role in doing what it can to take positive steps to reduce its reliance on carbon and has become a global role model.

The National Decarbonisation Plan

In 2018, Costa Rica announced a plan to move towards net-zero carbon emissions by 2050. In 2019, the National Decarbonisation Plan 2018–2050 was released. Net-zero emissions mean that emissions will be no greater than the capacity of the environment to absorb carbon. Essentially, it means the carbon cycle is returned to a balanced state. Costa Rica became a world leader with this plan by being the first country to make this type of commitment. Since then, other countries have followed suit—the UK announced a similar goal in 2019, with New Zealand following in 2020 and Australia declaring its intention in 2022.

The main plan included several smaller specific plans for each sector of Costa Rica's economy. These include tackling emissions in the transport sector, building sector, energy sector, agriculture, waste management and many other parts of the economy. These specific sector plans make Costa Rica's **decarbonisation** plan far more effective and more likely to succeed than those of other countries. The National Decarbonisation Plan is more ambitious than the commitments Costa Rica made under the Paris Agreement.

There will be many challenges for Costa Rica to achieve its plan to decarbonise its economy. Cost is the greatest. Moving towards electric transport, for example, will be very expensive. At present, many of Costa Rica's environmental initiatives are funded through taxation on fossil fuels. Ironically, a reduction in fossil fuel use will limit this revenue source. But Costa Rica has a superb track record in meeting ambitious environmental goals. It once had one of the world's highest rates of deforestation, but is now the leader in reforestation.

13.4.9 Wind farm in Tilaran in the Guancaste Province of Costa Rica

Renewable energy

Costa Rica is a world leader in the use of renewable energy sources. Since 2015, well over 95 per cent of the country's electricity has been produced from renewable sources. In 2018, it went 300 days straight using only renewable sources to meet its energy needs. By the mid 2020s, the country's electricity is expected to come from 100 per cent renewable energy. In comparison, by 2022 about 20 per cent of the USA's electricity came from renewable sources. In Australia, it's about 24 per cent and in France, it is 19 per cent. Seventy-four per cent of Costa Rica's renewable energy is derived from hydro-electric power, 15 per cent from wind power, 8 per cent from geothermal systems and about 1 per cent from biomass and solar (see Figure 13.4.9).

Decarbonisation of the transport sector

With so much of its electricity already derived from renewables, Costa Rica has set about reducing carbon emissions associated with transport. About 68 per cent of its current carbon emissions come from the transport sector.

In 2019, electric buses were introduced in the capital San José (see Figure 13.4.10). The government offers financial incentives for electric cars. These include exemptions from some taxes, special parking spaces and non-payment for parking meters. The government also supports infrastructure development such as recharging stations.

13.4.10 Costa Rica introduced electric buses in 2019.

Moratorium on oil extraction and exploration

In 2002, the Costa Rican government took the remarkable step of placing a moratorium—a temporary ban—on extracting or exploring for oil in its territory. The original moratorium was due to expire in 2021, but in February 2019 it extended it until 2050 as part of the National Decarbonisation Plan. Costa Rica imports around US$1 billion worth of oil each year, but with its extensive plans to decarbonise the transport industry, this is expected to fall.

Halting deforestation and developing ecotourism

In the 1940s, more than three-quarters of Costa Rica was covered in dense tropical rainforest. By 1983, this had fallen to 26 per cent due to extensive logging and land clearing, much of it illegal. At its height, more than 50 000 hectares of forest were being cleared a year. This is equivalent to more than 110 000 football fields.

Major policy changes dramatically altered this trend. Restricting and regulating logging permits, special payments to landowners to conserve forests, creating national parks and attracting foreign investment and support for reforestation projects were all included. By 1989, the rate of deforestation had more than halved, and by 1998 it fell to zero. By 2012, forest cover had increased by 52 per cent on 1983 levels and Costa Rica has set a target of a 70 per cent increase by the mid-2020s to help the nation achieve its carbon-neutrality goals (see Figure 13.4.11).

13.4.11 Costa Rica is actively halting deforestation with targets to increase forest cover.

13.4.12 Ecotourism is a major source of international revenue for Costa Rica.

Costa Rica was a pioneer in ecotourism, using its natural environment as a major tourism drawcard. Instead of logging its forests, Costa Rica now generates considerable income from its pristine environment and clean green policies. In 2018–2019, before the impact of the COVID-19 pandemic, Costa Rica generated almost US$4 billion from tourism with a significant share of this coming from the ecotourism sector (see Figure 13.4.12). This accounted for around 13 per cent of the country's GDP, the highest of any Central American nation.

Activities

Acquiring and processing geographical information

1 Describe Costa Rica's location.
2 Explain why Costa Rica is vulnerable to climate change.
3 Explain the concept of net-zero emissions.
4 Outline the key features of Costa Rica's National Decarbonisation Plan.
5 Explain why halting deforestation is important in addressing climate change.

Applying and communicating geographical understanding

6 Analyse Costa Rica's National Decarbonisation Plan. Explain the challenges of its implementation.
7 Create a table and summarise the key strategies Costa Rica is using to reduce its carbon emissions.
8 As a class, discuss the strategies used by Costa Rica to reduce its carbon emissions. Could some of these strategies be adopted in Australia? What advantages and disadvantages would they represent for Australia?

APPLICATION AND CONSOLIDATION TASKS

Task 1: Report writing: Climate change

Undertake research and analyse the extent and rate of climate change across the globe. In your report, include:

- the effect of greenhouse gases: carbon dioxide, methane, nitrous oxide
- the types of human activities that lead to an increase in these gases in the atmosphere
- the impacts of climate change on natural environments and human societies.

Include maps, graphs, illustrations and diagrams to support your findings.

Task 2: Podcast: Challenges, opportunities and responses to climate change in Australia

Prepare and record a podcast delving into the challenges, opportunities and responses to climate change in Australia. Include an assessment of the responses. Cover Australia and its response to the Paris Climate Change Agreement. The podcast should be 4–5 minutes long.

The podcast can include audio content, such as sound effects, music and snippets of interviews.

Task 3: Presentation: Cost Rica

Investigate and analyse the response of Costa Rica to climate change.

In your presentation:

- outline the international initiative
- explain whether or not the situation has improved since 2020 or its inception
- indicate if the initiative has been successful or if changes have had to be made
- include the future prognosis
- add visuals (maps, graphs, diagrams, photos).

Task 4: Infographic: Climate change strategies

Working in small groups, brainstorm strategies that could be adopted in your local community to mitigate climate change risks. Using the strategies your group proposes, create an infographic to inform your local community.

Ensure that your infographic is clear and engaging. It must explain why these are the best and most effective strategies for your local community. Consider including information refuting climate change sceptics.

CHAPTER

14 Contemporary hazard: Bushfires

Bushfires are becoming more frequent and destructive. They're moving faster, flaring up unpredictably and changing the weather. Places that have never burned before, such as the Arctic Circle, are going up in flames.

Fire was once used by many indigenous peoples around the world to manage the landscape. Today global warming is fanning fires that are increasingly difficult to control. During Australia's deadliest bushfire disaster—Victoria's Black Saturday in 2009—weather conditions were horrendous. Blistering high temperatures extended over three days, along with winds of over 100 kilometres per hour. In the 2019–2020 summer, drought and extreme heat turned much of the country into a tinderbox. Fires raged for months across multiple states and territories.

This chapter gives students the opportunity to enhance their understanding of the interactions between Earth's natural systems and human systems, by studying wildfires on a global scale. In doing so, they study the nature and spatial dimensions of wildfires, the natural processes and human interactions involved. They also assess the effectiveness of people and organisations in managing bushfires in the Australian context.

The sky turns hellish red and the haunting roar of terror mounts in the west. The heat builds and the smoke stings the eyes. The earth is scorched. A blackened landscape is all that remains. All is gone! A life's work and memories. In an instant our lives are changed forever.

Anonymous bushfire survivor

14.0.1 The flames of a fire front will ignite anything flammable it comes in contact with, NSW, 2018.

Chapter glossary

ambient temperature the temperature of the air (or other medium and surroundings) in any particular place

bushfire a uniquely Australian word used to describe any fire burning out of control in forest, scrub or grassland

combustible able to catch fire and burn easily

crown fires extremely hot fires that leap from treetop to treetop, burning whole trees

cultural burning the practice of Aboriginal and Torres Strait Islander peoples regularly using low-intensity fire to burn vegetation

fuel load the amount of fallen leaf matter, bark and small branches accumulated on the forest floor

fuel moisture the amount of moisture present in the fuel on the forest floor

gigafire an extraordinary fire event that devastates a large area, usually specified as more than 1 million acres or 404 685 hectares; as with megafires, such events are characterised by their intensity, size, duration and uncontrollable dimension

hazard reduction practices designed to reduce the severity of future bushfires

mechanical clearing the slashing, thinning and mowing of vegetation by machine

megafire an extraordinary fire event that devastates a large area (usually specified as more than 100 000 acres or 40 468 hectares). Such events are characterised by their intensity, size, duration and uncontrollable dimension

natural disaster a natural hazard, such as a flood, earthquake or cyclone, that causes great damage or loss of life

natural hazard an extreme event that occurs naturally and has the potential to harm humans and environments

prescribed burning a fire deliberately started to reduce the severity of future bushfires

pyrocumulonimbus fire-generated thunderstorms

QAnon a US-based right-wing group responsible for spreading disproven and discredited conspiracies

radiant heat heat transmitted by electromagnetic waves, in contrast with heat transmitted by conduction or convection

relative humidity the amount of water vapour present in air, expressed as a percentage of the amount needed for saturation at the same temperature

selective logging the practice of removing only some trees in a forest and leaving the rest intact, often considered a sustainable alternative to clear-felling forests

slope gradient a measure indicating how steep a slope is; the greater the gradient, the steeper the slope

surface fires a blaze largely limited to the undergrowth and debris on the forest floor

wildfire a fire burning out of control

UNIT 14.1
Bushfires

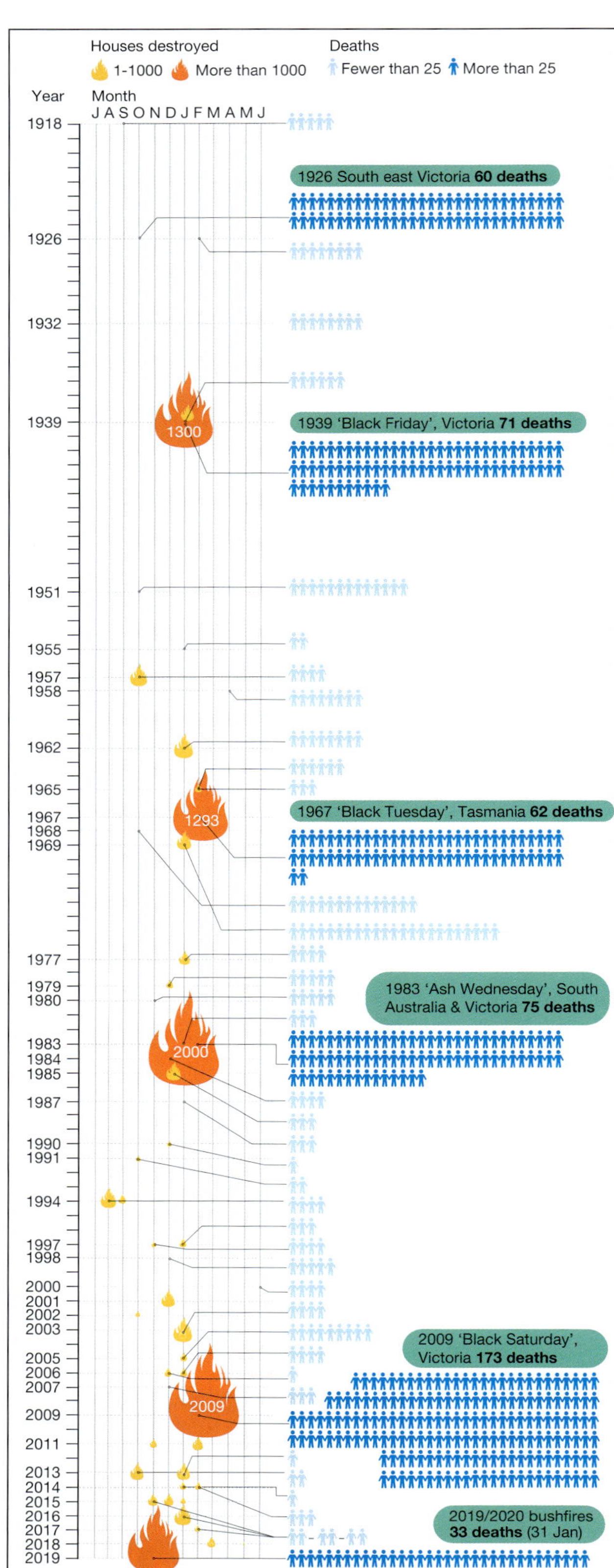

14.1.1 Australia's worst bushfires

While humans have historically tried to control the environment to meet their needs, there are elements of their surroundings that remain beyond their control. Extreme events within natural environments can devastate whole communities, disrupt communications and cause economic hardship. **Bushfires** are one such event.

Bushfire is a uniquely Australian word, used to describe any fire burning out of control. In other countries the term **wildfire** is used to describe such an event.

Bushfires can be classified in terms of the cause of ignition, their physical properties, the **combustible** material present and the effects of weather on the fire. They can be ignited by lightning strikes, accidents or arsonists. They are typically associated with periods of prolonged high temperatures, low humidity and/or very strong winds.

In Australia, bushfires occur as forest fires (bushfires with the trees, undergrowth and litter on the forest floor as the main fuel) and grass fires (bushfires with grass as the main fuel). Large areas of Australia suffer from the threat of bushfires, especially in the eucalyptus forests and woodlands of the south-east and south-west corners. Bushfires occurring in the north of the continent typically occur during the dry months of winter.

Bushfires damage property and kill livestock, blacken large tracts of forest and grasslands, and kill native animals and sometimes humans. Bushfire disasters put bushfires high on Australia's list of **natural hazards** (see Figure 14.1.1).

● **SPOTLIGHT**

Natural disasters

It is very easy to forget that the natural environment contains threats to human life and property. These extreme and unusual events are called natural hazards. They include major disturbances in the atmosphere and on Earth's surface. Australia's most common natural hazards include storms, cyclones, floods, droughts and bushfires. When natural hazards affect people, they are termed a **natural disaster**.

Global distribution of wildfires

Figure 14.1.2 shows the global distribution of wildfires. Such events typically occur in the world's grasslands and forests, especially those that experience seasonal and/or variable rainfall. Under such conditions, drying vegetation is readily ignited by lightning, by accident or intentionally. Forest-based wildfires are of two basic types:

- **surface fires:** largely limited to the undergrowth and debris on the forest floor. They kill seedlings and small trees, but spare most mature trees and allow most wildlife to escape. These occasional, low-intensity surface fires have several ecological benefits. They burn flammable material (fuel) that accumulates on the forest floor, reducing the risk of more destructive fires. They release nutrients trapped in slowly decomposing leaf litter. Releasing seeds from cone-bearing species stimulates the germination of other seeds. They also help control insects and tree diseases

- **crown fires:** extremely hot fires that leap from treetop to treetop, burning whole trees. They typically occur in forests that have not experienced a surface fire for several years. The absence of fire enables fallen branches, leaves and ground litter to accumulate. These fiercely burning fires can jump up to the forest canopy and destroy most vegetation, kill wildlife, increase soil erosion and destroy homes and infrastructure.

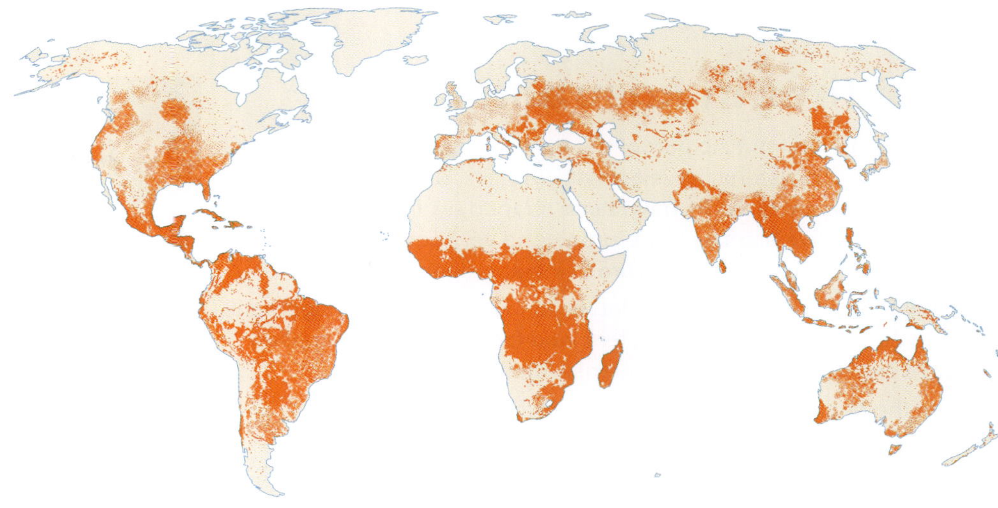

14.1.2 World distribution of wildfires

Sometimes these fires occur naturally, ignited by lightning strikes. However, many are the result of human carelessness or intent. Causes include arson, discarding lit cigarettes, and campfires and burn-offs getting out of control. Land clearing for cropping or grazing livestock is another cause.

The magnitude, frequency and duration of wildfires

The magnitude, frequency and duration of wildfires are all increasing, driven principally by the world's warming climate. In recent years, some 2.4 million hectares of Siberia have burned and more than a million hectares in Alaska. This is a massive loss for the Arctic region, which is already suffering disproportionally from global warming, as it is warming twice as fast as the rest of the planet. The frequency and magnitude of bushfires in Australia are also increasing.

Such fires contribute to a feedback loop of environmental change. As the world gets warmer, wildfires will become more prevalent. In their wake, more wildfires will leave less vegetation and higher carbon emissions. This will speed up global warming, particularly if those wildfires mainly occur in regions such as the Amazon or Arctic, which are powerful drivers of our climate system.

SPOTLIGHT

The increasing frequency and magnitude of bushfires in Australia

A recent study by researchers at the Australian National University (ANU) found that Victoria had only one **megafire** in 150 years of records. Since 2003 it has had three. They also noted the extensive and frequent re-burning of previously fire-damaged areas—sometimes with a gap as short as five or six years.

In the 2019–2020 fire season, bushfires burned approximately 1.5 million hectares in Victoria. That is roughly double the size of metropolitan Melbourne. This was the largest area affected since 1939 when 3.4 million hectares burned.

Of the 1.5 million hectares burnt in 2019–2020, more than 600 000 hectares have burned twice, and more than 112 000 hectares have burned three times over the past 25 years.

Based on their findings, the researchers concluded there was an urgent need to implement strategies for reducing the incidence of megafires, protect unburnt areas and manage repeatedly damaged ecosystems.

● SPOTLIGHT

Bushfires in Australia

Bushfires are an intrinsic part of the Australian environment. The continent's ecosystems have evolved with fire, and the landscape and its biological diversity have been shaped by fire. Many of Australia's native plants are fire-prone and very combustible, while numerous species depend on fire to initiate regeneration.

First Nations Australians have long used fire as a land management tool. It continues to be used to clear land for agricultural purposes and to protect properties from intense, uncontrolled fires.

From the earliest years of European occupation of Australia, bushfires caused a loss of life and significant property damage (see Figure 14.1.3). While naturally occurring bushfires cannot be prevented, their consequences can be minimised by implementing mitigation strategies in the most vulnerable areas. Figure 14.1.4 shows the bushfire risk zones of Australia, while Figure 14.1.5 shows Australia's bushfire seasons.

14.1.3 Following early European occupation, bushfires presented a major threat as the new residents failed to learn the land management skills of First Nations Australians.

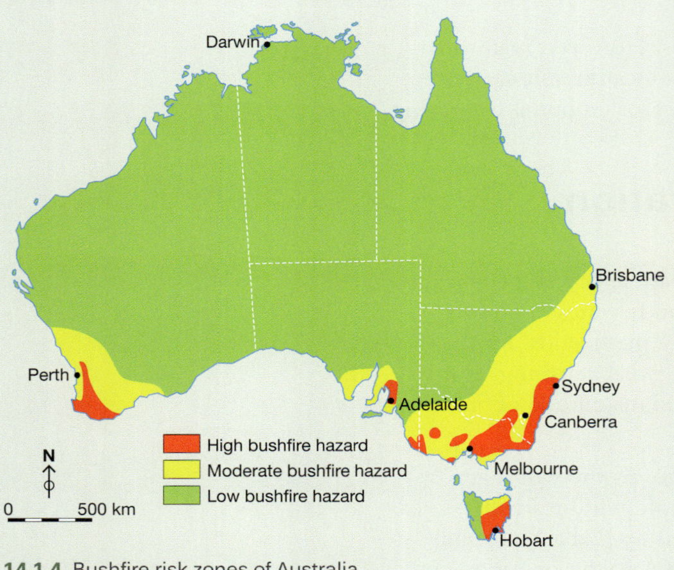

14.1.4 Bushfire risk zones of Australia

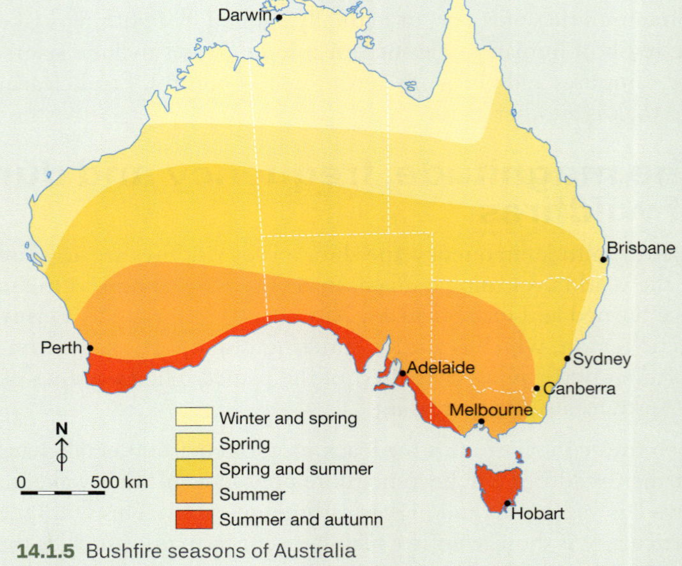

14.1.5 Bushfire seasons of Australia

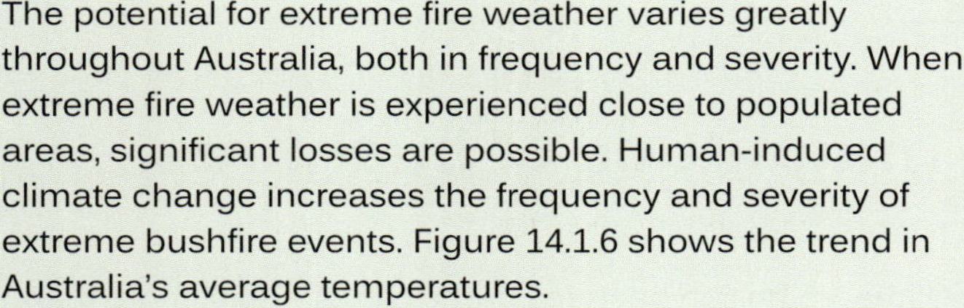

The potential for extreme fire weather varies greatly throughout Australia, both in frequency and severity. When extreme fire weather is experienced close to populated areas, significant losses are possible. Human-induced climate change increases the frequency and severity of extreme bushfire events. Figure 14.1.6 shows the trend in Australia's average temperatures.

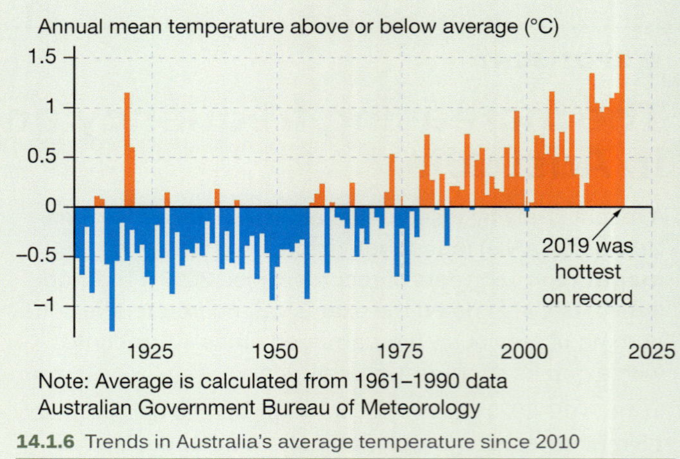

14.1.6 Trends in Australia's average temperature since 2010

The conditions under which bushfires occur

The conditions under which bushfires occur provide valuable insights into the characteristics of the natural environment. These include the physical processes and cycles that influence the nature and occurrence of bushfires.

The conditions or factors determining whether a bushfire will occur include the presence of fuel, oxygen and an ignition source. The intensity of the fire and the speed at which it spreads depend on **ambient temperature**, **fuel load**, **fuel moisture**, wind speed and the angle of a slope (see Figure 14.1.7).

Fuel load: High fuel loads can result in disastrous fires, should it be ignited.

Drought: Long periods of below-average rainfall dry out the fuel load.

Strong winds: Air provides the oxygen to keep fires burning. Stronger winds mean extra oxygen and more intense fires. They fan the fire and accelerate the speed at which the fire spreads.

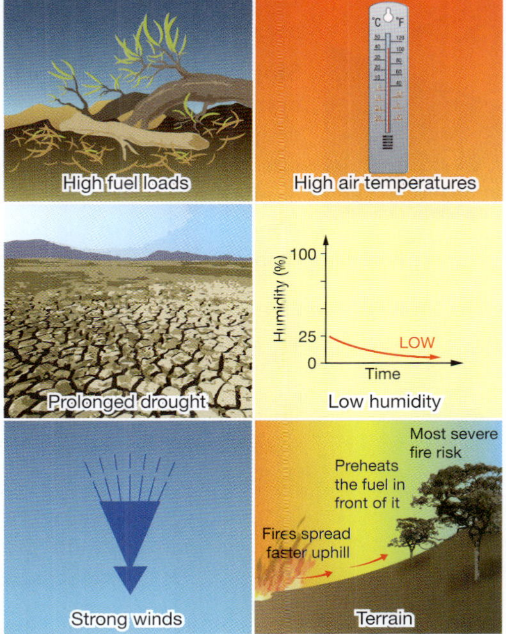

High temperatures: In Australia, summer temperatures reach the high 30s and can even exceed 40°C. High temperatures dry out forests and grasslands. This makes them easier to ignite.

Low humidity: Hot, dry air with humidity below 25 per cent creates dangerous bushfire conditions. Humidity is the amount of water vapour (or moisture) in the air.

Terrain: Fires spread more quickly up hillsides and slopes.

14.1.7 Factors affecting the ignition and spread of bushfires

Ignition source: Bushfires result from both human activity and natural causes. Lightning is the principal natural source, accounting for about 50 per cent of all ignitions in Australia. Fires resulting from human activity, either by accident or deliberately lit, account for the other 50 per cent.

Fuel load: The amount of fallen leaf matter, bark and small branches accumulating on the forest floor is referred to as fuel load. Fibrous and dry tree bark can carry fire up to the treetops. Generally, the greater the fuel load, the hotter and more intense the fire. Fuel that is loosely compacted will burn faster than heavily compacted fuel sources. Smaller pieces of fuel, such as twigs, leaf litter and branches, burn quickly, particularly when they are dry. Larger fuel sources, such as fallen trees, often burn later after the fire front has passed. The natural oils within eucalyptus trees promote the combustion of fuel.

Fuel moisture: This is the amount of moisture present in the fuel on the forest floor. Dry fuel will burn quickly, but damp or wet fuel may not burn at all. Consequently, the time since rainfall and the amount of rain received are important factors in assessing bushfire danger.

Wind speed and direction: Winds drive fires across landscapes by pushing the fire front (the leading edge of a fire) into fresh fuel sources and providing a continuous supply of oxygen. Wind also promotes the rapid spread of fire by spotting. This is the ignition of new fires lit by burning embers lifted, into the air by the wind. Spotting can occur up to 30 kilometres ahead of the fire front.

A threshold wind speed of around 12–15 km/h has been observed which makes a significant difference in the behaviour of bushfires. With wind speeds below this threshold, fires in areas with heavy fuel loads burn slowly. Above this threshold, fire behaviour

Did you know?

Bark shedding by many eucalyptus species helps to promote fires. The shed bark is low in nutrients and breaks down very slowly. When dry, it is very flammable.

becomes more erratic and its advance is more rapid. The width of a fire front also influences the rate of fire spread and any change in wind direction can quickly widen the forward fire front. Many people who die in bushfires get caught during or after a wind change. Wind also impacts the intensity of fires by feeding in a continuous supply of oxygen.

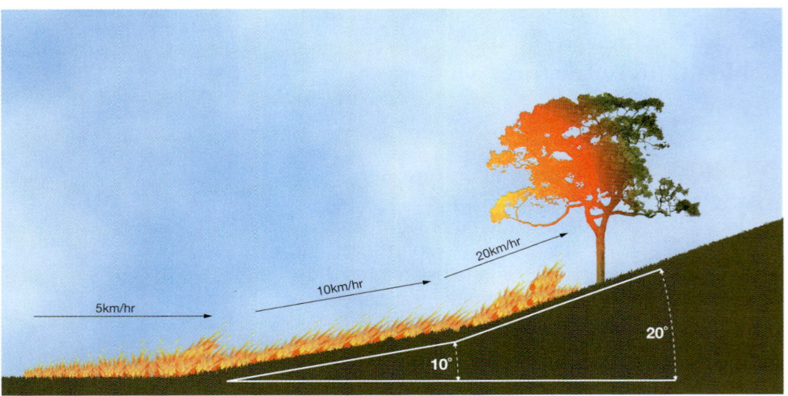

14.1.8 The relationship between slope gradient and the speed of bushfires

Ambient temperature and relative humidity: The higher the temperature, the higher the chance of ignition occurring, This is because the fuel load dries quickly, making it more likely to burn. Also, dry air promotes a greater intensity fire than does moist air, and plants become more flammable at a low humidity because they release their moisture more readily.

Slope gradient: Fires accelerate when travelling uphill and decelerate travelling downhill. The steepness of the slope plays an important role in the rate of fire spread. The speed of the fire front advance doubles with every 10° increase in slope. On a 20° slope, the speed of fire advance is four times greater than on flat ground (see Figure 14.1.8).

14.1.9 How bushfires move across the land

Fire attack

Figure 14.1.9 shows how bushfires move across the land. Fuel burning on the forest floor dries out the leaves on the trees making them more flammable. Once the crown of the tree catches fire, flames can readily spread from tree to tree until the whole canopy is consumed. Windborne embers ignite spot fires well ahead of the fire front. The speed and direction of the wind determine how quickly this spread occurs.

Bushfires pose a threat to life and property in several ways. This includes:

the fire front: The flames of the fire front ignite anything flammable that it comes into contact with. The most dangerous of all fire fronts burn in the crowns, or tops, of trees (crown fires). In eucalyptus forests, bushfires can advance at alarming speeds through the upper layers of the forest. The tops of the trees often appear to explode as the fire roars through. On hot days, the oils in eucalyptus leaves pass into the atmosphere as a vapour. This vapour is quite flammable. The trees themselves do not explode; it is the oil-rich vapour given off by the leaves that ignites in a fireball.

ember attack: Even if the fire front itself never reaches a certain location, spot fires can break out well ahead of the advancing blaze when hot embers fall from the sky. Embers are burning leaves, bark and small pieces of wood from tree branches. They can be carried great distances by strong winds. Ember attacks can occur before or during a fire and for a long period after the main fire has passed (see Figure 14.1.10).

heat: The amount of **radiant heat** coming from a large fire can be extreme (see Figure 14.1.11). It can melt metal and boil water in tanks. It may last for only a few minutes as the fire front passes, but it can be far longer when large logs, branches, grass tussocks and stump holes continue to burn and smoulder.

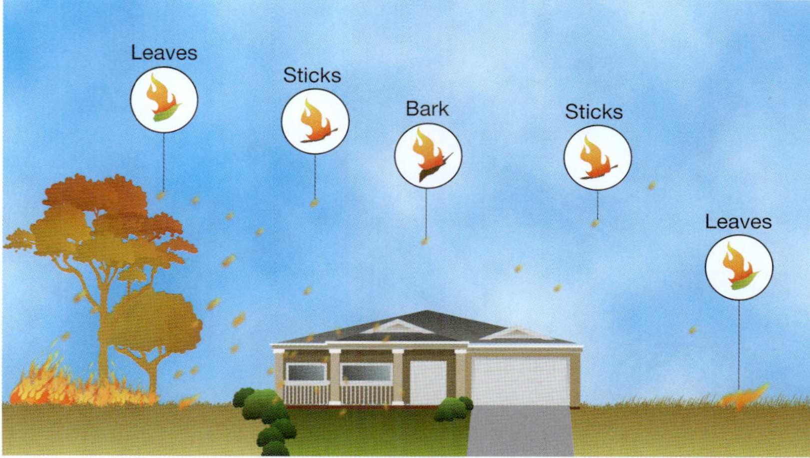

14.1.10 Ember attack and spot fires

14.1.11 Radiant heat can kill living things and damage property.

Role of drought in wildfires

Wildfires are more likely to occur during drought conditions when the land and vegetation have dried out. Australia's Bureau of Meteorology defines drought as a 'prolonged, abnormally dry period'. This distinguishes between 'serious' and 'severe' drought. 'Serious drought' is defined as rainfall over a three-month period being between the lowest 5 and 10 per cent recorded for that region in the past. 'Severe drought' occurs when, for three or more months, the rainfall is less than the lowest 5 per cent of recorded rainfall in that area over the long term. Such definitions cover drought, which is a relative term, as rainfall deficiencies need to be compared to typical rainfall patterns. This includes seasonal variations.

One of the outcomes of drought is the stress experienced by vegetation. Grasses die and trees shed leaves as they struggle to minimise water loss. The stressed vegetation, together with the dry fuel matter on the forest floor, makes it susceptible to ignition whether it be by a lightning strike, accident or intent.

When drought conditions develop after a period of above-average rainfall over an extended area, the once-thriving vegetation starts wilting as the dryer conditions take hold. If the dry conditions persist, the excess dried-out vegetation provides ample fuel for a fire.

Significantly, Australia has one of the most variable rainfall climates in the world. It tends to experience severe droughts every 10 to 15 years on average. Droughts may last for extended periods—in some cases, a decade or longer. These drought conditions predispose much of Australia to hazardous bushfire conditions.

Droughts over eastern and northern Australia are usually associated with the weather phenomenon known as El Niño. This is a reversal of the weather phenomenon known as La Niña, which brings rainfall to eastern and northern Australia.

Figure 14.1.12 illustrates an El Niño event. The main changes that occur in an El Niño period that result in low rainfall for Australia are:

- weaker trade winds from South America
- cool water off the east coast of Australia
- higher air pressure over eastern and northern Australia.

Figure 14.1.12 outlines a La Niña event. The main changes that occur in a La Niña period that result in high rainfall for Australia are:

- stronger easterly trade winds from South America
- warm water off the east coast of Australia
- low air pressure over eastern Australia.

El Niño and La Niña events are associated with a phenomenon known as the Southern Oscillation Index (SOI) (see Figure 14.1.13). It measures the monthly or seasonal

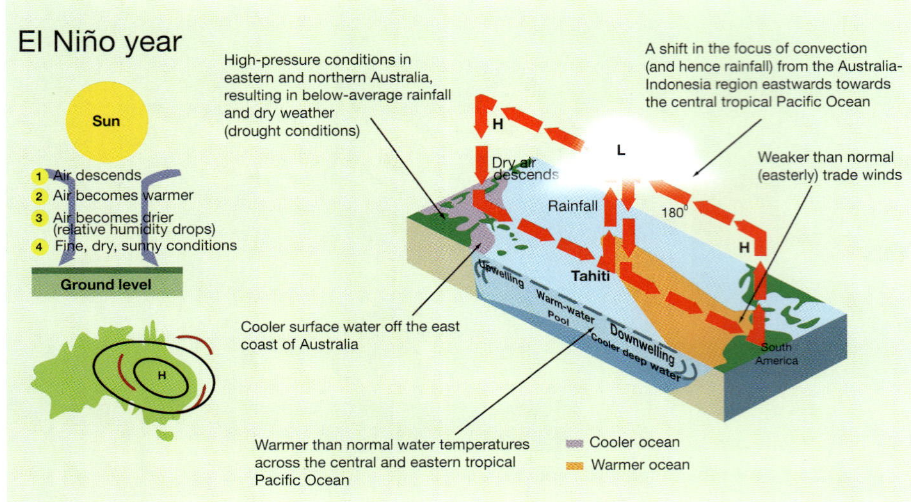

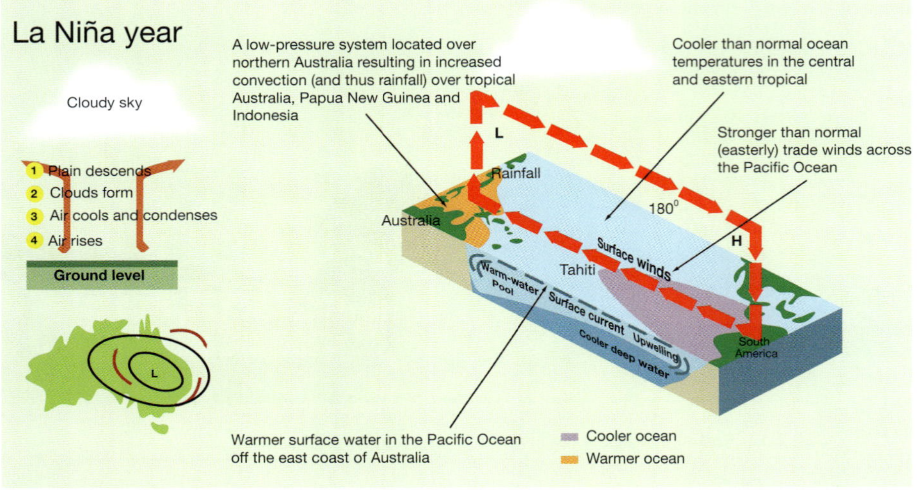

14.1.12 Impact of El Niño and La Niña

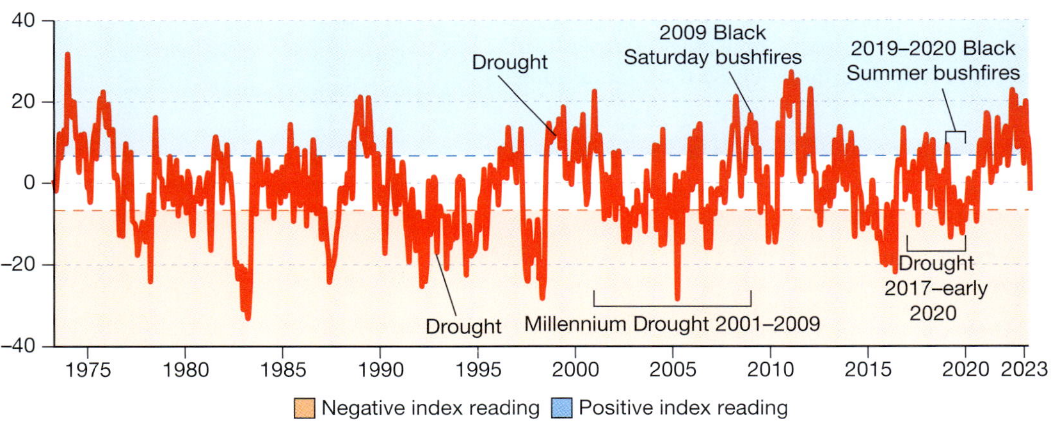

14.1.13 The Southern Oscillation Index—monthly, 1990–2023. Sustained positive values above +7 typically indicate La Niña, while sustained negative values below −7 typically indicate El Niño.

variations in the air pressure difference between Tahiti and Darwin. This, in turn, reveals broader changes in weather patterns across the Pacific Ocean. Long-term measurements show that:

- negative SOI values are linked to El Niño events and droughts
- positive SOI values are linked to a La Niña and high rainfall.

Bushfires: An ecological dimension

Fire is essential for some plant species to survive. The cones of coniferous pines need to be heated before they open and release their seeds. Chaparral is a shrubland plant community found in California. They require fire before seeds will germinate and their leaves include a flammable resin that feeds fire, helping the plants to propagate. Such plants depend on wildfires to pass through their regular life cycles. Some plants require fire every few years, while others survive on just one or two burns a century.

Many Australian eucalyptus trees have special fire-adaptive traits, including re-sprouting after fires. This is often referred to as epicormic sprouting and is very common (see Figure 14.1.14). Following a fire, a series of events trigger sprouting. One way is through damage to the tree's top, or crown. Hormones suppressing the sprouts may stop being generated by the tree's crown, triggering sprouting. The sprouting buds are often protected by a thick bark to help them survive a fire's intense heat.

Eucalyptus seed release is triggered by fire. When burned or subject to heat, tough woody capsules disperse their contents onto nutrient-rich ash on the land surface. All the understorey competition for light, water and nutrients has been removed by fire. Browsing animals are driven out for some time. The heated soil reduces the number of plant-eating insects and soil organisms during the short but crucial early growth period.

There are key differences between Australian eucalypt and rainforest trees. Eucalypts are adapted to, and take advantage of, major widespread disturbances of the forest canopy, especially those caused by fire. Rainforest species are less able to cope with fire.

Fires can also help keep ecosystems healthy. They can kill insects and diseases that harm trees. They often clear scrub and underbrush, making way for new grasses, herbs, and shrubs that provide food and habitat for animals and birds. Low-intensity fires can clean up debris and underbrush on the forest floor, add nutrients to the soil, and open up space to let sunlight through to the ground. All these changes nourish smaller plants and give larger trees room to grow and flourish.

14.1.14 Eucalyptus epicormic regrowth

Climate change impacts the frequency and severity of bushfires

Climate change impacts rainfall patterns. In Australia, 2019 was the warmest and driest year ever recorded.

Scientists are convinced that as climate change accelerates, southern parts of Australia will experience more spells of very dry weather. These conditions are likely to result in an increased incidence of extreme fire events. There is already evidence for such an increase.

Activities

Acquiring and processing geographical information

1 What are bushfires called in other countries?

2 Outline the various ways in which bushfires can be classified.

3 Outline where bushfires typically occur.

4 Distinguish between surface and crown fires.

5 Identify the principal causes of bushfires.

6 Explain why the magnitude, frequency and duration of bushfires are increasing worldwide.

7 Outline the factors that contribute to the ignition and spread of bushfires.

8 Outline how bushfires pose a threat to life and property.

9 Explain how the Bureau of Meteorology defines drought in Australia.

10 Outline the role of drought in the occurrence of bushfires.

11 Outline the ecological benefits derived from the periodic burning of fire-adapted species.

12 Describe the relationship between climate change and the changing incidence and severity of bushfires.

Applying and communicating geographical understanding

13 Distinguish between natural hazards and natural disasters.

14 Study Figure 14.1.1. Using data from the graph, write a report outlining Australia's worst bushfire since 1918.

15 Study Figure 14.1.2. Describe the global distribution of wildfires.

16 Study the box, Spotlight: The increasing frequency and magnitude of bushfires in Australia. Outline the evidence presented.

17 Study the box, Spotlight: Bushfires in Australia.

- **a** Outline the role of bushfires in shaping the Australian environment.
- **b** Using Figure 14.1.4, identify the parts of Australia that experience high and moderate bushfire risk.
- **c** Using Figure 14.1.5, describe the bushfire seasons experienced in different parts of Australia.
- **d** Using data from Figure 14.1.6, describe the trends in annual mean temperature. What are the implications of this trend for the magnitude, frequency and duration of bushfires?

18 Study Figure 14.1.7 Write a report outlining the factors that maximise the likelihood of the ignition and spread of wildfire/bushfire.

19 Study Figure 14.1.8. Describe the relationship between slope and the speed of wildfire/bushfire spread.

20 Study Figure 14.1.10. Describe the impact of ember attack and radiant heat on life and property.

21 Study Figure 14.1.13. Describe the relationship between variations in the Southern Oscillation Index and the incidence of drought and recent bushfire events.

■ CASE STUDY

Global wildfires

The following three examples highlight the extent to which wildfires are a global issue. They focus on the California wildfires of 2020 and the Oregon wildfires of 2020 in the USA, and the 2018 wildfires in Greece.

Californian wildfires

The 2020 Californian wildfires were the worst on record and have been described as the state's first-ever **gigafire**.

By the end of the year, 9639 fires had burned 1 779 730 hectares of land, the equivalent of 4 per cent of the state's total land area (see Figures 14.1.15 to 14.1.19). By the time the fires were extinguished, the blazes had destroyed over 10 000 structures. The damage costs ran over US$14 billion, including over US$10 billion in property damage. Thirty-one people died. At the peak of the emergency, 367 fires raged across the state.

14.1.15 Firefighters in California were overwhelmed by the wildfires.

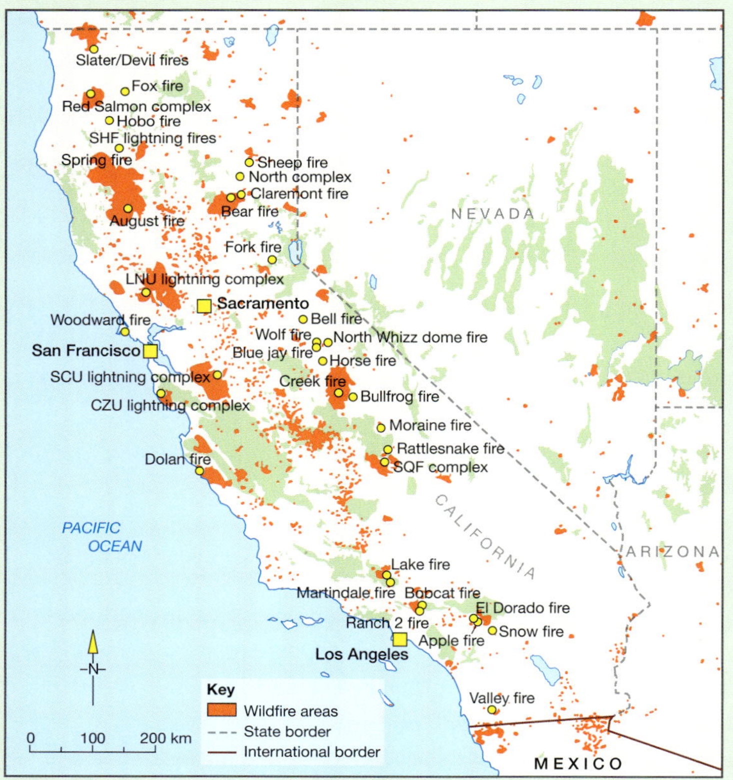

14.1.16 Vast areas of California burned during the 2020 gigafire.

The 2020 annual fire season was preceded by an unusually dry January and February, the driest on record. While the rainfall received during March and April was close to average it did little to moderate the severe drought conditions prevailing in Northern California. If the seasonal rainfall continues to arrive late in coming years, Central and Southern California face worsening fire conditions in the future.

The fires began in August. Many were sparked by intense thunderstorms during an extended heatwave. By early September, the combination of a record-breaking heatwave and strong winds resulted in an explosive spread of the existing fire fronts. Most notable was the Big Creek blaze, which would become California's largest recorded wildfire. It triggered mass evacuations and resulted in 16 deaths.

Thick smoke blanketed much of California for weeks. It made air quality in some areas among the worst in the world. In the San Francisco Bay area, skies glowed eerily orange for days. Authorities reported that the poor air quality arising from the smoke caused hundreds of extra deaths in cities and along the west coast from California to Washington and Oregon. The wildfires also destroyed groves of ancient redwoods, giant sequoias and Joshua trees. Many of these forests are lost forever.

Fire is part of the Californian ecology, yet the intensity and extent of the blazes were unprecedented. Scientists attributed the intensity of this fire season to climate change and poor forest management practices.

Like Australia, California is a landscape adapted to fire. Fires set intentionally by Native Americans or sparked by lightning, burned large areas each year before the arrival of Europeans. But decades of fire suppression policy removed it from the landscape, causing forests and scrublands to grow denser. As a result, forests now burn more severely.

Oregon wildfires

In 2020, Oregon also experienced one of the most destructive fire seasons on record. The fires killed at least 11 people, burned more than 400 000 hectares of land and destroyed thousands of homes.

In early September, unusually high winds and continued dry weather led to the rapid expansion of wildfires across the state. Rumours spread on social media falsely claiming that the alternatively far-right and far-left extremists were deliberately setting fires and preparing to loot evacuated properties. These conspiracy theories were magnified by **QAnon**. Some residents refused to evacuate, choosing to defend their homes against the supposed looters. Authorities pleaded with them to ignore the rumours, but some perished defending their homes.

14.1.17 The smoke plume of the 2020 US west coast wildfires could be seen from space.

Greek wildfires

In July 2018, a series of wildfires in Greece ignited during a European heatwave. The worst was in the coastal areas of the Attica peninsula, the region that includes Athens and its outer suburbs. The fires were the second-deadliest wildfire event in the twenty-first century, taking the lives of 102 people, including 11 children. Only the 2009 Black Saturday bushfires in Australia were more deadly (173 killed).

Among the worst affected areas were the communities north of the port city Rafina, most notably Kokkino Limanaki and Mati (see Figure 14.1.20). More than 700 residents had to be rescued from beaches and the sea. Tragically, authorities found the bodies of 26 people trapped, huddled together, just metres from the sea. Rescue boats also recovered bodies from the water. Ten people drowned when a rescue boat capsized. At least 4000 residents were left homeless by these wildfires.

The fire started near the town of Pentili due to negligence when a man burned wood in his garden. It quickly spread in the gusty conditions to the town of Kineta, where it burnt houses. It headed east towards the area's beachside settlements, where it started burning parts of Neos Voutzas, Kokkino Limanaki and Mati.

14.1.18 The Californian wildfires destroyed over 10 000 structures, including many homes.

14.1.19 In some areas of California, whole neighbourhoods were destroyed.

14.1.20 Mati's fleeing residents abandoned cars that became engulfed by the rapidly advancing flames.

Activities

Applying and communicating geographical understanding

1 Outline the causes and consequences of these catastrophic wildfire events. What do these suggest about the magnitude, frequency, duration and temporal spacing of wildfires?

UNIT 14.2

Bushfire mitigation strategies

Minimising the risks associated with fires provides a range of challenges, as well as some opportunities. Responses include **hazard reduction** interventions, planning and regulatory initiatives, and strategies to promote community awareness and preparedness. The land management strategies of Indigenous Australians could offer alternatives to other forms of hazard reduction.

Hazard reduction

Hazard reduction interventions aim to reduce the risks associated with bushfires. They typically focus on the actions that land managers can take to influence fire behaviour, mainly via managing the fuel load. Other factors influencing fire behaviour, such as weather and terrain, cannot be readily modified.

Fuel load management is achieved through three main processes:

- reducing the total mass of fuel
- altering its structure
- altering its composition.

These interventions can reduce the intensity and rate of a fire's spread, occurrences of spot fires and how far embers can travel ahead of a fire.

Such strategies are not intended to stop or prevent fires. They are designed to enhance and support the effectiveness of other complementary prevention, preparation and response measures. In particular, this includes fire suppression, but it also means urban planning, building regulations and community preparedness.

The most common fuel load hazard reduction techniques used in forest settings include:

- **prescribed burning**
- **mechanical clearing**, such as slashing, thinning and mowing
- chemical control or spraying, through both on-ground and aerial delivery
- grazing by animals.

Four factors determine which techniques are used and when. They include cost, practicalities, capabilities and risks. Sometimes a combination of these techniques may be most appropriate.

Prescribed burning

Fuel management strategies vary in terms of effectiveness. Prescribed burning is especially effective when undertaken in the forest–urban interface—areas where settlements meet or intermix with forest ecosystems. Prescribed burning in areas within the broader landscape, even those away from the forest–urban interface, can also significantly reduce the risk to settlements (see Figure 14.2.1).

The areas targeted for prescribed burning typically include high-ignition areas (e.g. elevated landforms susceptible to lightning strikes), areas where the topography and forest types create fire runs, ridges and other areas known to be associated with high-intensity crown fires. Prescribed burning can also reduce bushfire-related impacts on ecological assets and areas of high conservation value.

While prescribed burning can materially affect the extent of and severity of fires, its effectiveness varies depending on the nature of ecosystems and the prevailing climate. The impact of such practices is also short-lived. Fuel loads re-accumulate relatively quickly in Australian forests. This means prescribed burning must be undertaken regularly to be effective. In most instances, the benefits derived from prescribed burning last for just one to four years.

Weather has the greatest influence on bushfire behaviour. As weather conditions deteriorate, the role played by fuel load management declines. In Australia, the most severe bushfire impacts occur during extreme or catastrophic fire weather conditions. In such instances, fuel loads are of little significance.

Fuel reduction strategies such as prescribed burning are most effective in mediating the impact of relatively low-intensity surface fires. They are less effective in the case of extreme fire events where fire and the atmosphere interact with explosive effects.

Low-intensity fires typically have a well-defined fire front, with a relatively narrow band of fire activity that represents a line between the unburnt fuel ahead of the fire and the burnt fuel behind it. In extreme fires, fire behaviour is no longer solely a function of fuel load. Such fires generate their own behaviour by interacting with the surrounding atmosphere. This results in fire behaviour that is very difficult to predict and counter. Therefore, in extreme fires, fuel loads do not have a material impact on fire behaviour. As conditions deteriorate, fuel reduction is diminished in its effectiveness. Any evaluation of fuel management efforts needs to be viewed within this context.

14.2.1 Field officers of the NSW National Parks and Wildlife Service carry out hazard reduction burns ahead of the bushfire season.

Mechanical clearing

The mechanical clearing of fuel load involves slashing, thinning and mowing vegetation, typically undergrowth and grasses. This strategy is widely used to build buffer zones between housing and bushland. Slashing machines are used to clear scrub and tractor-mounted mowers are used to slash grasses. In some areas, topography means mechanical clearing tools are unable to operate effectively.

Larger-scale machinery can also be used to create firebreaks (see Figure 14.2.2). These serve to impede low-intensity fires and provide access for firefighters to respond to lightning strikes and spot fires.

14.2.2 Mechanical clearing of scrub

Thinning or selective logging

Thinning or **selective logging** is the practice of removing only some trees and leaving the rest of a stand of trees intact. It is often presented as an alternative to clear-fellling, in which a large swathe of forest is cut down, leaving little behind but wood debris and a denuded landscape (see Figure 14.2.3).

While thinning may prove effective in slowing the spread of low-intensity fires, it has little impact on fire behaviour in extreme fire events.

14.2.3 Selectively logged area in British Columbia, Canada

Chemical control by spraying

Chemical control of woody weeds and grasses can also be used to control the proliferation of fuels. Defoliants (chemicals that result in trees shedding their leaves) can be applied by ground-level or aerial delivery. Because of their adverse impacts on the health of both humans and fauna, their use is often restricted and even banned in some areas.

Grazing by animals

Grazing cattle in forested lands is sometimes promoted as a means to reduce fuel loads. The practice is, however, controversial. Those opposed argue that cattle grazing does not stop fires and introducing animals creates a host of other problems. They trample native plants, increase erosion, collapse riverbanks and pollute streams. They also spread fast-growing and highly flammable weeds, increasing fire risk and out-compete native species.

● SPOTLIGHT

Cultural burning practices of Aboriginal and Torres Strait Islander peoples

Fire has played an important role in the evolution and functioning of Australian environments. Aboriginal and Torres Strait Islander peoples have used fire as part of their land management processes for tens of thousands of years. As a result, many of the continent's plant species and indeed whole ecosystems have evolved to be dependent on regular or occasional exposure to fire.

After colonisation, fire regimes across the Australian continent changed significantly and resulted in the decline of these burning practices. Fire went from being a tool that was widely used to being seen as a threat to life and property that needed to be suppressed and excluded.

The catastrophic Australian bushfires of 2019–2020 prompted interest in Aboriginal and Torres Islander land management and burning practices. There are many fire experts who claim that Indigenous land management practices can play an important role in preventing catastrophic bushfires.

Cultural burning is a traditional land management practice that involves the use of a 'cool', gentle, creeping fire that burns gently through grasslands and links up with other fires lit on the forest floor to create a mosaic effect. Cultural burning is still widely practised in Northern Australia and Western Australia, where it has been for thousands of years. In the Great Western Woodlands of Western Australia, 16 million hectares of temperate woodland, heathland and mallee are cared for in part by the Ngadju people and the Ngadju Rangers Group. The Ngadju undertake cultural burns when the grass is green, a few days after rain, or when rain is coming. In the warmer seasons, the fire must be started in the morning before the sea breezes arrives.

In recent times, cultural burning has been reintroduced into some areas where the practice had been stopped long ago. In 2019, the Dja Wurrung people of Victoria lit fires in the bushland that surrounds the Tang and Thunder swamps in Central Victoria, north of Bendigo. In the 170 years that passed since cultural burning was regularly practised in the region, agriculture, urban development and mining have degraded this land and fragmented its ecosytems. The region was once covered in box-ironbark forests and woodlands. Now it is one of the most profoundly altered landscapes in Victoria.

Generations of land management expertise

For at least 65 000 years, Aboriginal and Torres Strait Islanders have used their knowledges of land management practices—including fire—to care for Country. The responsibility for such practices rests with the custodians of the land, or those who have been given their permission and guidance (see Figure 14.2.4). The traditional knowledge of how to practise cultural burning has been passed down from generation to generation, with the elders of the group instructing the young.

The actual type of fire used is specific to each location and its intended purpose. The burning typically has one or more interconnected objectives. These include protecting cultural or natural assets by maintaining the health of the surrounding Country; habitat protection and fuel reduction; and ceremonial purposes. Fuel reduction—targeted burning to reduce the amount or density of foliage is rarely the principal objective.

The use of 'cool' or low-intensity fire reduces the risk of extreme fire events that result in the burning of whole trees and forest canopies. Protecting the canopy is an important aim of cultural burning because the canopy supports resources important to Aboriginal and Torres Strait Islander peoples. For example, insects, birds' nests and shade. The loss of the canopy fundamentally alters the surrounding ecosystem—sunlight breaks through and dries out the soil.

Cultural burning differs from prescribed or hazard-reduction burning, the latter being the process of applying fire to a predetermined area to reduce the incidence and severity of bushfires. Prescribed burning is increasingly difficult to implement. Because of the warming climate, it is becoming increasingly difficult to find windows of time when prescribed burns can be safely lit. In such circumstances, cultural burning may prove beneficial where prescribed burning is considered to be too hazardous. Cultural burning offers an alternative.

Unlike hazard reduction burns, where fire often creates walls of flame, cultural burning focuses on the use of spot ignitions, creating a mosaic of burnt vegetation. This, in turn, leaves space for wildlife to escape. It also encourages the growth of native grasses and reduces the density of vegetation. Very hot fires are avoided.

14.2.4 For tens of thousands of years, Australian flora and fauna evolved in the presence of fire, indeed they depend on it for regrowth and regeneration. Much of this fire was lit by Indigenous Australians who, for at least 65 000 years, lit small fires as they moved around the landscape. This helped them to hunt for food, clear pathways and regenerate the bush.

Restraints around fuel reduction strategies

The forms of fuel management outlined come with costs and risks. These include resourcing, including the cost of labour, training and equipment. Prescribed burning can also go wrong. Fires can escape containment lines and result in the loss of property and in some cases, human lives. Indirect costs include respiratory-related health impacts from smoke exposure. Adverse environmental and heritage impacts may follow, such as the loss of amenity or biodiversity.

There is ongoing debate about the ecological consequences of prescribed burning. Some scientists express concern about the capacity of local biodiversity to recover from burns conducted too frequently.

Those downplaying the role played by climate change often blame the lack of prescribed burning for the fire-based disasters experienced in recent decades. These same interests single out public land management, especially national parks, as a factor in bushfire severity in Australia. They argue these management strategies are often driven by an ideological approach favouring minimal intervention.

Planning and regulatory responses

Drawing on the experiences of past extreme fire events, authorities embraced a range of planning and regulatory interventions aimed at reducing the impact of fire on property, infrastructure and people.

Land-use planning (zoning) and building regulations are tools deployed to minimise the risk to communities in fire zones. These laws and regulations can't be imposed retrospectively. They can, however, be imposed on communities as they rebuild after a fire disaster.

Typical planning interventions include:

- limiting development in fire-prone areas
- siting housing with due regard to topography and fire behaviour
- ensuring an adequate buffer zone exists between housing and adjacent bushland
- planning road links that are easily accessed by emergency vehicles and for efficiently evacuating residents
- ensuring vulnerable developments, such as schools, hospitals and aged care facilities, are built only in areas where speedy evacuation is possible
- protecting electricity and telecommunications infrastructure.

Historically, powerlines have often caused fires and require fire mitigation measures. It is vital to ensure that they are well maintained and their corridor is kept clear of fuel. Telecommunications infrastructure also needs protection, given its strategic importance in warning communities of impending danger.

Also relevant are laws governing vegetation clearing on private land, including where it can and cannot be done. Regulatory interventions that govern building design and the construction materials used are relevant. In Australia, requirements for the construction of buildings in bushfire-prone areas are specified by Standards Australia in AS3959–2009. The purpose of the standards is to reduce the risk of buildings igniting while a bushfire passes through. Construction requirements vary depending on the bushfire risk as assessed for the area. All aspects of construction are covered in the standards, including fire-resistant framing, roofing and cladding.

● **SPOTLIGHT**

The Resilient Building Council

The Resilient Building Council is an independent, not-for-profit organisation. Its members are scientists, civil engineers, structural engineers, fire safety engineers, bushfire architects, community risk management specialists and building regulatory experts. The council's interest is to improve the resilience of properties and communities.

The council has initiated a Bushfire Resilience Star Rating project. This aims to measurably reduce bushfire impacts, according to the evidence-based assessment conducted by independent bushfire experts. The rating recognises and rewards well-prepared properties and communities. Its purpose is to reduce the unsustainable financial impacts of bushfire disasters.

Community awareness

Alerting communities to an impending bushfire, especially those considered extreme, is a major focus of authorities seeking to minimise loss of life. The use of a fire danger rating system, combined with on-air radio and television warnings, plays an important role in the protection of people, livestock and property. During a bushfire emergency, community briefings are regularly held. Letterbox drops updating communities are used to keep people informed. Telephone and text-based warning systems are activated.

The Bush Fire Danger Ratings are based on the possible impacts of a fire, if one was to start. Fire authorities

The Australian fire danger ratings (AFDRS) levels:

MODERATE
Plan and prepare

HIGH
Be ready to act

EXTREME
Take action now to protect life and property

CATASTROPHIC
For your survival, **leave bushfire risk areas**

FIRE DANGER RATING	What you should do
CATASTROPHIC	• For your survival, leaving early is the only option. • Leave bush fire prone areas the night before or early in the day—do not just wait and see what happens. • Make a decision about when you leave, where you will go, how you will get there and when you will return. • Homes are not designed to withstand fires in catastrophic conditions so you should leave early.
EXTREME	• Leaving early is the safest option for your survival. • If you are not prepared to the highest level, leave early in the day. • Only consider staying if you are prepared to the highest level—such as your home is specially designed, constructed and modified, and situated to withstand fire, you are well prepared and can actively defend if a fire starts.
HIGH	• Leaving early is the safest option for your survival. • Well-prepared homes that are actively defended can provide safety—but only stay if you are physically and mentally prepared to defend in these conditions. • If you're prepared, leave early in the day.
MODERATE	• Review your bushfire survival plan with your family. Keep yourself informed and monitor conditions. Be ready to act if necessary.

14.2.5 Bushfire danger ratings

determine the risk level by assessing forecast cond tions such as temperature, humidity, wind and the dryness of forests and grasslands. The higher the fire danger rating, the more dangerous the conditions (see Figure 14.2.5).

Education campaigns are also used to inform residents of the actions they can take to protect their lives and properties. Figures 14.2.6 and 14.2.7 outline the steps residents can take to prepare their homes ahead of bushfire season and once the alarm is raised.

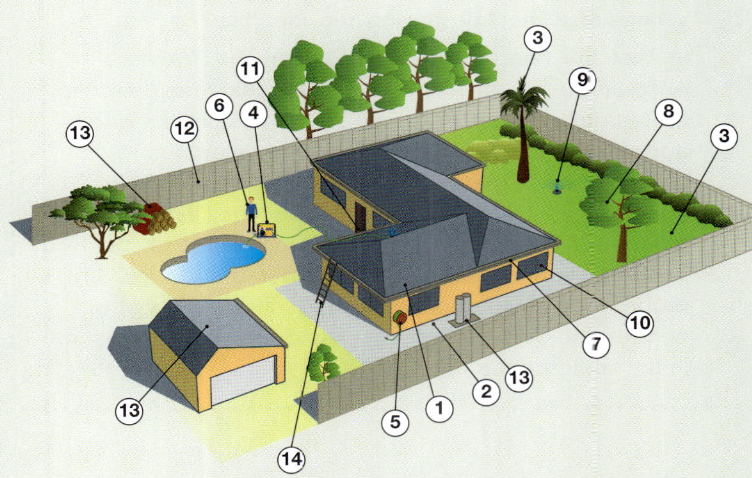

1 Clear leaves from gutters and cut back vegetation overhanging the house.
2 Seal any openings under the house or eaves. Fit wire screens to doors and windows.
3 Keep lawn and shrubs trimmed and rake up leaves.
4 Install non-electric sprinkler system that uses water in swimming pools and/or tanks.
5 Ensure hoses are in good order and long enough to reach all parts of the property.
6 Wear loose fitting clothing made from natural fibres such as wool, heavy cotton or denim.
7 Block downpipes and fill gutters with water. Hose down the house and surrounding areas.
8 Trim branches and trees.
9 Turn on sprinklers. Remove all flammable substances, such as gas cylinders and paints, from around the house.
10 Make sure that everyone (including pets) is inside.
11 Put wet towels against the spaces under doors. Close all windows, curtains, blinds and doors. Fill buckets, basins, baths and sinks with water to put out spot fires.
12 Install metal (not timber) fencing that shields property from an advancing fire front.
13 Store wood, gas, petrol and oil-based paints well clear of the house.
14 Keep ladders handy for roof access (inside and out).

14.2.6 Protecting a suburban home

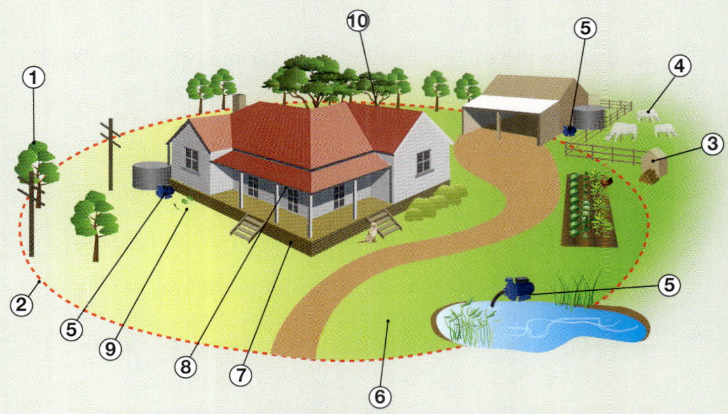

1 Trim tree branches away from power lines.
2 Create a firebreak around the homestead and other farm buildings.
3 Store firewood, petrol and gas well away from the house.
4 Move livestock to a well grazed paddock.
5 Use a diesel-powered pump to access water from farm dams, water tanks and swimming pools.
6 Keep lawns short and plant shrubs away from farm buildings.
7 Seal under-floor spaces to prevent embers entering.
8 Install gutter guards and keep gutters clear of leaf litter.
9 Have a hose ready to put out spot fires.
10 Block downpipes and fill gutters with water. Hose down the house and surrounding areas.

14.2.7 Protecting a farm-based home

● SPOTLIGHT

Spatial technologies and bushfire management

For those responding to and managing bushfire hazards, deploying spatial technologies can enhance their effectiveness. The US's Forest Service, for example, uses a model to assess fire risk developed by Esri, a US geographic information system software developer. This model analyses data on the distribution and types of trees and other ground cover, and the construction materials used in an area. The data are collected mainly by satellites and aircraft, but rangers and crews of firefighters contribute detail from the ground.

Another system developed at the University of California, Santa Barbara, called RHESSys, uses fuel and moisture data for 800 square kilometres of Californian wildland. It identifies where best to establish fire breaks, cut down trees or clear scrub.

The Los Alamos National Laboratory, New Mexico, designed another piece of sophisticated software called FIRETEC. It models how the flames of a planned burn—intended to clear vegetation in a controlled way—will be fed by the wind they generate. Its users include the forest services of Canada, France and the USA. The software lets them design precise patterns for planned burns and to clear surface vegetation without destroying tree canopies.

In Australia, researchers at the University of Western Australia (UWA) developed a new touchscreen device for mounting in fire trucks to help firefighters predict where and when a bushfire will spread. The system analyses comprehensive geospatial data in real-time, including geographical topography, vegetation types, fire-prone hotspots, time since the last burn, rate of spread, fuel accumulation and forecasted weather. Unlike the CSIRO's modelling software Spark, designed for use in control centres, the UWA technology puts critical fire information in the firefighters' hands.

Activities

Acquiring and processing geographical information

1 Explain the purpose of hazard reduction.

2 List the three processes involved in managing the fuel load. What is the intent of such interventions?

3 Name and describe the four most common forms of fuel load reduction techniques.

4 State where prescribed burning is considered to be most effective. What places are targeted? Outline the relationship between fuel reduction strategies and extreme weather.

5 Compare the behaviours of low-intensity and extreme bushfires.

6 Outline the costs and risks associated with fuel reduction strategies.

7 Explain the relationship between climate change and prescribed burning, and how it is being used as an ideological weapon.

8 Outline the planning interventions that can reduce the impacts of fire.

9 Outline the importance of raising community awareness of bushfire dangers. What forms does it take?

Applying and communicating geographical understanding

10 Study the box, Spotlight: Cultural burning practices of Aboriginal and Torres Strait Islander peoples. Complete the following activities.

- a Outline the ways in which Aboriginal and Torres Strait Islander peoples use fire as a land management strategy and the impact it has on plant species and environments.
- b What impact did the decline of cultural burning practices have?
- c Define Country.
- d How have First Nations Australians' culturally prescribed management practices, developed over 65 000 years, been passed from one generation to another?
- e What are the objectives of fire-based management processes?
- f Describe what is meant by the term 'cool' burn. Outline its characteristics.

11 Study Figures 14.2.7 and 14.2.8. Select the illustration that best reflects the type of home you live in. Assess how well prepared it is in the case of an extreme fire event.

12 Study the Spotlight: Spatial technologies and bushfire management. Investigate the technologies used by the NSW Rural Bushfire Service.

UNIT 14.3
The Black Summer fires of 2019–2020

Australia's massive Black Summer bushfires in 2019–2020 provide broad insights into the physical processes and cycles influencing the nature and occurrence of bushfires in Australia (see Figure 14.3.1).

14.3.1 Orroral Valley bushfire, Canberra, on 31 January 2020, burned nearly 25% of the ACT.

Australia's Black Summer

The summer of 2019–2020 was Australia's most catastrophic bushfire season. So much was lost and the impact so great that its consequences will be felt for decades. Almost 19 million hectares (186 000 square kilometres) were burnt, including 12.6 million hectares of forest and bushland. Thirty-four lives were lost, including six Australian firefighters and three US aerial firefighters. It destroyed 3094 homes and 65 000 people were forced to flee. Nearly 3 billion animals were killed or displaced. Many threatened species and ecological communities were extensively harmed. Estimates of the national financial impact of the Black Summer fires exceed $10 billion.

In NSW, the fires burnt 5.3 million hectares (6.7% of the state), including 2.7 million hectares in national parks (37% of the state's national park estate). More than 80 per cent of the World Heritage-listed Greater Blue Mountains Area and 54 per cent of the NSW components of the Gondwana Rainforests of Australia World Heritage property were impacted by fire.

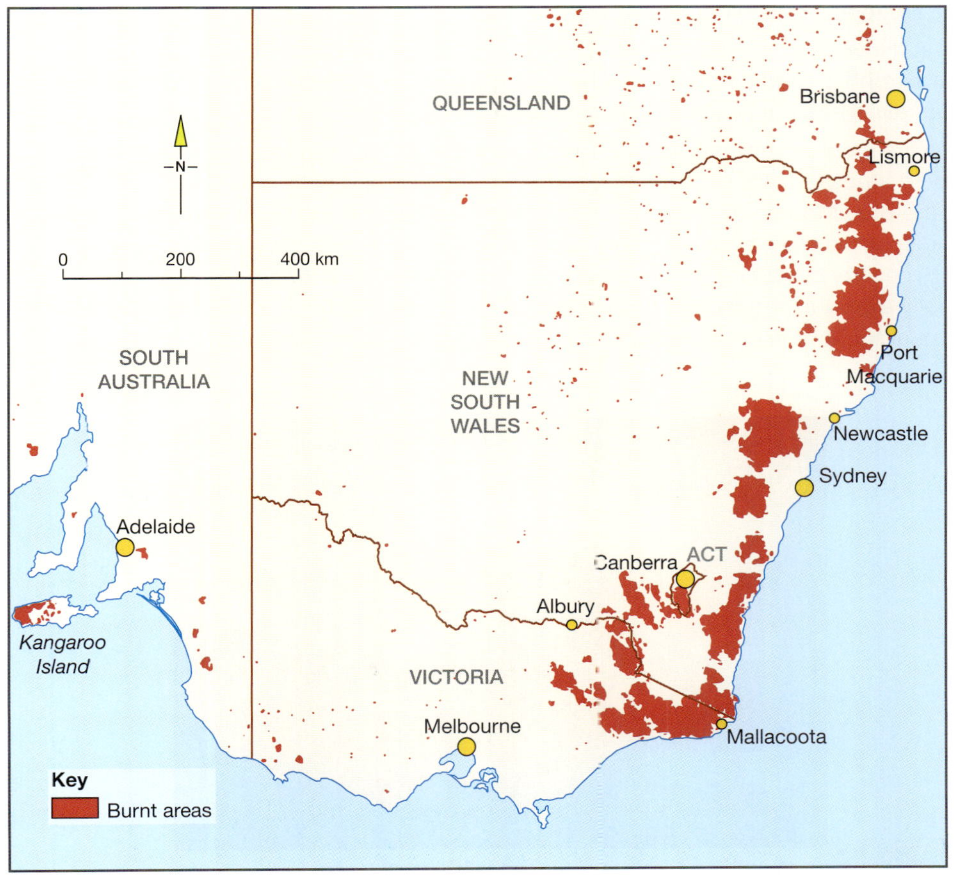

14.3.2 Distribution of burnt areas, 1 October 2019 to 5 February 2020, in south-east Australia

14.3.3 A NASA satellite image of the smoke plumes from the south-east Australian fires

According to early estimates, the bushfires probably released around 900 million tonnes of carbon dioxide into the atmosphere. This is roughly equivalent to the annual emissions from all commercial aircraft worldwide and is higher than the annual emissions of Germany. It is also far more than Australia's annual emissions of around 531 million tonnes in 2019–2020.

Lighting strikes, rather than arson or accidents, ignited the majority of the blazes. The extent of the fires is evident in Figures 14.3.2 and 14.3.3.

Major bushfire outbreaks

From September 2019 through to March 2020, fires swept across vast areas of south-east Australia. In eastern and north-eastern Victoria, large forest areas burned out of control for four weeks before the fires emerged into the surrounding grasslands in late December. Multiple states of emergency were declared across NSW, Victoria and the ACT. Three large fires are covered in Spotlight boxes to illustrate the extent and severity of the bushfire emergency as it devastated communities. They caused untold ecological damage and dominated global news services.

● **SPOTLIGHT**

The Gospers Mountain megafire

A megafire started from a single ignition point on the afternoon of 26 October 2019 (see Figure 14.3.4). A lightning bolt struck the ground near a disused airstrip at Gospers Mountain in the Wollemi National Park, NSW. It ignited a small fire on the floor of this heavily wooded area. By the day's end, the blaze had expanded to an area of 521 hectares. Still burning two weeks later, a sudden wind change turned the fire towards the south-east.

By 12 November, the Gospers fire was one among 300 that gripped the state in catastrophic conditions. Fanned by strong winds and temperatures in the mid-30s, Australia's eastern seaboard was ablaze from the Illawarra to Taree.

The fire would eventually burn through almost 500 000 hectares of forest (about seven times the size of Singapore). The fire zone extended from the western border of the Blue Mountains to the Central Coast hinterland, north to the Hunter Valley and south to the Hawkesbury and past Bells Line of Road. It eventually merged with several other fires to form a vast complex called a megafire. The fire was finally extinguished by rain on 10 February after burning for 77 days. Eighty-one per cent of the Blue Mountains World Heritage Area had been burnt.

14.3.4 Gospers Mountain Fire, NSW

SPOTLIGHT

The Currowan megafire

A fire started in Currowan, on the NSW South Coast, on 26 November, 2019 (see Figure 14.3.5). Over the next 74 days it blackened almost 500 000 hectares of forest and grassland, destroyed 312 homes and damaged another 173. Three people died.

This fire was ignited by a lightning strike somewhere in the rugged terrain of the Currowan State Forest. Fanned by a west-north-westerly wind, the fire spread rapidly.

A decade of dry conditions and two years of intense drought had preceded this fire. It sapped moisture from the land, greatly increasing the amount of fuel available to burn. Trees, stressed by drought, had dropped large amounts of leaves.

A combination of strong winds, low humidity, dry soil and high temperatures pushed the initial fire northwards to the Southern Highlands in less than six weeks. On the worst days, the fire advanced 12 to 15 kilometres. As the wind direction changed, it also spread to the east and south–east.

A week after the fire began, it started attacking coastal towns. Evacuation centres opened in Ulladulla and Batemans Bay, at that time considered to be relatively safe.

The region's distinctive escarpment and its hilly terrain, combined with strong winds, made the fires behave unpredictably. Spot fires, often many kilometres ahead of the fire front, made the situation even more hazardous.

Those living in coastal communities prepared for ember attacks as they watched the sky go black that afternoon. Bushfire-generated storms added to the sense of doom. The storms were generated when columns of heat rose with the smoke. Once they punched through a height of 15 to 16 kilometres, they formed a thunderhead that produced lightning storms. Once the system collapsed, it sent the wind in a 360-degree direction, making fires behave erratically on the ground and throwing spot fires kilometres ahead of the fire front.

As conditions worsened, authorities ordered a mass exodus of holiday-makers from the South Coast. The Princes Highway was soon clogged with traffic heading north towards Sydney. The congestion lasted up to 24 hours.

After 74 days, rainfall finally extinguished the Currowan blaze.

14.3.5 Smoke from the Currowan bushfire, 2019

● **SPOTLIGHT**

The Mallacoota evacuation

Mallacoota is a small coastal town in the far eastern corner of Victoria. The fire there is now seen as one of the defining moments of the 2019–2020 Black Summer bushfire season. A rapidly advancing fire front forced thousands of residents and holiday makers to huddle on the beach, where they were rescued by the Royal Australian Navy (RAN).

The Mallacoota disaster had its origin after Christmas 2019. On 28 December, lightning strikes started a small fire about 30 kilometres west, at Wingan River. This fire was soon out of control and generating its own weather. It began bearing down on Mallacoota.

On the morning of 30 December, local authorities alerted residents and visitors to the planned emergency procedures at a town meeting. There is only one road into the town, which meets the Princes Highway about 24 kilometres north-west of Mallacoota. The general advice was that those wanting to leave should do so immediately, but only by road to the north through Eden and then onto Canberra. The road to Melbourne had already been cut. The road north would, however, close within hours. About 1000 residents and 3000 visitors remained trapped in the town.

At 4:40 pm on 30 December, an emergency warning was issued for people to relocate to a safe evacuation location. This included the local wharf. Boats were launched to shelter on the lake—more than 50 boats carrying up to 300 people were on the water by the evening of 30 December.

Around 3500 people took shelter on the foreshore hoping to escape the wrath of the blaze. It was expected to reach the outskirts of the town at daybreak. Instructions were issued to get into the water if necessary, their only signal would be fire truck sirens. Mallacoota was now without power.

Early on 31 December, the blaze's 10-kilometre-wide fire front reached the town. Large parts of the settlement were lost during the following hours, including over 135 homes. The central area where people were sheltering, with most of the town's significant infrastructure, was saved as the fire front shifted north-east towards the NSW border.

On 2 January the RAN arrived bringing essential food and medical supplies. The following day, the Navy began evacuating 1000 people (see Figure 14.3.6). The Royal Australian Air Force evacuated a further 450 people. Road access was not possible until 18 January due to extensive fire damage.

Fortunately, no lives were lost. Images of people huddled on beaches and boats featured in media reports worldwide. They became symbolic of the devastation caused by the Black Summer blazes.

14.3.6 Evacuees board one of HMAS *Choules*' landing craft at Mallacoota to be ferried out to the ship, 3 January 2020.

Prevailing weather conditions

The prevailing environmental conditions in the months preceding the 2019–2020 bushfire season established the environmental conditions and circumstances that would cause any bushfire to swiftly get out of control.

The year 2019 was Australia's warmest and driest. Record low rainfall occurred over large areas of inland Australia that spring (see Figure 14.3.9). As a result, there were very low soil-moisture levels over most of the continent. The dry conditions continued through December—the driest on record, with rainfall below average across most of the continent.

In the northern tropics an absence of early wet-season rain caused the heat to build. Except for places near exposed coasts, many sites regularly reached 40°C in early December. High-temperature records were also set at several sites across the north of Western Australia, the Northern Territory and Queensland.

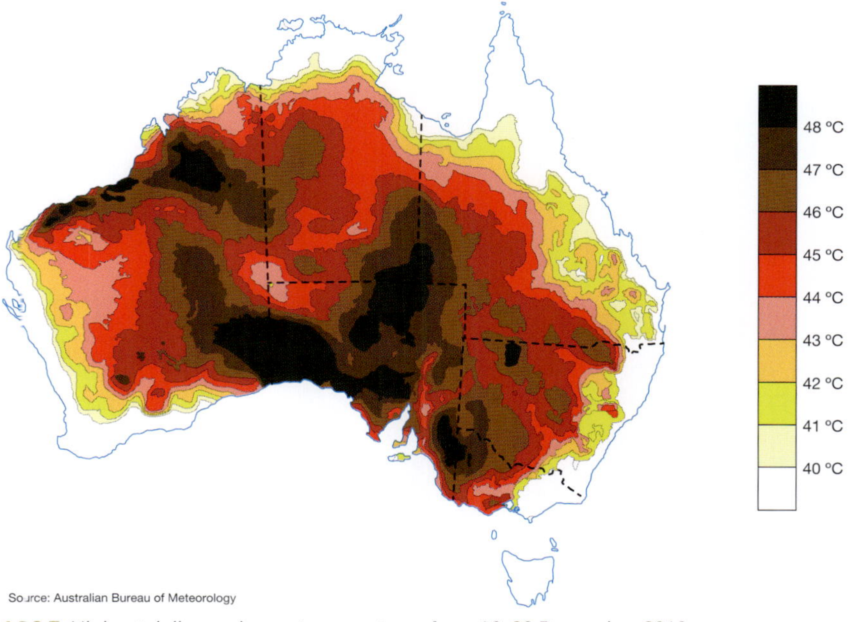

14.3.7 Highest daily maximum temperatures from 16–22 December 2019

The heat continued in northern Australia with only brief periods of relief. From 12 December, the mass of extreme heat moved into the south of Western Australia, then more broadly towards the east through southern Australia. While it had dissipated in southern Australia by 22 December, extreme heat continued in central and northern areas. It returned to south-eastern Australia from 28 December onwards peaking on 30 December (see Figures 14.3.7 and 14.3.8). Southern Australia experienced a further round of extreme heat in early January 2020, peaking in southern and Central NSW on 4 January. A final significant heatwave swept across south-eastern Australia at the end of January and the start of February.

A major feature of this heatwave was its extent. Australia experienced its hottest day ever recorded on 18 December, 2019. The average of 41.88°C was far above the previous record of 40.30°C set in January 2013. Six other days that month also exceeded the previous record temperatures. This high saw conditions send the bushfire risk rating to extreme.

In NSW the most extreme heat occurred on 4 January 2020. Temperatures reached 48.9°C in Penrith, the highest recorded in the Sydney basin. Temperatures this high are dangerously hot and place extreme thermal stress on people and the environment.

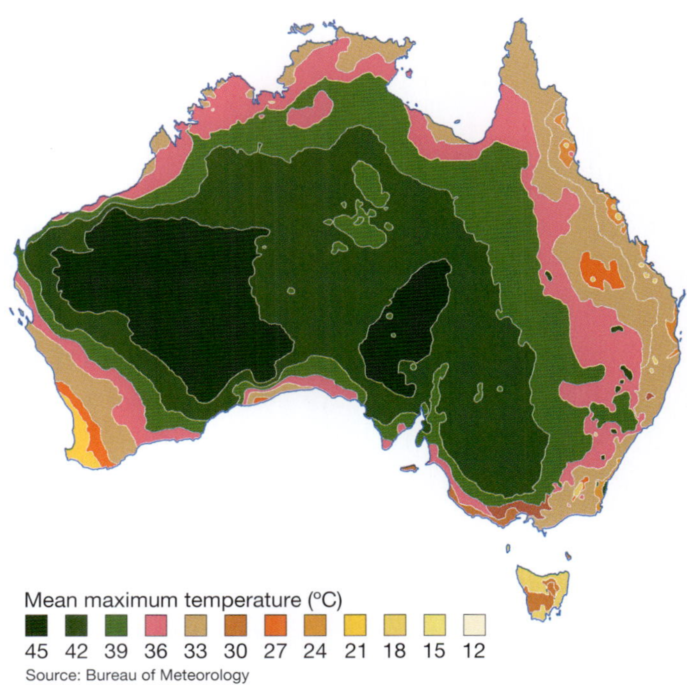

14.3.8 Maximum temperatures experienced, 29 December 2019

The heatwave conditions, combined with an already parched landscape and strong winds, produced elevated fire risk weather conditions from December 2019 into early January 2020. The severe rainfall deficiencies and hydrological drought exacerbated the fire weather conditions.

During this time, the Forest Fire Danger Index (FFDI) in south–east Australia was extreme. The FFDI is one common measure of fire weather conditions, reflecting longer-term rainfall and temperature patterns and shorter-term weather.

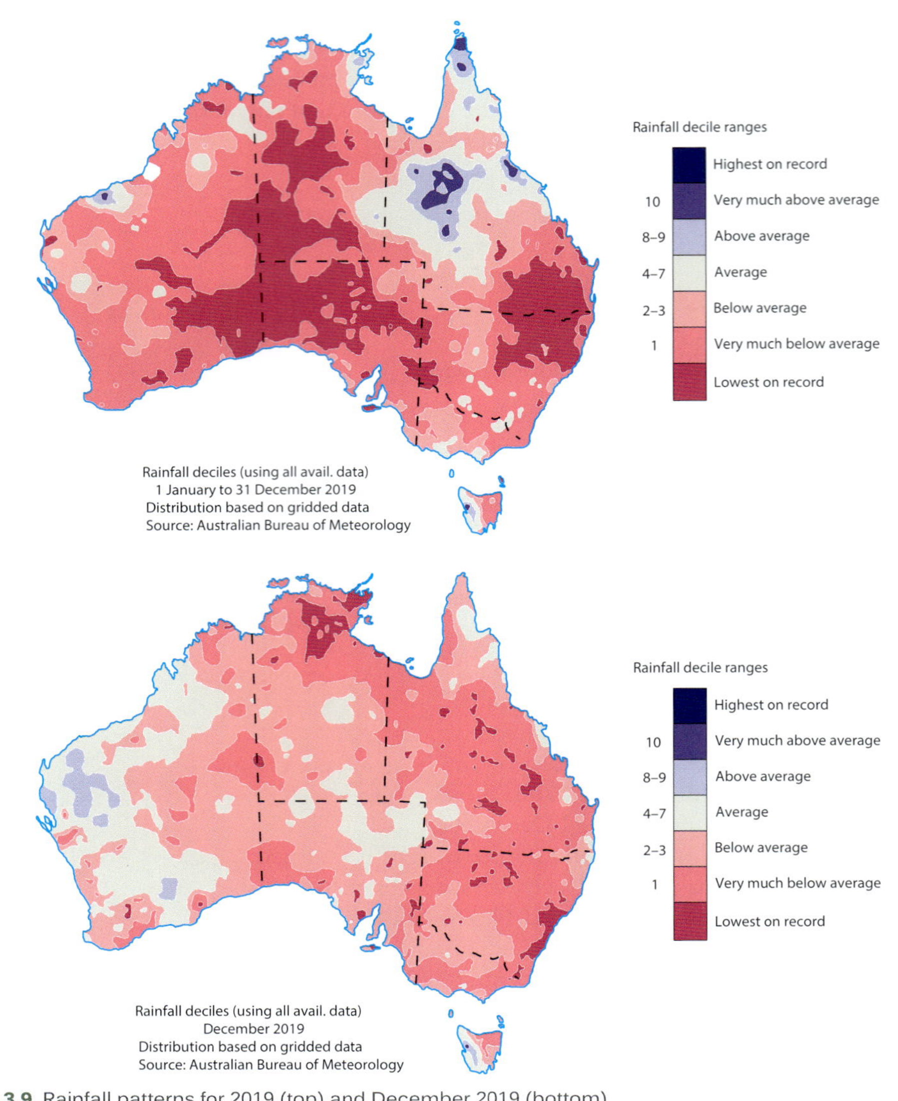

14.3.9 Rainfall patterns for 2019 (top) and December 2019 (bottom)

Impacts of human-induced climate change

In the wake of the Black Summer bushfires, scientists concluded that authorities needed to take more action in mitigating bushfire risk. They sought to improve the methods used to adapt to the now inevitable increase in fire risk in Australia, while also pursuing urgent global climate change mitigation efforts.

Scientists, including geographers, warn that fire disasters like Black Summer are made worse by human-induced climate change in numerous ways. Some are well understood and some require more research. Climate change impacts combine to increase the severity of bushfires, especially in south-east Australia.

These conclusions are consistent with the findings of both the NSW Bushfire Inquiry and the federal government's Royal Commission in the wake of the Black Summer disaster. They also concur with an open letter released at the height of Australia's Black Summer fire crisis, which was signed by more than 400 climate and fire experts from across the world. It warned that human-induced climate change is worsening fire weather and bushfires in southern and eastern Australia. They stated that the data shows a trend towards more frequent and extreme fire weather conditions during summer, along with an earlier start to the fire season, particularly in southern and eastern Australia. The letter concluded that Australia's dangerous fire weather will certainly worsen in the future and that ongoing human-induced climate change is making fire management increasingly challenging.

Given such findings, the overwhelming consensus of scientists is that urgent and ambitious action to reduce greenhouse gas emissions is needed if we are to pursue efforts to limit global warming to 1.5°C above pre-industrial levels. Global greenhouse gas emission reductions would curtail further climate change-related intensification of Australia's bushfire risk and give fire management and adaptation measures the best chance of success.

● SPOTLIGHT

Pyrocumulonimbus thunderstorms

During the Black Summer, many fires transitioned into extreme **pyrocumulonimbus** events, extremely dangerous fires that generate their own thunderstorms (see Figure 14.3.10).

Pyrocumulonimbus clouds can cause dangerous and unpredictable changes in fire behaviour, making the fire more difficult and hazardous to fight.

They form when the intense heat of a fire causes air to rise rapidly within the smoke plume. The rising hot air is turbulent and draws in cooler air from the surrounding air mass, helping to cool the plume as it rises.

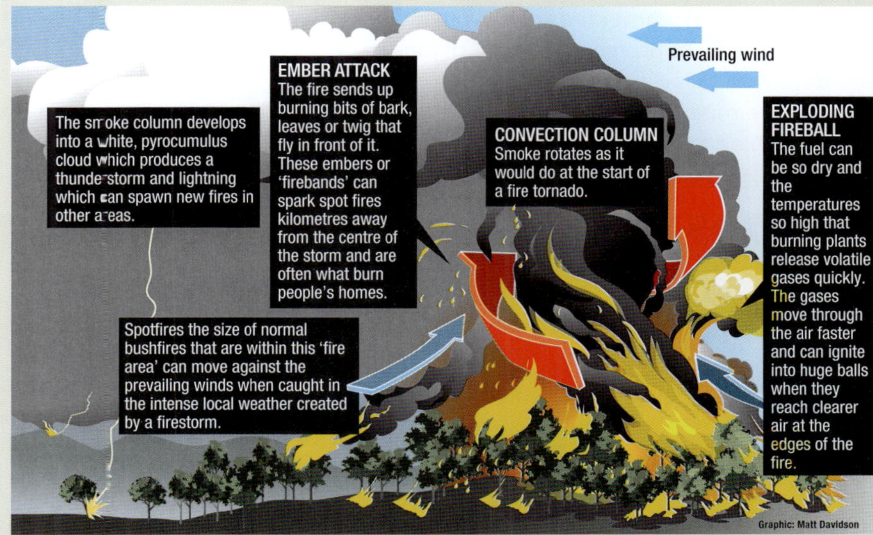

14.3.10 The formation of a pyrocumulonimbus thunderstorm

As the plume rises higher into the atmosphere, the air pressure declines, causing the plume air to expand rapidly and cool even further. If it cools enough, the moisture in the plume will condense and form cumulus clouds called pyrocumulus. The 'pyro' comes from the Ancient Greek word meaning fire.

The condensation process releases stored latent heat to be released. This makes the cloud warmer and more buoyant, causing the cloud air to accelerate up. Further expansion and cooling causes more moisture to condense and the cloud to accelerate up. In some instances, the air accelerates into the lower stratosphere before losing buoyancy. Collisions of ice particles in the very cold upper parts of these clouds cause a build-up of electrical charge, which is released as lightning. Having produced a thunderstorm, the cloud is now known as pyrocumulonimbus.

Ecological impacts of the fires

The Black Summer bushfires have been described as an ecological disaster. These fires extended across tens of millions of hectares of land, covering native forests and grasslands that serve as wildlife habitat. Over 330 threatened species and 37 threatened ecological communities protected under national environmental law were in the path of these bushfires. Six World Heritage sites were affected, including extensive areas of the Gondwana Rainforests of Queensland and NSW (54% burnt) and the Greater Blue Mountains Area in NSW (82% burnt).

At least 3 billion terrestrial vertebrates were displaced or killed by the fires. Reptiles made up two-thirds of the affected, and birds, mammals and amphibians comprised the other third. Ecologists feared some endangered species were driven to extinction.

It is still too early to be certain about the ultimate consequences of this bushfire season for Australian wildlife. But experts do predict serious, long-term, adverse effects on biodiversity.

Non-government wildlife organisations played a critical role in the aftermath of the fires. They rescued injured wildlife and ensured that as many species as possible were supported until the habitat on which they depend became viable.

● **SPOTLIGHT**

Saving the Wollemi Pine

The Wollemi pine is a critically endangered plant species dating from the Jurassic Period (199.6–145.5 million years ago). While successfully propagated commercially, fewer than 100 mature plants survive in the wild. All are in a small remote gorge in the Wollemi National Park. It is one of the world's greatest ecological treasures. Bushfire is the most significant threat to the species' survival.

During the Black Summer fires, the Wollemi pines were at risk of being lost. To protect the stand of pines, air tankers were deployed to lay a protective layer of fire retardant and a rescue mission was launched. A specialist team was assembled to fight the fire, consisting of national park staff, firefighters and Wollemi pine researchers.

Due to its remoteness and rugged terrain, access to the gorge is difficult. It can only be reached by air. Emergency firefighters had to be flown and winched in by helicopter. Their strategy was to install an irrigation system to increase the moisture content in ground fuels surrounding the pines, thereby reducing the severity and impact of the blaze. Helicopters were also deployed to drop water onto the fires as they approached the gorge. These measures protected these trees. They minimised the damage caused and preserved this unique species in the wild. See Figures 14.3.11 and 14.3.12.

14.3.11 Special firefighting efforts saved the wild Wollemi Pines from the Black Summer fires.

14.3.12 A NSW National Parks and Wildlife Service firefighter with some Wollemi pines he helped save. Their location remains a closely guarded secret.

● **SPOTLIGHT**

Impacts on air quality

Air quality in Sydney and much of eastern NSW was well above a level considered hazardous for much of December 2019 and January 2020 (see Figure 14.3.13). The air quality in New Zealand also suffered. By New Year's Day, 2020, a blanket of smoke from the Australian fires covered the whole South Island, giving the sky an orange-yellow haze. It moved over the North Island the following day, continuing to impact air quality until 6 January. By 7 January 2020, the smoke was carried another 11 000 kilometres across the South Pacific Ocean to Chile, Argentina, Brazil and Uruguay.

14.3.13 The bushfire crisis saw air quality in Sydney and much of eastern Australia range from bad to extremely hazardous.

● SPOTLIGHT

Forests struggle to recover

By the end of 2020, the fire-affected forests of south-east Australia were showing signs of recovery. But the rate of recovery was uneven. In some instances, there was little evidence that the original ecological complexity would return.

Several factors combined to make the Black Summer fires particularly devastating for Australia's eastern forests. The drought preceding the record-breaking blazes was so intense that the forests' capacity to bounce back to health was greatly reduced.

Because there were so few unburnt refuges for plants and animals to emerge from to repopulate the fire grounds, any recovery would be slow. Logging also took a toll. Tree culling reduces the overall condition of the forest because it involves removing ecologically significant large trees critical to the forest ecosystem.

The salvage logging of burnt forests also takes a toll. Soil structure is affected, as is the water quality of streams. Removing trees impacts the habitat of native animals.

Government authorities have sought to promote the recovery of the forest. They have, for example, collected millions of seeds to help regrow the wet eucalypt forests in Victoria. In the dry eucalypt forests, efforts have been made to stop the spread of invasive weeds and feral animals.

Scientists, including geographers, argue that the ecosystems of south-east Australia have not yet been able to adapt to the increasing frequency and severity of extreme bushfire events. In the first two decades of this century, there have already been three megafires on Australia's east coast—in 2009, 2014 and 2020. In the entire twentieth century, there was just one.

Some of the most lasting impacts occurred in the alpine ash forests of the Snowy Mountains and north-eastern Victoria. They evolved to cope with fires that occur no more than once every 75 years. But these forests burned in 2003 and again in 2020. When alpine ash burns too often, it does not produce the seeds necessary for the forest to regenerate.

There is a risk that the wetter forests across huge areas of Victoria and southern NSW won't be able to recover from the 2019–2020 fires. This is because most tree species in wet eucalypt forests grow from seeds. The lack of mature trees, which are only found in old-growth forests, is a key risk to forest recovery. These mature trees produce most of the seeds needed for forest regeneration. They also provide the habitat that hundreds of species of vertebrates rely on.

Trees in dry eucalypt forests, which are adapted to hotter and more frequent fires, don't shed seeds like the species found in the wetter forests. They sprout new growth from their trunks—a response known as epicormic growth (see Figure 14.3.14). In some areas, even these trees are struggling to regenerate because of repeated burning.

14.3.14 Within weeks of a fire regrowth begins to appear.

Response to the bushfire crisis

Authorities were overwhelmed by the scale of the emergency. The available firefighting resources proved to be inadequate, especially the availability of aerial water-bombing aircraft. In NSW, the Fire and Rescue NSW and the largely volunteer Rural Fire Service (RFS) formed the bulk of the primary response to the fires, mobilising thousands of firefighters and several hundred firefighting vehicles. Both organisations were supported in their firefighting efforts by the NSW National Parks and Wildlife Service and the Forestry Corporation of NSW, who are respectively responsible for national parks and forests across the state. Firefighting resources from other states were also deployed to assist in the response. In Victoria, the Country Fire Authority led the response. The Australian Defence Force supported state fire services in logistics and operational support.

Community-based volunteer organisations and charities also played an important supporting role. A sample of the organisations involved in the relief effort included:

- the Salvation Army and Australian Red Cross provided support at evacuation centres (see Figure 14.3.16).
- WIRES Wildlife Rescue mounted a rescue and treatment service for injured wildlife.
- the Animal Welfare League raised funds to assist the treatment of injured animals.
- Team Rubicon Australia supported communities by assisting with debris removal and the clean-up of fire-affected areas.
- Community Rebuilding Trust supported the rebuilding of communities by providing financial assistance to replace lost tools and office equipment for workers to return to work. They provided temporary business premises, sent in financial and business recovery expertise and rebuilt destroyed community facilities.
- Architects Assist, representing over 600 Australian architecture firms, provided their services free to individuals and communities affected by the bushfires.
- a diverse range of Australian charities raised an estimated half a billion dollars for bushfire relief.

Government-based relief responses included establishing the National Bushfire Recovery Agency in January 2020. Its role was to lead and coordinate the national response to rebuilding communities affected by the 2019–2020 bushfires. The Agency's responsibilities are outlined in Figure 14.3.15. Resilience NSW is the lead disaster management agency for the state. It is responsible for all aspects of disaster recovery and building community resilience to natural disasters.

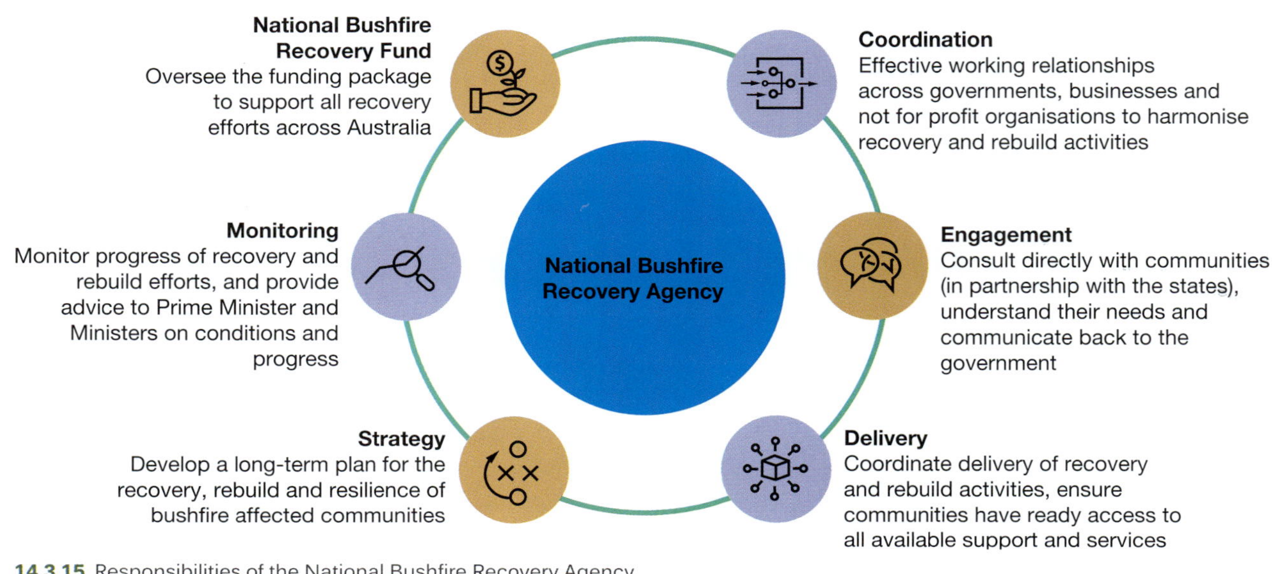

14.3.15 Responsibilities of the National Bushfire Recovery Agency

In the wake of the Black Summer fires, the then Australian Prime Minister, Scott Morrison, established a royal commission into the bushfires. Officially known as the Royal Commission into National Natural Disaster Arrangements, its terms of reference included an 'examination of the coordination, preparedness for—including hazard reduction burns—response to and recovery from disasters as well as improving resilience and adapting to changing climatic conditions and mitigating the impact of natural disasters.' While indicating support for most of the royal commission's recommendations, the federal government did take action on the call for creating a national aerial firefighting fleet. Today, the National Aerial Firefighting Centre (NAFC) contracts approximately 150 aircraft on behalf of state and territory governments.

14.3.16 The Red Cross distributing clothing at the Bega evacuation centre.

The NSW Government established its own independent expert inquiry into the 2019–2020 bushfire season. The inquiry made 76 recommendations and then-Premier, Gladys Berejiklian said her government would adopt all of them. Compulsory land clearing, night-time water bombing and aggressive hazard reduction burning were among the recommendations, as were strategies to protect fauna. Landowners in some areas will be legally required to ensure their properties are safe by clearing land and conducting hazard-reduction burns.

The scale of the Black Summer fires attracted the attention of the international media. This initiated a major international effort to assist Australian firefighters in their battle. Canada, Malaysia, New Zealand, the United Arab Emirates and the USA all sent experienced firefighters. Japan sent two water bombers and Singapore two heavy-lift helicopters. Fiji and Indonesia sent army engineers, while Papua New Guinea provided logistical support.

The response to the Black Summer bushfires was not without its critics. Those most seriously affected, including many who had lost their homes, complained that the promised assistance was slow in arriving. By mid-2020 many communities had received little in terms of relief. By mid-2023 many people were still living in temporary (often makeshift) accommodation, waiting on the processing of insurance claims, bank loans and building approvals. Without these, people were unable to start rebuilding their lives.

Businesses also struggled to survive in the aftermath. Those in the tourism sector lost the revenue generated by the peak summer holiday season. Many businesses and farmers faced delays in the distribution of much-needed money to assist in their recovery. This was despite the federal government setting aside hundreds of millions of dollars to help with the recovery. The paperwork required to access the funds was considered excessive.

Australia's federal system of government proved a hindrance to both the fighting of the blazes and the distribution of disaster relief. Disaster management is primarily a state responsibility, but the Commonwealth also has a role, particularly when fires cross state lines and those states need additional resources. There is no clear power in the Australian Constitution to declare a national emergency. Some important functions are in the hands of the Commonwealth Government or funded by it. Notably, these are aerial firefighting, telecommunications infrastructure, emergency broadcasts on the ABC, electricity grids and assistance from the military. Commonwealth agencies, such as the CSIRO and the Bureau of Meteorology, have important roles in predicting the likely impacts of bushfires and how their effects might be mitigated.

Looking to the future

The Black Summer bushfires of 2019–2020 proved to be a watershed moment in managing Australia's growing fire risk. It raised several questions. How can Australia's landscapes be better prepared for such events? How might fires best be responded to when they occur? How might the recovery from such events be better managed?

The events of 2019–2020 revealed that current fire management strategies will not, or are unlikely to, sustain the full range of ecosystem processes and biodiversity. Nor will they reduce the impact of bushfires on people, property, infrastructure, forests and ecological communities, biodiversity and wood resources to an acceptable level. Finding sustainable, affordable and socially acceptable approaches to managing fire will require a cohesive and considered, evidence-based approach.

Hazard reduction, especially prescribed burning, is favoured by many. But there are limits to such practices, especially as a warming climate provides a much narrower window of time for prescribed burning to take place. Land-use planning and zoning, combined with using fire-resistant building materials, have the potential to reduce the loss of property and life. Indigenous land management techniques provide one possible way ahead.

Unfortunately, the frequency and severity of bushfires is likely to increase as the climate warms. Ultimately, humans need to tackle climate change if the risk associated with extreme bushfire events is to be minimised.

Activities

Acquiring and processing geographical information

1 Outline the nature and extent of the losses associated with the Black Summer bushfires.

2 List the causes of most of the Black Summer of 2019–2020 fires.

3 Describe the prevailing weather conditions that preceded and sustained the Black Summer bushfires.

4 Explain the relationship between human-induced climate change and the increasing frequency and severity of bushfires.

5 Outline the ecological impacts of the Black Summer fires.

6 Describe the impact of the Black Summer fires on air quality.

7 Outline how authorities responded to the bushfire crisis and helped facilitate the recovery from the fires.

8 Outline the key findings of the Royal Commission into National Natural Disaster Arrangements and the New South Wales bushfire inquiry.

9 List the criticisms of the bushfire response and recovery effort.

Applying and communicating geographical understanding

10 Study Figures 13.3.2 and 13.3.3. Describe the spatial extent of the Black Summer bushfires. Why were they largely confined to the Great Dividing Range and the coastal strip?

11 Study the boxes, Spotlight: The Gospers Mountain megafire, Spotlight: The Currowan megafire and Spotlight: The Mallacoota evacuation. Complete the following tasks.

- a What do each of these fires have in common?
- b Describe the extent of the damage caused by the Gospers Mountain megafire.
- c What factors made the Currowan blaze so unpredictable?
- d How were people affected by the Currowan fire?
- e Explain why the Mallacoota blaze was so newsworthy.

12 Study Figure 14.3.7. Describe the distribution of the extreme hot weather experienced from 16 to 22 December 2019.

13 Study Figure 14.3.8. What parts of Australia experienced temperatures in excess of 40°C?

14 Study Figure 14.3.9. What parts of Australia experienced the lowest rainfall on record in all of 2019 and in December 2019?

15 Explain the process by which pyrocumulonimbus thunderstorms develop.

16 Study the box, Spotlight: Saving the Wollemi Pine. Outline the efforts made to protect this vulnerable species.

17 Study the box, Spotlight: Forests struggle to recover.

- a Outline the factors that made the Black Summer particularly devastating for Australia's eastern forests.
- b State the impact of logging on forest recovery.
- c State the evidence cited that confirms the increasing incidence of megafires. What impact is this having?
- d Explain the difference in the prospects for recovery of wet and dry eucalypt forests.

18 Study Figure 14.3.15. Using information from the infographic, describe the responsibilities of the National Bushfire Recovery Agency.

19 Conduct a class debate on the topic, 'The Federal Government should immediately establish a national fleet of aerial water bombers'.

APPLICATION AND CONSOLIDATION TASKS

Task 1: Infographic: El Niño and La Niña

Prepare an infographic explaining the impact of El Niño and La Niña events on the weather experienced in Australia. Start by studying Figure 14.1.13.

Using visuals, clearly outline the different impacts of these weather events. Your infographic should include information on these aspects:

- environmental
- economic
- social.

Consider past El Niño and La Niña events and their impacts in your response. Include the role of the Bureau of Meteorology in long-range forecasting. How does this enable public officials, farmers and others prepare for these events?

Task 2: Report: Resilient Building Council

Investigate the role of the Resilient Building Council, an independent, not-for-profit organisation. In your report include:

- an overview of the organisation
- its role
- its guidelines for new buildings
- how these guidelines are being implemented.

Your report can include visuals (maps, graphs, diagrams, photos) and comments from experts in the field.

Alternatively, assess your home and/or school with reference to bushfire resilience guidelines. Prepare a report of what would be required to meet the new guidelines.

Task 3: Poster: Bushfire plan

Prepare a bushfire survival plan for your family.

In your bushfire plan:

- identify the bushfire risk where you live
- include information about where to find out about bushfires, for example, apps for your phone, news sites, RFS
- assess the fire preparedness of your home.

Prepare a step-by-step guide of what needs to be done:

- before
- during
- after a bushfire.

CHAPTER

15 Contemporary hazard: COVID-19

In late 2019, reports began to emerge from China of a previously unknown, highly contagious respiratory disease. The disease outbreak was centred around the megacity of Wuhan in Hubei Province. The disease is a type of **coronavirus** and within months had spread globally creating a **pandemic**. COVID-19 is highly infectious and authorities around the world responded by restricting travel and, in many cases, forcing people to remain in their homes for long periods to reduce the threat of infection.

This chapter investigates the human–environment interactions that contributed to the outbreak of COVID-19 as an example of a contemporary ecological hazard. In doing so, it studies the nature and spatial dimensions of the pandemic and the factors that contributed to its nature and occurrence. It explores the human–environment interactions relevant to the disease, along with the challenges, opportunities and responses stemming from the disease. Also covered is how effectively people and organisations managed the outbreak.

Pandemic is not a word to use lightly or carelessly. It is a word that, if misused, can cause unreasonable fear, or unjustified acceptance that the fight is over, leading to unnecessary suffering and death.

Dr Tedros Adhanom, World Health Organization Director-General

15.0.1 Coronaviruses include the common cold and COVID-19

Chapter glossary

asymptomatic when a person is infected with a disease but does not display the typical symptoms of the illness

community transmission the spread of disease to people in the wider community outside of those in quarantine

contact tracing steps taken by health officials to link cases of COVID-19 infection to find the sources of the infection and limit its spread

coronavirus a family of viruses that use RNA instead of DNA as their genetic material. These viruses often cause respiratory illnesses in humans and other animals, including influenza and the common cold

COVID-19 the official name given to the newly discovered coronavirus that was the cause of the pandemic; it is an acronym for **co**rona**vi**rus **d**isease and 19 because it was first reported in 2019

epidemiological relating to the branch of medicine that deals with the prevalence, distribution and patterns of disease, often referred to as the geography of disease

herd immunity the idea that if enough people become immune to a disease, either through immunisation or from being infected, the disease will die out due to a lack of hosts

hotel quarantine a system whereby people arriving from overseas were required to isolate in special locked-down hotels for 14 days after arriving in the country

novel coronavirus a temporary name given to a new variant of a coronavirus until it is officially named

pandemic a situation where an infectious disease spreads over a large region or the entire world

patient zero the first person infected with a disease

quarantine placing people who are infected, suspected of being infected or may have been in contact with infected people, into isolation for a period of time to prevent them from spreading the disease to others

Spanish flu a strain of influenza (flu) that caused a global pandemic in 1919 and into 1920; there were many similarities between the Spanish flu pandemic and the COVID-19 pandemic

super-spreader a person who transmits a disease to a disproportionately large number of people

virus a very small and simple infectious agent that can multiply only in the living cells of plants, animals or bacteria

zoonosis a disease that can be transmitted from animals to humans

UNIT 15.1

Spatial dimensions and COVID-19

15.1.1 Deserted streets of Wuhan, China, January 2020

COVID-19 is a highly infectious disease that was first recognised in November 2019. As winter settled across China, residents of the city of Wuhan began to report severe colds that caused people to have trouble breathing. By December 2019, Wuhan's hospitals were filling with people suffering from a pneumonia-like illness. On 31 December 2019, the Chinese government formally announced an outbreak of a new type of coronavirus. Authorities responded by placing the population of more than 11 million people into lockdown, meaning movement outside their homes was severely restricted (see Figure 15.1.1). On 23 January 2020, Professor Edward Holmes, an Australian scientist from the University of Sydney, released the virus's genomic sequence confirming it was a new disease.

By mid-2023 almost 770 million cases of COVID-19 had been diagnosed since the pandemic was announced. These are just the confirmed cases. Millions more are expected to have been infected who were never tested. This is especially the case in developing nations. For example, data shows many more infections in developed countries per capita (see Figures 15.1.3 and 15.1.4), but this is far more likely due to comprehensive testing and reporting in these countries.

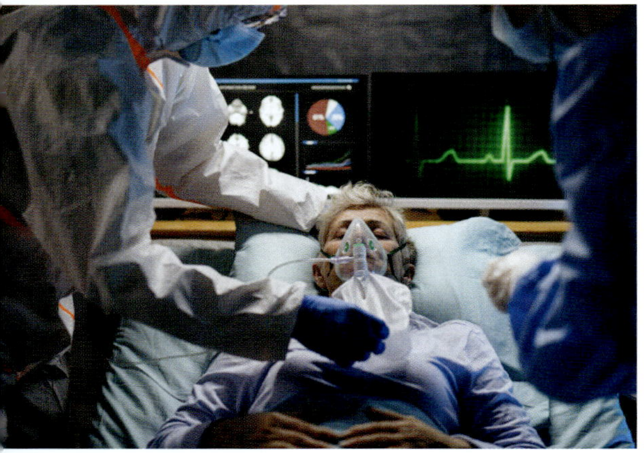

15.1.2 A COVID-19 patient in Beragamo, Italy

By mid-2023 almost 6.95 million people had died from the virus. Many more deaths associated with other illnesses, some previously undiagnosed, were made worse as a result of COVID-19. Lockdowns became common around the globe, including Australia. Significant restrictions were placed on people's movements to curtail the spread of the disease.

In mid-March 2020, the WHO declared a global pandemic. This term describes a situation where an infectious disease becomes very prevalent across a large region (see Figure 15.1.2). WHO declared a global pandemic, meaning it was impacting the entire world. At the time of the announcement, the Director-General of WHO said the pandemic would impact every sector of human society and that every sector must be involved in the fight against the disease.

Early in the pandemic, the world-renowned US-based John Hopkins University started to keep comprehensive statistics on COVID-19. On New Year's Day 2021, just over a year after the first cases were reported in China, they reported on the disease. By then it was found in 191 countries worldwide, and more than 85 million people had been infected (see Figure 15.1.3). Very few places globally entered 2021 free of the disease, with small island nations being notable exceptions. By mid-2023, only two countries had reported zero cases to WHO, Turkmenistan and North Korea. Both countries have authoritarian regimes with strict controls on media coverage and the release of information. So it is more than likely that no countries were free of infections.

Did you know?

Even in Antarctica COVID-19 cases were reported, with 36 staff at the Chilean General Bernardo O'Higgins Riquelme Research Base diagnosed with the illness in late December 2020.

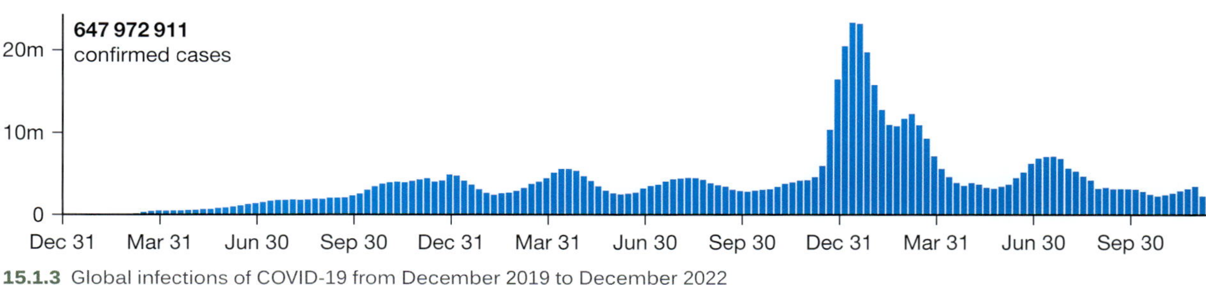

15.1.3 Global infections of COVID-19 from December 2019 to December 2022

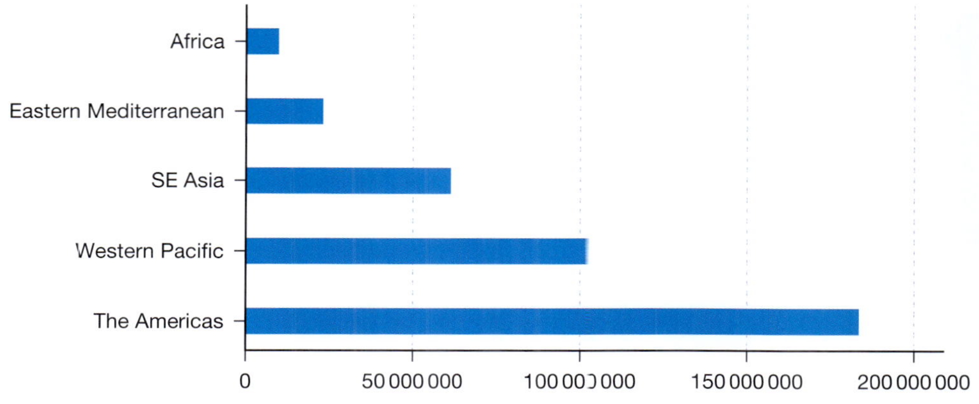

15.1.4 Infections of COVID-19 from December 2019 to December 2022 by region

The nature of COVID-19

COVID-19 is a **virus**, which is a tiny sub-microscopic infectious agent. A virus can only survive and multiply in the living cells of animals, plants or bacteria. Once outside the infected cells, a virus will eventually die unless it can be transmitted to another living cell. In the case of COVID-19, it is typically transmitted from one person to another by respiratory droplets. These can be spread when a person coughs or sneezes. Smaller aerosol droplets are emitted when people breathe. The virus in these droplets can be taken in through the nose, mouth or eyes. This is far more likely if someone is within one metre of the infected person. For this reason, one of the key strategies for reducing the disease's spread was to encourage people to maintain social distancing, staying at least 1.5 metres away from others, and wearing masks to minimise the transmission of respiratory droplets and aerosols (see Figure 15.1.5).

15.1.5 Masks reduce the transmission of COVID-19.

When COVID-19 emerged, some people sought to downplay the seriousness of the pandemic. They argued that it was simply another strain of the influenza virus and nothing too serious. This notion arose from it being part of the coronavirus family, which includes influenza and the common cold. There are many more viruses in the coronavirus family, their common feature being a genetic structure made up of RNA, ribonucleic acid, instead of DNA, deoxyribonucleic acid. However, being a new form of coronavirus that humans had no previous exposure to, there was no resistance to the disease in our bodies. This meant it was more severe and far more infectious than existing coronaviruses.

● **SPOTLIGHT**

One disease, many names

Scientifically, the disease is officially referred to as COVID-19. This is an acronym for coronavirus disease, with the 19 indicating it was first reported in 2019. The WHO gave it this designation in February 2020. Until then the official name had been 2019n-CoV.

But the disease has been referred by many other terms. Coronavirus is one of the most common, especially in the media from the USA. However, this refers to a family of diseases rather than this specific one. Another commonly used term is the '**novel coronavirus**'—meaning 'new' coronavirus.

Some names were used for purely political purposes. For example, former US president Donald Trump adopted the term, 'China Flu' in attempts to downplay the disease's significance. It also gave credence to theories that the disease was deliberately spread by Chinese authorities.

15.1.6 Live ducks sold in a neighbourhood wet market in Shanghai, China

Source of COVID-19

The source of COVID-19 is the subject of considerable debate. **Patient zero**—the first person infected with the disease—was never successfully identified. This has made it very challenging to identify the original source of the disease. Most scientists believe it is a **zoonosis**, a disease transmitted from animals to humans. This has been a common form of transmission for numerous other respiratory coronavirus diseases that emerged in the twentieth and twenty-first centuries.

One of the leading theories is that COVID-19 arose from a disease originating in bats, then mutated so it could infect humans. How long this process took is not yet known, but it may have happened long ago. For example, after extensive research into the HIV-AIDS virus, it was determined it was a disease found in chimpanzees. It was transmitted to a human who was butchering an animal for bushmeat in the 1920s. Not until the 1980s had the disease spread enough among humans to a level it could be readily identified as a new disease.

Several other theories have also been suggested, including the possibility of COVID-19 originating in various mammal species. The debate around this highlights the challenge of researching the origins of diseases.

A seafood market in Wuhan is considered the epicentre of the disease's spread into the human population. Like many other markets in Asia, a 'wet market' is a food market where animals are sometimes sold alive. These markets can sell a wide variety of exotic land-based species, not just sea creatures (see Figure 15.1.6). Whether the disease spread from an animal at the market to a human or whether a human at the market was already infected is not known. However, it does appear someone at the market was a **super-spreader**, a person who transmitted the disease to many others.

Identifying the exact source of the virus is extremely difficult but vital work. By knowing the source of the virus, scientists can begin to take active steps to cut off the source and deal with future outbreaks. Many in the West have called on the Chinese government to close the country's wet markets. There is, however, considerable cultural pressure from the Chinese people not to do so.

Throughout the pandemic, there were allegations that the disease was the result of an intended escape from an infectious disease laboratory in China. Part of the reason was that the Wuhan Institute of Virology is in the city and was thought to be the possible source of the disease. This was initially dismissed as a conspiracy theory, but it lingers and the reluctance of Chinese authorities to participate in a transparent investigation into the source of COVID-19 has added weight to the theory.

The truth may never be known. Based on research into past pandemics it is clear the source of the disease is more likely to be zoonotic than anything else.

Activities

Acquiring and processing geographical information

1 Outline the spread of COVID-19.
2 Explain what the term pandemic means.
3 Describe what a virus is.
4 Describe the methods of transmission for COVID-19.
5 Differentiate between the terms coronavirus and COVID-19.
6 Explain the term zoonosis.

Applying and communicating geographical understanding

7 Using Figures 15.1.3 and 15.1.4, describe the overall global trends of infections of COVID-19 over the first year of the pandemic.
8 Take on the role of a public health professional. With a partner, prepare a short film or brochure explaining the main COVID-19 transmission methods and strategies for reducing transmission.
9 Write a short report outlining the presumed source of COVID-19.

UNIT 15.2

The natural environment and COVID-19

A virus is a naturally occurring infectious agent that has ancient origins. They are simple structures, just a string of genes. Although there are many types of viruses, they all work in roughly the same way by taking over cells within a living plant, animal or bacteria. The virus then hijacks the cells' molecular systems to create copies of itself, allowing it to spread to new cells by repeating the process.

Viruses can jump from one host to the next through various transmission methods. Some, such as the HIV-AIDS virus, use blood and other bodily fluids. Many others, such as coronaviruses—including COVID-19—use respiratory droplets and aerosols. Blocking this transmission from one host to next is the simplest method of containing a virus.

Zoonotic diseases

COVID-19 is most likely a zoonosis virus—a disease that began in animals and then evolved to infect humans. Such diseases are often referred to as zoonotic. These diseases can be transmitted from animals to humans through physical contact, via air or water. Other means are by humans consuming meat or other animal products from the infected animals or being bitten or scratched by infected animals.

Zoonotic diseases are common and are naturally occurring. In many instances, the infected host animal may not be affected by the disease. WHO estimates that more than 60 per cent of all known human diseases are zoonotic in their origin. These diseases account for considerable mortality (death) and morbidity (illness) throughout the world.

A key aspect of research into COVID-19 is examining its zoonotic origins. This will help better understand how it was transmitted from animals to humans. Research has focused on bats as the most probable disease vector, with the horseshoe bat the most likely origin (see Figure 15.2.1). Researchers have identified bats found in caves near Mojiang in the southern region of China which carry a virus very similar to COVID-19. Chinese authorities have placed enormous restrictions on the research, confiscating samples taken from caves in the area in late 2020. This action was seen as China attempting to minimise links between COVID-19 and the country, particularly its trade in animals.

15.2.1 A horseshoe bat

● SPOTLIGHT

Hendra Virus

Bats, also known as flying foxes, are known to cause several other zoonotic diseases, including two deadly diseases associated with Australian bats—Hendra virus (see Figure 15.2.2) and Australian Bat Lyssavirus (ABLV). In January 2021, an alert was issued for ABLV in areas of Brisbane after an infected bat, which later died, was found in a bushland reserve in the city's north.

Like many zoonotic diseases, Hendra virus uses another species as an intermediate host, first infecting horses, which then transmit the disease to humans. The horse becomes infected after eating grass that an infected Australian flying fox has urinated on. After close contact with the infected horse, a human can become infected. Since being first discovered in 1994, seven people have been infected with Hendra virus, resulting in four deaths. In 2012, a vaccine for horses was developed and no further cases in humans have been detected since.

15.2.2 Flying foxes are vectors of the deadly Hendra virus.

15.2.3 Mosquito nets reduce the chance of being bitten by malaria-carrying mosquitoes

Malaria

Malaria is one of the most significant zoonotic diseases, one which infects the liver and blood of humans and causes intense fevers, sweating and often seizures. It is caused by a parasite that lives in the gut and saliva glands of mosquitoes. When a human is bitten by the mosquito, the parasite is transmitted from the mosquito into the blood of the bitten person. This type of transmission is known as a vector transmission—the disease uses another organism (in this case the mosquito) to carry and transmit the disease.

Malaria is found in more than 90 countries, predominantly developing nations within, or close to the tropics. In 2019 it infected 209 million people, resulting in around 409 000 deaths. Active steps to reduce the mosquito vectors include removing standing water, such as containers of water (where they breed), and using mosquito nets and insect repellent (see Figure 15.2.3). These practices help reduce the chances of being bitten. Together with the use of anti-malarial drugs, they have significantly reduced the rates of infection and death. WHO estimates that such measures have reduced infections by more than 1.5 billion over the last 20 years, saving more than 7.6 million lives.

WHO states that since 2010, about three-quarters of all new diseases, including COVID-19, are zoonotic diseases. Many of these diseases are poorly understood and often misdiagnosed. They are more prevalent in developing countries where contact between animals and humans is more common. The spread, severity and impact of the pandemic has heightened the attention given to zoonotic diseases and the threats they pose to human health.

Pandemics through history

The COVID-19 pandemic and the restrictions that accompanied it, were, for many, a once-in-a-lifetime experience. Yet pandemics are not a rare occurrence in human history. They have featured periodically and can be considered a natural occurrence.

One of the first recorded pandemics took place in the second century CE across the Roman Empire. The Antonine Plague raged between 165 CE and 180 CE. It is thought to have been a measles pandemic and was spread by traders and soldiers.

Bubonic plague

The 'Black Death' is considered to be the greatest pandemic in recorded history. It ravaged Medieval Europe in a series of waves of infection. These began in the mid-1300s with the last great infection occurring in 1400, although there had been smaller outbreaks previously. The first of these was recorded in the sixth century and there were several later. The Black Death is thought to have originated in Central Asia and spread along trade routes into Europe. The actual disease—the bubonic plague—was caused by bacteria carried by fleas. These fleas infected rats, which were prevalent across Europe. From the rats it spread to humans. The pandemic wreaked havoc across Europe, reducing the continent's population from around 80 million to just 30 million in less than half a century. It had a profound impact on the economy, development and social structures of Europe that lasted centuries.

Spanish flu

The last great pandemic prior to COVID-19 was the **Spanish flu**. It emerged in the wake of World War I and was to claim at least 50 million lives globally, although the real figure may have been as high as 100 million (see Figure 15.2.4). This is far more than were killed in the devastating war. Despite attempts to **quarantine** Australia, the disease began to emerge in early 1919. An estimated 40 per cent of the population contracted the illness and around 15 000 people died from its effects.

Although called the Spanish flu, the disease did not emerge in Spain. As a neutral country during the war, it was one of a few countries with an uncensored press. So it was first widely reported there, hence the disease was attributed to Spain.

There are some similarities between the Spanish flu and COVID-19. Both are respiratory diseases and the responses were also similar. Lockdowns and quarantines were put in place. In Australia, tensions between states emerged as each sought to restrict the movement of people, similar to the COVID-19 response. The widespread use of face masks and social distancing was also adopted then (see Figure 15.2.5).

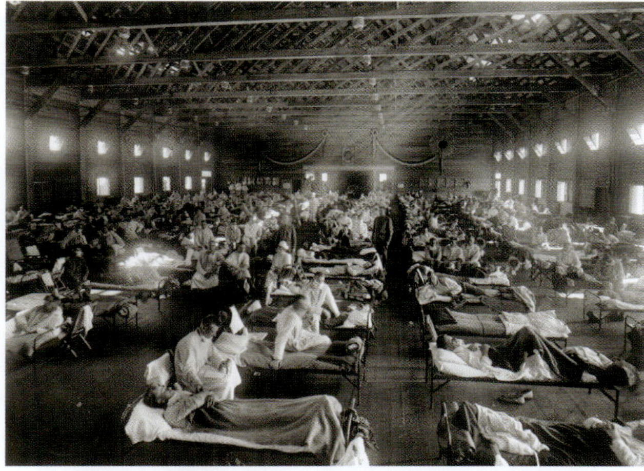

15.2.4 During the Spanish flu outbreak, medical facilities were overwhelmed and makeshift wards were established, like this one in an army camp in Kansas, USA, 1918.

15.2.5 Face masks were widely used during the Spanish flu pandemic in 1919.

Impact of weather on COVID-19

When COVID-19 first emerged, there was speculation that weather was a factor in the spread and prevalence of the disease. Then-US President, Donald Trump, infamously stated in February 2020, 'Looks like by April, you know in theory when it gets a little warmer, it miraculously goes away.'

Some believed that because the virus appeared to be susceptible to heat outside of the body, that warmer spring temperatures would kill it off. The early spread of the disease was greater in regions that had colder winter temperatures. These included Central China, Northern USA and parts of Europe, adding credence to such a claim. Anecdotally, this confirmed what many already believed about common coronaviruses, such as influenza, which is far more prevalent during winter months.

Figure 15.2.6 shows the number of cases of COVID-19 reported in the USA through much of 2020. The prediction that the Northern Hemisphere summer (June to August) would see the end of the virus was clearly incorrect. While the rates stabilised a little in spring, this coincided with changes in government policies. These significantly reduced, and often eliminated, travel restrictions and social distancing requirements, so there was a spike in cases during summer. Similarly, Australia saw a significant spike in cases in late 2022, despite warm summer temperatures. By mid-December there were 1606 people in hospital with COVID-19 and 74 deaths from the disease. However, this spike coincided with the removal of the last of the COVID-19 restrictions. Widespread mass domestic and international travel was returning, pre-Christmas celebrations meant social interactions were increasing and community attitudes towards precautions, such as mask-wearing, had waned considerably (see Figure 15.2.6 and 15.2.7).

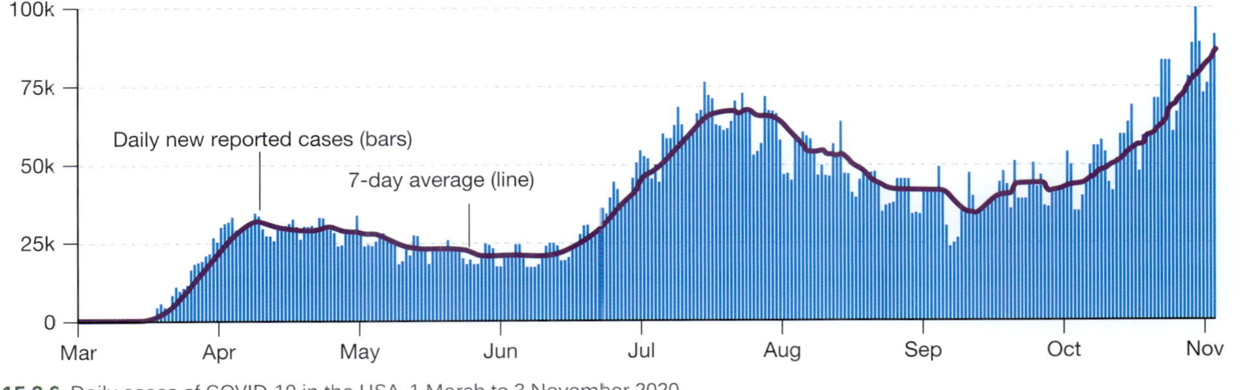

15.2.6 Daily cases of COVID-19 in the USA, 1 March to 3 November 2020

15.2.7 The late 2022 COVID-19 spike in Australia was partly associated with travel restrictions being lifted.

The link between weather and COVID-19, as well as other coronaviruses, has more to do with human activities than the disease itself. Colder temperatures drive people inside bringing them into closer contact with others, creating more opportunities for transmission.

Activities

Acquiring and processing geographical information

1. Describe the structure and operation of a virus.
2. Explain how zoonotic diseases infect humans.
3. Outline the link between bats and COVID-19.
4. Name the cause of the fourteenth-century Black Death.
5. Outline the impacts of the fourteenth-century Black Death pandemic.
6. Explain why some people linked the spread of COVID-19 to cold weather.
7. Outline the impact of weather on COVID-19.

Applying and communicating geographical understanding

8. Write a paragraph comparing the Spanish flu and COVID-19 pandemics.
9. Examine Figure 15.2.6 and complete these activities:
   - **a** Outline the trends shown on the graph.
   - **b** July marks the start of the long summer holidays in the USA. How might this have impacted COVID-19 case numbers?
   - **c** Explain how this graph could help dispel the idea that COVID-19 increases are linked to cold weather.

UNIT 15.3

Human activity and COVID-19

The impact on global travel

Prior to the impact of COVID-19, global travel was at record highs. In 2018, 1.4 billion people travelled internationally for tourism purposes, creating a US$1.7 trillion travel industry that employed hundreds of millions of people globally. Figure 15.3.1 shows the rapid growth of international travel from the second half of the twentieth century into the twenty-first century. Just 25 million people travelled internationally in 1950, less than 2 per cent of the 1.4 billion that were travelling in 2018.

The economic, social and cultural benefits of the rise of modern global travel have been enormous. However, the ease of global travel also helped facilitate the rapid spread of the disease, bringing about the pandemic. One of the earliest actions many countries, including Australia, did to reduce the spread of COVID-19, was to restrict international travel.

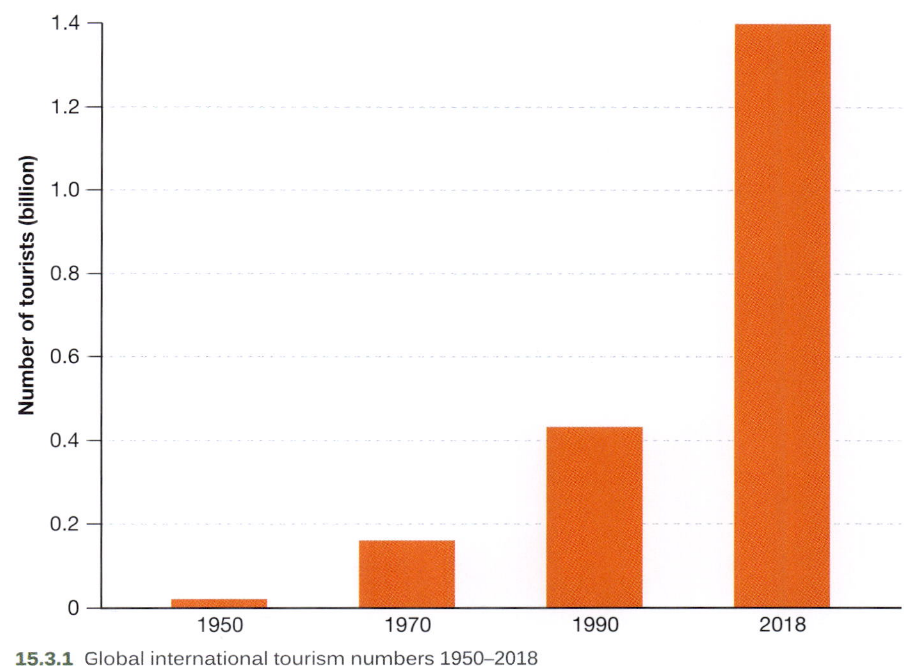

15.3.1 Global international tourism numbers 1950–2018

Figure 15.3.2 highlights the speed with which COVID-19 spread globally. After first being reported in Central China in December 2019, by 20 January 2020, four countries, all bordering China, had also reported infections. Within 25 days, the virus was confirmed in 28 countries, including the USA, Australia and several in Europe. These infections were transmitted initially by international arrivals in these countries. Forty days later, on 25 March, 183 countries on every continent outside Antarctica, reported cases. Virtually every country on Earth was infected within three months. Only the most isolated nations were free of the disease, including some small Pacific island countries and apparently some nations, principally in Africa, that were unwilling or unable to test for and report on cases.

Not only did infected travellers transmit the disease, but the nature of international travel acted as a significant infectious environment. The disease easily spread via hundreds of people sharing an airline cabin, given the extended periods in very close contact. Similarly, cruise ships carrying thousands of passengers in close quarters were an early source of transmission, until authorities effectively stopped cruising by barring ships' entry to ports.

Growth of urban areas

The world's population has become increasingly urbanised. More than 55 per cent of all humans now reside in cities. The UN predicts this will rise to almost 70 per cent by 2050. By their very nature, cities bring large groups of people into close contact with each other, often using shared resources, such as public transport. For diseases like COVID-19, which rely on transmission from one host to the next, via respiratory droplets and aerosols, such close contact between people greatly aids its transmission.

Map 1

Map 2

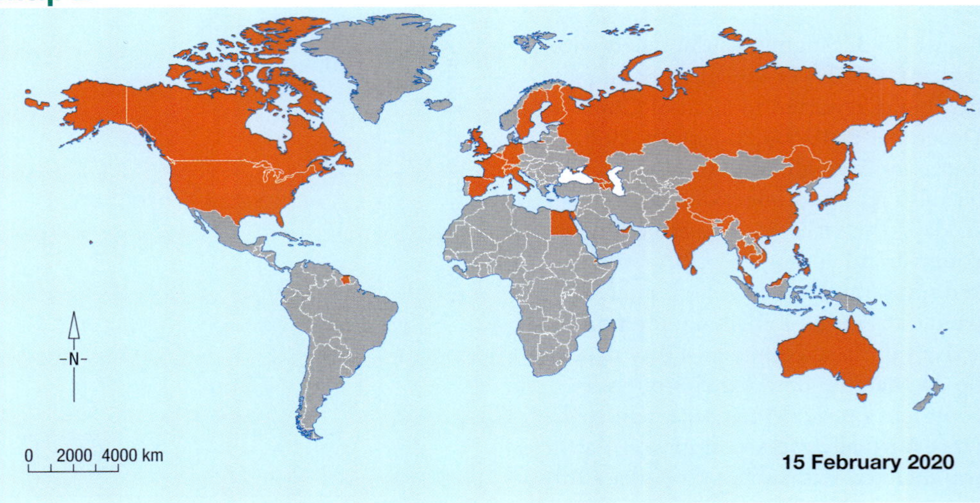

Map 3

15.3.2 The number of countries with COVID-19 changed swiftly.

Epidemiologists and urban geographers have found a strong link between urban density and the spread of COVID-19. Large cities bear a disproportionate number of infections around the world compared to more rural areas. However, density is not the only factor. For example, some less-urban areas in Switzerland reported very high rates of infection. These were typically areas with very significant tourist numbers in alpine regions. Equally, some more rural areas also reported high death rates. This tended to be in areas with higher proportions of older and more vulnerable individuals.

Urban growth has also increased the exposure of more people to zoonotic diseases (see Figure 15.3.3). As urban areas push into rural and less-developed areas, contact between people and wild animals has grown, bringing them into competition for the same spaces. For example, in northern Australia urban development into bushland has increased the prevalence of Q Fever, a zoonotic disease carried by bandicoots, wallabies and kangaroos, in places including Townsville, Queensland.

15.3.3 Urban spread into rural areas brings people and wild animals like foxes into closer contact, increasing the risks of zoonotic diseases.

Density, wealth and COVID-19

Urban geographers conducted a study into the experience of COVID-19 infections in New York City that highlights the complexity of this disease's links with urban density. Researching case numbers across the city, they found an important link between wealth, density and infection rates.

New York City is made up of five boroughs or districts. The most densely populated borough is Manhattan, with more 1.6 million people living in less than 60 square kilometres (see Figure 15.3.4). This creates a density of around 27 500 per square kilometre. Despite this, Manhattan's rate of infection was lower than in the other boroughs. This includes the very suburban area of Staten Island which has a density of a little more than 3000 people per square kilometre (see Figure 15.3.5).

Manhattan is not just densely populated, it is also the most expensive part of New York City. People living there are generally wealthier, and many occupy highly paid professional roles. The researchers found that far more Manhattan residents were able to work from home, avoiding the need to use public transport, workplaces and other shared city amenities, compared to those in the other boroughs. Many people living in areas such as Staten Island were front-line workers: nurses, bus drivers and retail workers. They were far less able to 'shelter at home' during the peak of the pandemic, so were forced to travel to work and leave their homes for daily needs, increasing the risk of being exposed to COVID-19.

Impact of cultural events

Many places responded to the COVID-19 pandemic by introducing restrictions on mass gatherings. This led to widespread cancellation of significant events. The 2020 Tokyo Olympics were delayed, and many other major sporting events were cancelled.

15.3.4 and 15.3.5 New York City's Manhattan (left) has a much higher population density than Staten Island (right).

The 2020 Sydney–Hobart Yacht Race was cancelled after an outbreak in Sydney saw city-wide restrictions imposed. This was the first time the race was cancelled in its 76-year history.

The annual hajj pilgrimage to Mecca in Saudi Arabia is one of the most sacred acts for people of the Islamic faith. Each year, over two million people from all over the world undertake the pilgrimage, which lasts just a few days. In 2020, only 1000 people were allowed to participate due to strict social distancing rules. With the introduction of vaccines, 60 000 pilgrims were allowed to visit in 2021. All had to be fully vaccinated. By 2022, the restrictions had eased considerably and more than 1 million visitors made the pilgrimage.

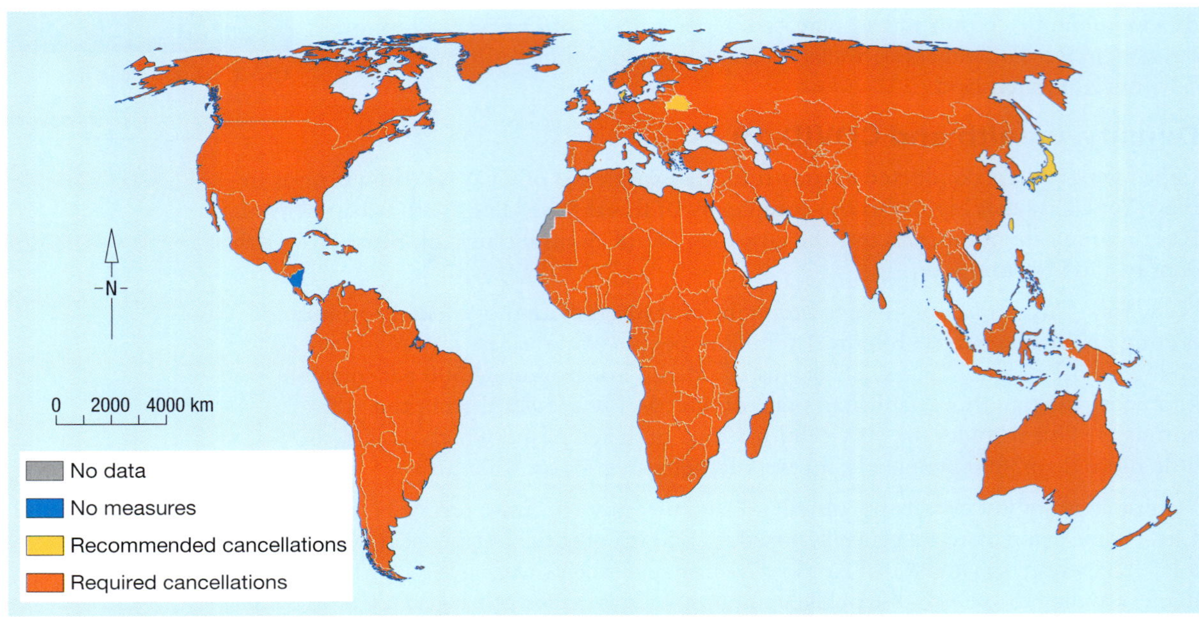

15.3.6 Countries with policies to cancel public events due to COVID-19 in mid-April 2020

Figure 15.3.6 depicts countries around the world where governments implemented widespread policies to cancel public events. This map shows data for mid-April 2020 when such policies were most widespread and typically at their most stringent. The map reveals this to be an almost universal policy globally.

Policies to reduce public gatherings were an important strategy to help limit the spread of the virus. It was also challenging to make people understand and comply with these policies. Events that bring families together such as Christmas (for Christians), Eid (for Muslims), Diwali (for Hindus) and Thanksgiving (for the USA) would all become periods impacting the spread of COVID-19. Such events brought people into closer contact than usual. They also saw widespread movements of people. Subsequently, the prevalence, scale and spread of the disease increased.

Climate change and zoonotic diseases

The weather plays no significant role by itself in spreading COVID-19. Instead, variations are more associated with people's activity associated with the seasons. For example, cold weather brings people inside and into closer contact with others. Equally, climate change does not have an immediate effect on this disease. However, there are very strong links between the spread of zoonotic diseases and human-induced climate change. Those diseases associated with the tropics, such as malaria, are becoming more widespread because the world is warming. The risk of future pandemics, which are often zoonotic, is heightened by climate change.

● SPOTLIGHT

Thanksgiving holiday and COVID-19

Thanksgiving is an important tradition in the USA (and to some extent, Canada). In the USA, Thanksgiving is celebrated on the fourth Thursday in November. It comes from the tradition of holding a harvest feast and dates back to the early colonisation of North America by English settlers. It is usually a time when families reunite, so travel across the USA peaks around that time. In 2019, more than 31 million people took flights in the USA during Thanksgiving.

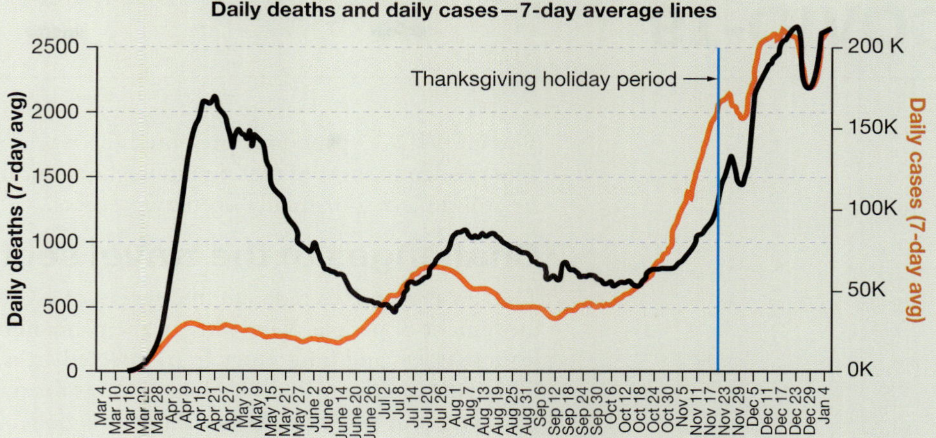

15.3.7 Impact of Thanksgiving on COVID-19 in the USA, 2020

Despite the USA being one of the countries hardest hit by COVID-19, many Americans still travelled during the 2020 Thanksgiving holiday period. Close to 10 million people took flights and millions more drove long distances. At that time there were very few travel restrictions in place. During the following fortnight, COVID-19 cases in the USA increased markedly compared to the period immediately prior to the holiday (see Figure 15.3.7). Case numbers rose by 20 per cent, hospitalisation rates increased by 21 per cent and deaths increased by 39 per cent. Over December, immediately after the holiday, there were 63 000 deaths from COVID-19, overwhelming hospitals across the country.

Activities

Acquiring and processing geographical information

1 Explain how global travel impacted the spread of COVID-19.

2 Outline why aircraft and cruise ship travel increase risks of infection.

3 Explain the link between urban living and infection rates.

4 Outline the impact urban growth has on exposure to zoonotic diseases.

5 Explain how cultural events could lead to increased COVID-19 exposure.

6 Outline the link between climate change and the spread of some zoonotic diseases.

Applying and communicating geographical understanding

7 Use Figure 15.3.1 and information from the text to prepare a short report on the rise of global tourism and the impact that this had on the spread of COVID-19.

8 Study the maps in Figure 15.3.2. Using these maps, and an atlas if you need to, describe the spread of COVID-19 globally over the period.

9 Write a report explaining the findings of research in New York City and links between wealth, urban density and exposure to COVID-19.

10 Describe the patterns shown in Figure 15.3.6.

11 Using Figure 15.3.7 and the text information in the box, Spotlight: Thanksgiving holiday and COVID-19, describe and account for the impact of the holiday on COVID-19 cases and deaths in the USA.

UNIT 15.4

Challenges and opportunities created by COVID-19

The COVID-19 pandemic presented the world with many unprecedented challenges. The pandemic showed that despite technological, cultural and economic advances, we are still subject to naturally occurring events.

Challenges to the travel sector

The key response to reducing the spread of COVID-19 was to restrict travel and the movement of people. Very early in the pandemic, governments moved to restrict travel from overseas and limit entry from infected areas. In January 2020, the Australian Government announced travel restrictions for travellers from some destinations and on 18 March this was extended to the entire globe. The Australian Government's travel advice to Australians from 24 March 2020 was 'Do not go overseas. A travel ban is in place' (see Figure 15.4.1).

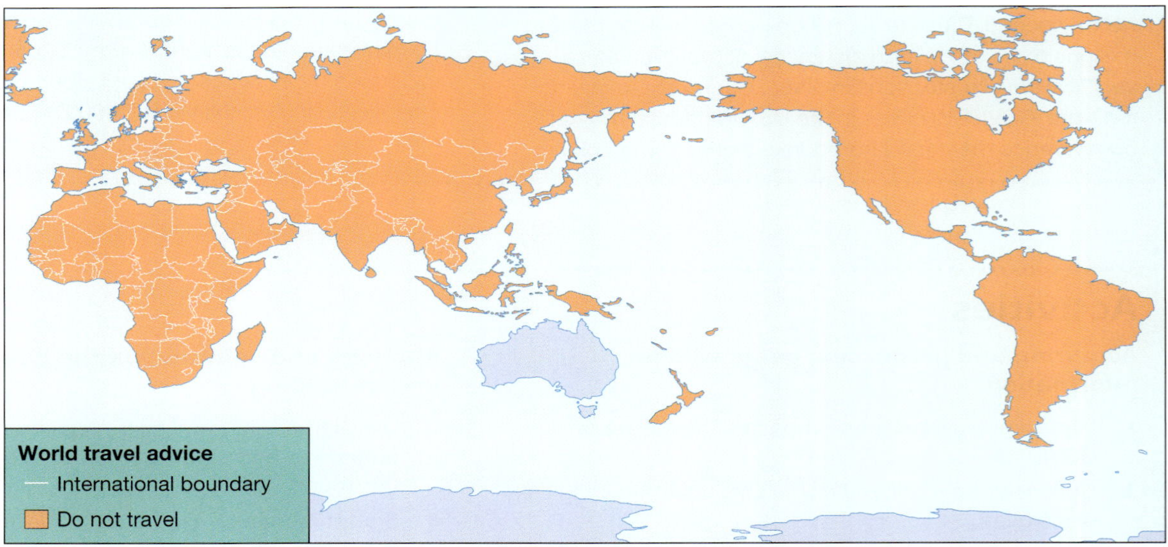

15.4.1 Australia's travel advice on 24 March 2020, showing the entire world as 'Do Not Travel'

Figure 15.4.2 highlights the extent of travel restrictions globally by mid-April 2020. Total border closures were the norm, with nearly all remaining countries opting for highly restrictive policies. Many countries, especially in Europe, did gradually reduce these restrictions across 2020, although many were re-applied as subsequent waves of the virus took hold.

The global travel restrictions had an immediate impact of the tourism and travel industries. The US Tourism Association estimated the US travel industry lost US$425 million per day, resulting in around 12 million job losses in 2020. Airlines were immediately impacted with mass cancellations of flights and airlines around the world placed their aircraft into long-term storage. On 2 April 2020, for the first time in aviation history, the same number of aircraft were in storage around the world as were operational.

Several airlines faced a severe financial crisis, including Australia's second biggest airline, Virgin Australia. The company was placed into administration with enormous debts. It was unable to repay these and sold to overseas financiers. In 2018, there were

54 519 flights per year between Sydney and Melbourne, an average of 149 flights per day. This made it the second busiest domestic air route in the world. In November 2020, there were just 10 flights per week between the cities and at some points in the year there were none at all.

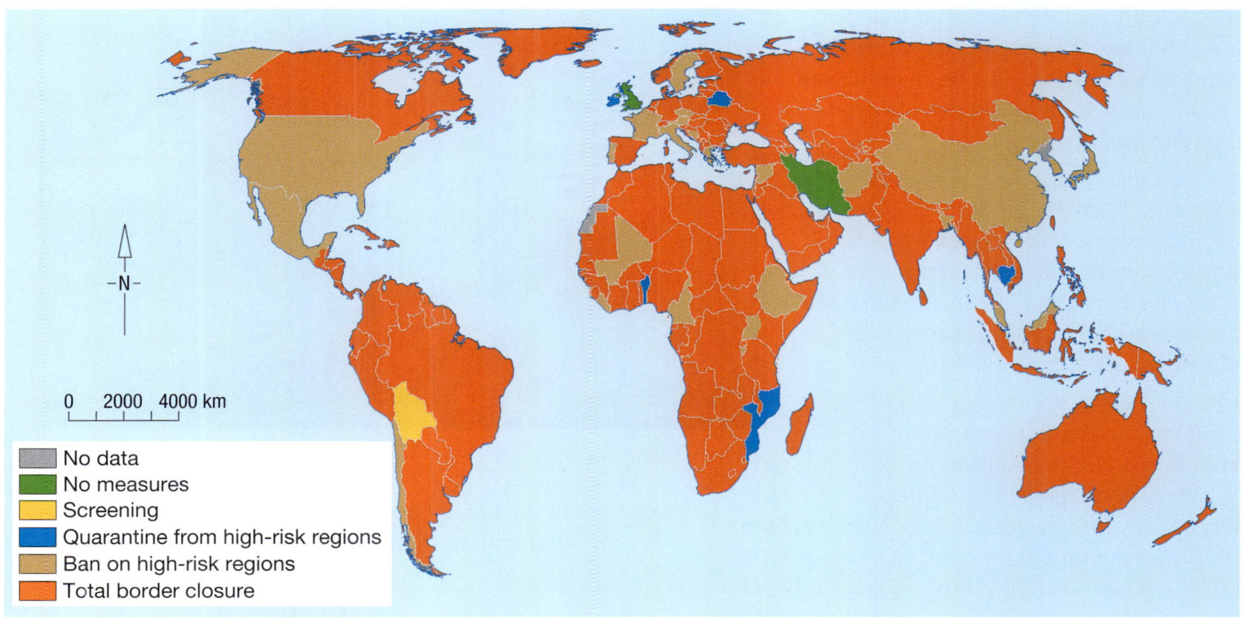

15.4.2 International travel restrictions during the COVID-19 pandemic, mid-April 2020

The International Air Transportation Association estimated the world's airlines lost US$314 billion in 2020. In late 2020 and into 2021, many airlines around the world began to fly their aircraft without passengers, using the luggage bays for cargo to gain some revenue.

Data from Tourism Australia shows that international arrivals to Australia in September 2020 totalled 3720. This was a 99.5 per cent decrease in the number of international arrivals from September 2019, which saw almost 695 000 arrivals (see Figure 15.4.3).

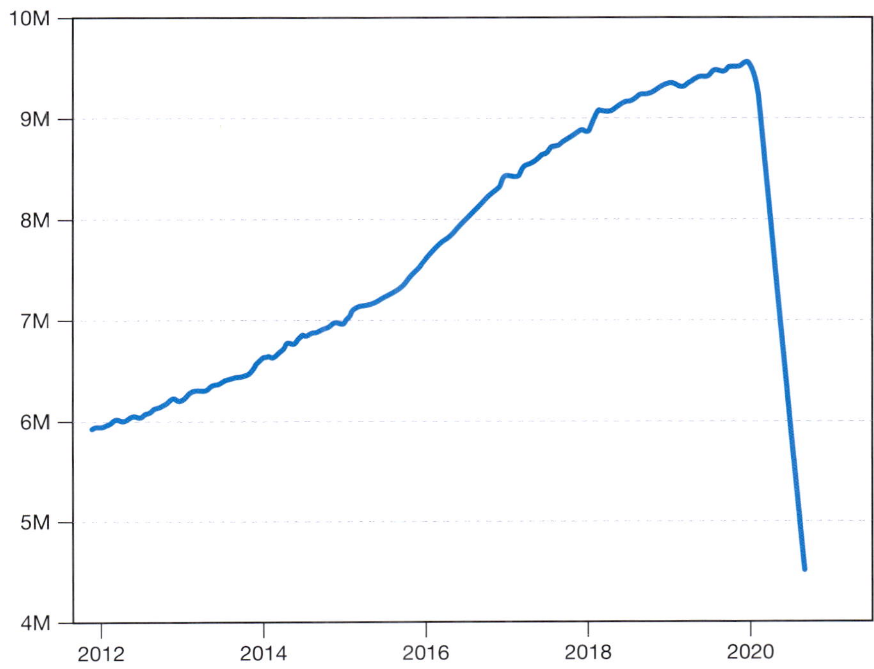

15.4.3 International arrivals to Australia for each year ending September, 2012–2020

● **SPOTLIGHT**

Cruising sunk by COVID-19

In its 2019 annual report, the Cruise Line International Association (CLIA) predicted that 32 million people would take a cruise in 2020. There was good reason for this—there had been continual growth since 2009, when 17.8 million people cruised, a 44 per cent growth in a decade.

The 2020 reality was all major cruise lines posted revenue declines of at least 75 per cent. Outbreaks of COVID-19 onboard some cruises saw restrictions on cruise ship movements in many locations. CLIA estimates that the industry lost 520 000 jobs in 2020 as cruise ships were left with skeleton crews of engineers and sailors just to maintain them.

In a bid to stimulate their economies, many governments around the world announced large-scale spending on infrastructure during the pandemic. This saw a significant increase in demand for steel. Many cruise ships became worth more as scrap metal. Consequently, 2020 saw the scrapping of dozens of previously popular cruise ships, many long before their anticipated end of life (see Figure 15.4.4).

15.4.4 Cruise ships at a scrap yard in Turkey being broken down for scrap steel

● **SPOTLIGHT**

Tourism recovery post-pandemic

By mid-2021 much of the world was beginning to relax restrictions, especially in Europe and the USA. By 2022, the travel industry was beginning to focus on recovery from the COVID-19 lockdowns. The United States Travel Association reported that by November 2022 domestic air travel had returned to 95 per cent of pre-pandemic levels. In Australia, domestic air travel was around 75 per cent of pre-pandemic levels by late 2022 (see Figure 15.4.5).

The recovery of international travel was considerably slower. Official statistics by Tourism Australia found the number of international visitors to Australia in September 2022 was 47 per cent lower than in September 2019. However, the extent of COVID-19's impact is more apparent in the September 2022 figure when it was a staggering 8332 per cent higher than in September 2021.

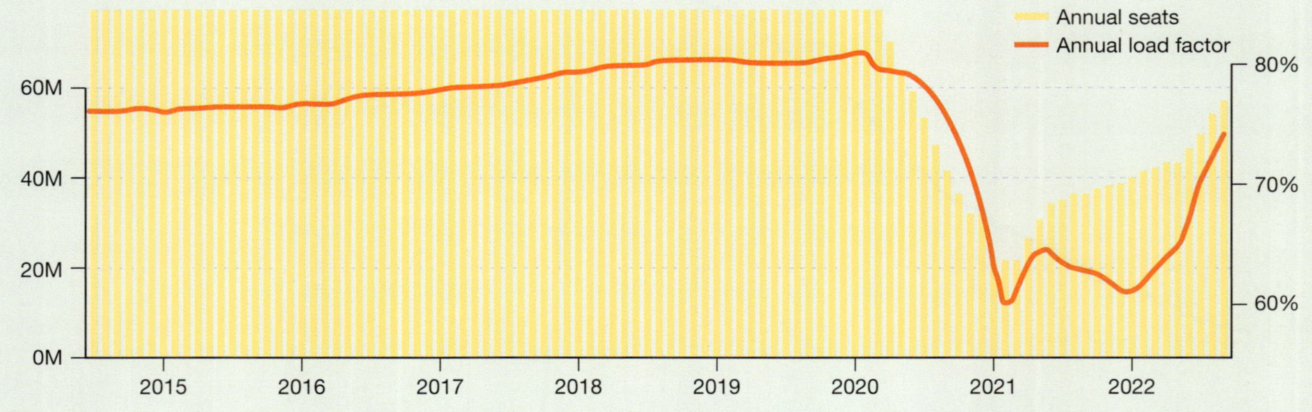

15.4.5 Domestic air travel in Australia (annual seats indicate the actual seats available for travel; the load factor is how many are actually used)

COVID-19 waves

One of the most iconic phrases of the COVID-19 pandemic was 'flatten the curve'. This described the goal to reduce the number of infections. It would be reflected in graphs with relatively flat curves rather than steep lines showing spikes in infections. By 'flattening the curve', medical experts hoped to keep the rate of infection low enough that hospitals could cope with admissions and deaths would be minimised.

Throughout the first half of 2020 efforts were focused on this as a goal. Extensive lockdowns, restrictions on travel, work-from-home rules, wearing masks and an emphasis on hygiene were all strategies to achieve this. Figure 15.4.6 shows the number of infections in Australia between January 2020 and January 2021, the first year of the pandemic. The effects of the extensive lockdowns in April and June 2020 are visible with the flattened curve.

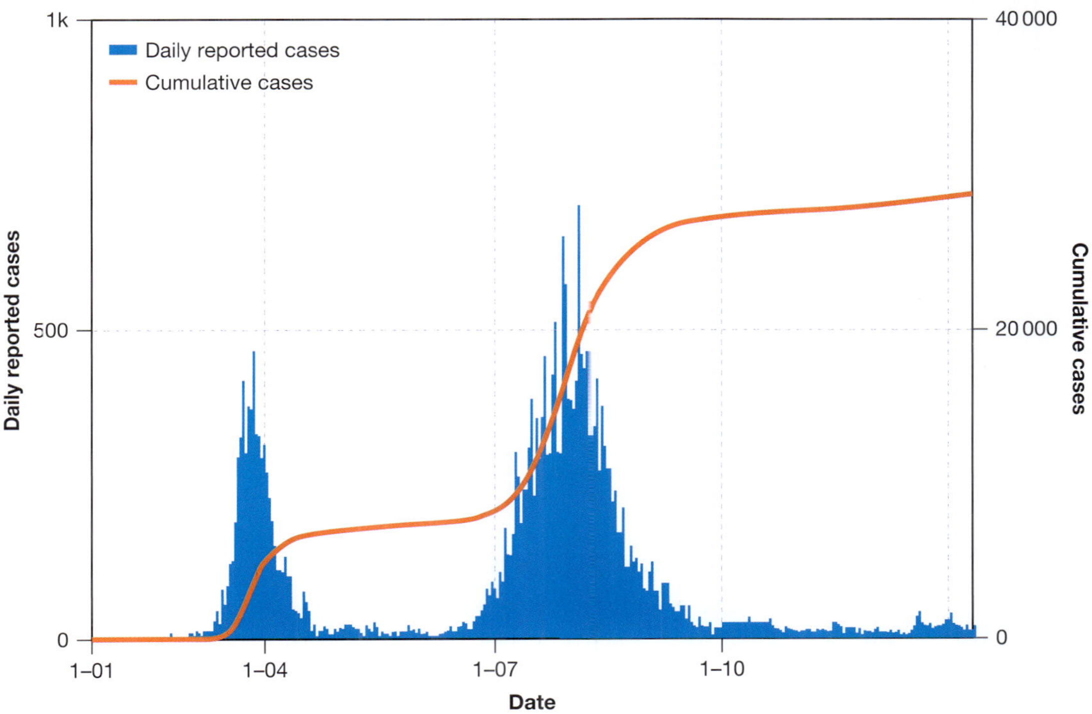

15.4.6 Confirmed COVID-19 cases in Australia, January 2020 to January 2021, indicates its pattern of peaks and troughs

Figure 15.4.6 also highlights the spike in cases that occurred in July and August 2020. This was associated with an outbreak after workers at a hotel quarantine site contracted COVID-19 from a returned traveller. Widespread **community transmission** ensued, resulting in a so-called second wave as the previously flattened curve spiked again with infections. In Victoria's second wave, an extensive lockdown period was reinstated, which lasted over four months.

Incidences of multiple waves of infection were common around the world. In the UK, after getting infection rates under control in mid-2020, a second wave of infections began emerging in September. Limited restrictions led to some flattening, but by November cases soared across the country. This pattern of spikes and troughs continued, even after vaccines were introduced (see Figure 15.4.7). The vaccines tended to reduce the severity of infection but didn't prevent people from becoming infected.

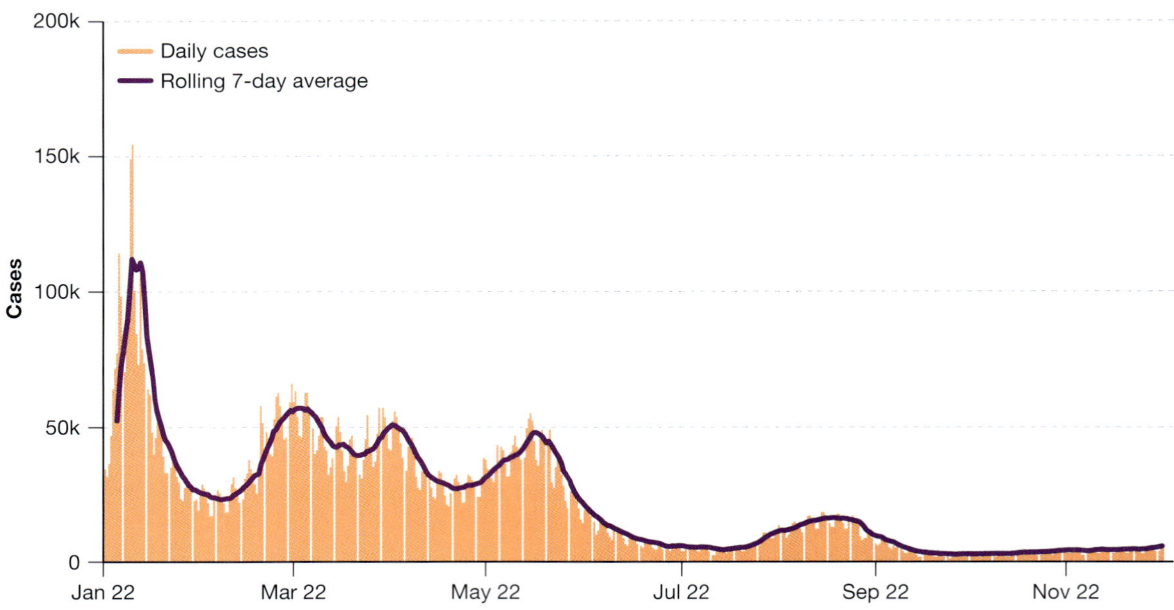

15.4.7 Confirmed COVID-19 cases in Australia, January 2022 to December 2022. The pattern of peaks and troughs, or 'waves' remains evident.

The second and subsequent waves of infection highlighted the challenges faced by medical officials in containing COVID-19. Extensive restrictions on the population had positive effects on flattening the curve, but came with great economic and social consequences. Governments had to deal with a constant tension between the positive and negative effects of lockdowns and restrictions. The disease's highly infectious nature and its relative ease of transmission meant that efforts to contain the virus could be quickly undone.

The impact of lockdowns

Lockdowns were the most common strategy used to reduce the spread of COVID-19. It was not until there were widespread vaccinations that lockdown restrictions were eased. Melbourne experienced the most and longest lockdowns of any city in Australia. In 2020 and 2021, the city experienced six lockdown periods totalling a staggering 262 days. Schools, businesses, universities—virtually everything was forced to stay closed. It was not until late October 2021 that the last lockdown ended.

Sydney also experienced significant lockdowns in 2020 and 2021. The longest began in the second half of 2021 running through to October, lasting more than 100 days. Again, this saw much of the city, including schools, closed. Figures 15.4.10 and 15.4.11 highlight this impact visually. Martin Place is usually one of the busiest areas of the CBD full of pedestrians, but was deserted during the lockdowns.

Across the world, 'lockdown' became one of the most widely used words in the media and on social media sites. In Australia, lockdowns divided the nation, with state borders closed to each other. 'Hard' borders were established and patrolled by police. Lockdowns had profound impacts on families, and changed the nature of schooling and work. Their impacts will be felt long into the future.

● SPOTLIGHT

Age and COVID-19

Like most infectious diseases, COVID-19's impact was most keenly felt by the elderly and other vulnerable individuals with existing illnesses. They were more likely to become very ill, be hospitalised or die. Figure 15.4.8 demonstrates that only a tiny number of Australians aged under 50 died from the illness in 2020, before anyone was vaccinated in Australia.

Figure 15.4.9 also reveals that the rate of infection among younger people in Australia in 2020 was much greater than in older people. People aged between 20 and 29 were the most likely to contract COVID-19, but they were also among the age group least likely to die from the illness. This created a unique challenge for health authorities as they attempted to contain the illness. This age group is more likely to socialise and gather in groups than older groups.

While younger people have a low risk of death from illnesses like COVID-19, the challenge is that they can transmit it to older people and others who have a very high risk of dying. This led to education campaigns in Australia targeting young people about their responsibility to help limit the spread and risk of the disease to their parents and grandparents.

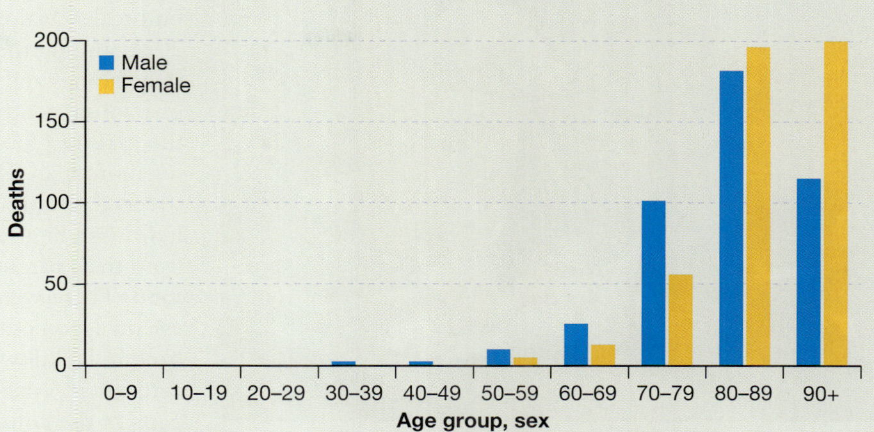

15.4.8 Deaths caused by COVID-19 in Australia by age group and gender in 2020

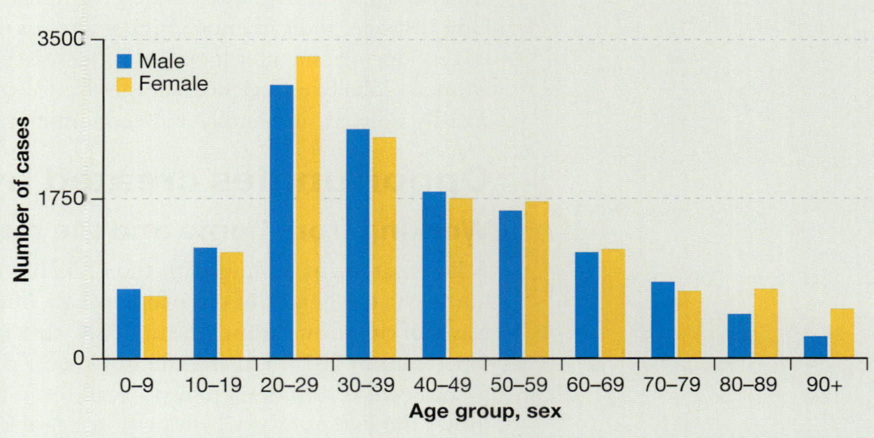

15.4.9 The number of COVID-19 cases in Australia by age group and gender in 2020

15.4.10 Martin Place, Sydney, mid-week in August 2021 during the city's longest lockdown.

15.4.11 Martin Place, Sydney, mid-week in 2018

15.4.12 A worker outside a residential building as Shanghai enters another lockdown, March 2022

By 2022 most countries began to see vaccination as the key strategy for dealing with COVID-19 outbreaks, so lockdowns were slowly wound back. However, in China the authorities continued to pursue strict lockdowns in communities with outbreaks. Major cities remained in lockdown throughout 2022, among the strictest used. People were confined to their homes and those infected were removed and taken to quarantine centres. One such event was in May 2022, when the massive global city of Shanghai went into strict lockdown for more than 2 months. Police and health officials patrolled streets to ensure no one left their home, even for essential items (see Figure 15.4.12). This saw young children separated from their parents, people going without medicine and even food. There were shortages as authorities struggled to provide essentials to communities. By December 2022, Shanghai was back in lockdown again as officials continued with a policy of zero-covid. Mass protests, rare in China, broke out in many parts of the country due to the extensive lockdowns.

One of the most profound long-term effects of lockdowns has been the significant increase in mental health conditions associated with isolation. Research by mental health advocacy groups, including the Black Dog Institute, indicate much higher rates of mental health distress among all groups in society during and after lockdowns. Research by the WHO suggests that 45 per cent of women globally experienced domestic violence during the lockdowns as they became socially isolated, and family and economic pressures mounted.

Opportunities created by COVID-19

Working from home and the return of neighbourhoods

A key strategy in dealing with the COVID-19 pandemic was a shift to working from home. At the height of the pandemic in Australia, around 50 per cent of the workforce was working from home. In the USA, that number was around 35 per cent, up from 4 per cent in 2019. During the UK's second wave in early 2021, on any given workday 57 per cent of London's massive workforce was working from home. Figure 15.4.13 shows the extent of work-from-home restrictions across the world in April 2020. Most countries had at least some restrictions in place. Many places had extensive restrictions on workers to avoid attending their workplace in person.

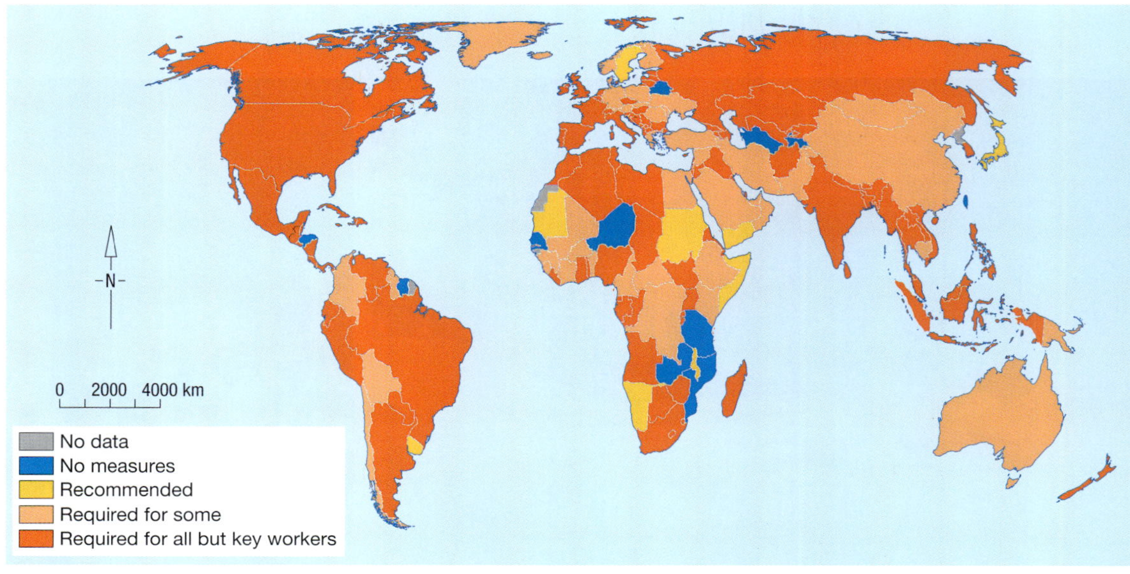

15.4.13 Work from home requirements around the world in April 2020

The move to working from home turned traditional CBDs into virtual ghost towns overnight. This had severe repercussions for businesses that rely on city workers, such as cafes and shops (see Figure 15.4.14). Another consequence of this radical change in work patterns was to revitalise many suburban areas. People often commented that they rarely experienced their neighbourhoods during weekdays. The move to working from home saw many people 'rediscovering' their neighbourhoods.

In Sydney, many suburban cafes and other business reported 2021 as one of their busiest years. While similar businesses located in the CBD floundered due a lack of customers, their customers were spending in their local neighbourhoods instead. Estimates reveal the increase in spending in suburban cafes and restaurants was over 30 per cent that year. At the height of the 2021 lockdowns, well over 40 per cent of workers in Sydney were working from home full-time. Many more were working from home at least some of the time. Prior to the pandemic, the figure was around 8 per cent.

15.4.14 Deserted CBDs had a devastating effect on businesses.

Research by the NSW Department of Planning, Industry and Environment found a significant increase in Sydneysiders accessing local parks and other amenities as restrictions on travel came into force. Sydney Western Parklands recorded a 100 per cent increase in visitations during lockdown periods. Authorities noted that many people discovered walking trails, bike tracks and hidden parks across their suburbs, encouraging greater use of these facilities.

The long-term consequences of the work-from-home requirements are yet to be fully understood. Some urban geographers have predicted that demand for traditional office space will decline in the coming decade. This is because workers and employers now understand the benefits of working from home, at least some of the time. The pandemic reinforced the ready availability of technologies, such as video conferencing, which are effective. By the end of 2022, around one in four Americans were still working from home at least some days each week. Numbers in Australia and other developed nations were similar. There are considerable differences between jobs, and it depends on the nature of the work, with office workers and professional occupations dominating work-from-home statistics.

Positive environmental consequences

With pandemic restrictions significantly reducing economic activity, one consequence was a sharp fall in greenhouse gas emissions. Emissions of carbon dioxide alone fell by 7 per cent in 2020, almost 2.5 billion tonnes less than in 2019. It was the biggest reduction in emissions since widespread fossil fuel use began in the mid-nineteenth century.

Much of the reduction in emissions came from the transport sector. Far fewer vehicle trips were made as many nations moved to work-from-home arrangements (see Figure 15.4.15). In the USA and Europe, work-from-home arrangements were widespread. These places saw the biggest reductions, 12 per cent and 11 per cent respectively. In China, shorter periods of restriction saw only around a 1.7 per cent carbon dioxide emission reduction in 2020. The widespread grounding of aircraft worldwide was another significant contributing factor.

15.4.15 A deserted freeway in Los Angeles, USA, at the height of the pandemic lockdown

To reduce emissions adequately to avoid the worst effects of climate change, the percentage reductions seen in 2020 would need to be maintained forever. Lockdowns and travel restrictions are not a long-term solution for dealing with climate change. Nonetheless, one of the key benefits of the pandemic was to highlight how changes around personal transport can have significant positive benefits for emission-reduction targets.

● **SPOTLIGHT**

Venice's clear water

Early in the COVID-19 pandemic, Italy was one of the countries hardest hit, with tens of thousands of infections and deaths. A comprehensive lockdown was implemented, so the Italian tourism industry ground to a halt. Venice, usually awash with visitors, fell quiet. The lack of boat traffic helped minimise sediment disturbance. For the first time in living memory, the waters of Venice cleared (Figure 15.4.16). Fish and other marine life not seen in the city's canals returned within just weeks of the boat traffic reduction, highlighting the resilience of nature. Unfortunately, decades of pollution meant that water quality did not substantially improve.

15.4.16 Venice, normally crowded by tourists, was largely deserted at the height of the pandemic.

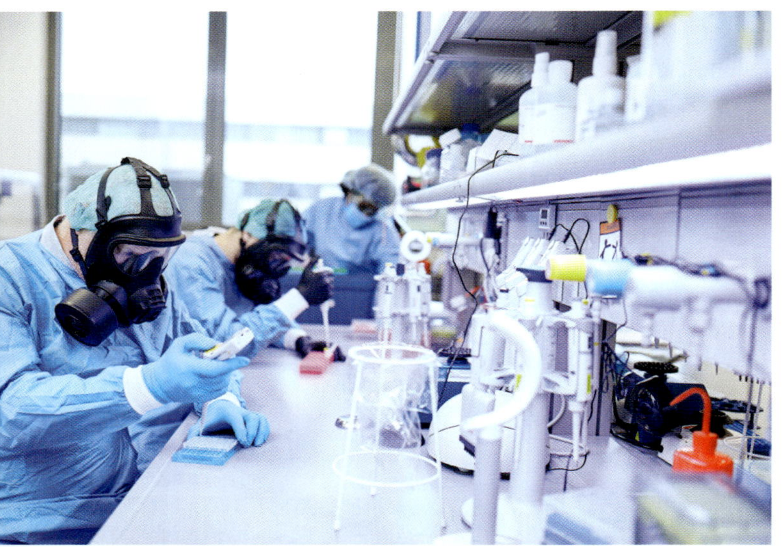

15.4.17 Laboratory team working on COVID-19 vaccine manufacturing.

Research and innovation

Times of great crisis often result in periods of significant innovation. The COVID-19 pandemic saw a period of intense medical and virological research. Scientists quickly identified the virus's genomic code, which led to swiftly developing tests for the disease. Testing was one of the most significant steps in containing its spread.

The focus was on developing effective vaccines. In record time, scientists working in teams in various countries were able to develop vaccines that helped to further limit the spread of the disease (see Figure 15.4.17). Within 12 months of COVID-19 first being identified, vaccinations began. Mass vaccination programs were well underway across many countries by mid-2021.

By late 2022, 70 per cent of the world's population had received one vaccination, around 64 per cent receiving two doses (the level considered to be fully vaccinated). In many developed nations, many people received three, four and in some cases five doses by late 2022. In Australia, over 97 per cent of people over 16 had had one dose by December 2022 and 96 per cent had received two doses, one of the highest rates globally. The speed of the vaccine's development and rollout was unprecedented.

Many countries, including Australia, offered incentives for having the vaccine, such as fewer restrictions. For some jobs, such as those in hospitals and schools, vaccination was mandatory. This was controversial, but was seen as being an essential part of the ability to ease restrictions.

The COVID-19 pandemic directed global attention to the risks posed by zoonotic diseases. GLEWS+ (The Global Early Warning System) is a joint initiative of WHO, the World Organisation for Animal Health (OIE) and the UN Food and Agriculture Organization. Formed in 2006, GLEWS evolved into GLEWS+ with a focus on rapidly identifying and assessing the risk of and responding to zoonotic diseases. This type of joint approach brings together experts who will be better able to deal with novel zoonotic diseases more swiftly in the future.

Varying perspectives on COVID-19

15.4.18 Protestors opposing the COVID-19 vaccine roll out in Melbourne, 2021

Many different perspectives emerged on how best to deal with the COVID-19 pandemic. At the most extreme end were those who claimed the disease was a hoax. Jair Bolsonaro, the then-president of Brazil, called it a 'little flu'. He later became infected, but recovered and continued to downplay the virus. By early June 2023, 702 900 Brazilians had died from COVID-related causes.

Health officials and most world leaders understood the gravity of the pandemic, but differed widely on approaches to mitigating its worst effects. Australia and New Zealand were among the countries to take very strict measures. They aimed to contain the virus by implementing extensive restrictions on movement (see Figure 15.4.18). Other countries, such as the USA, had a mix of policies, but tended towards fewer restrictions fearing the economic consequences.

A few countries, most notably Sweden and initially the UK, adopted a policy of **herd immunity**. This is based on the concept that people will develop an immunity to further infection once they have recovered from a disease. The idea in Sweden was that if enough people developed immunity the disease would die out. However, this relies on large numbers of people being infected. It assumes they will not die or face long-term consequences and immunity will follow the infection. Unfortunately, neither proved correct and Sweden had one of the highest rates of infection and deaths in Europe. Herd immunity is usually achieved through extensive vaccination programs.

Activities

Acquiring and processing geographical information

1 Describe the impact of travel restrictions on the Australian and global tourism industries.

2 Outline the strategies adopted to flatten the curve of COVID-19 infections.

3 Explain how the second and third waves of COVID-19 developed.

4 Explain why working from home was a strategy implemented to restrict the spread of COVID-19.

5 Describe some of the impacts of the lockdowns.

6 Outline some of the positive consequences of widespread work-from-home arrangements.

7 Describe the positive environmental consequences of the pandemic.

8 Outline how COVID-19 enhanced research and innovation.

9 Outline some of the different perspectives on the COVID-19 pandemic.

10 Explain the concept of herd immunity.

Applying and communicating geographical understanding

11 As a class, discuss the use of travel restrictions during the COVID-19 pandemic. Brainstorm the advantages and disadvantages of these restrictions and record your thoughts using a graphic organiser.

12 Examine Figure 15.4.2. Describe the data shown on the map.

13 Write a short report, using statistics, outlining how airlines and cruise ship operators were impacted by COVID-19.

14 Using Figure 15.4.5 and the text, describe the recovery of the tourism sector from COVID in 2022.

15 Examine Figures 15.4.8 and 15.4.9. Complete the following activities.

a Which age groups suffered the greatest number of deaths from COVID-19 in Australia?

b Why do you think more women in the 90+ age bracket died from COVID-19?

c Outline the data shown in Figure 15.4.9.

d Write a short discussion on why you think the rates of infection were higher in younger people. What were the health implications for the wider community?

16 Examine Figures 15.4.10 and 15.4.11. Use these photographs and the information in the text to write a short report on the extent of work-from-home regulations.

UNIT 15.5

COVID-19 management in Australia

By mid-January 2021, around a year after the pandemic began, 28 614 Australians had been infected with COVID-19 and 909 had died as a result of the disease. This represented an infection rate of 0.1 per cent of the population. At the same time, countries with higher rates were the USA, with an infection rate of 6.8 per cent, and the UK at 4.7 per cent. New Zealand had one of the lowest rates in the world at 0.04 per cent, with 1866 infections and 25 deaths.

Figure 15.5.1 details the extent of infections and deaths as at 11 January 2021 for selected countries and their estimated populations. Australia had a relatively low infection and death rate compared to countries with similar economic development and wealth. These are useful statistics as they occurred before any nations could introduce a vaccine. They represent the pandemic's first year when the spread of the disease was at its most rapid.

Country	Total infections	Total deaths	Total population
USA	22 699 938	381 480	330 317 000
UK	3 118 518	81 954	65 220 000
France	2 737 501	67 368	65 495 000
Italy	2 276 491	78 555	61 558 000
Germany	1 921 024	40 686	75 588 000
Canada	668 181	17 086	35 652 000
Japan	286 752	4044	125 667 000
South Korea	69 114	1140	51 525 000
Australia	28 614	909	26 121 000
New Zealand	1866	25	4 595 000

15.5.1 COVID-19 infections and deaths for selected countries, 1 January 2020–11 January 2021

Australia's geographical advantages

Australia's strategies dealing with the COVID-19 pandemic were aided by its geography. Sharing no land borders with other countries meant island nations like Australia and New Zealand could more effectively seal off their borders to international travellers. While most countries closed their borders across 2020, those with complex land borders faced greater challenges.

In Europe, vast quantities of freight moves between countries by truck. In December 2020, France closed its border with the UK due to a highly infectious new strain. Freight trucks were unable to use the ferries across the English Channel or the Eurotunnel train service underneath it. They began to pile up in southern England (see Figure 15.5.2). Australia was spared such extensive border problems, simply closing its international airports and taking careful quarantine measures at ports.

15.5.2 Thousands of trucks parked after France closed its border to the UK

Australia's relatively low population density was another factor in its favour. Most cases were confined to large capital cities, with limited outbreaks of infection in regional and remote areas. With their concentration of health services, these cities were better able to cope with infections and smaller regional hospitals were spared the impact.

Australia's response to COVID-19

Responses from authorities at state and federal levels combined with the actions of individuals and business, worked together synergistically. This enabled Australia to experience a better containment of the pandemic compared to other economically and culturally similar countries.

Contact tracing

A feature of the COVID-19 pandemic in Australia was the use of **contact tracing**. Using **epidemiological** methodologies, contact tracing aims to identify the source of an infection and the people who may have been exposed to it. Australia was one of the first countries to adopt comprehensive contact tracing early on. Figure 15.5.3 shows the status of contact tracing on 26 January 2020. This is at the pandemic's very start, long before the worst effects of the outbreak were being experienced. Australia was one of only a handful of countries that implemented comprehensive tracing.

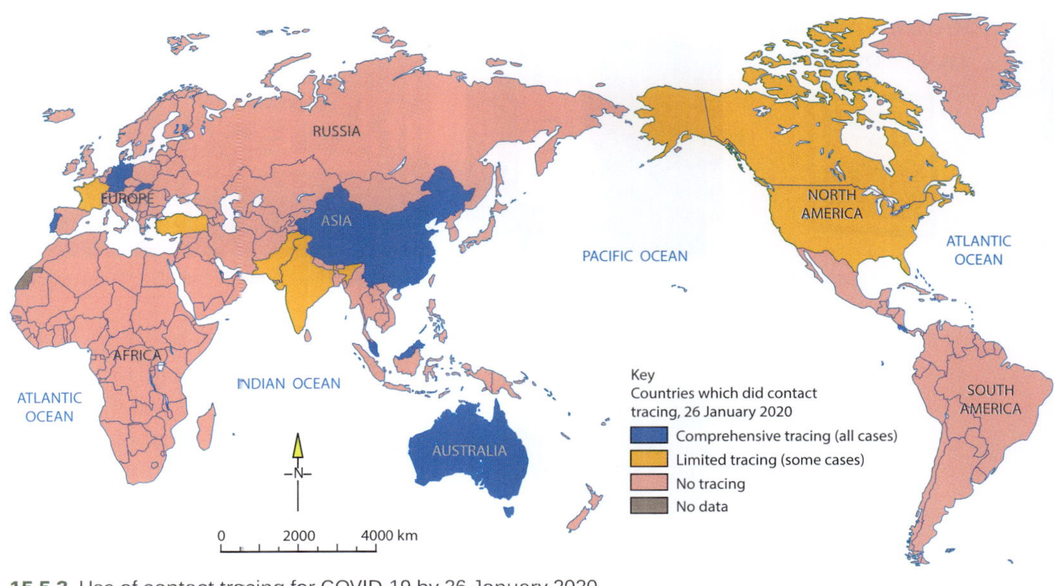

15.5.3 Use of contact tracing for COVID-19 by 26 January 2020

Contact tracing proved to be one of the most important strategies for limiting the spread of COVID-19. Teams of contact tracers were employed to interview infected people and determine where they had been and who they had been in contact with. This information was used to contact others in contact with an infected person. They had to isolate until after the incubation period had ended.

The Australian Government developed the COVIDSafe app to assist with contact tracing and implemented a major public health education campaign (see Figure 15.5.4). By June 2020, six million Australians had downloaded the app onto their smart devices. The app was designed to constantly run in the background and noted which other devices were close by.

The COVIDSafe app was controversial and was considered to have limited effect. State governments instituted a more comprehensive system of requiring people to sign in at many hospitality and entertainment venues, on public transport and other stores. This was typically done through a QR code (see Figure 15.5.5), which assisted greatly in identifying close contacts quickly. Contact tracing also played a key role in allowing for economic activity to continue (at least partially) through much of the pandemic period. By having patrons sign in, cafes and restaurants were able to remain open during most of the pandemic period.

15.5.4 Adversting campaign to encourage the use of the COVIDSafe app

Widespread testing

Testing for COVID-19 was crucial for limiting the spread of the illness and ensuring appropriate treatment. Early in the pandemic, health authorities across Australia implemented widespread testing. Testing was free and clinics were established in multiple locations (see Figure 15.5.6). In comparison, tests in some US cities could cost up to US$850, discouraging people with symptoms to get tested. Where infection hotspots were identified, emergency testing centres were established. Although waiting times could be high, this response helped Australia deal with the virus.

15.5.5 NSW required people to sign in using a QR code at many locations, to assist in contact tracing.

15.5.6 A drive-through COVID-19 testing centre in Sydney

Authorities also adopted special testing protocols of sewage checking for COVID-19. In Sydney, the Sewage Surveillance Program was implemented by NSW Health across 60 individual sites. The program tested for fragments of the COVID-19 virus in untreated sewage to check on its prevalence in the population. A feature of the disease is that many people are **asymptomatic**; they do not display symptoms despite being infected. Testing sewage proved a useful strategy for identifying whether there were possible unidentified cases in an area. Authorities then issued calls for residents of these areas to be tested.

Individual actions

Another key aspect of Australia's pandemic management was the willingness of nearly all Australians to follow the advice of health authorities. Australians proved themselves to be quite compliant with the restrictions placed on them. Although governments instituted considerable penalties for breaches, relatively few were issued. In January 2021 in Sydney, Brisbane and Melbourne, for example, there were some poorly attended protests in response to mask-wearing mandates. Overall, there were extremely low levels of non-compliance. Similarly, sign-in protocols were widely adopted across the country.

Activities

Acquiring and processing geographical information

1. Outline how Australia's relative isolation assisted in containing the virus.
2. Describe the process of contact tracing.
3. Explain why contact tracing was effective in limiting the spread of COVID-19.
4. Describe how technology was used to assist in contact tracing.
5. Outline how COVID-19 testing was implemented in Australia.
6. Explain why widespread testing reduced the spread of COVID-19.
7. Outline the role of individuals in dealing with COVID-19.

Applying and communicating geographical understanding

8. Using the data in Figure 15.5.1, write a report comparing COVID-19 infections and death rates in Australia to other similar countries.
9. Using Figure 15.5.3 and the text, explain Australia's adoption of contact tracing.
10. During the pandemic, some people thought government regulations like mask-wearing and requirements to sign in at venues were breaches of civil liberties and were unreasonable. As a class, discuss this idea and whether these steps were reasonable. Use a graphic organiser to record the key points of the discussion.

APPLICATION AND CONSOLIDATION TASKS

Task 1: Presentation: An overview of the COVID-19 pandemic

Prepare a digital presentation on the origins, transmission methods and impacts of the pandemic. In your presentation, ensure you:

- explain the nature and spatial dimensions of the COVID-19 pandemic
- describe how naturally-occurring and human activities impacted the spread of COVID-19
- assess the challenges and opportunities created by the COVID-19 pandemic.

Your presentation should include maps, graphs, illustrations and diagrams to support your findings.

Task 2: Speech: Analysing Australia's response

Undertake research and analyse Australia's response to the COVID-19 pandemic and write a speech. In it, include a comparison between the approaches taken by Australia and two other countries to mitigate the COVID-19 pandemic and assess their effectiveness.

Although you are not presenting your speech, include the visuals that would accompany one, including maps, graphs and other illustrative material.

Task 3: Report writing: A significant pandemic

Working with a partner, conduct research into a significant pandemic in history, excluding COVID-19, and write a report providing an overview of the pandemic. In your report outline:

- the type of pandemic and its causes
- the impacts in the short-, medium- and long-term
- what happened.

Your report should include visuals to support your report, including maps, graphs, illustrations and diagrams.

Task 4: Podcast: Living during the time of COVID-19

Conduct a series of short interviews of people who were working and those who were in school during the COVID-19 pandemic. Record the interviews and then present them as a podcast. In preparing your interviews, include information on the common experiences people faced during the pandemic and outline what restrictions they faced.

The podcast can also include audio content, such as sound effects and music.

SECTION

4 Geographical investigation

Geographical investigations lie at the heart of the discipline of geography. Geographers investigate places and the relationships between people and their environments. In doing so, they explore both the physical properties of Earth's surface and the human societies spread across it.

Geographers also study how human culture interacts with the natural environment. It reveals the ways in which locations and places can have an impact on people. These investigations help geographers to better understand where things are found, why they are there, and how they develop and change over time. In brief, geographical investigations help everyone make sense of the world around them.

Part of the Year 11 Geography curriculum is to plan and conduct a geographical investigation. This investigation's aim is to help develop an understanding of the nature of geographical inquiry. It will facilitate the exploration of the value of geographical inquiry in the contemporary world through practical research and the application of geographical concepts, skills and, most particularly, tools.

Content focus

In the section of the text, students focus on the nature of geographical inquiry and how it can contribute to the understanding and effective management of places and environments. They also examine the nature of engagement with geographical challenges by research institutions, in political debate, in the media and by individuals. The section also provides students with a step-by-step guide on how to conduct a geographical investigation.

In this section

CHAPTER

16 Contemporary geographical inquiry

Geographers are interested in many different topics—environments, people, cultures, politics, settlements, biodiversity, landforms and more. What is unique in their approach to studying these topics is the perspective adopted. A geographer's perspective is determined by the concepts central to the study of discipline—place, space, environment, interconnection, scale, sustainability and change. These shape the nature of the lens that geographers use to investigate phenomena and issues.

Geographers ask spatial questions. How and why are things distributed or arranged in particular ways on Earth's surface? They study these distributions and arrangements at a variety of scales, from local to global. Geographers investigate how human and natural activities on Earth's surface interact to shape the characteristics of the world we live in.

This chapter focuses on the nature of geographical inquiry and how it can contribute to our understanding and effective management of places and environments. It examines what research institutions, government and advocacy groups, the media and individuals do to engage with geographical challenges.

The study of geography is about more than just memorising places on a map. It's about understanding the complexity of our world, appreciating the diversity of cultures that exists across continents. And in the end, it's about using all that knowledge to help bridge divides and bring people together.

Barack Obama, former President of the United States

16.0.1 Kayaking in Neko Harbour, Antarctica

Chapter glossary

advertorial an advertisement in the form of editorial content

bias inclination or prejudice for or against a particular point of view

clickbait an online heading or content whose main purpose is to get the reader to click; the headlines are usually sensationalist or misleading

data a set of ideas or values recorded and interpreted to produce information

sockpuppeting the use of an online identity used for purposes of deception

Twitter bot a computer program that automatically performs Twitter (now known as X) actions

UNIT 16.1

The nature of geographical inquiry

Inquiry-based learning is at the heart of any geographical investigation. It is central to the study of geography, whether it is undertaken as an independent study or as an organised fieldwork activity.

In inquiry-based learning:

> ... students have an opportunity for independent learning based on a set of problems or hypotheses, for which they can find the information and formulate answers. The hypothesis (or equivalent) drive[s] the desire to collect evidence in order to find answers.
>
> Margaret Roberts, *Geography Through Enquiry*, 2013

Inquiry-based investigations occur when students decide (often with teacher guidance) what they are going to investigate and then find their own evidence. In doing so, they determine their learning pathway. Figure 16.1.1 illustrates an inquiry sequence from the perspective of the learner.

When developing a geographical investigation, remember that an inquiry is a set of sequential (and increasingly complex) processes that involves a progression from demonstrations of competence to confidence. Over time, students progress from what can be described as closed investigative tasks to a framed inquiry/investigation, and then a fully independent inquiry/investigation. At each stage, students engage in learning in ways that demonstrate a higher level of independence. Figure 16.1.2 sets out this progression and the growing independence students will display. By Year 11, it's expected that students can demonstrate a reasonably high level of independence.

16.1.1 An inquiry sequence from the perspective of the learner

	Closed tasks	Framed inquiry/ investigation	Independent inquiry/ investigation
Ask questions	A pre-planned investigative task is presented to students. Questions are not made explicit.	The teacher formulates the investigation's questions and they are made explicit.	Students determine the investigation's questions, albeit framed by teacher guidance.
Collect data	The teacher determines the nature of the investigation to be undertaken and decides which methodologies will be used. The data is presented as authoritative.	The teacher makes all major decisions regarding the investigation's procedures. Data are presented as information for students to interpret.	Students actively engage in determining the methodologies employed and data sources.
Make sense	The teacher devises the activities to achieve predetermined objectives. Students follow the teacher's instructions.	Students can choose their preferred method of representation. Analysis is independent.	Students analyse the data collected and make decisions or reach conclusions independently.
Reflect	Investigation outcomes are largely predetermined. Students are given limited, if any, opportunity to reflect on the nature of the investigation undertaken.	Students engage in discussions about what they have learnt. Different outcomes are possible.	Students reflect on the data-collection methods used and the validity of the data collected.

16.1.2 Geographical investigation progression and increasing independence

The investigations undertaken by students in Australian schools typically emphasise only part of the inquiry sequence—observation, collection and recording of **data**, and its representation and analysis. In Year 11, students have the opportunity to conduct an inquiry/investigation independently using the following stages:

- topic selection
- formulating research questions and identifying the best method/s to gather the required data
- collecting and analysing data
- engaging in the evaluation stage of the inquiry sequence
- applying the knowledge and understandings acquired through the investigation undertaken.

The geographical tools and skills that students master throughout their geography studies are used to acquire, process, apply and communicate geographical information (see Figure 16.1.3).

Developing geographical inquiry skills

The geographical inquiry skills developed in this investigation include those associated with:

Acquiring geographical information

Developing geographical questions and collecting primary data and secondary information.

Processing geographical information

Representing, interpreting, synthesising and analysing data and information to make generalisations and inferences; proposing explanations; drawing conclusions and predicting future trends. Students evaluate data and information for reliability, bias, validity and usefulness.

Applying geographical understanding

Evaluating and ranking options, proposing actions and predicting outcomes, and developing an implementation plan. Students apply their knowledge and understanding of other geographic contexts.

Communicating geographical understanding

Communicating conceptual understanding, findings and/or conclusions to a range of audiences, including stakeholders. Students reflect on the process and effectiveness of geographical inquiry undertaken.

16.1.3 Developing geographical inquiry skills

Geographical inquiries

Investigating the challenges facing people and places is central to the study of geography. Through such investigations, geographers can engage with geographical issues and phenomena at a variety of scales and contexts. They learn to appreciate the complexity of our world and the diversity of its environments, economies and cultures. They use this knowledge to promote a more sustainable way of life and an awareness of social and spatial inequalities. Engaging with geographical challenges extends beyond schools, universities and research institutions. Geographers are subject to considerable political attention and feature prominently in the media. They provide opportunities for individuals to engage in active and informed citizenship.

Research institutions and areas

Research institutions, including Australia's 43 universities and numerous government agencies (e.g. the CSIRO, the Australian Bureau of Meteorology, the Australian Bureau of Statistics and Geoscience Australia) play an important role in investigating geographical issues.

In 2018 the Australian Academy of Science's National Committee for the Geographical Sciences delivered its landmark report, *Geography: Shaping Australia's Future*. It identified 10 broad Australia-based research areas in which geographers play a leading role:

- environmental change and human response
- land, water and food
- health and wellbeing
- the economy
- the Asia–Pacific region
- natural hazards
- rural and regional Australia
- Australia's cities (see Figure 16.1.4)
- coastal and marine environments
- geographical information systems and science.

These topics demonstrate the nature and scope of geographical research in Australia.

16.1.4 Unchecked urban sprawl poses a threat to the sustainability of Australian cities.

● SPOTLIGHT

Research example: Australian cities

Compared with other developed countries, a high proportion of Australia's population lives in urban areas, and most of this urban population is concentrated in cities with populations of more than 1 million. Two-thirds of residents live in state and territory capital cities, which are growing at nearly twice the rate of the rest of the country.

These cities have a wide range of environmental, economic, infrastructure, social and cultural issues, which are the subject of extensive geographical research. These follow the geographical themes of space, interconnection and environment, and managing space and place through planning. A concern for social equity and justice is a common theme. There are many broader areas of research interest, including global and national city hierarchies and networks, cities' internal structures, employment, transport and mobility, infrastructure, housing, diversity, the urban physical environment, urban sustainability and planning.

An example of a research initiative is investigating the causes and consequences of Australia's population being concentrated in five large cities, both positive and negative. Another is determining strategies to shift urban population growth from Melbourne and Sydney to smaller cities and regional centres. One research study might investigate ways to improve the physical quality of city neighbourhoods, looking for strategies to improve mobility, or how to use diversity to enhance the quality of urban life. Another could seek to discover how to make cities more liveable, ecologically sustainable and economically viable. Or determine how various age groups, including the elderly and children, experience and negotiate the city, and how their experiences could be improved. During their studies of sustainable places, HSC students will have the opportunity to study a range of these issues.

Research in geography

Geography is a wide-ranging, dynamic subject concerned with exploring issues that affect the wellbeing of people and places. The concept of wellbeing, along with its emphasis on access to employment and general economic wellbeing, includes the core values of environmental sustainability, equity and social justice.

The research geographers undertake is based on a set of core concepts that guide the choice of research topics, identify significant questions and suggest explanations. Geography's core concepts of space, place and environments are supplemented by a set of complementary concepts—interconnection, geographical scale, change (time) and sustainability. These seven concepts provide the lens through which geographers study geographical issues and phenomena.

Activities

Acquiring and processing geographical information

1. Describe what makes geography unique. What concepts inform the perspective adopted by geographers?
2. List the key spatial questions posed by geographers.
3. Outline the key features of inquiry-based learning.
4. Describe the nature of the inquiry that students should be able to undertake by Year 11.
5. Identify the four geographical tools and skills that students will master throughout their studies of geography.

Applying and communicating geographical understanding

6. How does Barack Obama's quote at the beginning of this chapter inform our understanding of the role and contribution of geography?
7. Working in groups, formulate a definition for each of geography's seven key concepts. Share these with the rest of the class.
8. Working in groups, study Figure 16.1.1. Select a contemporary geographical issue, event or phenomenon and work your way around the inquiry learning cycle, thinking about how you might investigate it.

UNIT 16.2

Political debate, advocacy groups and the media

Geographical issues form an important element of contemporary political discourse. Debates feature prominently in the media. Consider, for example, the attention given to natural hazards. These include droughts, floods and bushfires, and the issues of climate change. More specific are issues such as the state of the Murray–Darling River system and threats to the Great Barrier Reef. Also contested are levels of immigration, urban congestion and infrastructure planning.

Contemporary issues are those discussed and widely debated in the community because their impacts will be felt in our own lifetime and beyond. Truly geographical issues are those with a spatial and temporal dimension. This means that they occur in a particular context and at a particular time. One might be a local community-based issue that is a focus of people's attention for just a short period—an example is a proposal to rezone a local area to accommodate high-density housing. Another might affect the whole planet, such as an environmental issue of concern for generations (e.g. climate change).

People often have different worldviews that shape their perspective or point of view on issues and on the best way to use resources or manage the environment (see Figure 16.2.1). Sometimes people disagree on how best to address such issues. Disagreement is healthy, provided we respect the rights of others to express their views, even though we might not necessarily agree with them. Being able to identify and evaluate such points of view greatly enhances our decision-making processes. Any conclusion we draw about an issue must be based on sound knowledge and understanding of the processes and forces involved.

Newspaper, television, radio and social media play important roles in bringing issues to the public's attention. They are also the main means by which politicians communicate with the public and provide a forum for all to comment.

16.2.1 A fracking protest in Darwin

Advocacy groups and individuals

Advocacy groups are organisations that seek to influence, either directly or indirectly, the decision-making processes of governments or large corporations. Commonly called interest groups, they do this by trying to exercise direct influence over decision-making processes or by attempting to shape the demands that other groups and the public make on decision-makers.

Advocacy groups are formed when people with a common cause or concern unite with the aim of having their concerns addressed. They play an important role in democratic societies, such as Australia. Some are very effective and influence decisions at the highest level of government. Others are weak and have little impact.

How much influence an interest group has is not necessarily related to its size. Some small groups can exercise great power and operate through direct contact with government. Others need to engage in direct action, such as demonstrations, to be successful. Often an interest group's influence depends on the group's ability to draw the media's attention to the issues and causes it promotes.

Advocacy groups with a geographical focus may be concerned with a single issue, such as the protection of an area of bushland in a residential area. Others may focus on long-term objectives, such as reducing greenhouse gas emissions.

One advocacy group may come into existence as a reaction to a particular government decision or action. Another may result from a long-standing commitment to a particular viewpoint or cause. A group may be highly organised, with a paid professional staff, or it may be forced to rely on volunteers. It may be well-resourced or have limited means.

Collective action has several advantages. Individuals often feel powerless to influence decision-making processes. By acting collectively with like-minded people, they are more likely to have an impact. Collective action is harder to ignore.

Advocacy groups are quite different to political parties. While both are organisations that communicate the views and concerns of citizens to governments, political parties are accountable to the electorate and seek to control the functions of government. Advocacy groups lack this type of accountability. They are, instead, a political force that seeks to influence government policies.

Environmental advocacy groups are concerned with protecting the physical and human environment. They are now represented in Australia's parliaments and are usually referred to as the Greens or Teals. Examples of environmental advocacy groups are Greenpeace, the Australian Conservation Foundation, Friends of the Earth and The Wilderness Society.

Using the media critically

Students must exercise caution when using the media as a secondary source of information. The media plays an important role in shaping the nature of political discourse and debate in a democracy. But it no longer merely seeks to 'report' the news. It is increasingly opinion-based. There is a polarisation in the worldviews presented.

Conservative media outlets in the Australian context include *The Australian*, *Daily Telegraph*, *Courier-Mail* and the *Herald Sun* newspapers and Sky News. They promote a conservative and increasingly right-wing political agenda. Alongside are the conservative talkback radio networks, such as Sydney's 2GB, with its line-up of right-wing shock jocks. In the political centre are the Nine newspapers (*Sydney Morning Herald* and Melbourne's *The Age*), the commercial (free-to-air) television networks and the ABC. On the progressive left is the online newspaper, *The Guardian*, the print and online magazine, *The Monthly*, and web-based news sources including Crikey and The Conversation.

Being able to identify media **bias** is an increasingly important skill that can be applied to a wide variety of secondary sources.

● SPOTLIGHT

Left wing and right wing

The terms left wing and right wing are a way of classifying the political or ideological positions held by individuals and parties. Left-wing (or progressive) thinking is characterised by an emphasis on ideas such as freedom, equality, rights, progress, reform and a commitment to global institutions (see Figure 16.2.2). Right-wing (or conservative) thinking is characterised by a focus on 'family values', authority, order, duty, tradition and nationalism. In the Australian political context, the Greens would be classified as left-wing, Labor as centre-left and the Liberal–National Coalition as centre-right. In the USA, the Republican Party would be classified as right-wing and the Democrats as centre-left. In the USA, the political left are often referred to as 'liberals'.

	Partisan left	Leans left	Centre	Leans right	Partisan right
High quality		The Conversation Crikey *Saturday Paper* *The Monthly*	Four Corners SBS News ABC News & 7.30 *The Sydney Morning Herald* *The Age*		
Mixed quality	*The Guardian* Opinion New Matilda	*Eureka Street* *The Guardian* *Independent Australia* Q&A	Inside Story The New Daily The Drum Insiders	*The Australian* Sky News *Australian Financial Review* 7 News 9 News news.com.au *Herald Sun* *The Daily Telegraph* Eyewitness News	
Poor quality	Green Left Junkee	The Feed The Project Pedestrian		60 Minutes A Current Affair	*The Australian* Opinion Sky News Opinion *The Spectator* *Quadrant* *Daily Mail Australia*

← Left — Bias — Right →

16.2.2 Australian media outlets by quality and political ideology

● SPOTLIGHT

Talkback radio

Early morning talkback radio plays a role in setting the daily political agenda, even though its listening audience is relatively small compared with the total population.

There is little evidence these broadcasters have much influence or that they change how people vote. This is because their audience largely consists of those with fixed conservative worldviews seeking affirmation of their opinions. There is, however, a perception that these broadcasters have the power to change votes and this means they can exercise a disproportionate degree of influence over decision-making processes. For example, some may champion the cause of climate change denial, making it difficult for governments to build the community-wide support needed to tackle climate change.

There is commercial interest in appealing to these sections of the Australian community. Ratings success attracts the advertising revenue that funds these radio stations. This is why progressive activist groups promote advertising boycotts against certain media outlets and individual broadcasters. They aim to inflict commercial losses and expenses on media outlets by forcing advertisers to withdraw their commercials.

Contemporary issue investigation scaffold

When planning an investigation, students may consider the range of contemporary issues highlighted in the media, particularly those featuring prominently in public discourse and debate. They can spark ideas for potential topics. The following scaffold list may be useful to use as a guide. These points may help develop knowledge and understanding of an issue, as well as determine whether it is geographical and worth investigating.

- Read or view the material dealing with the issue.
- Name and briefly outline the issue.
- Identify the main people and/or organisations involved in the issue.
- Identify the scale at which the issue is relevant. Is it a global, national, regional or local issue?
- Explain why the issue is of interest to geographers.
- List the main sources of information about the issue.
- State whether the sources used present different points of view or perspectives on the issue.
- State whether the sources of information are reliable.
- Outline the actions that people could take to address the issue.
- Consider how the media have influenced the study of the issue.
- Explain how the study of the issue has affected students' views on the issue.
- Reflect on the actions that can be taken to address the issue.

SKILLS BUILDER

Identifying media bias

When seeking to identify media bias there are several questions we should consider:

- Who are the sources? Is there, for example, an over-reliance on 'official' sources, such as government officials, corporate representatives or spokespeople of established think tanks?
- Is there a lack of diversity? What, for example, is the race and gender of those selecting and presenting the news, compared to the community it serves?
- From whose point of view is the news reported? Media coverage often focuses on how an issue impacts corporate interests or governments rather than those who are directly affected by the issue.
- Are there double standards? Some media outlets hold individuals and groups to one standard while applying a different standard for other individuals or groups.
- Public policy think tanks such as The Australia Institute, the Evatt Foundation and The McKell Institute are often badged as Labor-leaning. Conservative think tanks, such as the Centre for Independent Studies, the Institute of Public Affairs and its spin-off the Sydney Institute, are aligned with conservative Australian politics. For example, young people committing property crimes are often portrayed differently than corporate executives engaged in white-collar criminal activities.
- Do stereotypes skew coverage? Does the reporting of crime in Australian cities, for example, emphasise the racial or cultural origins of some offenders while ignoring that of others?
- What are the unchallenged assumptions? To what extent does a media report promote an uncontested claim that is accepted without proof and is not argued against?
- Is the language loaded? Does the media report use loaded terminology to shape public opinion? For example, do conservative media commentators use the term 'political correctness' or 'woke' to dismiss progressive policies and worldviews? Progressive media outlets will use terms such as 'reactionary' and 'right-wing extremism' to stigmatise conservative policies and worldviews.
- Do the headlines and stories match? Rarely is a story's headline created by the journalist writing the story. Many readers just skim the headlines, so a misleading headline can have a significant impact.
- Are stories on important issues featured prominently? Consider where the story appears. Newspaper articles on the front page are the most widely read, as are stories at the top of a news website homepage. Lead stories on television and radio news have the greatest impact on public opinion.

Remember that news reporting and presentation are highly selective processes. What news is to be reported? What stories or issues are to be downplayed or ignored? Students need to be critical consumers when it comes to such secondary sources.

Activities

Acquiring and processing geographical information

1 Explain what is meant by a contemporary issue.

2 Explain why it is important to exercise caution when using the media as a secondary source of information.

Applying and communicating geographical understanding

3 Study Figure 16.2.2. Identify the media sources you are personally familiar with.

  a Where do these fit on the progressive–conservative continuum?

  b To what extent may this reflect and/or influence your worldview?

4 Distinguish between left-wing (progressive) and right-wing (conservative) worldviews.

5 Select one Australian-based environmental organisation (e.g. Australian Conservation Foundation, Bush Heritage Australia, Greening Australia, etc.). Briefly outline the nature of their campaigns. What strategies do they use to influence public discourse and debate?

6 Select a geographical issue featured in a recent media report. Use the contemporary issue investigation scaffold to investigate and report on the issue. Present your findings as a written report.

7 Select a geographic issue in the news. It needs to be a well-publicised issue that has been in the news for a long period. This makes it possible to source material from a variety of sources. For example:

- energy policy in Australia
- Australia's carbon dioxide emissions target
- bushfire reduction practices
- urban congestion taxes
- urban consolidation.

  a Select the media organisations for comparison. Choose at least three, one from each column.

Leans left	Centre	Leans right
The Guardian The Feed The Project *The Saturday Paper* *The Monthly*	ABC News *Sydney Morning Herald* *The Age* The Conversation The New Daily 10 News 7.30 Insiders	*The Australian* *Australian Financial Review* *The Daily Telegraph* *Herald Sun* *The Courier-Mail* Sky News 7 News 9 News

  b Find articles or news stories about your chosen topic, read/watch and make notes. Take in these points:

- Who is the author or journalist responsible for the information provided?
- Describe their point of view/argument.
- Do they use facts and/or reference experts? How?
- Describe the language used (emotive, anecdotal, assumptions, attacks, accusations, expert opinion, generalisations, hyperbole, metaphors, similes, inclusive language etc.)
- Are images, cartoons and graphics used? Describe what they are and how they influence the story.
- How and where is the story/article presented? For example, it is on the front page, at the top of the webpage, at the beginning of the news etc.
- Identify the bias of the story/article/post and explain how the bias influences the reporting.

● SKILLS BUILDER

Identifying fake news

Fake news

Fake news is disinformation, false news and conspiracy theories deliberately spread via traditional media sources (print and broadcast) or, more typically, via online social media. The intent of those spreading fake news is to mislead. They aim to manipulate the political views and thinking of individuals and/or influence the nature of public discourse and debate.

Fake news is often created and published with the intention of damaging political candidates or a particular party, or to advance a particular worldview or ideology. Fake news typically uses sensationalist, dishonest or fabricated headlines to increase readership. Those generating fake news hope people will share the misinformation via social media.

With advances in technology, fake news can sometimes be harder to identify. Manipulated images and deepfake video and audio content can look like genuine evidence. Although they can usually be identified as fake under close inspection, advances in the technology used to create this kind of content may make it harder to identify in the future.

The use of fake news to influence the outcome of elections is becoming more widespread. In the 2019 Australian federal election, some incorrect information about a Labor Party policy escalated into a fake news story. The misinformation that the Labor Party intended to reintroduce death duties (a tax on the estates of the deceased) appeared first on social media and was later amplified in the traditional media. In the 2016 federal election, Labor was accused of distributing material alleging that the Coalition government was going to privatise Medicare. The so called 'Mediscare' had no factual basis. In the USA, Fox News agreed, in an out-of-court-settlement, to pay Dominion (a voting systems manufacturer) US$787.5 million in damages for Fox's alleged promotion of the idea that Dominion was part of conspiracy to rig the election against Donald Trump in 2020.

Australia's bushfires in the news

During the Australian bushfire crisis of the summer 2019–2020, social media was used to promote the false claim that most fires resulted from arsonists' activities. Many used a media report that incorrectly claimed there had been 183 arrests for arson-related crimes made in conjunction with the bushfires. Despite a correction being made, that this was the number of people arrested since the beginning of 2019, the story continued to be spread.

Figure 16.2.3 and 16.2.4 are two images taken out of context and tweeted around the world, supposedly showing the impact of the fires. The tweet of the fake 'satellite image' was shared thousands of times. For the uninformed, it might have appeared that the entire continent was on fire. Figure 16.2.4 was taken in Dunalley, Tasmania in 2013. The photo was shared on Instagram in January 2020 in a post that said it was 'happening right now' with the hashtag #mallacoota.

Giving further support to the false claims of arson, the hashtag #ArsonEmergency started trending on Twitter (now known as X), in early 2020. The hashtag was used over 27 000 times. It was used in posts that denied the evidence that climate change was linked to the severity of the bushfire crisis. Although it was suspected that **Twitter bot** accounts might be spreading the hashtag, researchers from the University of Adelaide found that most of the posts came from accounts that were verified as genuine people.

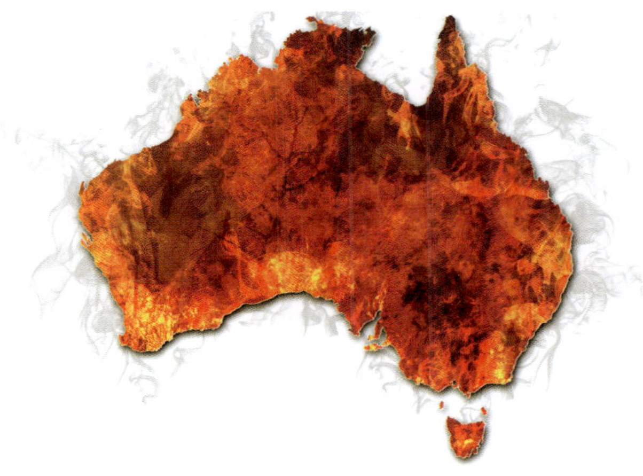

16.2.3 A 3D graphic visualisation of the bushfires, not a NASA satellite image of the 2019–2020 Australian bushfire disaster

16.2.4 Tammy Holmes and her five grandchildren take refuge from a bushfire in water, Tasmania, 2013. Tammy's husband took the photo to show his daughter that her children were safe.

How to identify fake news

Those generating fake news seek to take advantage of people's emotions or to make money. They do this by using persuasive headlines with eye-catching images to gain attention. Such posts are more likely to be spread on social media. Every time someone clicks on the link, the creators make money.

There are 10 steps students can take to determine whether a site can be classified as fake news (see Figure 16.2.5).

16.2.5 Sometimes we need to look closely to determine whether news is fake or real.

Step 1: Check the website's address. Any web address (URL) that is unfamiliar should be regarded as suspicious. For example, the ABC's URL is www.abc.net.au. Be wary of lookalike sites such as www.abc.net.com.

Step 2: Investigate the site's name. Fake news websites often have names that sound authentic, but many have already been identified as fictitious. If students search for the site's URL they may discover it has already been identified as promoting fake news.

Step 3: Look for indicators that it is an established credible news site. For example, check if the post has the date it was posted and a by-line for the journalist responsible for the story. Also, sources are often provided using links to previous reports on the issue or event. Finally, check for consistency. Does the content of the post match the headline and any accompanying images?

Step 4: Check the website's 'about' page. For legal reasons, many fake news sites have a disclaimer on their 'about' page alerting readers that their stories are fabricated.

Step 5: Check the website's registration. When and where was it registered? Copy and paste the URL into whois.icann.org or who.is. These sites will tell you when the site was launched. Be wary of relatively new sites. Often, they may have been launched in response to a particular event. The country of registration can also offer a clue about the site's authenticity.

Step 6: Fact-check any claims made in the post. Use a search engine to check whether a credible news organisation has reported the post you are reading. Credible news stories are typically reported by multiple news agencies.

Step 7: Be alert for overly sensational headlines. Fake news stories and conspiracy theories seek to elicit an emotional response because they know readers are more likely to share these via social media.

Step 8: Do a reverse image search. Copy and paste the image's URL into google.com/images. If the image is old or manipulated, the search will show the same image used multiple times on different websites.

Step 9: Look for visual clues identifying replicated websites. Fake news distributors may replicate websites resembling authoritative media organisations. They match the format and design elements, images and logos.

Step 10: Look for excessive advertising. A website showing excessive advertisements, pop-ups or flashing banners indicates its owner is more likely to be interested in generating revenue than providing objective, quality content.

It is also worthwhile checking if the post is an **advertorial**. Print publications often present an advertisement designed to look like a legitimate and independent news story. Check to see if the story is accompanied by an 'Advertisement' tag in small print.

Be aware of Twitter bots being used to distribute fake news. Bot software automatically performs actions such as tweeting, re-tweeting, liking, following, unfollowing or directing messaging to other accounts. There are legitimate uses, such as broadcasting useful information, or automatically generating interesting content. Improper usage includes violating user privacy, spamming and **sockpuppeting** (use of online identities to deceive).

Did you know?

The Shovel and the Bettoota Advocate are Australian satirical online newspapers. One of the Betoota Advocate's stories about a Sydney parking ranger accidentally giving himself a parking ticket was taken up as a 'factual' news story on one of the commercial television channels.

Clickbait

Clickbait is a form of false advertising on social media that uses a hyperlink to encourage users to read, view, or listen to a linked piece of online content that is deceptive or misleading. Some clickbait has relatively harmless content, containing celebrity gossip or anonymous stories taken from social media. This kind of content is created to get site visits that earn advertising revenue. Other clickbait is more insidious and the aim is to have the recipient share the disinformation via social media.

The spread of fake news, especially in the form of conspiracy theories, reflects an increased distrust in scientific expertise, media scepticism and a rejection of liberal democratic authority in some elements of society.

Activities

Applying and communicating geographical understanding

1. Define fake news. Why is it published?
2. Undertake internet research to locate fake news related to a contemporary geographical issue. For example, climate change and the increased frequency of natural hazards such as bushfires, drought and other extreme weather events.
3. Using online media sources, locate an article or post suspected of being fake news. Use the 10-step process to assess the authenticity of the article or post.

CHAPTER

17 Undertaking a geographical investigation

A geographical inquiry is a process students use to learn about and deepen their understanding of geography. It involves investigations that start with geographical questions and proceed through the collection, interpretation, analysis and evaluation of information to the development of conclusions and proposals for actions. In undertaking a geographical inquiry, students have the opportunity to apply their geographical skills and use geographical tools to acquire, process and communicate geographical information and formulate proposals. Where appropriate, they can take action. Inquiries may vary in scale and geographical context. Fieldwork-based investigations provide an opportunity to be involved in an active inquiry outside the classroom.

This chapter focuses on the curriculum requirement to plan and conduct a geographical investigation to develop an understanding of geographical inquiry and its value in the contemporary world. The investigation is designed to provide opportunities for students to apply geographical concepts, skills and tools through practical research in their local area or another accessible area for primary data collection.

As early as 1500 BCE, Polynesian navigators in the Pacific Ocean used complex maps made of tiny sticks and shells that represented islands and ocean currents they would encounter on their voyages. Today, satellites placed into orbit ... [have mapped] almost the entire surface of Earth ... with remarkable accuracy, and much of this information is available instantly on the internet. One of the most remarkable of these websites is Google Earth™, which 'lets you fly anywhere on Earth to view satellite imagery, maps, terrain, 3D buildings, from galaxies in outer space to the canyons of the ocean.' In essence, anyone can be a virtual Christopher Columbus from the comfort of home.

National Geographic

17.0.1 A school lesson on water quality

Chapter glossary

evidence facts and information used as the grounds for belief or disbelief

inquiry question a well-worded question that focuses on a researchable issue whose answer takes the form of a claim supported by evidence, information and reasoning

interval scale variables are in order, with known differences in values

nominal scale variables are simply named or labelled, with no specific order

ordinal scale variables placed in a specific order, beyond just naming them

primary data data collected by an investigator

qualitative data data that records subjective qualities, for example, opinions, attitudes and beliefs

qualitative methodologies research methods that focus on obtaining data through open-ended questions and conversation-based communication, focusing on not only what people think, but also why they think so

quantitative data data that records quantities, for example, numbers, sizes or frequencies

quantitative methodologies those methods used to collect quantitative data and the statistical, mathematical or numerical analysis of data, collected through polls, questionnaires and surveys or by manipulating pre-existing statistical data using computational techniques

ratio scale variables are in order with known differences in values and have an absolute zero point

research action plan a plan that lists what steps must be taken to investigate a specific inquiry question, its purpose being to clarify what resources are required to reach the goal, formulate a timeline for completing specific tasks and determine the resources required

secondary data data collected by someone other than the investigator, including published data (such as census results and records of rainfall), historical data (such as old photos) and data collected by other students

spatial phenomena the distribution of the arrangement of a feature across Earth's surface

UNIT 17.1

Planning and conducting a geographical investigation

Undertaking a geographical investigation involves applying a well-established inquiry sequence. This sequence is expressed here in terms of a **research action plan** and is shown in Figure 17.1.1. As students develop their investigation, they will customise and add additional details to this inquiry sequence.

When developing their research action plan, students must:

- identify the focus of their investigation
- select the methodology they will use to collect, present and analyse data
- decide how the investigation is to be sequenced
- determine the time allocated to the various steps
- decide how to present the inquiry's findings and reflect on how they will evaluate their study
- determine whether any action is appropriate.

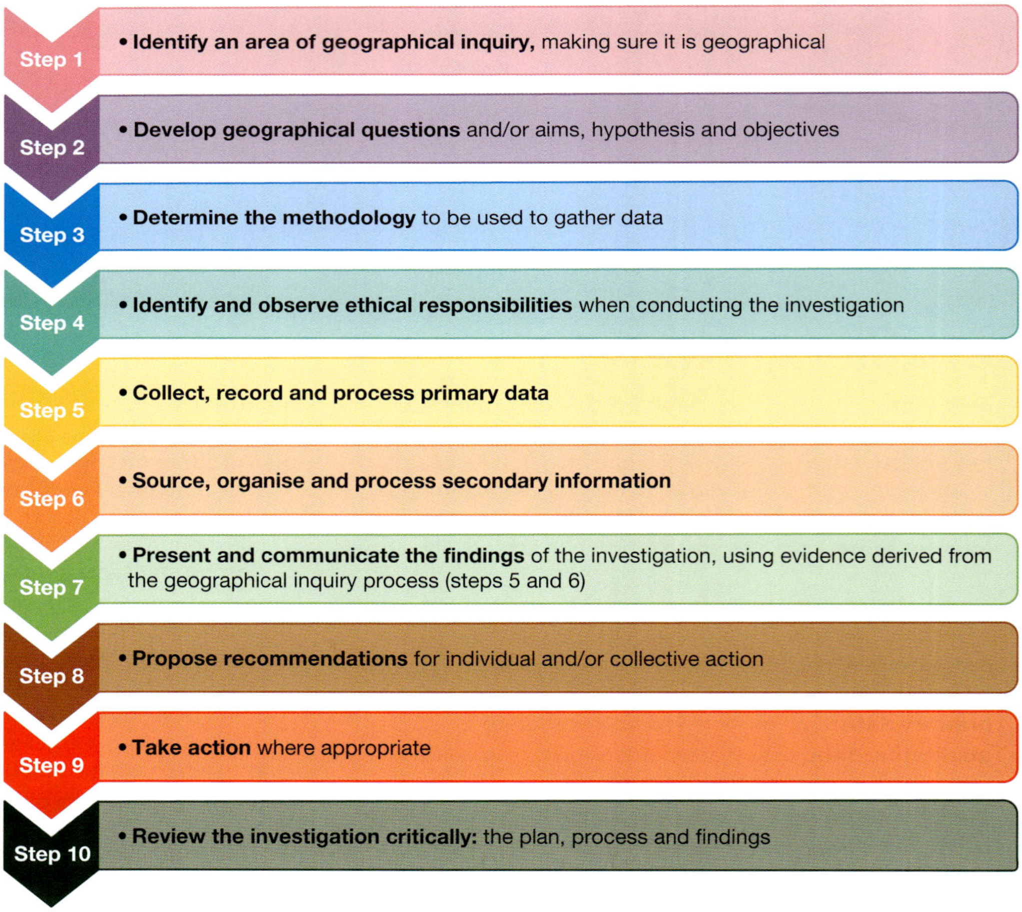

Step	
Step 1	• **Identify an area of geographical inquiry,** making sure it is geographical
Step 2	• **Develop geographical questions** and/or aims, hypothesis and objectives
Step 3	• **Determine the methodology** to be used to gather data
Step 4	• **Identify and observe ethical responsibilities** when conducting the investigation
Step 5	• **Collect, record and process primary data**
Step 6	• **Source, organise and process secondary information**
Step 7	• **Present and communicate the findings** of the investigation, using evidence derived from the geographical inquiry process (steps 5 and 6)
Step 8	• **Propose recommendations** for individual and/or collective action
Step 9	• **Take action** where appropriate
Step 10	• **Review the investigation critically:** the plan, process and findings

17.1.1 Research action plan

Identify an area of geographical inquiry

Given the diverse nature of the discipline, students have a wide choice of topics to choose from when selecting an area of geographical inquiry. There are, however, some restraints. The key question students need to ask themselves is: Does the chosen topic have a geographical focus?

Geographical investigations involve asking and answering questions about **spatial phenomena**. The key geographic questions asked are:

- Where is it located?
- Why is it there?
- What is the significance of the location?

As students develop their geographical spatial understanding, they need to pose additional questions: What is this place like? What is it associated with? What are the consequences of its location and associations?

The geography studied in Year 11 provides an insight into the nature of topics that students might elect to investigate. The syllabus has three content-based components: Earth's natural systems, People, patterns and processes, and Human–environment interactions.

In Earth's natural systems, students investigate Earth's uniqueness and diversity. They explore the diversity of Earth's landscapes and their physical characteristics. They examine the cycles, circulations, interconnections and spatial patterns that combine to form Earth's integrated system. Students also explore natural cycles and human modifications that change Earth's land cover, including investigations into major processes and changing global land cover through studies of deforestation, desertification or melting glaciers and ice sheets.

In People, patterns and processes, students investigate the extent of the human footprint on Earth, as evidenced through human phenomena, patterns and processes. They investigate the drivers of international integration and how this human transformation is shaping the patterns and extent of the human footprint on Earth by studying population change and resource consumption.

Students investigate the unique character of places. They explore the extent to which various human processes are shaping places through studies of human resilience in diverse environments, place and cultural change, or political power and contested spaces.

In Human–environment interactions, students investigate the global interactions of Earth's natural and human systems through the study of a natural hazard, a contemporary ecological hazard or climate change.

Drawing on these broad areas of study, students have the opportunity to identify a topic that interests them. They then develop their research questions(s). This determines the methodology they adopt and implement. These steps form the basis of the student's research action plan. Figure 17.1.2 provides a list of potential topics, but there are many other topics students might like to investigate.

Time management

There is one note of caution. Many students discover, often too late, that the scope of the topic they have selected is too ambitious to be completed in the time available. It is quite often the small, seemingly straightforward research activity that achieves the best outcome. Most vital is that students demonstrate they have a sound understanding of the research process. That means they can develop, apply and evaluate a research framework. Figure 17.1.3 gives an example of the criteria a teacher might use in assessing a geographical investigation.

The NSW Stage 6 Geography Syllabus assigns just 20 hours to the geographical Investigation that students are required to do, out of the 120 hours allocated to the study of geography in Year 11. While students will usually spend many more hours engaged in their investigation, it's worth noting the study must be achievable within an acceptable timeframe.

Potential geographical investigation topics
Earth's natural systems
Investigating a local landform feature
Investigating the processes of weathering, erosion and deposition
Investigating the factors affecting vegetation patterns in a local area
Investigating plant succession on sand dunes
Investigating plant succession on sandstone plateaus
Investigating the hydrology of a local catchment
Investigating the connection to Country of a First Nations group
People, patterns and processes
Investigating spatial patterns and how they have changed over time
Investigating the changing demographic profile of an Australian community
Investigating the impacts of population pressure on a selected resource
Investigating environmental degradation associated with a selected resource
Investigating the impacts of cultural integration in a selected community
Investigating the impacts of economic integration on a selected place
Investigating changing patterns of employment is a selected place
Investigating the changing retail landscape
Investigating a local example of the adaption, adoption and diffusion of popular culture
Human–environment interactions
Investigating the nature of human–environment interactions in a geographic region
Investigating a natural hazard's impact within a local context
Investigating a contemporary ecological hazard with a local context
Investigating an economic activity and its environmental interactions
Investigating the impacts of climate change in a local context
Investigating contrasting perspectives of the issue of climate change in a local community

17.1.2 There are many topics that can be investigated in geography

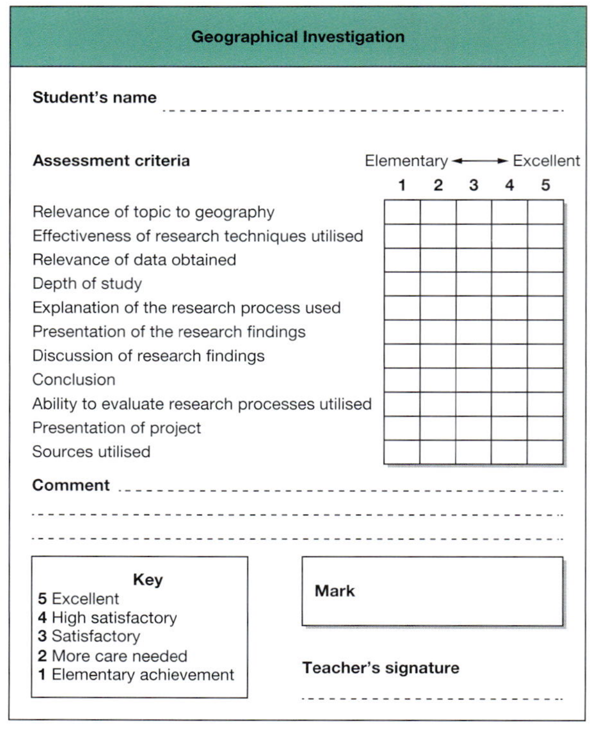
Geographical Investigation

Student's name ....

Assessment criteria — Elementary ⟷ Excellent

Criterion	1	2	3	4	5
Relevance of topic to geography					
Effectiveness of research techniques utilised					
Relevance of data obtained					
Depth of study					
Explanation of the research process used					
Presentation of the research findings					
Discussion of research findings					
Conclusion					
Ability to evaluate research processes utilised					
Presentation of project					
Sources utilised					

Comment ....

Key
5 Excellent
4 High satisfactory
3 Satisfactory
2 More care needed
1 Elementary achievement

Mark

Teacher's signature ....

17.1.3 Sample geographical investigation marking criteria

Develop geographical questions

Once students have settled on an area of geographical inquiry, they need to develop an **inquiry question** (or questions) to frame their investigation. While some students find the inquiry question an adequate starting point, others find it useful to further frame their investigation using aims, hypotheses and objectives. Some students combine both approaches. That is, they have an overarching inquiry question and then use aims, hypotheses and objectives to add a degree of precision to their investigation.

Inquiry questions

In developing an inquiry question, students should appreciate that the body of geographical knowledge they studied in Years 7 to 11 was constructed by geographers in response to questions that sparked their curiosity. In developing their inquiry questions, students are encouraged to ask themselves, 'What sparked my curiosity?' This will help them to identify a geographical phenomenon, process or issue worth investigating. From this, they develop a question they want answered.

Well-crafted inquiry questions have several important qualities. They:

- promote thinking and a genuine curiosity about the world
- encourage us to think about something in a way never considered before
- encourage us to think critically and creatively
- lead to more good questions
- are open-ended—typically there is no definitive, correct answer
- point to important, transferable ideas within and across disciplines.

Examples of inquiry questions related to the topic Bushfire hazards include:

- What factors are responsible for the increasing frequency and severity of bushfires?
- Under what environmental conditions is the bushfire risk greatest?
- How can the threat of bushfires to life and property be minimised?

Aims, hypothesis and objectives

The aims of any investigation are concise statements of what students have set out to investigate. An aim is usually expressed in terms such as to investigate, to discover, to identify, to explain, to analyse or to evaluate. It is possible to have more than one aim. The aim/s should be derived from the overarching inquiry question the student seeks to answer.

Examples of possible aims for the topic Bushfire hazards are:

Aim 1: To identify the factors responsible for the increasing frequency and severity of bushfires in the study area.

Aim 2: To investigate the bushfire risk at a particular point in time in the area studied.

Aim 3: To evaluate the strategies used to minimise the threat of bushfires to life and property in the area studied.

These aims form the basis for developing the hypothesis that is central to the investigation being undertaken.

A hypothesis is a tentative proposal used to explain an observation or a fact. It needs to be investigated before it can be confirmed. Hypotheses are the expected answers to the issues addressed in the investigation's aims and are based on students' prior learnings, observations and experiences. They are only informed guesses, so they may be wrong. This does not matter; researchers have often found that unexpected outcomes of their research have led to important new discoveries.

For Bushfire hazards, examples of appropriate hypotheses might be:

Hypothesis 1: The activities of people have increased the frequency and severity of bushfires.

Hypothesis 2: The level of bushfire risk can be determined by applying specific criteria.

Hypothesis 3: The risk to life and property can be reduced by developing a bushfire survival plan.

The number of hypotheses should match the number of aims.

Sometimes, people undertaking an investigation also nominate objectives. These are the steps taken to achieve the stated aims. They may be quite specific and closely related to the methodology developed to test the selected hypotheses.

For the same topic, examples of objectives related to Aim 1 might be:

Objective 1: To determine whether land-use practices in the study area have increased bushfire risk.

Objective 2: To determine whether climate change (rising temperatures and a reduction in average rainfall) has increased bushfire risk in the study area.

Determine the methodology

To conduct a successful geographical investigation, students need to develop their knowledge and understanding of the various research methodologies that geographers use. These are the approaches or techniques used to conduct research and gather data in the field. Specifically, students need to develop skills in collecting, presenting and analysing data.

Research methodologies exist on a continuum, from **quantitative methodologies** to qualitative ones. Quantitative methodologies measure (or quantify) data. **Qualitative methodologies** collect less measurable and, therefore, less quantifiable data. Few methodologies are exclusively quantitative or qualitative. Most have elements of both approaches.

Quantitative research is a research approach or data collection process that enables information to be collected, measured and compared. Quantitative research methods include surveys and structured interviews.

Qualitative research is a non-statistical approach. Geographers use qualitative research when they want to understand the actions of people in particular situations. The focus of qualitative research is on how participants (rather than the researcher) interpret their experiences. Such research methods include unstructured interviews, focus groups, open-ended questionnaires and participant observation. A qualitative survey is based on the use of an unstructured questionnaire. The interviewer uses a conversational style of interaction with the respondents. The responses are in the respondents' own words and often have emotional content. The characteristics of quantitative and qualitative methodologies are set out in more detail in Figure 17.1.4.

Students also need to identify the types of **primary** and **secondary data** they will need to answer their inquiry question or prove their hypothesis. They must determine how this is to be collected and analysed. That is, what methodology will they use?

The primary data provides the information needed to answer the inquiry question or prove the hypothesis. It is normally collected during the investigation's fieldwork stage. Data collection tools include questionnaires, interviews, focus groups, measurements and observations.

Secondary data is existing data already collected and published by others, such as in newspapers, magazines, journals and government reports or on the internet. It may be available in the same form as primary data (e.g. measurements not yet graphed or analysed), but, unlike primary data, it has been sourced by someone else, not the student. Often, secondary data is used before the study commences. It helps the researcher to develop their background knowledge or understanding of the context in which the investigation takes place.

Field-based geographical investigations principally involve collecting primary data. They may also draw on secondary data to assist the research process, especially in the planning phase and when analysing the information collected. Data collection methods are shown in Figures 17.1.4 to 17.1.8.

Quantitative research methodologies	Qualitative research methodologies
■ Results are dominated by numerical data, which are often presented in tables, graphs and diagrams ■ Survey questionnaires and measuring processes are used to gather data ■ Specific data from large populations can be collected using surveys ■ Results are easy to compare with other studies ■ Rely on the researchers' organisational skills in writing research questions, developing questionnaires, and collating and analysing the data collected	■ Results are often descriptive, with questions used to convey people's ideas ■ Personal interviews and observations are the principal data-gathering tools ■ Detailed data can be gathered, but only from small populations ■ Results are difficult to compare with other studies ■ Rely on the researcher's interpretive skills to understand what the data collected is showing

17.1.4 Characteristics of quantitative and qualitative research

Type of data	Source	
	Primary	**Secondary**
Qualitative	Photographs	Historical photographs, satellite images, aerial photographs, graphs and published statistics
	Sketch maps	Topographic maps, Google Maps
	Interviews and questionnaires	Newspaper articles, television programs (e.g. documentaries), internet-sourced information, journal articles
	Qualitative surveys	Published evaluative materials and reports
	Observations	Government reports
Quantitative	Statistical surveys and counts	Government reports, Australian Bureau of Statistics (ABS) census data and surveys
	Population surveys	ABS census data
	Quantitative questionnaires	Newspaper and magazine articles, internet-sourced information
	Land-use mapping	Topographic maps, land-use maps, zoning maps
	River channel surveys	Hydrological maps, data collected by catchment management authorities
	Weather-related measurements	Bureau of Meteorology data

17.1.5 Examples of primary and secondary sources for quantitative and quantitative data

Research approach	Explanation
Structured interview	An interview method where each interviewee is presented with the same questions in the same order
Unstructured interviews	An interview method where questions can be changed or adapted to suit the respondent's understanding or beliefs, knowledge or role in an issue or event
Focus group	A small cross-section of people brought together to provide feedback on an issue
Survey	A systematic collection, analysis and interpretation about a topic or feature being studied
Open-ended questionnaire	A type of questionnaire containing survey questions that encourage people to talk openly about issues
Participant observation	The process by which the researcher immerses themselves in the subject being studied, usually over a long period

17.1.6 Summary of research approaches

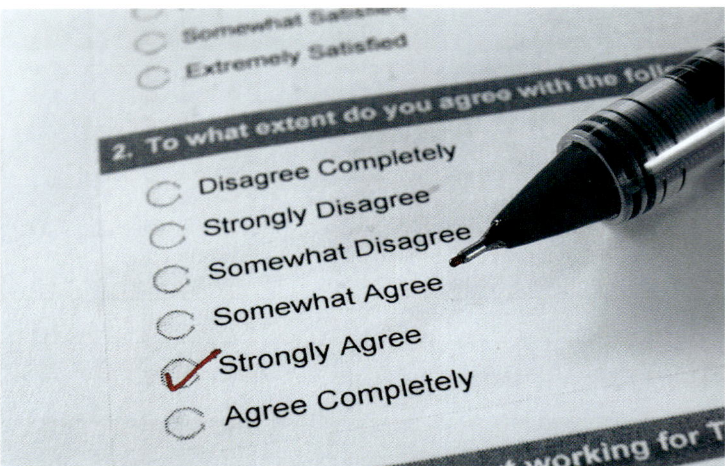

17.1.7 A questionnaire-based survey is an effective tool for gathering primary data.

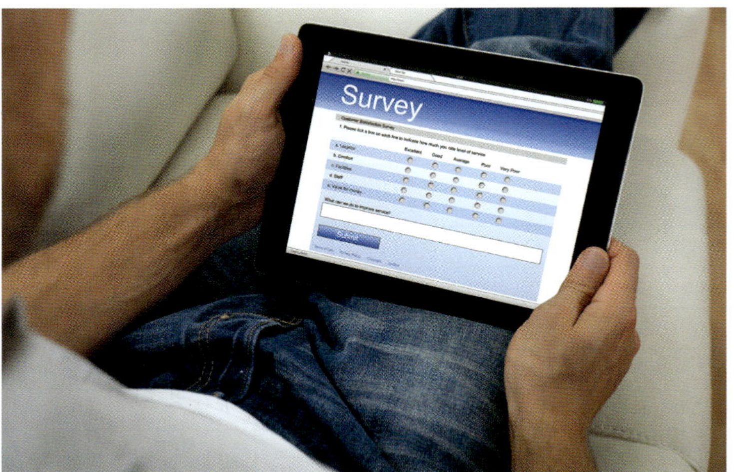

17.1.8 Online surveys are an increasingly popular way of gathering primary data.

17.1.9 Drones are an increasingly popular way of gathering primary data.

Identify and observe relevant ethical responsibilities

When undertaking a geographical investigation, students should conduct themselves according to accepted ethical standards. This includes the following:

- Obtain the permission and respect the privacy of people providing you with information and/or data.
- Communicate honestly when obtaining information and/or data from people, avoiding any deceptive or coercive behaviour.
- Do not put anyone in danger at any time.
- Do not enter private property without authorisation.
- Avoid environmental damage.
- Do not take someone's work and copy it or pass it off as your own.
- Credit where data, ideas or information come from, using an appropriate referencing style.
- Respect cultural protocols and sensitivities.
- Store data in a secure place (such as a password-protected laptop), especially if it is sensitive.

Collect, record and process primary data

Collecting and processing the data obtained during an investigation is the hands-on part of the investigative process. It might involve making direct observations (e.g. identifying dune vegetation types), conducting a survey, interviewing members of the public, measuring, taking photographs, constructing field sketches or collecting samples. Data collected directly from sources—as opposed to information sourced from the research of others—are called primary data (see Figure 17.1.9).

Once collected, the primary data needs to be used in a manner that best addresses the inquiry questions. It must also achieve the study's stated aims and/or prove or disprove the hypotheses.

Recording and processing the information collected must be done carefully to ensure accuracy. If the data is subjected to mathematical analysis, make sure the calculations are carefully checked. Measurement-based data can be tabulated or presented in a range of graphic forms to help with analysing it.

Survey results can be tallied on a blank copy of the questionnaire or entered into a spreadsheet. Spreadsheet data can be readily converted into a variety of graphic forms if the software has graphing functions. The graphs that spreadsheets generate are often easier to interpret than the original data tables.

Presenting the data graphically is one way to present data that makes the statistical analysis easier. Text-based forms of presentation and statistical tables are alternatives. Photographic images can be set out so they can easily be compared. For example, if the fieldwork aims to study change over time, order the photographs sequentially.

Source, organise and process secondary data

Secondary data is information collected by someone other than the investigator. Common sources include census, government department and authority information (e.g. published reports and online databases), international bodies (e.g. the UN and the World Bank) and geospatial databases (e.g. Google Earth™). Others are media, topographic maps and photographs, and a range of NGOs, such as policy think tanks and environmental groups.

Using secondary data can save time that would otherwise be devoted to its collection. With quantitative data, it can provide larger and higher-quality databases than is practical for an individual researcher to gather.

Secondary data use is typically associated with quantitative databases or for analysing verbal or visual materials created for another purpose. It is still considered a legitimate element of qualitative studies. Geographers investigating social and economic change consider that secondary data sources are essential, given that it is impossible to conduct one's own survey to adequately capture past change and/or developments.

While some investigations are based solely on the analysis of secondary data, it is more commonly used to better understand the context or circumstances in which the investigation is being conducted.

Secondary data, as with primary data, can be presented in a variety of formats: text-based, oral, graphical or multimedia.

Present and communicate findings

After analysing the primary and secondary data collected, students can present and communicate their findings. This involves drawing on the **evidence** derived from the geographical inquiry process to answer the inquiry question.

In communicating their findings, students are required to develop and submit a well-structured oral, written, and/or multimedia presentation. They need to use relevant geographical terms and concepts and a recognised form of referencing.

When outlining their results, students describe any trends, patterns or relationships identified and make comparisons between different locations or groups of people. Some useful descriptors are:

- trends: increase, decrease, no change
- patterns: equally distributed, unequally distributed, dispersed, concentrated, spaced
- relationships: direction of correlation (positive, negative, none), the strength of correlation (strong, weak), statements of proportion (e.g. 'A is 40% higher than B').

Propose recommendations

Often the investigation will highlight a particular environmental or social issue. Students may feel strongly about this issue and want to initiate some form of active citizenship. Any such action would aim to raise the community's awareness of the issue and encourage the relevant authorities to take appropriate action.

Take action

Everyone can engage with the geographical challenges facing people and places by either taking individual action or joining like-minded people involved in issue-based organisations. Sometimes called 'interest groups', these organisations help individuals raise public awareness of an issue and influence decision-making processes.

An interest group is a group of people who campaign on a cause or interest they have in common. Interest groups are major players in the political life of all democratic countries. People with similar views often work together to achieve a common objective.

Strategies that students might like to use to raise the broader community's awareness of the issue are outlined in Figure 17.1.10.

Strategy	Description
Petitions to parliament	As part of parliamentary procedures, petitions (written requests by people concerned about a particular issue) are presented and read out in parliament.
Deputations to members of parliament or local government representatives	Citizens often seek to gain the support of a member of parliament or local government representative through a deputation: a small group representing the view of a larger group, meets with the member or representative. They in turn approach the relevant minister or raise the issue in parliament or a local government forum.
Direct appeals to a minister or the head of the relevant local government authority	Many citizens attempt to lobby ministers (or mayors) or their senior advisers. Sometimes individual citizens or groups of like-minded citizens use the services of a professional lobbyist, a person paid by a group to present its point of view.
Letter writing	Letter-writing campaigns are used to persuade decision-makers there is widespread community concern about a particular issue. Letters written to the editors of newspapers are also used to help shape public opinion.
Advertising	Some individuals and groups use various forms of advertising to promote their concerns. However, this can be very expensive. Consider inexpensive forms of advertising to use.
Social media campaigns	Social media (including Facebook, TikTok, Instagram, YouTube, Twitter (X) and blogs) is used to draw attention to particular issues and influence public opinion.
Talkback radio	Participating in talkback radio programs helps to influence public opinion.
Protests and demonstrations	Demonstrations and other forms of peaceful protest are used to attract media attention. Individuals and groups hope the publicity generated will help shape public opinion.
Public meetings	Public meetings can be used to educate and inform the public about a particular concern.
Other strategies	As an activity, students can design issue-specific bumper stickers, posters and T-shirts with slogans.

17.1.10 Strategies for active citizenship

Alternatively, students may elect to modify their behaviour or lifestyle to live more sustainably, thereby adopting practices that minimise their environmental footprint.

Examples include:

- using public transport in favour of their own car
- shopping at local farmers' markets or growing their own vegetables and herbs
- introducing energy-saving technologies at home (e.g. energy-efficient appliances and light bulbs, insulation, double glazing)
- developing energy-saving habits (e.g. minimising air conditioning, turning off lights and unplugging appliances)
- choosing foods with less packaging and avoiding plastics
- saving water (e.g. taking shorter showers, washing full loads and planting drought-tolerant plants).

Review the investigation critically

The investigation's evaluation identifies the strengths and weaknesses of the study undertaken. Students mention what went well and what did not go so well. They then suggest how the study could be improved if it were carried out again. They also provide ideas for further investigating the same topic.

Students could start their evaluation with, 'I learnt a lot about ... from undertaking this fieldwork study. I also learnt how important it was to ...'.

They could continue with, 'I had some difficulty collecting information because ...'. Finally, they provide suggestions of how they could extend their investigations. For example, 'It would be wise to do further study on ...'

In evaluating their study, students reflect on whether the study's conclusion is supported by data obtained using a methodology based on sound reasoning. Also, of relevance are the issues of validity and reliability. Did, for example, the questions asked (or the measurements taken) provide an expected or readily substantiated answer? Reliability refers to the extent to which any measurements are consistent.

● SKILLS BUILDER

Methods of collecting data

Data can be collected via observation, questioning (surveys and interviews) and measurement (in the case of statistical data).

Observation (qualitative)

Observational data is typically recorded using audio and/or video recordings. There are three types of observation methods:

- **Naturalistic:** observing people in their natural environment
- **Participant:** observing behaviour as a participant in a group
- **Controlled:** observing behaviour(s) under certain conditions and/or time parameters.

Important!

Consider the issue of validity. Does the investigator's presence influence the behaviour observed? Have the participants consented? This is an ethical consideration.

Questioning (quantitative and qualitative)

Closed questions: answers are limited to single words, numbers or listed options. The collated data is quantitative. Example: 'Does the bushfire danger increase in summer?' Answers: Yes / No / Don't know.

Statements: answers are limited to positions on a scale. The collated data is quantitative.

Example: 'Climate change has contributed to an increase in the incidence and severity of bushfires.' Answers: Strongly agree / Agree / Don't know / Disagree / Strongly disagree.

Open questions: answers are detailed text. The data is qualitative.

Example: 'How might the threat of climate change be addressed without impacting on our material standard of living?'

Questionnaires and interviews

Questionnaires can cover relatively large samples. They are useful for gathering data on people's perceptions and experiences of place.

Interviews often involve a smaller sample than questionnaires. They can gather more in-depth data. Focus groups are a type of interview involving multiple interviewees in a focused discussion.

Sampling

Sampling is the process of collecting data from some sites or people in order to gain insight into a whole population. It applies to qualitative and quantitative methodologies.

- **Sample:** a limited number of things, such as a group of 100 community members or testing water quality at 10 points along the course of a river.
- **Population:** the total number of things, such as all the inhabitants of a particular place.
- **Representative:** the extent to which a sample matches the characteristics of the whole population.

Students need to explain how representative a sample is. It determines how the findings can be generalised.

Probability sampling involves selecting a sample that is representative of the population as a whole. There are three ways of doing this:

- **Random sampling:** each member of the population is equally likely to be included.
- **Stratification sampling:** a proportionate number of observations is taken from each part of the population. For example, each age group is included in the sample according to the percentage of the population being accounted for.

- **Systematic sampling:** observations are taken at regular intervals, such as every 10 metres or every tenth person. All surveys conducted by the ABS employ stratification sampling. Household surveys (such as the Monthly Population Survey and the Household Expenditure Survey) use a geographic stratification sample. Business surveys use variables such as state and industry stratification sampling and also a size measurement (e.g. the number of employees) to form size strata.

Non-probability sampling can be used for some qualitative methods, such as interviews, where it may not be possible to identify a representative sample. In non-probability sampling, the investigator selects the sample through their own subjective judgement. There are three ways of doing this:

- **Convenience and haphazard sampling:** selecting people who are easy to reach. For example, surveying the first 20 people encountered during the investigation of bushfire preparedness. This includes magazine and newspaper questionnaires and phone-in polls. These surveys also tend to ask questions that are loaded or have biased wording. This makes them subject to biased or unrepresentative samples as only people who feel strongly about the topic will respond.
- **Judgement or purposive sampling:** selecting a 'representative' sample by an expert in the field of study. Judgement sampling is subject to unknown biases, but may be justified for very small samples.
- **Snowball sampling:** the selection of at least two people who are subsequently asked to help identify other potential interviewees. The process is continued until the desired sample size is achieved.

Measurement (quantitative)

Those things to be measured are called variables. There are two types:

- **Independent variables** are not affected by other things. Each is said to be independent of the other variables.
- **Dependent variables** are affected by other things. Each is dependent on other variables.

While an independent variable causes a change in a dependent variable, a dependent variable cannot cause a change in an independent variable.

There are four measurement scales for variables:

- **Nominal:** variables that are not numerical (e.g. categories such as gender and ethnicity).
- **Ordinal:** variables where the order has meaning, but the difference between values is not important (e.g. the rank assigned to a value, such as first, second and third).
- **Interval:** variables where the order and exact difference between values is known (e.g. actual numbers, such as the temperature in degrees Celsius or a measurement in centimetres).
- **Ratio:** variables that allow for comparisons such as ratios, percentages and averages.

Measures of central tendency

- **Normal distribution:** the symmetrical bell-shaped distribution derived from a large series of measurements plotted on a frequency histogram. The mean is in the middle, with an equal number of smaller and larger values either side of it.
- **Mean:** all the measurements collected are added together and then divided by the number of measurements taken. The mean can only be used if the data approximates a normal distribution and has a numeral measurement scale.
- **Median:** the data is arranged in order with the middle value being defined as the median. It can be used for data that are not normally distributed. Suitable for variables with an ordinal scale.
- **Mode:** the value which occurs most often. Suitable for variables with a nominal scale.

Measures of dispersion

- **Dispersion:** the spread of data around the average, usually expressed as the mean ± interquartile range or mean ± standard deviation.
- **Range:** the distance between the highest and lowest value.
- **Interquartile range:** that part of the range covering the middle 50 per cent of the data.

If the variable has a numeral scale and if the data are normally distributed, use the interquartile range or the standard deviation. Otherwise, use the median and range to show dispersion.

● SKILLS BUILDER

Using surveys and questionnaires

Surveys are used to collect primary data about a population, their characteristics, behaviours and opinions. Examples of surveys include:

- **a census:** collecting information about every member of a population. In Australia, the ABS conducts a census every five years.
- **sampling:** collecting information on only a sample of a population. In Australia, the ABS uses sampling for a range of surveys, including its regular household surveys, the General Social Survey, Monthly Population Survey and its Survey of Income and Housing.

Surveys of human populations are common in opinion polling, market research and data collection by government departments. Depending on its purpose, a survey can ask for opinions or seek factual information.

When the investigator administers the questions, the survey is referred to as a structured interview. When the person being surveyed is free to answer a predetermined set of questions (e.g. a census form), it is referred to as a self-administered questionnaire.

Surveys have several advantages. They are an efficient, relatively easy way to collect information from an entire population or a representative sample. Surveys enable a wide range of information to be collected. They can be used to study attitudes, values, beliefs and past behaviours. The data collected can be analysed using a range of statistical processes (e.g. calculating the average and/or median values).

Surveys also have a number of disadvantages (e.g. they depend on the participant's motivation, honesty, memory and ability to respond). The respondents (those from whom the information is being collected) may not always be willing to give accurate answers. They may be motivated to give answers that cast themselves in a favourable light. This problem can be addressed by seeking anonymous survey responses.

Surveys that use randomly selected individuals may provide inaccurate findings if the sample is not large enough and/or representative of the whole population. Some people choose not to complete surveys. Those groups motivated to respond may differ from those who do not respond. This can produce biased findings.

Questionnaire-based surveys are particularly useful for collecting a range of geographical data. This includes:

- basic demographic and sociological data for a sample population (e.g. age, gender, occupation, education and ethnic origin)
- data about behaviour patterns (e.g. pedestrian and traffic movements, recreational activities and shopping preferences)
- data about spatial patterns (e.g. holiday destinations, journeys to work and origin of visitors to a destination).

Types of questions

As noted above, two types of questions can be used in questionnaires: open-ended or closed questions.

- **Open-ended questions:** interview-style questions. Participants are asked to construct their responses rather than choose between predetermined answers.
- **Closed questions:** participants are provided with a choice of answers to a specific question. Respondents are asked to tick boxes and/or rank or score responses.

Types of questionnaires

There are two types of questionnaires: those carried out directly (face-to-face) and those given to the respondent to complete in their own time, such as drop-and-collect or electronic questionnaires.

Direct questionnaires

Direct (face-to-face) questionnaires are interview-style surveys. These are usually conducted in public spaces, such as in the street or at a resident's front door. The questions asked are generally closed, requiring only a short response.

Drop-and-collect questionnaires

Drop-and-collect questionnaires are posted or delivered to the participant. It is a good idea to include a stamped, self-addressed envelope with these questionnaires to encourage participants to return them quickly. These questionnaires are usually accompanied by a cover letter explaining the survey's purpose and the processes involved in its completion and collection. This is the method previously used for the Australian Census, which is conducted by the ABS (see Figure 17.1.11), but now, most Australians complete the Census online. Participants complete these questionnaires without input or guidance from the investigator. They are self-administered questionnaires.

Drop-and-collect questionnaires have a range of advantages and disadvantages. They are less expensive to conduct than interviews. They don't need a large staff of skilled interviewers and they can be administered to an entire population, or a large sample, at the same time. They also generate information that can be collated and analysed using computers. Anonymity and privacy encourage the person completing the questionnaire to be more candid and honest when responding. Additionally, interviewer bias is absent from the data collection process and there is less pressure on the respondent.

Self-administered questionnaires also have several disadvantages. Respondents are, for example, more likely to become distracted and stop answering questions, meaning that they do not complete the survey. The respondents are not able to ask for clarification. Often, those motivated to complete such surveys are people who have extremes of opinion. This can result in skewed responses. Response rates also tend to be relatively low.

Electronic questionnaires

Many surveys are now conducted electronically, usually distributed as email messages sent to potential respondents. Electronic questionnaires eliminate the costs associated with printing and distributing paper-based questionnaires. Collecting data in an electronic format reduces the time and costs required for data processing and may increase public confidence in the anonymity of their responses. There are many free online survey tools.

Tips for effective questionnaires

When using questionnaires as an inquiry tool, students should not assume that people will always supply accurate details. Some people are prone to exaggeration or inclined to give the answer they think will receive the most positive response.

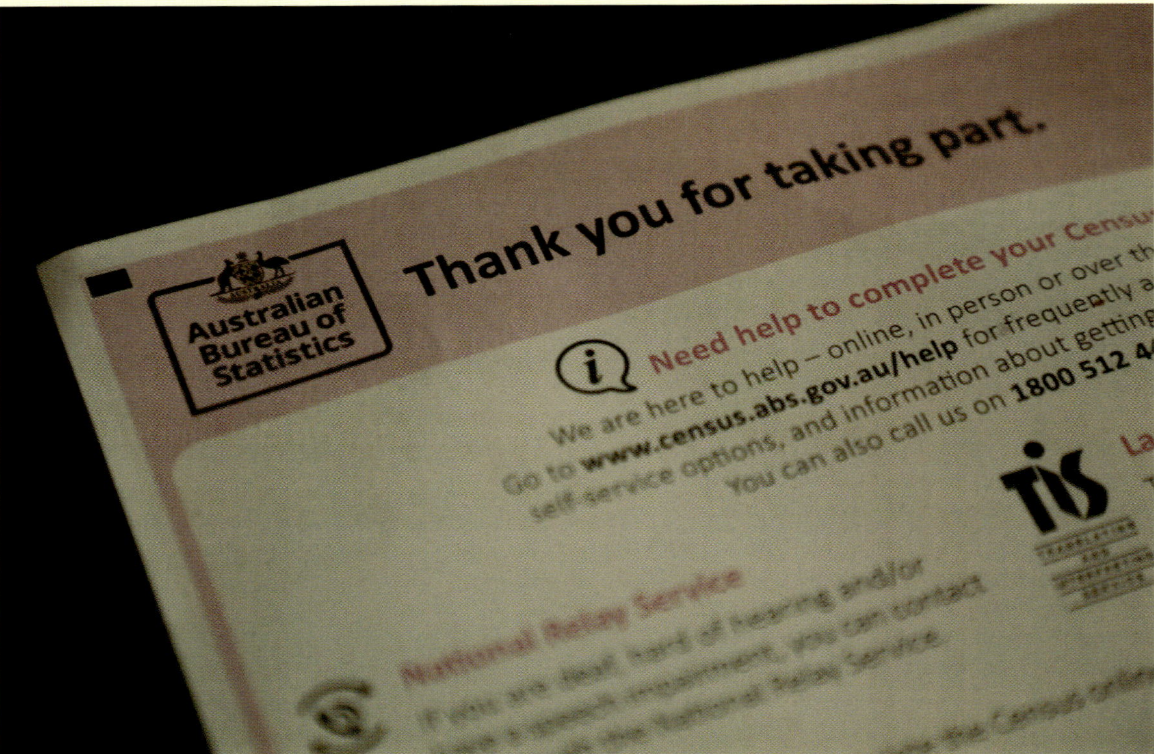

17.1.11 The Australian Census is the largest survey in Australia and is conducted every five years.

Students need to survey a variety of people and select people from a range of ages, ethnicities and occupations and both genders.

The criteria used to select people to survey will depend on the aim of the investigation being undertaken.

The questionnaire should be designed to fulfil an investigation's specific objectives. A questionnaire is an impersonal means of collecting information, so it must contain clear questions. The questions need to be worded as simply as possible to avoid any confusion or ambiguity, because the researcher will probably not be present to explain what any particular question means.

To make sure the questionnaire is effective, students should write questions following these tips:

- Ensure they are relevant to the investigation's inquiry question and/or aims.
- Do not ask too many questions.
- Avoid questions that require extensive thinking on the respondent's part.
- Ensure the questions can be answered briefly.
- Use simple and direct language.
- Avoid starting with 'Do you agree with ... ?' as this will often encourage the participant to give just a 'yes' or 'no' response.
- Prepare the questions carefully to avoid bias.
- Avoid questions that are of a personal nature and/or are likely to offend.
- Ensure the questions are given in a logical sequence.

Students should always treat the respondents respectfully and assure them that their information will remain confidential. Remember, these people are doing a favour by responding, so thank everyone who responds.

Interviews

Face-to-face interviews are a useful way to collect data that is not available from other sources. They are especially useful when aiming to investigate attitudes and values. Interviews can provide **quantitative** or **qualitative data** (or both), depending on the nature of the questions asked. It is, however, more difficult to carry out a quantitative analysis of the information gathered via an interview.

Interviewing is a skill that needs to be practised. Students need to be particularly clear about the nature of the information they wish to collect. They should also be aware of any potential for upsetting people by asking controversial or sensitive questions.

Interviews have several advantages over questionnaires. They are, for example, more flexible. The questions asked can be longer, more detailed and even open-ended. Interviews allow people to express their feelings and opinions. Unlike questionnaires, interviews do not force people to choose between categories or responses constructed by others. Finally, the interviewer can follow up on a response by asking supplementary (additional) questions. Sometimes, the interviewee (the person being interviewed) provides unexpected information, which adds greater substance to the research.

Tips for surveying people

Students are encouraged to follow these guidelines when surveying people through an interview or a direct questionnaire. They need to:

- introduce themselves politely and state clearly what they are doing. This will help them to obtain the cooperation of potential survey participants
- remember some people will not be interested in participating in the survey
- explain why they are collecting information and that all responses will be anonymous
- think carefully about the sequence of the questions. They should test them on friends first
- ensure the questions focus on the topic being surveyed and the demographics of the people being surveyed
- take care of the 'tone' they use when asking questions
- be neutral to the responses of the interviewee, even if they don't agree with them
- record the information from each respondent on a separate recording sheet
- dress appropriately, ideally in their school uniform
- always be polite and thank the participant for their cooperation.

When using surveys in an investigation, it is a good idea for students to employ both prearranged interviews and self-administered questionnaires. They need to be safety-conscious—they should always work in pairs or groups, and they should not give out their personal information to strangers.

When surveying a random sample of people (e.g. people at a shopping centre) students should follow these tips:

- Avoid approaching people who are engaged in an activity or are obviously in a hurry.
- Make the survey as brief as possible to minimise disruption to the people being interviewed; restrict it to a minute or two.
- Ensure the sample of people is broad enough to include a representative cross-section.

● SKILLS BUILDER

Presenting data graphically

After collecting the data, they need to be compiled and displayed in a way that highlights the essential features clearly to an audience. The statistical analysis can only be performed if it is properly presented. There are three modes of presentation of data. These are text-based presentations, table-based presentations and graphical representations (see Figure 17.1.12).

Some common approaches to the graphical representation of data include:

- **bar graph** (categorical data): a pictorial representation of data that uses bars to compare different categories of data. These are used to compare discrete variables. There are spaces between bars. The width of the bars is the same
- **histogram** (interval or ratio data): a graphical representation that displays data using bars to show the frequency of numerical data. It shows the distribution of non-discrete variables. There are no spaces between bars. The width of the bars need not be the same
- **pie chart** (nominal or ordinal data): area of circle segments representing the proportion. Multiple pie charts can be used with the radius of the circle having meaning
- **line graph** (ordinal, interval or ratio data): both axes are numerical. If time is one of the variables, always plot it on the X-axis. Only join up the points if the data is continuous
- **scattergraph:** needs one independent variable (on the x-axis) and one dependent variable (on the Y-axis). Both axes need interval or ratio data, and both must be continuous data. Do not join up each point. Use a line of best fit instead

17.1.12 A graphical representation is one of the best and most visually appealing ways of presenting data as it caters for the widest audience.

● SKILLS BUILDER

Writing reports

Most students will elect to submit a written report. If this is the preferred method of presentation, it is important for students to follow generally accepted guidelines.

Elements of the report

Written reports typically include all of the following elements.

Title page

The first page of the report is the title page. It should include the title of the geographical investigation, the student's name, class and the date of submission. The title's wording should convey to the reader a concise understanding of what the research investigation is about.

Table of contents

The next page is the table of contents. This should give the headings for all sections of the report and the page numbers they begin on. By referring to the table of contents, a reader should be able to see at a glance that the investigator has used the correct report structure.

List of tables and figures

The list of tables usually precedes the list of figures. In each case, the titles should be listed and numbered sequentially. Figures include photographs, graphs, maps, sketches and diagrams.

Inquiry question(s) and aims, hypotheses and objectives

Showcase the inquiry question and list the aims, hypotheses and objectives (if included) in point form or as discrete (separate) statements. It is best to space these out, so each point is clear, or use a format that draws the reader's attention to them.

Methodology

In the methodology section of the report, students describe how they investigated the geographical phenomenon, process or issue they studied and how they tested each of the hypotheses. They describe the processes they went through to collect, record and process primary data, and source, organise and process secondary data and explain why the particular approach was adopted.

Where appropriate, students include information, such as:

- how the primary data was collected, recorded and processed
- how any secondary data was sourced, organised and processed
- an explanation of why these methods were selected
- a note of the size of the sample, if a survey was used
- an outline of how this sample was selected
- a statement on the duration of the research.

Research findings

The research findings section represents the report's body. It outlines the results of the investigation. This can be done using appropriate graphs, tables, diagrams, photographs and maps, but should also include a written description of the results. Students refer to the tables and figures used to illustrate the results. This can be done by using statements such as:

'The results illustrated in Figure 1.2 demonstrate ...'

'Figure 1.3 shows that ...'

'As shown in Table 2.1 ...'.

Students need to take care with graphs, maps and other illustrative material. They need to use colour and make sure the generally accepted conventions are followed. They should ensure each figure is numbered and has a caption or title that identifies what it represents.

Conclusion

In the conclusion, students analyse the research results and evaluate them in terms of the inquiry question and (where appropriate) the stated aims and hypotheses.

Students discuss how the findings reflected, or differed from, what was expected. They also explain why particular results were achieved and outline any unexpected or interesting outcomes from the investigation. If appropriate, they reflect on the geographical implications of the research.

Evaluation

Many teachers regard the evaluation section to be the most important part of the geographical investigation. This is an opportunity for students to demonstrate their understanding of research processes by critically evaluating the framework developed for the project. Disappointing research results can still be used to gain a good grade. Students use the results to identify what went wrong and how mistakes could be prevented in a future similar research project.

The evaluation should outline both the limitations and successes of the geographical investigation. Students state the problems encountered and how these affected the final outcome. Common problems encountered by students include:

- the non-return of survey forms
- poorly constructed questions that failed to obtain the information required
- a sample that proved to be too small
- a lack of appropriate resources
- a reluctance by individuals and institutions to provide access to the required information
- an inquiry question that was too ambitious for the time and resources available.

Students balance the discussion by highlighting the successes of the research and the benefits derived from completing the project. These might include:

- an understanding of the research process
- an enhanced understanding of geographical concepts and terminology
- communication skills
- skills in data presentation
- greater self-confidence
- a sense of achievement from completing the task
- students conclude this section by describing how to improve the project if it were to be repeated.

References

In this section of the report, students cite the secondary resources they used while undertaking the geographical investigation (see the box, Skills Builder: Referencing sources on page 466.)

Acknowledgements

The acknowledgements section is an opportunity for students to acknowledge and thank those individuals and organisations who have assisted them in their investigation.

Appendix

The appendix is where background information is placed that is relevant to the investigation. It might include newspaper cuttings that deal with the topic, copies of questionnaires used and tables of data used to construct the graphs that illustrate research findings.

SKILLS BUILDER

Preparing oral reports

Students might find it appropriate to use an oral presentation to communicate their investigation's findings. Oral presentations are often supported by multimedia presentations or other visual aids. Developing and practising public speaking skills will help students do well in this aspect of the task.

The fear of speaking in front of a large group is very common, and there are ways of overcoming it. One is to carefully plan and prepare the presentation. Other good advice includes:

- students need to know their audience. The more they know about their audience, the more appropriate and focused their oral report will be. For example, if the audience already has a good knowledge of the topic investigated, the student may choose to leave out familiar information.
- students must ensure their information is well organised. The most successful oral reports have a structure similar to a written report; that is, they have an introduction, a body and a conclusion.

Using speaker's notes

Rather than simply reading a prepared speech, it is more effective for students to use a series of words or phrases that act as cues for their presentation. They should make sure, however, that the words and phrases are written in large print and that there are not too many points on a page. It is also a good idea for students to number the points, so that they do not get out of sequence. Different colours and highlighter pens may prove useful.

Delivering an oral presentation

To avoid stage fright, students should check these points before delivering their report:

- Know the material thoroughly.
- Use numbered points on speaking cards.
- Practise the speech several times.
- Do not rush the presentation.
- Use appropriate gestures and maintain good posture.
- Use a range of visual aids.
- Make eye contact with members of the audience.
- Vary their voice throughout the presentation.
- Stand still.

Structure of an oral report

Introduction

The first part of the oral report tells the audience what the topic is and outlines the investigation's inquiry question(s). It will also provide an insight into the methodologies used to gather, process, present and analyse the data on which the study's findings have been based. Students should seek to do this in a way that gains the audience's attention. A good way to do this is to show some visual material featured in a PowerPoint® presentation or in a video.

Body

The body of the oral report is the focus of the student's presentation. It provides an opportunity of drawing the audience's attention to the key findings of the study, as informed by the primary and secondary data collected, and what their geographical implications are. Perhaps there are recommendations the investigator can make based on these findings. It is also an opportunity to evaluate the methodology used in the study. Again, visual material, featured in a PowerPoint® presentation or in a video, can be used to enhance the audience's level of engagement and understanding.

Conclusion

Students as presenters should let the audience know when they are nearing the end of the presentation by using phrases such as 'In conclusion' or 'To sum up'. In their conclusion, students briefly highlight the key findings and recommendations of the study and outline any actions they intend to take.

● SKILLS BUILDER

Preparing multimedia presentations

In today's geography classroom, oral reports are, more often than not, accompanied by a multimedia presentation. Such presentations combine various types of media, including text, graphics, clip art, digital photographs, video, and perhaps even sound effects and music.

The most widely used multimedia presentation tool is Microsoft PowerPoint®. Apple's Keynote® is another popular tool. PowerPoint® and Keynote® are powerful software tools used for presenting information in a slide-show format. Text, charts, graphs, sound effects and video are just some of the elements students can incorporate into a slideshow presentation. This is a great way to communicate geographical information, including the results of the student's geographical investigations.

Guide for a successful slideshow presentation

General guidelines

- Plan the presentation carefully.
- Use the same design template throughout the presentation.
- Limit the number of slides used—too many can bore and confuse the audience.
- Include only essential information.
- Standardise the position, colours and styles of headings, text and images.
- Use colours that contrast—for example, black text on a light-coloured background. Yellow or white text on a dark blue background also works well.
- Be consistent with sound effects, transitions and animations.

Text guidelines

- Use the same font throughout the presentation.
- Do not include too much text on each slide.
- Have no more than six lines of text per slide, with no more than six words a line.
- Avoid long sentences.
- Select a suitable font size—in the range from 18 point to 48 point.
- Use a larger font to highlight key points.
- Avoid fancy fonts as they can be hard to read.
- Avoid words in all capital letters as they are hard to read.
- Avoid abbreviations and acronyms, such as 'approx.' and 'WTO'.

Sound

Sound effects can be used when text and/or objects appear in each slide, as well as during slide transitions. Students must take care as too much sound can distract the audience from their oral presentation.

Clip art, photos and graphics

- Use a variety of illustrations to make slides interesting and informative, including clip art, photos and maps.
- Ensure these illustrations balance the slide and enhance and complement the text, rather than overwhelm it.
- Present any data as a graphic.
- Include no more than two graphics per slide.

Delivering the multimedia presentation

- Practise and time the presentation.
- Distribute a printout of the presentation with spaces for note-taking.
- Speak confidently and clearly.
- Make the presentation available on the internet. PowerPoint® allows students to turn their presentation into a ready-made web page. Go to 'Save as'. From the 'Save file as type' list, select 'Web page' and then click 'Save'.

● SKILLS BUILDER

Referencing sources

Referencing is how students acknowledge the sources of the secondary information used in their investigation. They must include a reference whenever they draw on someone else's words, ideas or research, and also provide references (citations) for any graphics-based information used (see Figure 17.1.13).

The most widely used referencing procedure in the social sciences is the APA (American Psychological Association) style. The APA style involves both in-text citations and a list of references at the end of the work.

Referencing books

In-text citations

Within the text of an essay or report, students need to include two pieces of information about a source being used:

- the name of the author (or authors)
- the year of publication

Single author:

... (Kleeman, 2023)

or

Kleeman (2023) suggests (argues, contends etc.) that ...

Multiple authors:

... (Kleeman, Hamper & Rhodes, 2023)

or

(Kleeman, Hamper & Rhodes, 2023) suggest (argue, contend etc.) that ...

For multiple authors, second and subsequent use:

(Kleeman et al., 2023)

List of references

In the list of references at the end of the report, students should include the:

- author's surname, and initial(s)
- year of publication (in brackets)
- title of the publication (in italics and with minimal capitalisation)
- edition (if applicable, abbreviated as 'edn'),
- place of publication
- publisher.

Single author:

Kleeman, G. (2022). *Skills in geography* (3rd edn.). Port Melbourne: Cambridge University Press.

17.1.13 Acknowledging the sources you use is an important part of any investigation.

Multiple authors:

Kleeman, G., Hamper D. & Rhodes, H., (2023). *Global interactions Year 11* (4th edn). Melbourne: Pearson.

Edited work:

Kleeman, G. (Ed.). (2022). *Geography Skills Unlocked* (Second Revision). Sydney: Australian Geography Teachers Association.

Referencing newspapers and magazines (electronic sources)

In-text

If there is no author, list the name of the article and year:

(Win for endangered animals at global wildlife conference, 2019)

If there is an author:

(Farge, 2023)

List of references

If there is no author, include the:

- name of the article
- year, month and date of publication (in brackets)
- name of newspaper or magazine (in italics)
- URL from which the article was retrieved.

Win for endangered animals at global wildlife conference (2019, 30 August). *Sydney Morning Herald*. Retrieved from https://www.smh.com.au/environment/conservation/win-for-endangered-animals-at-global-wildlife-conference-20190830-p52mc5.html

If there is a named author, include:

- author's surname and initial(s)
- year, month and date of publication (in brackets)
- name of the article
- name of newspaper or magazine (in italics)
- URL from which the article was retrieved.

Farge, E. (2023, April 22) Glaciers 'already lost': Extreme melt has contributed to 10cm rise in ocean levels. *Sydney Morning Herald*. Retrieved from https://www.smh.com.au/world/europe/glaciers-already-lost-extreme-melt-has-contributed-to-10cm-rise-in-ocean-levels-20230422-p5d2gz.html

Referencing government publications (electronically sourced)

In text

If there is no obvious author or editor (which is common), students cite the relevant government department or agency as the author:

(Australian Bureau of Statistics [ABS], 2022)

List of references

If there is no named author, include:

- the relevant government department or agency
- year of publication
- name of article (in italics) including any relevant coding
- URL from which the article was retrieved.

Australian Bureau of Statistics (2022). *Australian demographic statistics*, September 2022 (Cat. no. 3101.0). Retrieved from http://www.abs.gov.au.

Index

Note: G1 and G2 page numbers refer to digital chapters

D

E

H

I

J

K

L

S

Acknowledgements

The following abbreviations are used in this list: t = top, b = bottom, l = left, r = right, c = centre.

123RF: bennymarty, p. 443; Braydenstanfordphoto, p. 17; Ozbandit, p. 258t; rushay, p. 454b.

AAP: Peter Eve, p. 384; Tim Holmes, p. 443; Dean Lewins, p. 399; NSW National Parks and Wildlife Service, pp. 353b, 396l, 396r; PKKP and PKKP Aboriginal Corporation, p. 251.

Agence VU: JR Image, p. 314.

Alamy: Abaca Press, p. 212b; Peter Adams, Jon Arnold Images Ltd, p. 257; Dario Bajurin, p. 211b; Ian Beattie, p. 250; Ricardo Beliel/ BrazilPhotos, BrazilPhotos, p. 126; Joerg Boethling, p. 175b Rosemary Calvert, p. 346; CBW, p. 140t; Mary Clarke, p. 218; Ashley Cooper, Global Warming Images, pp. 5, 141, 363t; Simon Dack Archive, p. 303b; Bob Daemmrich, p. 172t; Ethan Daniels, p. 23t; Danita Delimont Creative, p. 129; Beth Dixson, p. 306t; Jean-Paul Ferrero/ Auscape, Auscape International Pty Ltd, p. 111; Lincoln Fowler, p. 361t; FoxTree gfx, pp. 379, 389; FreeProd, agefotostock, p. 174b; Geopix, p. 381t; GL Archive, p. 409t; Gondong/BSIP, BSIP SA, p. 408; Peter Horree, p. 270; imageBROKER, p. 269c; Michael Interisano/ AgStock - RM, Design Pics Inc, p. 171bl; David Jackson, p. 288c; Barbara Jean, p. 383c; Inge Johnsson, p. 21; Interfoto/History, p. 174t; JG Photography, p. 175t; Robert Kneschke, p. 424b; Eric Lafforgue, p. 253; Lakeview Images, p. 174c; Lebrecht Music & Arts, p. 8; frans lemmens, p. 171c;

Andrew Lloyd, p. 354; Paul Lovelace, p. 421bl; LWM/NASA/Landsat, p. 146; Loren McIntyre, Stock Connection Blue, p. 124l; McPhoto/ Ingo Schulz, p. 3; Buddy Mays, p. 162; Regis Martin, p. 438; MB_Photo, p. 161; MediaWorldImages, p. 311; van der Meer Marica, Arterra Picture Library, p. 171t;

Richard Milnes, p. 462; Tuul and Bruno Morandi, p. 172b; Newscom, p. 281; Christian Offenberg, p. 212tl; Sean Pavone, pp. 23c, 237; Photo12/Ann Ronan Picture Library, Photo 12, p. 180t; Photo Researchers, Science History Images, p. 64; Planet Observer, Universal Images Group North America LLC, p. 51b;

Pulsar Imagens, p. 169tr; Dave Reede, All Canada Photos, p. 288t; Michel Roggo, Nature Picture Library, p. 139t; ronmolnar.com.au, p. 260; searagen, p. 178; Shawshots Stock Photo, p. 180bl; Christina Simons, p. 397; Sorge, Agencja Fotograficzna Caro, p. 294b; Eayne Stanley/ZUMA Press Wire, p. 192; Alexey Stiop, p. 171br; Keren Su/ China Span, p. 24; Timothy Swope, p. 423b; Technology And Industry Concepts, p. 383b; Trekkerimages, p. 166tr; US Navy Photo, p. 322b; Vintage_Space, p. 409c; Sebastian Wasek/Loop Images, Loop Images Ltd, p. 22;

WaterFrame_dpr, WaterFrame, p. 358t; Jan Wlodarczyk, pp. 10-11; Xinhua, p. 83b;

ZUMA Press, ZUMA Press, Inc., p. 381br.

Australian Bureau of Meteorology: pp. 393, 394.

Australian Bureau of Statistics: p. 417.

Australian Government, Department of Health and Aged Care, Weekly COVID-19 reporting: pp. 418, 419, 420.

Bayarcal, Carmelo: 166b.

Breiehagen, Per: p. 99.

Catton, Alex & Turrel, Mark: p. 145b.

The COVID Tracking Project at The Atlantic: p. 415.

CSIS Asia Maritime Transparency Initiative/Maxar Technologies: pp. 322t (both).

The Economist: p. 179.

European Space Agency: 2 Mission, p. 359: C. Yakiwchuck, p. 336b.

FairfaxPhotos: Chris Hopkins, p. 289; Fairfax Media, p. 169tl; Justin McManus, p. 331r.

Food and Agriculture Organization of the United Nations 2014: John Latham, Renato Cumani, Ilaria Rosati and Mario Blois, FAO Global Land Cover (GLC-SHARE). Reproduced with permission, pp. 80, 81tr, 81br, 82tr, 82br, 83tr, 84tr, 84br, 85tr, 85br, 86tr, 86br, 104, 105, 110.

von Gerhardt, Rachel: p. 139bl.

Getty Images: Anadolu Agency, p. 428tr; The Asahi Shimbun, p. 321b; Auscape, p. 6; Airphoto Australia, p. 169br; Tolga Akmen, p. 304; Ted Aljibe/AFP, pp. 321t, 323b;

Raul Arboleda/AFP, p. 317; Australian Department of Defence/Helen Frank/Handout/Anadolu Agency via Getty Images, p. 392; John W Banagan, p. 10b;

Patrick Baz/AFP, p. 315b; Ezequiel Becerra, p. 367b; Milos Bicanski, p. 303c; Bloomberg, p. 149; John Carnemolla, p. 264; Fabrice Coffrini, p. 310b; Ashley Cooper, p. 15t; De Agostini Editorial, p. 263t; Graham Denholm, p. 262; DimItar Dilkoff, p. 308b; Josh Edelson/AFP, p. 380; William Edwards, p. 426; Ryan Evans, p. 266b; Future Publishing, p. 167t; David Grey, p. 390b; Bartosz Hadyniak, p. 190b; Michael Hall, p. 331l; Martin Harvey, p. 335; Hiroshi Higuchi, p. 13r; Hulton Deutsch, p. 263c; Icon Sportswire, p. 427b; Kyodo News, p. 358b; Thomas Langer, p. 269t; Mark Leffingwell/Digital First Media/Boulder Daily Camera, p. 8; Feng Li, p. 279; Ricardo Lima, p. 49; Chris McGrath, p. 418; Jake McPherson, p. 299; Emilio Maranon III, Philippines, pp. 326 - 7; James D. Morgan, p. 353t; Dan Mullan, p. 283; Hoang Dinh Nam/ AFP, p. 323t; Marilyn Nieves, p. 214; NurPhoto, p. 191; Andrea Pattaro, p. 424t; Monty Rakusen, p. 19; Jason Redmond/AFP, p. 182; Sarah Reed/AFL Photos, p. 284; Benjamin Ridgway, p. 391; Karim Sahib, p. 294t; Heikki Saukkomaa, p. 313 Qilai Shen, p. 422; Alexander Spatari, p. 266t; Paul Souders, p. 329; Justin Sullivan, p. 381c; Universal History Archive, p. 263bl; Philip Wallick, p. 381bl; Lisa Maree Williams/Stringer, p. 383t; Trevor Williams, p. 120; Winfried Wisniewski, p. 14tl; vladimir zakharov, p. 211t.

GRID-Arendal: pp. 133-4.

Jennings, BJ: p. 178.

JFDP13: p. 183.

John Hopkins University and Medicine, Coronavirus Resource Centre: pp. 409b, 412 (all);

Lliboutry, Louis, p. 145bt.

LIMA/US Geological Survey: p. 91l.

MERICS: p. 193.

NASA Earth Observations: pp. 332 (all), 336t.

NASA Earth Observatory: pp. 9, 18, 93 (all), 123; Jeff Schmaltz, MODIS Rapid Response Team, NASA GSFC, 139br.

National Gallery of Victoria: John Longstaff, Gippsland, Sunday night, February 20th, 1898, oil on canvas,144.8 × 198.7 cm, National Gallery of Victoria, Melbourne,

Purchased, 1898, photo: National Gallery of Victoria, Melbourne, p. 374.

National Geographic: p. 181.

National Oceanic and Atmospheric Administration: pp. 75, 142, 358.

Natural Resources Canada, Government of Canada: p. 51.

Our World in Data, Oxford University: p. 414, 417t, 421t, 427t.

Royal Australian Historical Society: Darling Harbour and Pyrmont Bridge, 26 Nov 1937 [RAHS Adastra Aerial Survey Collection, p. 169bl.

Science Photo Library: Theirry Berrod, Mona Lisa Production, pp. 15.

Shutterstock: 4H4 Photography, p. 303t; Emrah C. Adalioglu, p. 362t;

Adwo, p. 421br; Aerial-motion, p. 263br; alextrp, p. 140c; AlinaKruchkoff, p. 11t; Am/Kobal , p. 204; Uwe Aranas, p.18 (pump station); arjma, p. 48; Atstock Productions, p.68 (ecosystems); Altrendo Images, p. 295bl; Toni Aules, p. 68 (community); Avigator Fortuner, p. 86tl; Salvador Aznar, p. 244; Roman Babakin, p. 413br; CarlosBarquero, p. 187; William Barton, p. 423t; Mohammad Bash, p. 58;

V. Belov, p. 12t; Necole A Berry, p. 106cl; Betelejze, p. 194b; thala bhula, p. 140b;

Bill45, p. 4bl; Lukas Bischoff Photograph, p. 27; BMJ, p. 87t; Boris-B, p. 268t; Boulenger Xavier, p. 119; Peter Braakmann, p. 20t; Bumihills, p. 249; CarlsPix, p. 94; Rich Carey, pp, 87b, 114, 240, 350; CatalinT, p. 18t; ChameleonsEye, p. 410; Li Hui Chen, p. 23b; Maksym Chub, p. 308c; Cloudcatcher Media, p. 54t;

Claudio Giovanni Colombo, p. 173; Corona Borealis Studio, p. 403; Rob Crandall, p. 293; Creative Family, p. 295tl; Darkydoors, p. 84tl; Kenneth Dedeu, p. 301; Madeleine Deaton, p. 131r; Colin Dewar, p. 4br; Designua, p. 55; Roberto Destarac Photo, p. 4tl; Elena Dijour, p. 305; DimaBerlin, p. 339; Craig Dingle, p. 407b; ESB Professional, p. 460; Esmeralda Edenberg, p. 13; Efired, p. 7t; Efimova Anna, p. 14r; NicoElNino, p. 158; elRoce, p. 390t; emperorcosar, p. 268b; Dirk Ercken, p. 241t; Erlantz P.R, p. 42; Everett Collection, p. 12; Juergen Faelchle, p. 348 ;

FiledImage, pp. 68b, 425; Sheila Fitzgerald, p. 306b; Ana Flasker, p. 15b;

frees, p. 300; GagliardiPhotography, p. 39; Germhess, p.179; Angelo Giampiccolo, p. 117t; Pavlo Glazkov, p. 18b; goodluz, p. 447; Charles Goudy, p. 65c; Mark Green, p. 122b; Jess Gregg, p. 406; Oliver Griffiths, p. 32; Igor Grochev, p. 320; Ground Picture, pp. 272b, 404b, 467; guentermanaus, p. 124r; Guitar photographer, p. 14bl; Jamie Hall, p. 413t; Menno van der Haven, p. 17b; Karl Hofman, p. 371; www.hollandfoto.net, p. 11b; Homo Cosmicos, p.7c; humphery, p. 241b; M. W. Hunt, p. 355b; iofoto, p. 95; ImagineStock, p. 10t; Chris Ison, p. 10; JaySi, p. 82tl; Ben Jeayes, p. 436; Jenso, pp. 5br, 180br; Tanya Jones, p. 265 (t); rodrigo_ jorda, p. 117b; JRJfin, p. 44; Julia700702, p. 19; Dmitry Kalinovsky, p. 9; kamilpetran, p. 330; Anjo Kan, p. 315t; Dmitriy Kandinskiy, p. 85bl; Dimitrios Karamitros, p. 350; Khunjompol, p. 83tl; King Ropes Access, p. 247; kleinstofzuigertje, p. 362b; Kletr, p. 106b; Sergey Krasnoshchokov, p. 68 (population); Nicolaj Larsen, p. 85tl; Chintung Lee, p. 265 (b); Leeborn, p. 31; Lesovik, p. 151; leungchopan, p. 212tr; Loco Photography, p.12b ; LongJon, p. 295tr; LukaKikina, p. 347; jon lyall, p. 363b; Richard A McMillin, p. 4tr; MAGNIFIER, p. 197; Eder Maioli, p. 51t; maloff, p. 18 (Afsluitdijk); Matt Makes Photos, p. 91r; mattymeis, p. 145t; Maximillian cabinet, p. 43r; metamorworks, pp. 194t, 272t; Mick2770, p. 81bl; Alexandros Michailidis, p. 364; evgenii mitroshin, p. 3; Luciano Mortula-LGM, pp. 288b, 296; Karla Mtz, p. 352; Nature, Parks/Outdoor, p. 92; Naufal MQ, p. 291; nechaevkon, p. 355t; Nemo2023, p. 103; Daria Nipo, p. 428tl; noina, p. 167b; R.M. Nunes, pp. 13l, 136; Nuntiya, p. 74; Manoej Paateel, p.14; paintings, p. 190t; hal pand, p. 135; Sean Pavone, pp. 5bl, 273; Dmitry Pichugin, p. 9; Arnold O. A. Pinto, p. 13; Pitroviz, p. 106cr; Pit Stock, p. 295br; Patrick Poendl, p. 130; Ekaterina Pokrovsky, p. 63; Andrey_Popov, p. 454c; paul prescott, p. 190c;

Ondrej Prosicky, p. 82bl; Daniel Prudek, p. 57; pysto_photography, p. 413bl; Torsten Pursche, p. 5; Quintanilla, p. 54b; Rawpixel.com, p. 5tl; Andis Rea, p. 3;

Alf Ribeiro, p. 86bl; Prachaya Roekdeethaweesab, p. 307; RossHelen, p. 269b;

Jane Rix, p. 34; robert_s, p. 68t; Mabelin Santos, p. 367t; Sathit, p. 259; Robin Scholte, p.16; Eleanor Scriven, p. 433; Giviryak Sergey, p. 4; sittitap, p. 106t;

Vitaly Sosnovskiy, p. 430; SRStudio, p. 50; Andrei Stepanov, p.155; Alexey Stiop, p. 368t; Boris Stroujko, p. 215; Ihor Sulyatytskyy, p. 404t; Suriya99, p. 5tr;

Teo Tarras, p. 81tl; testing, p. 280; thanasus, p. 148; Joseph Thomas Photography, p. 131l; Tobetv, p. 147; Travelpixs, p. 361b; Travel Stock, p. 65t; TR Stok, p. 20;

AlecTrusler2015, p. 349; Victoria Tucholka, p. 69; TungCheung, p. 282; turtix, p. 60;

Joost van Uffelen, p. 20b; T.W. van Urk, p. 17c; Sergey Uryadnikov, p. 113; View Apart, p. 200; Margus Vilbas, p. 368b; Jose L Vilchez, p. 14; Stefan Voswinkel, p. 150b; Frank Wagner, p. 84bl; Ashish_ wassup6730, p. 43l; Anton Watman, p. 221;

Natsicha Wetchasart, p. 396b; Richard Whitcombe, p. 79; PaulWong, p. 310t;

worldclassphoto, pp. 71, 122t, 242b; John Wreford, p. 7b; Xiangli Li, p. 166tl;

Drazen Zigic, p. 405; zimmytws, p. 454t; ZRyzner, p. 242t; Rudmer Zwerver, pp. 150t, 407t;

SmartTraveller, Department of Foreign Affairs and Trade, Commonwealth of Australia: p. 416b.

Tandberg, Ron: p. 223.

United Nations, DESA, Population Division, 2018: p. 164 (both).

World Health Organization: pp. 404, 405, 411